Cover Page Designed by:

Ms. Gowri Vemuri

First Edition

May, 2019

Author:

MATH-KNOTS

Index

Preface

The word "Science"brings to mind many different pictures: a fat textbook, bunch of equations, space shuttle, medicines and microscopes etc., simply discovering and learning about the nature around us. There are many branches of science and physics the most basic area of sciences, which deals with energy, matter, space, time and their interactions. The word physics is derived from the Latin word *physica*, which means "natural thing."

Physics as a fundamental science helps to uncover the laws of nature. The language of expression is mathematics. During ancient times, humans studied Stars and celestial bodies that lead to the development of disciplines like astronomy and mathematics. Based on size and speed, Physics as a subject is categorized several categories, which you can find from the following picture.

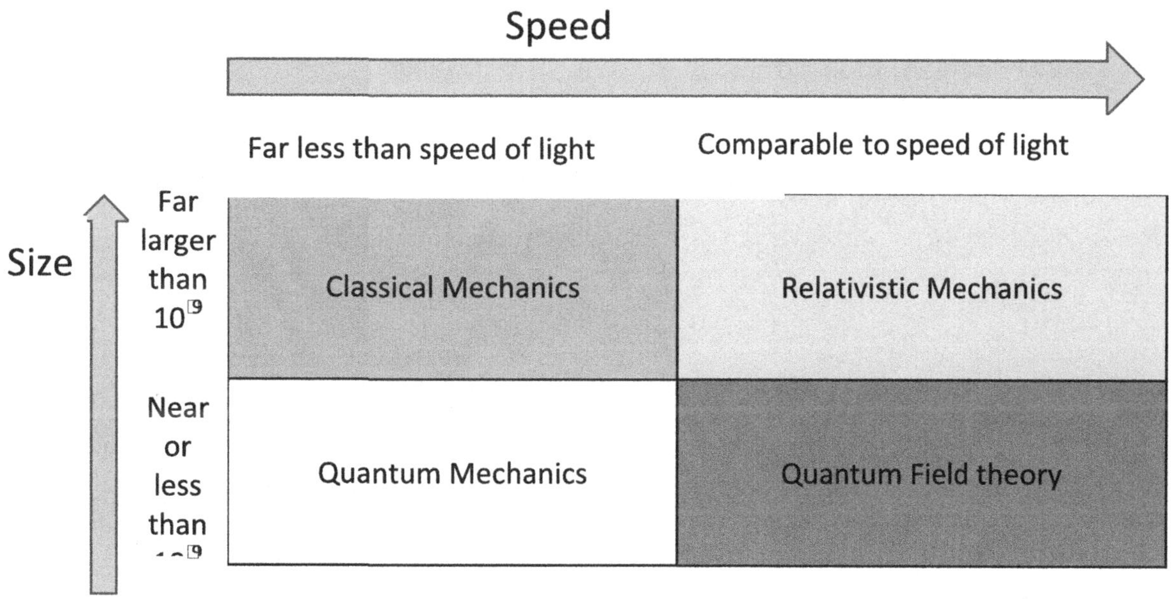

There are many ways to learn physics, but often students face difficulty in solving a problem by the concept they have learnt in the class. Our book is designed to help students in this area. Each chapter in the book, provides a summary of a concept, along with many practice problems.

Chapter - 1

MECHANICAL PROPERTIES OF SOLIDS

MECHANICAL PROPERTIES OF SOLIDS

SYNOPSIS

1. **Rigid body :** A body whose shape and size cannot be changed, however large the applied force may be, is called rigid body.

   There is no perfectly rigid body in nature.

2. **Deformation force :** A force which changes the size or shape or both of a body without moving it as a whole is called deformation force.

3. **Restoring force :** The force which restores the size and shape of the body when deformation forces are removed is called restoring force. ***Deformation force and restoring force are not action reaction pair.***

4. **Elasticity :** The property of a body by virtue of which it regains its original size and shape when deformation forces are removed is called elasticity.

   Elasticity is a molecular phenomenon.

5. **Elastic body :** A body which shows elastic behaviour is called elastic body

   Eg. Steel, rubber

   Nearest approach to perfectly elastic body is quartz.

6. **Plastic body :** A body which does not show elastic behaviour is called plastic body.

   Eg : Putty, chewing gum, lead solder, wax.

7. A body which is more difficult to strain is more elastic.

Eg: i) Steel is more elastic than rubber

ii) Springs are made of steel but not of copper because steel is more elastic than copper.

8. Hammering, rolling and addition of impurity increases elasticity.

9. Annealing and rise in temperature decreases elasticity.

10. For invar steel, the elastic property does not change with temperature.

11. **Stress :** The restoring force developed per unit area of the deformed body is called stress

$$\text{Stress} = \frac{\text{Restoring force}}{\text{cross sectional area}} = \frac{F}{A}$$

F = External force applied

$A = \pi r^2 =$ area of cross section

r = radius of the wire

Unit : $\frac{N}{m^2}$ or pascal , $\frac{dyne}{cm^2}$

Dimensional Formula : $M^1L^{-1}T^{-2}$

Stress is a tensor

12. **Stress is of three types :**

    i) Longitudinal stress

    ii) Tangential stress

    iii) Bulk stress

i) **Longitudinal stress :** If the restoring forces are perpendicular to the area of cross section and are along the length of the wire, the stress is called longitudinal stress.

During longitudinal stress, the body undergoes change in length but no change in shape and volume.

ii) **Tangential stress (or shearing stress) :** If the restoring forces are parallel to the surface, the stress is called shearing stress.

During tangential stress the shape of the body changes, size and volume remain same.

iii) **Bulk stress (or volume stress) :** If a body is subjected to equal forces normally on all the faces the stress involved is called bulk stress.

13. **Strain :** The deformation produced per unit magnitude is called strain

a) longitudinal strain $= \frac{\text{change in length}}{\text{original length}} = \frac{\Delta l}{l}$

b) Bulk strain $= \frac{\text{changein volume}}{\text{originalvolume}} = \frac{\Delta v}{v}$

c) Shearing strain = θ =

$$\frac{\text{lateral displacement between two layers}}{\text{perpendicular distance between the two layers}}$$

d) Transverse strain $= \frac{\text{change in radius}}{\text{original radius}} = \frac{\Delta r}{r}$

e) Shearing strain = 2 x longitudinal strain

f) Bulk strain = 3 x longitudinal strain

g) Longitudinal strain : shearing strain : Bulk strain = 1 : 2 : 3

14. The maximum value of the stress within which the body regain its original size and shape completely is called elastic limit.

15. Hooke's law : Within the elastic limit, stress is directly proportional to strain.

$$\frac{\text{stress}}{\text{strain}} = E = \text{constant}$$

E is modulus of elasticity

Unit of E : $\frac{N}{m^2}$ or pascal , $\frac{dyne}{cm^2}$

i) within the elastic limit, stress - strain graph is a straight line passing through the origin

Slope of the straight line is E

ii) A spring balance works on the principle of Hooke's law.

iii) Modulus of elasticity does not depend upon the dimensions of the body but it is a property of the material of the body.

iv) *Within the elastic limit, the load-extension graph is a straight line passing through the origin.*

16. Young's modulus : $Y = \frac{\text{longitudinal stress}}{\text{longitudinal strain}}$

$$Y = \frac{Fl}{Ae}$$

If load is M, then F = Mg and $\boxed{Y = \frac{Mgl}{\pi r^2 e}}$ where r is radius of the wire.

i) The stress required to double the length of a wire or to produce 100% longitudinal strain is equal to Young's modulus of the wire.

ii) Y of a perfectly elastic material is infinite and that of a perfectly inelastic material is zero.

iii) If a load 'M' produces an elongation e in a wire, the rise in temperature required to produce the same elongation is Δt. Then

$$\boxed{Mg = YA\alpha\Delta t}$$

where A is area of cross section of the wire and α is coefficient of linear expansion of the material of the wire.

iv) when a metal bar is fixed between two walls has a rise in temperature, the bar tries to expand and exerts force on the walls. This force is called ***thermal force***.

Thermal force $\boxed{F = YA\alpha\theta}$

where α is co-efficient of linear expansion of the bar
θ is rise in temperature.
Y is young's modulus,
A is area of cross section.

Thermal force is independent of length of the bar.

v) **Thermal stress :** Thermal force per unit area is called thermal stress

$$\text{Thermal stress} = \frac{\text{Thermal force}}{\text{area}} = \frac{YA\alpha\theta}{A} = Y\alpha\theta$$

Thermal stress is independent of area of cross section and length of the bar.

vi) A long wire suspended vertically can elongate due to its own weight.

- Elongation of a wire due to its own weight

$$\boxed{e = \frac{\ell^2 d g}{2Y}}$$

l = length of the wire, d = density of the wire

Y = young's modulus of the material of the wire

g = acceleration due to gravity

vii) Two wires made of ***same material*** having lengths l_1 & l_2 and radii r_1 and r_2 are subjected to the ***same force***. Then the ratio of their elongations is

$$\frac{e_1}{e_2} = \frac{l_1}{l_2} \times \frac{r_2^2}{r_1^2} \quad (\text{since } e \propto \frac{l}{r^2})$$

viii) Two wires made of ***same material*** having lengths l_1 and l_2 and masses m_1 and m_2 are subjected to the ***same force***. Then the ratio of their elongations is

$$\frac{e_1}{e_2} = \frac{l_1^2}{l_2^2} \times \frac{m_2}{m_1} \quad (\text{since } e \propto \frac{l^2}{m})$$

ix) Two wires of ***same material*** and ***same volume*** having areas of cross section A_1 & A_2 are subjected to the ***same force***. Then the ratio of their elogations is

$$\frac{e_1}{e_2} = \frac{A_2^2}{A_1^2} \quad (\text{since } e \propto \frac{1}{A^2})$$

x) Two wires of ***same material*** and ***same volume*** having radii r_1 and r_2 are subjected to the ***same force***. Then the ratio of their elongations is

$$\frac{e_1}{e_2} = \frac{r_2^4}{r_1^4} \quad (\text{since } e \propto \frac{1}{r^4})$$

viii) When a body of mass 'm' and density 'd_B' is suspended from a wire its elongation is 'e' when it is in air. If it is completely immersed in a non-viscous liquid of density d_L then its new elongation is

$$e^1 = e(1 - \frac{d_L}{d_B})$$

ix) Two wires of ***same length*** and ***radius*** have youngs modulii Y_1 & Y_2 are subjected to the ***same force***. Then the ratio of their elongations is

$$\frac{e_1}{e_2} = \frac{Y_2}{Y_1} \quad (\text{since } e \propto \frac{1}{Y})$$

x) Two wires of same length and radius are joined end to end and loaded. If the youngs modulii of the materials of the wires are Y_1 & Y_2 and the combination behaves as a single wire then its young's modulus is

$$Y = \frac{2Y_1Y_2}{Y_1+Y_2}$$

xi) Two wires of same length and radius are joined in parallel and loaded. If the youngs moduli of the materials of the wires are Y_1 & Y_2 and the combination behaves as a single wire then its young's modulus is

$$Y = \frac{Y_1+Y_2}{2}$$

17. Rigidity modulus, $n = \frac{\text{Tangential stress}}{\text{shearing strain}} = \frac{F}{A\theta}$

where θ is shearing strain

i) If n is low for a wire then it can be twisted very easily.

ii) Since phosphorbronze has very low rigidity modulus, it is used as suspension wire in moving coil galvanometer.

iii) For a fluid (liquid or gas) the rigidity modulus is zero.

18. A rod of length 'l' and radius 'r' is fixed at one end. If the other free end is twisted through an angle 'θ' then the angle of shear ϕ is given by

$$\ell\phi = r\theta$$

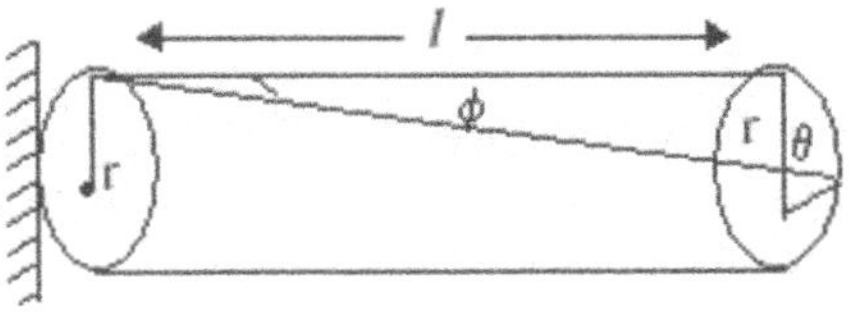

19. One end of the rod is fixed. The other free end is twisted through an angle 'θ' by applying a torque 'τ' then the work done on the rod (or) energy stored in the rod is

$$W = \frac{1}{2}\tau\theta$$

where θ is in radians.

20. Bulk modulus $K = \frac{\text{volume stress}}{\text{volume strain}} = \frac{V\Delta P}{\Delta V}$

i) If a block of coefficient of cubical expansion γ is heated through a rise in temperature of θ, the pressure to be applied on it to prevent its expansion, $P = K\gamma\theta$ where K is its bulk modulus.

ii) When a rubber ball of volume V, bulk modulus K is taken to a depth h in water, then decrease in its volume $\Delta V = \frac{hdgV}{K}$

(d = density of material)

21. Solids posses Y, n and K, but liquids and gases posses only K.

22. Isothermal bulk modulus of a gas = pressure of the gas (P)

Adiabatic bulk modulus of a gas = γP

where γ is the ratio of two specific heats.

23. **Compressibility :** The reciprocal of bulk modulus is called compressibility, C = 1/K

For incompressible substances, C = 0; K = ∞.

24. Possion's ratio (σ) =

$$\frac{\text{lateral contraction strain}}{\text{longitudinal elongation strain}} = \left(\frac{-\Delta r/r}{\Delta l/l}\right)$$

i) Poisson's ratio has no units and no dimensions

ii) Theoretical limits of σ : -1 to 0.5

iii) Practical limit of σ : 0 to 0.5

iv) If σ = 0.5 then the substance is perfectly incompressible.

v) The Poisson's ratio of a given material is 'σ'. When the load is applied on it, the longitudinal elongation strain is $\frac{\Delta l}{l}$. Then the fractional change in its volume is

$$\frac{\Delta V}{V} = (1-2\sigma)\frac{\Delta l}{l}$$

25. i) If large deformation occurs between the elastic limit and the breaking point, the material is ductile.

<u>**Ex :**</u> copper, silver, gold etc.,

ii) If the deformation between the elastic limit and the breaking point is very small or if the wire breaks immediately after crossing the elastic limit, the material is brittle.

<u>**Ex :**</u> glass, ceramic etc.

26. **Breaking stress :**

i) The breaking stress of a wire is the maximum stress the material can withstand.

ii) Breaking stress = $\frac{\text{Breaking Force}}{\text{initial area of cross section}}$

iii) Breaking force = Breaking stress x area of cross section.

iv) Breaking stress a) depends only on the nature of material of the wire b) independent of the length and area of cross section of the wire.

v) Breaking force a) is independent of length of the wire b) depends on the area of cross section and nature of material of the wire.

vi) Breaking force is proportional to area of cross-section.

vii) If we cut a cable that can support a maximum load of W into two equal parts, then each part can support a maximum load of W.

viii) ***A very long wire suspended vertically can break due to its own weight***

The maximum length of a wire that can hung without breaking under its own weight is

$$l = \frac{\text{Breaking Stress}}{dg}$$

27. Elastic fatigue :

a) The state of temporary loss of elastic nature due to continuous strain is called elastic fatigue.

b) Due to elastic fatigue

i) a wire can be broken within the elastic limit.

ii) a wire can be cut into pieces without using instruments.

iii) Railway tracks and bridges are declared unsafe after long use.

iv) spring balances show wrong readings after long use.

28. Elastic after affect: Within the elastic limit , the delay in regaining the original length after the removal of deforming force by a material is called its elastic after effect.

i) For a perfectly elastic body, elastic after effect is zero.

ii) For a perfectly inelastic body, elastic after effect is infinite.

29. Work done in stretching a wire,

$w = \frac{1}{2}$ x stretching force x extension.

i) $w = \frac{1}{2} Fe = \frac{1}{2}\frac{YAe^2}{l} = \frac{1}{2}\frac{F^2 l}{AY} = \frac{1}{2}\frac{F^2 l}{\pi r^2 Y}$

ii) $w = \frac{1}{2}$ x stress x strain x volume of the wire

iii) $\frac{\text{work done}}{\text{volume}}$ (a strain energy per unit volume or elastic potential energy per unit volume) $= \frac{1}{2}$ x stress x strain

$= \frac{1}{2}$ x Y x (strain)^2 $= \frac{1}{2}\frac{(\text{Stress})^2}{Y}$

iv) The total work done per unit volume in stretching an object upto the limits of perfect elasticity is known as its resiliance.

v) Area under F-e graph gives the work done or the strain energy stored in the wire.

Area $= \frac{1}{2}$ F.e = W

30. Springs :

i) A spring that obeys Hooke's law has force constant or spring constant

$Y = \frac{Fl}{Ax}$; $F = \frac{YA}{l}x$ and F = Kx

x is elongation

Where K is called the force constant or spring constant. It is the force needed to stretch a wire or spring over unit distance. i.e., $K = \frac{YA}{l}$

ii) $K \alpha Y, K \alpha A, K \alpha \frac{1}{l}$

iii) If a spring (or a wire) of force constant K is cut into 'n' equal parts, the force constant of each part of the wire is 'nk'.

iv) Potential energy of a stretched spring $= \frac{1}{2}$ $Fx = \frac{1}{2} Kx^2 = \frac{1}{2}\frac{F^2}{K}$

v) Two springs have force constant K_1 and K_2

a) When they are stretched by the same force and if their elastic energies are E_1 and E_2 then $\frac{E_1}{E_2} = \frac{K_2}{K_1}$

b) When they are extended by the same length, $\frac{E_1}{E_2} = \frac{K_1}{K_2}$

c) When they are extended till their energies are same then $\frac{F_1}{F_2} = \sqrt{\frac{K_1}{K_2}}$

d) The potential energy of a spring increases, whether it is stretched or compressed.

e) Springs in series, $K_{eff} = \frac{K_1K_2}{K_1+K_2}$

f) Springs in parallel , $K_{eff} = K_1 + K_2$

g) The reciprocal of spring constant is called compliance.

31. i) When a spring is stretched strain involved is shearing strain

ii) When a wire is stretched modulus of elasticity involved is Young's modulus

iii) When a wire is twisted modulus of elasticity involved is rigidity modulus.

32. Torsion of a cylinder :

When one end of a cylinder of length 'l' and radius 'r' is clamped and the other end is given a twist 'φ', then the restoring torque is given by

$$\tau = \frac{\pi n r^4 \varphi}{2l}$$

Where 'n' is the modulus of rigidity of the material of the cylinder.

EXERCISE - I

THEORITICAL QUESTIONS

1. A spiral spring is stretched by a force the resultant strain produced in the spring is

1) Volume strain 2) Shearing strain
3) Tensile strain 4) None of the above

2. Reason for the deformation of a regular body

1) Bulk strain 2) Shearing strain
3) Linear strain 4) Lateral strain

3. When a spring is stretched by a force, the resultant strain in known as

1) Longitudinal strain 2) Sheariing strain
3) Volume strain 4) Bulk strain

4. For a gas elastic limit

1) Exists 2) Does not exist
3) Exists only at absolute zero
4) Exists for a perfect gas

5. Hooke's law states that

1) Stress is inversely proportional to strain
2) Stress is directly proportional to strain
3) Stress is independent of strain
4) Stress is proportional to elastic modulus

6. Within elastic limit, if stress is doubled the strain becomes

1) Half 2) Twice
3) Thrice 4) None

7. The law which governs the working of a spring balance is

1) Kepler's law 2) Robert Hooke's law
3) Newton's law 4) Young's law

8. According to Hooke's law of elasticity the ratio of stress to strain

1) Does not remain constant 2) Remains constant
3) Increases 4) Decreases

9. A body subjected to strain a number of times does not obey Hooke's law due to

1) Yield point 2) Breaking stress
3) Elastic fatigue 4) Permanent set

10. The modulus of elasticity is dimensionally equivalent to

1) Stress 2) Surface tension
3) Strain 4) Coefficient of viscosity

11. Graph between the applied force and the change in length of a wire within elastic limit is a

1) Straight line with negative slope
2) Straight line with zero slope
3) Straight line with positive slope 4) None of the above

12. Which of the following substances has the highest elasticity

1) Rubber 2) Steel 3) Copper 4) Wood

13. A steel wire is stretched by 5 kg wt. If the radius of the wire is doubled its Young's modulus

1) Remains unchanged 2) Becomes double
3) Becomes half 4) Becomes 1/4 times

14. A copper wire and a steel wire of the same diameter and length are connected end to end and a force is applied which stretches their combined length by 1 cm. Then the two wires will have

1) The same stress and strain
2) The same stress but different strain
3) The same strain but different stresses
4) Different stresses and strain

15. The effect of temperature on the value of modulus of elasticity for different substances in general

1) Increases with increase in temperature
2) Decreases with increase in temperature
3) Independent of change in temperature
4) Does not change with temperature

16. A stone is suspended in a tub of water with copper wire. Another stone of equal mass is suspended in kerosene with equal length of copper wire then

1) Young's modulus is more in the first case
2) Young's modulus is more in the second case
3) Elongation will be less in the former case
4) Elongation will be more in the former case

17. In the given figure if the dimension of the wires are the same and materials are different Young's modulus is less for

1) A
2) B
3) Both
4) None

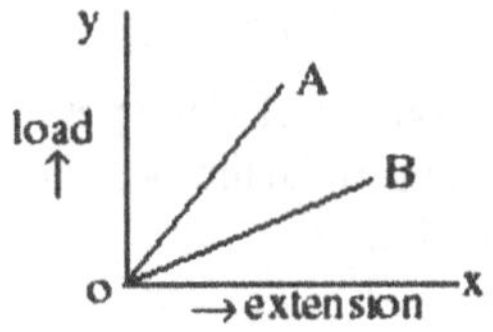

18. A heavy mass is attached to a thin wire and is whireled in a vertical circle. The wire is most likely to break

1) When the mass is at the lowest point
2) When mass is at the highest point
3) When wire is horizontal
4) When mass is at an angle of $\cos^{-1}\left(\frac{1}{\sqrt{2}}\right)$ from upward vertical

19. Steel is preferred for making springs over copper because

1) Young's modulus of steel is more than that of copper
2) Steel is cheaper
3) Young's modulus of copper is more
4) Steel is less likely to be oxidised

20. The possible value of Poission's ratio is

1) 0.9 2) 0.8 3) 0.4 4) 1

21. The Poisson's ratio can not have the value

1) 0.7 2) 0.2 3) 0.1 4) – 0.5

22. The Poission's ratio (s) should satisfy the relation

1) $-1 < \sigma < 0.5$ 2) $-0.5 < \sigma < 1$
3) $0.5 < \sigma < 1$ 4) $-1 < \sigma < -0.5$

23. The breaking stress of a wire depends on

1) Material of a wire 2) Shape of cross section
3) Length of the wire 4) Radius of the wire

24. Breaking stress per unit area of cross section of a wire is called

1) Yield point 2) Breaking stress
3) Ductibility 4) Tensile stress

25. The property of metals whereby they could be drawn into thin wires beyond their elastic limit without breaking is

1) Ductility 2) Malleability
3) Elasticity 4) Hardness

26. When an elastic material with young's modulus 'y' is subjected to a stretching stress 'S' the elastic energy stored per unit volume of the material is ---

1) $\frac{YS}{2}$ 2) $\frac{S^2Y}{2}$ 3) $\frac{S^2}{2Y}$ 4) $\frac{S}{2Y}$

27. A wire of length 'L' and cross sectional area 'A' is made up of a material of young's modulus 'Y'. If the wire is stretched by 'X' the work done is

1) $\frac{YAX}{L}$ 2) $\frac{YAX^2}{2}$ 3) YAX^2 4) $\frac{YAX^2}{2L}$

28. Strain energy per unit volume in a stretched string is

1) $\frac{1}{2}\times \text{stress}\times \text{strain}$ 2) stress x strain
3) stress x $(\text{strain})^2$ 4) $\frac{\text{stress}}{\text{strain}}$

29. When a wire is stretched to double it's length

a) Radius is halved b) Strain is unity
c) Stress is equal to Youngs modulus
d) Stress is unity

1) 'a' and 'b' are correct 2) 'b' and 'c' are correct
3) 'c', 'd' are correct 4) 'd' is correct

30. The formula for strain energy per unit volume

a) 1/2 (stress)(strain) b) 1/2 Y $(\text{strain})^2$
c) $\frac{1}{2}$ x y (stress) d) $\frac{1}{2}$ x $\frac{(\text{strain})}{Y}$

1) a, b are correct 2) a, c are correct
3) c, d are correct 4) a, b, c are correct

31. Regarding elasticity which of the following statements are correct?

a) the point where the elasticity ends and plasticity starts is called yield point
b) breaking stress depends on the area of cross section
c) the value of poisions ratio can not be greater than 0.5
d) air is more elastic than water

1) a, b are correct 2) a, c are correct
3) b, d are correct 4) a and d are correct

32. A spring of young's modulus 2×10^{11} pa is suspended vertically and subjected to a load of 5 kg and elongation is 2 mm. when theload is doubled – match the following (g =9.8 m/s²)

Section - A	**Section - B**
a)Elongating force	**e) 4 mm**
b)Stress	**f) Un changed**
c)Elongation	**g) 98 N**
d)Young's modulus	**h)Doubled**

1) a–h; b–g; c–f; d–e
2) a–g; b–h; c–e; d–f
3) a–e; b–f; c–g; d–h
4) a–f; b–e; c–g; d–h

33. Section - A | Section - B

Section - A	Section - B
a) Young modulus	e) resistance to change in shape
b) Rigidity modulus	f) resistance to change in linear dimension
c) Shear modulus	g) resistance to change in volume
d) Stress	h) resistance to against defermation forces

1) a–f; b–g; c–e; d–h 2) a–g; b–h; c–e; d–f
3) a–f; b–e; c–g; d–h 4) a–g; b–e; c–h; d–f

34. Of the following four wires of same material which will have maximum increase in length and which will have minimum increase in length if all are under the action of same force? Arrange the wires in the descending order of increase in length.

i) Natural length=0.5m, diameter=0.5mm
ii) Natural length=1m, diameter=1mm
iii) Natural length=1.5m,diameter=1.5mm
iv) Natural length=2.0m, diameter=2mm

1) i, ii, iii, iv 2) ii, i, iv, iii
3) i, iii, iv, ii 4) iv, iii, ii, i

35. Four springs of force constants K_1=1000N/m, K_2=1500 N/m, K_3=2500 N/m and K_4=2000 N/m are subjected to different loads producing same extension. Arrange the springs with workdone in descending order

1) 1,2,3,4 2) 3, 4, 1, 2
3) 3, 2, 1, 4 4) 3, 4, 2, 1

36. Four wires made of Al, Cu, Fe and steel and of lengths l_1 = 100m, l_2=150m, l_3=250m and l_4= 350m, break due to their own weights. Arrange these wires with their breaking stresses in ascending order

d_{Al} = 7500 kg/m^3 d_{cu} = 7800 kg/m^3
d_{Fe} = 9000 kg/m^3 and d_{steel} = 11000 kg/m^3

1) St, Fe, Al, Ci 2) St, Fe, Cu, Al
3) Al, Cu, Fe, St 4) Fe, St, Al, Cu

Instruction For Questions :

In each of the following questions, a statement of Assertion (A) is followed by a corresponding statement of reason (R) just below it. Of the statement mark the correct answer.

1) Both 'A' and 'R' are true and 'R' is the correct explanation of 'A'
2) Both 'A' and 'R' are true and 'R' is not the correct explanation of 'A'
3) 'A' is true but 'R' is false
4) 'A' is false but 'R' is true

37 . A) : Stress is Restoring force per unit area
R) : Interatomic forces in solids are responsible for the property of elasticity.

38. A) : Steel is more elastic than rubber.
R) : Under given deforming force, steel is deformed less than rubber.

39. A) : Bulk modulus of elasticity (k) represents incompressibility of the material.
R) : Bulk modulus of elasticity is proportional to change in pressure.

40 . A) : A steel wire of natural length 'l' breaks, when elongated by length 'x'. If original length is halved and elongated by x/2 it breaks.
R) : Breaking force is directly proportional to the length of the wire

41 . A) : Silver is a ductile material
R) : For a ductile material yield point and Breaking polnt are seperated by larger distance than for Brittle materials on thestress–strain curve.

42. A) : The stretching of a coil is determined by its shear modulus.
R) : Shear modulus change only shape of a body keeping its dimensions unchanged.

43. A) : Strain is a unitless quantity.
R) : Strain is equivalent to force.

44. A) : The bridges declared unsafe after a long use.
R) : Elastic strength of bridges losses with time.

45. A) : Two identical solid balls, one of ivory and the other of wet-clay are dropped from the same height on the floor. Both the balls will rise to same height after bouncing.
R) : Ivory is more elastic than wet-clay.

46. A) : Young's modulus for a perfectly plastic body is zero.

R): For a perfectly plastic body, restoring force is zero.

47. A) : Identical springs of steel and copper are equally stretched. More work will be done on the steel spring.

R) : Youngs modulus of steel is higher than copper.

48. Consider the statements A and B and identify the correct answer given below.

A) If the volume of a body remains unchanged, When subjected to tensile strain, the value of Possion's ratio is 1/2

B) Phosphor bronze has low Young's modulus and high rigidity modulus [2003 M]

1) A and B are correct
2) A and B ar wrong
3) A is correct and B is wrong
4) A is wrong and B is right

49. Work done per unit volume is stretching a wire
1) is equal to half the product of the stress and strain
2) is equal to the product of the stress and the strain
3) is euqal to the ratio of the stress to the strain
4) is equal to the ratio of the strain to stress

50. The breaking stress of wire depends upon [AIEEE 2002]

1) length of wire 2) radius of wire
3) material of wire 4) shape of cross section

51. Which of the following affects the elasticity of a substance [AIEEE 2002]
1) hammering and annealing
2) change in temperature
3) impurity in substance 4) all of these

52. Consider an ideal monoatomic gas of volume at pressureP. The bulk modulus at constant temperature is [UPSEAT 2000]

1) $\frac{p}{2}$ 2) P 3) γP 4) $\frac{PdP}{dV}$

53. If k is the force constant of a spring the work done [against restoring force] in increasing its extension from l_1 to l_2 will be

1) $K(l_2 - l_1)$ 2) $K/2(l_2 + l_1)$

3) $K(l_2^2 - l_1^2)$ 4) $\frac{K}{2}(l_2^2 - l_1^2)$

54. A wire of length L and cross -section 'A' has youngs modulus of material Y. It is stretched by an amount 'x'. The work done against restoring force is [MP PET 1996]

1) $\frac{YAx}{2L}$ 2) $\frac{YAx^2}{L}$

3) $\frac{YAx^2}{2L}$ 4) $\frac{2YAx^2}{L}$

ANSWERS

1) 2	2) 2	3) 2	4) 2	5) 2
6) 2	7) 2	8) 2	9) 3	10) 1
11) 3	12) 2	13) 1	14) 2	15) 2
16) 3	17) 2	18) 1	19) 1	20) 3
21) 1	22) 1	23) 1	24) 4	25) 1
26) 3	27) 4	28) 1	29) 2	30) 1
31) 2	32) 2	33) 1	34) 1	35) 4
36) 3	37) 2	38) 1	39) 3	40) 3
41) 1	42) 1	43) 3	44) 1	45) 4
46) 1	47) 1	48) 3	49) 1	50) 3
51) 4	52) 2	53) 4	54) 4	

EXERCISE - II(A)

YOUNG 'S MODULUS

1. A copper wire of 1 mm diameter is stretched by applying a force on 10N. Find the stress in the wire.

1) $1.273 \times 10^7 N/m^2$ 2) $1.373 \times 10^7 N/m^2$
3) $1.473 \times 10^7 N/m^2$ 4) $1.573 \times 10^7 N/m^2$

2. A tungsten wire of length 20 cm is stretched by 0.1 cm. Find the strain on the wire.

1) 0.002 2) 0.005 3) 0.001 4) 0.004

3. In the Searle's method to determine the Young's modulus of a wire, a steel wire of length 156 cm and diameter 0.054 cm is taken as experimental wire. The average increase in length for 1 1/2 kg wt is found to be 0.050 cm. Then the Young's modulus of the wire.

1) $3.002 \times 10^{11} N/m^2$ 2) $1.002 \times 10^{11} N/m^2$
3) $2.002 \times 10^{11} N/m^2$ 4) $2.5 \times 10^{11} N/m^2$

4. The area of cross-section of a wire is $10^{-5}m^2$ when its length is increased by 0.1% a tension of 1000N is produced. The Young's modulus of the wire will be ---

1) 10^{12} Nm^{-2} 2) 10^{11} Nm^{-2}
3) 10^{9} Nm^{-2} 4) 10^{10} Nm^{-2}

5. The following four wires are made of the same material. Which of these will have the largest elongation when the same tension is applied

1) l = 50cm and diameter 0.5mm
2) l = 100cm and diameter 1.0mm
3) l = 200cm and diameter 2.0mm
4) l = 300cm and diameter 3.0mm

6. A copper wire and an aluminium wire have lengths in the ratio 3 : 2, diameters in the ratio 2 : 3 and forces applied in the ratio 4 : 5. Find the ratio of the increase in length of the two wires. ($Y_{Cu}=1.1x10^{11}Nm^{-2}$, $Y_{Al}=0.70x10^{11}Nm^{-2}$)

1) 110 : 189 2) 180 : 110
3) 189 : 110 4) 80 : 11

7. There are two wires of same material. Their radii and lengths are both in the ratio 1 : 2. If the extensions produced are equal, what is the ratio of the loads ?

1) 1 : 2 2) 2 : 1 3) 1 : 4 4) 4 : 1

8. If stress is numerically equal to young's modulus the elongation will be

1) $\frac{1}{4}$ the original length
2) $\frac{1}{2}$ the original length
3) Equal to the original length
4) Twice the original length

9. An aluminium rod has a breaking strain 0.2%. The minimum cross-sectional area of the rod in m^2 in order to support a load of 10^4N is if Youngs modulus is $7x10^9Nm^{-2}$

1) $1.7x10^{-4}$ 2) $1.7x10^{-3}$
3) $7.1x10^{-4}$ 4) $1.4x10^{-4}$

10. A string of cross section $2cm^2$ is doubled in length by the application of a longitudinal force $2x10^5$ dynes. The youngs modulus is -

1) 10^7 dy/cm^2 2) 10^5 dy/cm^2
3) 10^4 dy/cm^2 4) 100 dy/cm^2

11. A metallic ring of radius 'r' and cross sectional area A is fitted into a wooden circular disc of radius R (R > r) . If the Young's modulus of the material of the ring is Y, the force with which the metal ring expands is : [2004 E]

1) $\frac{AYR}{r}$ 2) $\frac{AY(R-r)}{r}$
3) $\frac{Y(R-r)}{Ar}$ 4) $\frac{YR}{AR}$

12. When the tension on a wire is 4N its length is ℓ_1. When the tension on the wire is 5N its length is ℓ_2. Find its natural length.

1) $5\ell_1-4\ell_2$ 2) $4\ell_1-5\ell_2$
3) $10\ell_1-8\ell_2$ 4) $8\ell_1-10\ell_2$

13. When a metal sphere is suspended at the end of a metal wire its extension is 0.4mm. If another metal sphere of the same material with its radius half as the previous is suspended then extension would be

1) 0.05mm 2) 0.02mm
3) 0.01mm 4) 0.32mm

14. A wire whose cross - sectional area is 4 mm^2 is stretched by 0.1 mm by a certain load. If a similar wire of double the area of cross-section is under the same load, then the elongation would be

1)0.5mm 2) 0.05 mm
3) 0.005mm 4) 5mm

15. When a tension 'F' is applied, the elongation produced in uniform wire of length l, radius r is e, when tension 2F is applied, the elongation produced in another uniform wire of length 2l and radius 2r made of same material is [2000 M]

1) 0.5e 2) 1.0e 3) 1.5e 4) 2.0e

16. Two wires A and B have Young's modulii in the ratio 1:2 and ratio of lengths is 1:1. Under the application of same stress the ratio of elongations is ---

1) 1:1 2) 1:2 3) 2:1 4) 1:4

17. A wire is stretched by 0.01m by a certain force 'F' another wire of same material whose diameter and lengths are double to original wire is stretched by the same force then its elongation will be ....

1) 0.005m 2) 0.01m 3) 0.02m 4) 0.04m

18. A steel wire is 1m long and $1mm^2$ in area of cross-section. If it takes 200 N to stretch this wire by 1mm how much force will be required to stretch the wire of the same area and of same material having length of 10m to 1002 cm

1) 100 N 2) 200 N 3) 400 N 4) 2000 N

19. Two bars A and B of circular cross section and of same volume made of same material are subjected to tension. If the diameter of A is half that of B and if the force applied to both the rods is the same and it is in the elastic limit the ratio of extension of A to that of B will be ---

1) 16 2) 8 3) 4 4) 2

20. An iron wire of length 4m and diameter 2mm is loaded with a weight of 8kg. If the Young's modulus 'Y' for iron is $2x10^{11}Nm^{-2}$ then the increase in length of the wire is ---($g = 10\ ms^{-2}$)

1) 0.2mm 2) 0.5mm 3) 2 mm 4) 1 mm

21. Two wires of the same material have masses in the ratio 3:4 . The ratio of their extensions under the same load if their lengths are in the ratio 9:10 is

1) 5 : 3 2) 27 : 40 3) 6 : 5 4) 27 : 25

22. Two exactly similar wires of steel ($Y = 20 \times 10^{11}$ dy/cm^2) and copper ($Y=12x10^{11}dy/cm^2$) are stretched by equal forces. If the total elongation is 1cm elongation of copper wire is

1) 3/5 cm 2) 5/3 cm 3) 3/8 cm 4) 5/8 cm

23. Two wires of same length and thickness are joined end to end. Their Young's moduli are $15x10^{10}$pa and $20x10^{10}$pa. If the combina-tion is stretched by a certain load, the elongations of these wires will be in the ratio

1) 3:4 2) 4:3 3) 1:1 4) 1:2

24. A steel wire of length 20 cm is stretched to increase its length by 0.2 cm. Find the lateral strain in the wire if the Poisson's ratio for steel is 0.19.

1) 0.0019 2) 0.0008 3) 0.019 4) 0.008

25. Two rods of different materials having coefficients of linear expansion α_1 and α_2 and youngs modulii Y_1 and Y_2 respectively are fixed between two rigid massive walls.The rods are heated such that they undergo the same increase in temp. There is no bending of rods. If $\alpha_1 : \alpha_2$ = 2:3. The thermal stresses developed in the two rods are equal if $Y_1 : Y_2$ is equal to

1) 4 : 9 2) 3 : 2 3) 9 : 4 4) 2 : 2

26. A tungsten wire, 0.5 mm in diameter, is just stretched between two fixed points at a temperature of 40°C. Determine the tension in the wire when the temperature falls to 20°C. (coefficient of linear expansion of tungsten $= 4.5 \times 10^{-6}/^0C$; Young's modulus of tungsten $= 3.45 \times 10^{10}Nm^{-2}$)

1) 6.097N 2) 3.097N 3) 5.097N 4) 7.097N

*27. A uniform steel rod of length 1m and area of cross section $20\ cm^2$ is hanging from a fixed support. Find the increase in the length of the rod.

($Y_{steel}=2.0x10^{11}Nm^{-2}, \rho_{steel} =7.85x10^3kgm^{-3}$)

1) 1.923×10^{-5}cm 2) 2.923×10^{-5}cm
3) 1.123×10^{-5}cm 4) 3.123×10^{-5}cm

RIGIDITY MODULUS & BULK MODULUS :

28. A 2.0 cm cube of some substance has its upper face displaced by 0.15 cm, by a tangential force of 0.30N fixing its lower face. Calculate the rigidity modulus of the substance.

1) $1 \times 10^4N/m^2$ 2) $2 \times 10^4N/m^2$
3) $3 \times 10^4N/m^2$ 4) $0.5 \times 10^4N/m^2$

29. A spherical ball of volume $1000\ cm^3$ is subjected to a pressure of 10 atmosphere. The change in volume is $10^{-2}cm^3$. If the ball is made of iron find its bulk modulus. (Atmospheric pressure $=1x10^5Nm^{-2}$)

1) $1 \times 10^{11}\ N/m^2$ 2) $2 \times 10^{11}\ N/m^2$
3) $3 \times 10^{11}\ N/m^2$ 4) $4 \times 10^{11}\ N/m^2$

30. A fractional change in volume of oil is 1 percent when a pressure of $2x10^7\ Nm^{-2}$ is applied. Calculate the bulk modulus and its compressibility

1) $3x10^8\ Nm^{-2}$, $0.33x10^{-9}\ m^2N^{-1}$
2) $5x10^9\ Nm^{-2}$, $2x10^{-10}\ m^2N^{-1}$
3) $2x10^9\ Nm^{-2}$, $5x10^{-10}\ m^2N^{-1}$
4) $2x10^9\ Nm^{-2}$, $5x10^{-9}\ m^2N^{-1}$

31. Determine the pressure required to reduce the given volume of water by 2%. Bulk modulus of water is $2.2x10^9Nm^{-2}$.

1) $4.4 \times 10^7\ N/m^2$ 2) $2.2 \times 10^7\ N/m^2$
3) $3.3 \times 10^7\ N/m^2$ 4) $1.1 \times 10^7\ N/m^2$

32. When 200ml of water is subjected to a pressure of $2x10^6$ Pa. The decrease in its volume is 0.2ml. The compressibility of water is ---

1) $5x10^{-8}\ m^2N^{-1}$ 2) $5x10^{-10}\ m^2N^{-1}$
3) $5x10^{-12}\ m^2N^{-1}$ 4) None

33. A cubical ball is taken to a depth of 200m in a sea. The decrease in volume observed to be 0.1%. The bulk modulus of the ball is – ($g = 10\ ms^{-2}$)

1) $2x10^7$ Pa   2) $2x10^6$ Pa
3) $2x10^9$ Pa   4) $1.2x10^9$ Pa

34. A material has poissions ratio 0.5 . If a uniform rod of it suffers a longitudinal strain of 2×10^{-3}, then the percentage increase in its volume is

1) 0%   2) 10%   3) 20%   4) 5%

35. The increase in length of a wire on streching is 0.025%. If its poisson ratio is 0.4, then the percentage decrease in the diameter is :

1) 0.01   2) 0.02   3) 0.03   4) 0.04

36. A material has normal density ρ and bulk modulus K. The increase in the density of the material when it is subjected to an external pressure 'p' from all sides is

1) $\frac{P}{\rho K}$ 2) $\frac{KP}{\rho}$ 3) $\frac{P\rho}{K}$ 4) $\frac{K\rho}{p}$

37. A wire can be broken by 400 kg.wt. The load required to break the wire of double the thickness of the same material will be -

1) 800 kg wt   2) 1600 kg wt
3) 3200 kg wt   4) 6400 kg wt

38. A cable breaks if stretched by more than 2mm. It is cut into two equal parts. By how much either part can be stretched without breaking

1) 0.25mm   2) 0.5mm   3) 1mm   4) 2mm

39. The breaking stress of steel is $7.9x10^9\ Nm^{-2}$ and density of steel is $7.9x10^3\ kgm^{-3}$ and g = $10ms^{-2}$. The maximum length of steel wire that can hang vertically without breaking is

1) 10^3 m   2) 10^4 m   3) 10^2 m   4) 10^5 m

STRAIN ENERGY

40. A metal rod of Young's modulus $2\times10^{10}\ Nm^{-2}$ undergoes an elastic strain of 0.02% the energy per unit volume stored in the rod in joule/m³ is

1) 400   2) 800   3) 1200   4) 1600

41. A brass rod has length 0.2m area of cross section $1.0cm^2$ and Young's modulus $10^{11}Nm^{-2}$. If it is compressed by 5kg. wt along its length then the change in its energy will be

1) An increase of 2.4×10^{-5} J
2) A decrease of 2.4×10^{-5} J
3) An increase of 2.4×10^{7} J
4) A decrease of 2.4×10^{7} J

42. A metal wire of length 2.5 m and area of cross section $1.5x10^{-6}m^2$ is stretched through 2 mm. Calculate the work done during stretching. ($Y=1.25x10^{11}Nm^{-2}$)

1) 0.15 J   2) 0.51 J   3) 1.5 J   4) 5.1 J

43. A steel wire of length 1.5 m and area of cross section 1.5 mm² is stretched by 1.5cm. then the work done per unit volume. ($Y=2x10^{11}Nm^{-2}$)

1) 1 x 10^7J/m³   2) 2 x 10^7J/m³
3) 3 x 10^7J/m³   4) 4 x 10^7J/m³

44. When a steel rod of length 2m is compressed its length decreases by 0.1cm. Find the work done against the compressive stress. ($Y_{steel}=2.0x10^{11}Nm^{-2}$)

1) 2.5 x 10^4J   2) 1.25 x 10^4 J
3) 5.0 x 10^4J   4) 7.5 x 10^4J

45. The youngs modulus of the material of a rod is $20x10^{10}$ pascal. When the longitudinal strain is 0.04%, The energy stored per unit volume is

1) 4 x 10^{-3} J/m³   2) 8 x 10^3 J/m³
3) 16 x 10^{-3} J/m³   4) 16 x 10^3 J/m³

46. The work done in stretching a wire by 1mm is 2J. The work necessary for stretching another wire of same material but with double the radius of cross-section and half the length by 1mm is in Joules [1991 M]

1) 16   2) 8   3) 4   4) 1/4

47. Two wires of same radius and length are subjected to the same load. One wire is of steel and the other is of copper. It the Young's modulus of steel is twice that of copper, the ratio of elastic energy stored per unit volume in steel to that of copper wire is [UPSEAT 2000]

1) 1:2   2) 2:1   3) 1:4   4) 4:1

48. A spring of force constant 800N/m has an extension of 5cm. The work done in extending it from 5cm to 15cm is [AIEEE 2002]

1) 16J   2) 8J   3) 32J   4) 24J

ANSWERS

1) 1	2) 2	3) 3	4) 2	5) 1
6) 3	7) 1	8) 3	9) 3	10) 2
11) 2	12) 1	13) 1	14) 2	15) 2
16) 3	17) 1	18) 3	19) 1	20) 2
21) 4	22) 4	23) 2	24) 1	25) 2
26) 1	27) 1	28) 1	29) 1	30) 3
31) 1	32) 2	33) 3	34) 1	35) 1

36) 3 37) 2 38) 3 39) 4 40) 1
41) 1 42) 1 43) 1 44) 1 45) 4
46) 1 47) 1 48) 2

EXERCISE - II(B)

YOUNG 'S MODULUS

1. A brass wire of diameter 1 mm and of length 2m is stretched by applying a force of 20N. Then the stress in the wire is

1) $2.546 \times 10^7 N/m^2$ 2) $3.146 \times 10^7 N/m^2$
3) $2.346 \times 10^7 N/m^2$ 4) $2.246 \times 10^7 N/m^2$

2. If an iron wire is stretched by 1% what is the strain on the wire ?

1) 0.01 2) 0.005 3) 0.02 4) 0.03

3. A brass wire of diameter 1 mm and of length 2m is stretched by applying a force of 20N. If the increase in length is 0.51 mm, Then the Young's modulus of the wire is

1) $8.848 \times 10^{10} N/m^2$ 2) $7.984 \times 10^{10} N/m^2$
3) $6.984 \times 10^{10} N/m^2$ 4) $9.984 \times 10^{10} N/m^2$

4. Force required to increase the length of wire of area of cross section 'A' by one percent, if 'Y' is young's modulus, is

1) AY 2) $\frac{AY}{10}$ 3) $\frac{AY}{100}$ 4) $\frac{AY}{1000}$

5. Two wires of same length and material but with radii 0.1mm and 0.2mm are loaded with the same weight. Then

1) Thick wire extends more
2) Thin wire extends more
3) Both undergo same extension 4) None

6. A wire of length 'L' and radius 'r' is fixed at one end and a force 'F' is applied to the other end to produce an extension 'l'. The extension produced in another wire of same material of length '2L' and radius '2r' by a force '2F' is --

1) l 2) $2l$ 3) $l/2$ 4) $4l$

7. Two wires of different material have same lengths and areas of cross section. What is the ratio of their increases in length when the forces applied are the same ?

$(Y_1=0.90 \times 10^{11} Nm^{-2},\ Y_2=3.60 \times 10^{11} Nm^{-2})$

1) 4 : 1 2) 1 : 4 3) 2 : 1 4) 1 : 2

8. The stress required to double the length of a wire of Young's modulus 'E' is ---

1) E 2) 2E 3) 4E 4) E/2

9. The length of a brass wire is 1m. The strain produced by a load is 0.001. The change in its length is

1) 1cm 2) 1mm 3) 10 cm 4) 2 cm

10. The Young's modulus of a material having a cross-sectional area of $1 cm^2$ is 2×10^{12} dyne/cm^2. If the length is to be doubled the force required is

1) 4×10^7 N 2) 2×10^7 N 3) 10^7 N 4) Zero

11. A metal ring of inner radius r_1 and cross-sectional area 'A' is fitted onto a wooden disc of radius $r_2 > r_1$. If 'Y' is the young's modulus of the metal then the tension in the ring is

1) $\frac{AYr_2}{r_1}$ 2) $\frac{AY(r_2-r_1)}{r_1}$

3) $\frac{Y(r_2-r_1)}{Ar_1}$ 4) $\frac{Yr_1}{Ar_2}$

12. The length of a metal wire is l when the tension in it is 'F' and 'xl' when the tension is 'yF'. Then the natural length of the wire is

1) $\frac{(x-y)\,l}{x-1}$ 2) $\frac{(y-x)\,l}{y-1}$

3) $\frac{(x-y)\,l}{x+1}$ 4) $\frac{(y-x)\,l}{y+1}$

13. Two wires made of the same material have lengths 3 m & 4 m and masses 18 gm and 16 gm respectively . When they are subjected to the same force, then the ratio of their elongations is

1) 1 : 2 2) 4 : 3 3) 16 : 18 4) 9 : 4

14. Two wires of same material and same length but diameters in the ratio 2:3 are stretched by the same length. The ratio of the forces applied on two wires is ----

1) 4:3 2) 4:9 3) 2:5 4) 9:4

15. When a uniform wire of radius 'r' is stretched by a 2kg weight, the increase in its length is 2.00 mm. If the radius of the wire is r/2 and other conditions remaining the same, the increase in its length is [2000 E]

1) 2.00mm 2) 4.00mm
3) 6.00mm 4) 8.00mm

16. The extension of a wire by application of load is 0.3cm. The extension in a wire of same material but of double the length and half the radius of cross section by the same load will be in (cm)

1) 0.3 2) 0.6 3) 0.2 4) 2.4

17. A wire whose cross sectional area is $2mm^2$ is stretched by 0.1mm by a certain load, If a similar wire of triple the area of cross section is stretched by the same load the elongation of the second wire would be

1) 0.33mm 2) 0.033mm
3) 3.3mm 4) 0.0033mm

18. A brass wire of length 300 cm when subjected to a force F produces an elongation "a". Another wire of twice the diameter and of same length and material , when subjected to the force F produces an elongation b. Then the value of 'a/b' is

1) 1 : 1 2) 4 : 1 3) 2 : 1 4) 1 : 2

19. A hollow cylinder of inner radius 3 cm and outer radius 5 cm and a solid cylinder of radius 2 cm are subjected to the same force. If they are made of same material and of same length, then the ratio of their elongations is

1) 1 : 1 2) 1 : 2 3) 1 : 4 4) 2 : 3

20. A wire of length 10 m and of cross -section 1×10^{-4} sq.m is fixed at one end and a load of 1200 kg is attached at the other end. Then the extension produced is ($Y = 9.8 \times 10^{10} N/m^2$)

1) 12×10^{-3} m 2) 12×10^{-5} m
3) 12×10^{-4} m 4) 12×10^{-2} m

21. Two wires of the same material but of lengths in the ratio 2:3 and diameters in the ratio of 3:2 are stretched with equal force. Then the ratio of elongations produced is

1) 9 : 4 2) 1 : 1 3) 8 : 27 4) 9 : 16

22. An iron wire and copper wire having same length and cross-section are suspended from same roof Young's modulus of copper is 1/3rd that of iron. Then the ratio of the weights to be added at their ends so that their ends are at the same level is

1) 1:3 2) 1:9 3) 3:1 4) 9:1

23. Two wires A and B of same material have lengths in the ratio 1:2 and diameters in the ratio 2:1. If they are stretched by same force then the ratio of the increase of their lengths will be ----

1) 2:1 2) 1:4 3) 8:1 4) 1:8

24. Find the fractional increase in the volume of a wire of circular cross section if its longitudinal strain is 2%. $(\sigma = 0.4)$

1) 0.004 2) 0.002
3) 0.001 4) 0.005

25. Two rods of different materials are clamped at their ends rigidly. When they are heated for the same rise in temperature, same thermal stresses are produced in them. If their Young's modulii are in the ratio x:y then ratio of coefficients of their linear expansion is ....

1) x : y 2) y : x 3) $x^2 : y^2$ 4) $y^2 : x^2$

26. A steel wire of cross sectional area $2\ mm^2$ is just stretched horizontally between two fixed points at a temperature of 35°C. The temperature of the wire changed to 25°C. Then (Coefficient of linear expansion of steel = $11 \times 10^{-6}/^0C$; $Y_{steel} = 2 \times 10^{11} Nm^{-2}$)

1) the thermal strain is 11×10^{-5} ;

2) the thermal stress is $2.2 \times 10^7\ N/m^2$

3) the tension in the wire at temperature 25°C is 44N

4) All the above

27. A rope of rubber of density $1.5 \times 10^3\ Kg/m^3$ and Young's modulus $5 \times 10^6\ N/m^2$, 8m in length is hung from the ceiling of a room. Then the increase in length due to its own weight is

1) 96×10^{-2} m 2) 12×10^{-4} m
3) 96×10^{-3} m 4) 12×10^{-3} m

RIGIDITY MODULUS & BULK MODULUS :

28. A tangential force of 1000N is applied on upper surface of area $2\ mm^2$ which is 0.05 m from its fixed face. Find the shift of the upper surface with respect to its fixed face if its modulus of rigidity is $0.30 \times 10^{11}\ Nm^{-2}$.

1) 0.833 mm 2) 0.734 mm
3) 0.433 mm 4) 0.533 mm

29. A copper cube of each side of length 1 cm is subjected to a pressure of 100 atmospheres. Find the change in its volume if the bulk modulus of copper is $1.4 \times 10^{11} Nm^{-2}$. (Atmospheric pressure=$1 \times 10^5 Nm^{-2}$)

1) $0.7143 \times 10^{-10}\ m^3$ 2) $0.5143 \times 10^{-10}\ m^3$
3) $0.4143 \times 10^{-10}\ m^3$ 4) $0.8243 \times 10^{-10}\ m^3$

30. If a rubber ball is taken down to a 100 m deep lake its volume is decreased by 0.1%. If g = 10 m/sec^2 . Then the bulk modulus of elasticity is

1) 10^9Pa 2) 10^{10}Pa
3) 10^{11}Pa 4) $3x10^9$Pa

31. Find the pressure to be applied to increase the density of water by 1% if the bulk modulus of water is $2.2x10^9Nm^{-2}$.

1) 2.2×10^7 N/m^2 2) 1.2×10^7 N/m^2
3) 3.2×10^7 N/m^2 4) 4.2×10^7 N/m^2

32. A sphere contracts in volume by 0.01% when taken to the bottom of sea 1km deep. If the density of water is 1gm /cc Then the bulk modulus of water is

1) 9.8×10^5 Nm^{-2} 2) 9.8×10^8 Nm^{-2}
3) 9.8×10^{10} Nm^{-2} 4) 9.8×10^6 Nm^{-2}

33. A ball falling in a lake of depth 200m shows a decrease of 0.1% in its volume at the bottom. What is the bulk modulus [AFMC 1997]

1) 19.6×10^8N/m^2 2) 19.6×10^{-10}N/m^2
3) 19.6×10^{10}N/m^{-2} 4) 19.6×10^{-8}N/m^{-2}

34. A uniform cube is subjected to volume compression. If each side is decreased by 1%. Then the bulk strain is

1) 0.01 2) 0.06 3) 0.02 4) 0.03

35. The longitudinal strain in a metal bar is 0.05. If the Poission's ratio for the metal is 0.25, the lateral strain is

1) 0.0125 2) 0.4 3) 0.6 4) 0.025

36. Estimate the change in the density of water in ocean at a depth of 500 m below the surface. The density of water at the surface = 1030 kgm^{-3} and the bulk modulus of water = $2.2x10^9Nm^{-2}$

1) 2.363 kg/m^3 2) 1.363 kg/m^3
3) 4.363 kg/m^3 4) 3.363 kg/m^3

37. The breaking force of a wire is 'F'. When the thickness of the wire is doubled, the breaking force will be

1) 2F 2) 4F 3) F 4) 8F

38. A wire can support a load Mg without breaking. It is cut into two equal parts. The maximum load that each part can support is

1) Mg/4 2) Mg/2 3) Mg 4) 2Mg

39. Breaking stress for steel is $8x10^6$ Nm^{-2}. The density of steel is $8x10^3$ kgm^{-3} and g = 10ms^{-2}. What will be the maximum length of steel wire which will hang without breaking under its load.

1) 200m 2) 100m 3) 50m 4) 25m

STRAIN ENERGY

40. A rod of Young's modulus 20 GPa undergoes a linear strain of $6x10^{-4}$. Then increase in its energy density is ---

1) 3600 Jm^{-3} 2) 7200 Jm^{-3}
3) 1800 Jm^{-3} 4) None

41. A wire is suspended vertically from one of it's ends is stretched by attaching a weight of 200N to the lower end. The weight stretches the wire by 1mm. Then the elastic energy stored in the wire is [AIEEE 2003]

1) 0.2J 2) 10J 3) 20J 4) 0.1J

42. A brass wire of length 1m and area of cross section 1 mm^2 is stretched by applying a force of 20N. then the work done per unit volume of the wire is($Y_{brass}=0.91x10^{11}Nm^{-2}$)

1) 2198 J 2) 1049 J
3) 4422 J 4) 4521 J

43. When a copper wire of area of cross section 1 mm^2 is stretched by a force of 10N. Then the work done in stretching the wire is ($Y=1.2x10^{11}Nm^{-2}$)

1) 416.7 J 2) 208.35 J
3) 833.4 J 4) 316.7 J

44. A steel wire of length 80 cm and of area of cross section 1.5 mm^2 is stretched. Find the work done in increasing its length from 80.2 cm to 80.3 cm. ($Y=2.0x10^{11}Nm^{-2}$)

1) 0.9375 J 2) 0.5375 J
3) 0.6374 J 4) 0.8695 J

45. A metal wire 4m long and $2x10^{-7}$ sq.m in cross-section is stretched by a force of 30N. If the work done in stretching that wire is $4.5x10^{-2}$J the young's modulus of the wire is

1) $2x10^{11}$ Pa 2) $4x10^{11}$ Pa
3) $2x10^{12}$ Pa 4) $4x10^{12}$ Pa

46. A long spring when stretched by 2cm its potential energy is U. If the same spring is stretched by 8cm the potential energy of the spring is ----

1) 8U 2) 16U 3) 4U 4) U

47. Two wires of same material and size but with lengths in the ratio 5:3 are stretched by the same force. The ratio of work done in two cases is

1) 5:8 2) 8:5 3) 5:3 4) 3:5

48. A man of mass 100 kg gets into a car. If the centre of gravity of the car system is depressed by 5mm. Spring constant of the system is ---

1) 1.96×10^5 Nm^{-1} 2) 10^4 Nm^{-1}

3) 1.2 Nm^{-1} 4) Infinity

ANSWERS

1) 1	2) 1	3) 4	4) 3	5) 2
6) 1	7) 1	8) 1	9) 2	10) 2
11) 2	12) 2	13) 1	14) 2	15) 4
16) 4	17) 2	18) 2	19) 3	20) 1
21) 3	22) 3	23) 4	24) 1	25) 2
26) 4	27) 3	28) 1	29) 1	30) 1
31) 1	32) 3	33) 1	34) 4	35) 1
36) 1	37) 2	38) 3	39) 2	40) 1
41) 4	42) 1	43) 1	44) 1	45) 1
46) 2	47) 3	48) 1		

EXERCISE - III(A)

1. What mass must be suspended from the free end of a tungsten wire of length 50 cm and diameter 0.25 mm to stretch it by 1 mm (Y of tungsten = 3.448×10^{11} Nm^{-2})

1) 3.455 kg 2) 4.445 kg

3) 1.445 kg 4) 2.445 kg

2. The length of a rubber cord is l_1 meters when the tension is 4N and l_2 meters when the the tension is 5N. The length in meters when the tension is 9 N is

1) $5l_1 - 4l_2$ 2) $5l_2 - 4l_1$

3) $9l_1 - 8l_2$ 4) $8l_2 - 5l_1$

3. A wire of length 1m and radius 1mm is subjected to a load. The extension is 'x'. The wire is melted and then drawn into a wire of square cross-section of side 1mm. What is its extension under the same load?

1) $\pi^2 x$ 2) πx^2 3) πx 4) π/x

4. A steel wire of length 1 m and cross sectional area 1.5 mm^2 is hung from a rigid support, with a stone of volume 2000 cm^3 hanging from the other end. Find the decrease in the length of the wire, when the stone is completely immersed in water.

($Y_{steel} = 2 \times 10^{11} Nm^{-2}$, $\rho_{water} = 10^3 kgm^{-3}$)

1) 0.06533 mm 2) 0.02533 mm

3) 0.04533 mm 4) 0.01533 mm

5. Two wires of different materials, each 2m long and of diameter 2mm, are joined in series to form a composite wire. What force will produce a total extension of 0.9 mm.

($Y_1 = 2 \times 10^{11}$ pa & $Y_2 = 6 \times 10^{11}$ Pa)

1) 282.6 N 2) 212 N

3) 319.8 N 4) 382.6 N

6. An aluminium wire and a copper wire of the same length and cross section are joined end to end. The composite wire is hung from a rigid support and a load is suspended from the free end. if the increase in length of the composite wire is 1.8mm, find the increase in the length of each wire.

($Y_{Al} = 7 \times 10^{10} Nm^{-2}$, $Y_{Cu} = 1.1 \times 10^{11} Nm^{-2}$)

1) 1.1 mm, 0.7 mm

2) 0.7 mm, 1.1 mm

3) 2.2 mm, 1.4 mm

4) 1.4 mm, 2.2 mm

7. Two wires A & B are identical in shape and size and are stretched by same magnitude of force. Then the extensions are found to be 0.2% and 0.3% respectively. Find the ratio of their Young's modulii

1) 2 : 3 2) 3 : 2 3) 4 : 9 4) 9 : 4

8. One end of a uniform wire of length L and of weight W is attached rigidly to a point in the roof and a weight W_1 is suspended from its lower end. If 'S' is the area of cross- section of the wire, then find the stress in the wire at height (3L/4) from its lower end?

1) $\left(\frac{w_1 + (3w/4)}{S}\right)$ 2) $\left(\frac{w + (3w_1/4)}{S}\right)$

3) $\frac{w_1 + w}{4}$ 4) $\left(\frac{3w_1 + (w/4)}{S}\right)$

9. The adjacent graph shows the extension $(\Delta \ell)$ of a wire of length 1 m suspended from the top of a roof at one end with a load W connected to the other end. If the cross sectional area of the wire is $10^{-6}m^2$, calculate the young's modulus of the material of the wire

1) $2\times10^{11} N/m^2$
2) $2\times10^{-11} N/m^2$
3) $3\times10^{-12} N/m^2$
4) $2\times10^{-13} N/m^2$

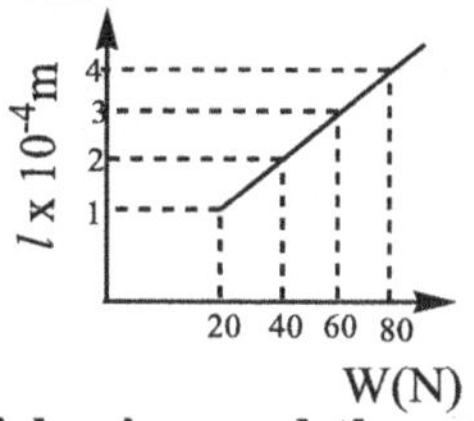

10. A lift is tied with thick wires and the mass of the lift is one metric ton. If the maximum acceleration of the lift is $1ms^{-2}$ and the maximumstressof the wire is $1.4x10^8Nm^{-2}$. If g = $10ms^{-2}$ the minimum diameter of that wire is ----

1) 10m 2) 10^{-1}m 3) 10^{-2}m 4) None

11. A light rod of length 200cm is suspended from the ceiling horizontally by means of two vertical wires of equal length tied to its ends. One of the wires is made of steel and is of cross-section 0.1sq.cm and the other is of brass of cross-section 0.2 sq.cm. Find the position along the rod at which a weight may be hung to produce (i) equal stresses in both wires and (ii) equal strains in both wires.

Y_{brass} = $10x10^{11}$ dyne/cm^2 and
Y_{steel} = 20 x 10^{11} dyne/cm^2

1) 133.3cm; 100cm 2) 167cm; 50cm
3) 200cm; 100cm 4) none

12. A wire 3m in length and 1mm in diameter at 30^0 c and kept in a low temperature at 170^0c and is stretched by hanging a weight of 10kg at one end calculate the change in the length of the wire. Given $\alpha=1.2x10^{-5}/^0C$ $Y=2x10^{11}$ Nm^{-2} $g=10ms^{-2}$

1) $2x10^{-3}$ m 2) $7.2x10^{-3}$ m
3) 95.6 x 10^{-5} m 4) 5.2 m

13. A long steel wire of length 'L' is suspended from the ceiling of a room. A sphere of mass 'm' and radius 'r^1' is attached to the lower end of the wire. the height of the ceiling is ($L+2r^1+l$). When the sphere is made to oscillate as a pendulum, its lowest point just touches the floor. The velocity of the sphere at the lowest point will be (L >> r^1,l and r is the radius of the wire)

1) $\sqrt{\frac{\pi r^2 ye}{m}-Lg}$ 2) $\sqrt{\frac{\pi r^2 y}{me}-Lg}$ 3) $\sqrt{\frac{y\pi r^2 e}{m}}$ 4) None

14. A block of mass 2 kg is attached to one end of a wire of cross sectional area 1 mm^2 and is whirled in a vertical circle of radius 40 cm. The speed of the block at the bottom of the circle $5ms^{-1}$. Find the elongation of the wire when the block is at the bottom. ($Y=2x10^{11}Nm^{-2}$)

1) 0.2892 mm 2) 0.054 mm
3) 0.1446 mm 4) 0.0017 mm

15. A stone of mass 2 kg is fastened to one end of a steel wire of cross sectional area 2 mm^2 and is whirled in a horizontal circle of radius 25cm. If the breaking stress of steel is $1.2x10^9Nm^{-2}$ find the maximum number of revolutions the stone can make per minute without the string breaking.

1) 661.5 rpm 2) 331.5 rpm
3) 221.5 rpm 4) 441.5 rpm

16. A metal cube of side length 8.0cm has its upper surface displaced with respect to the bottom by 0.10 mm when tangential force of 4×10^4N is applied at the top with bottom surface fixed. The rigidity modulus of the material of the cube is

1) $4\times10^9Nm^{-2}$ 2) $5\times10^9Nm^{-2}$
3) $8\times10^9Nm^{-2}$ 4) 10^8Nm^{-2}

17. A metal rope of density $6000kgm^{-3}$ has breaking stress $9.8x10^8$ Nm^{-2}. This rope is used to measure the depth of the sea. Then the depth of the sea that can be measured without breaking is -----

1) $10x10^3$ m 2) $20x10^3$ m
3) $30x10^3$ m 4) $40x10^3$ m

18. The upper end of a wire of radius 4mm and length 100cm is clamped and its other end is twisted through an angle of 30^0 then the angle of shear is

1) 0.012^0 2) 0.12^0 3) 1.2^0 4) 12^0

19. A uniform metal rod of 2mm^2 cross section is heated from 0^0C to 20^0C. The coefficient of linear expansion of the rod is $12\times10^{-6}/^0C$. Its youngs modulus of elasticity is 10^{11} Nm^{-2}. The energy stored per unit volume of the rod is

1) 2880 Jm^{-3} 2) 1500 Jm^{-3}
3) 5760 Jm^{-3} 4) 1440 Jm^{-3}

20. Two springs of spring constants 1500N/m and 3000 N/m respectively are stretched by the same force. The potential energy possessed by the two will be in the ratio [MP PMT 1998]

1) 4:1 2) 1:4 3) 2:1 4) 1:2

21. A boy has a catapult made of a rubber cord of length 42cm and diameter 6.0mm. The boy stretches the cord by 20cm to catapult a pebble of mass 20gm. The pebble flies off with a speed of 20m/sec. Find the young's modulus for rubber. Ignore the change in the cross-section of the cord in stretching

Hint : $\frac{1}{2}$ Fe = $\frac{1}{2}$ mV^2

1) 5.97×10^6 N/m² 2) 2.97×10^6 N/m²
3) 0.97×10^6 N/m² 4) 9.7×10^6 Nm²

22. One end of a long metallic wire of length L is tied to the ceiling. The other end is tied to massless spring of spring constant K.A mass (m) hangs freely from the free end of the spring. Th area of crossection and youngs modulus of the wire are A and Y respectively. If the mass is slightly pulled down and released it will oscillate with a time period T equal to

1) $2\pi\sqrt{\frac{m}{k}}$ 2) $2\pi\sqrt{\frac{m(YA+KL)}{YAK}}$

3) $2\pi\sqrt{\frac{mYA}{KL}}$ 4) $2\pi\sqrt{\frac{mL}{YA}}$

23. On loading a metal wire of cross section 10^{-6} m² and length 2m by a mass of 210 kg it extends by 16mm and suddenly broke from the point of support. If density of that metal is 8000 kgm^{-3} and its specific heat is 420 Jkg^{-1} K^{-1} the rise in temperature of wire is (g = 10 ms^{-2})

1) 2.5^0c 2) 5^0c 3) 6^0c 4) 10^0c

ANSWERS

1) 1	2) 2	3) 1	4) 1	5) 2
6) 1	7) 2	8) 1	9) 1	10) 3
11) 1	12) 3	13) 1	14) 1	15) 1
16) 2	17) 2	18) 2	19) 1	20) 3
21) 2	22) 2	23) 1		

EXERCISE - III(B)

1. A steel wire of length 1.5 m and diameter 0.5 mm is stretched by applying a weight of 1 kg wt. Then find the increase in the length of wire ($Y=2 \times 10^{11}$ Nm^{-2})

1) 0.2743 mm 2) 0.3743 mm
3) 0.1743 mm 4) 0.4743 mm

2. The length of a rubber cord is l_1 meters when the tension 4N and l_2 meters when the the tension is 5N. The original length of wire in meters is

1) $5l_1 - 4l_2$ 2) $5l_2 - 4l_1$
3) $9l_1 - 8l_2$ 4) $8l_2 - 5l_1$

3. The elongation of a steel wire, strectched by a force is e. If a wire of the same material, double the length and half the diameter is subjected to double the force, its elongation will be [1999 E]

1) 16e 2) 4e 3) (1/4)e 4) (1/16)e

4. A brass wire of cross sectional area 2 mm² is suspended from a rigid support and a body of volume 100 cm³ is attached to its other end. If the decrease in the length of the wire is 0.11mm, when the body is completely immersed in water, find the natural length of the wire.

($Y_{brass}=0.91\times10^{11}Nm^{-2}$, $\rho_{water}=10^3kgm^{-3}$)

1) 2.043 m 2) 1.043 m 3) 4.043 m 4) 3.043 m

5. Two wires of the same length and radius are stretched by the same load and the difference of their elongations produced is found to be 0.25 cm . If their modulus of elasticity are in the ratio 5:3. Then the individual elongations are

1) 0.375cm, 0.625 cm 2) 0.625 cm, 0.650 cm
3) 0.325 cm, 0.350cm 4) 0.675 cm, 0.325cm

6. An aluminium wire and a steel wire of the same length and cross section are joined end to end. The composite wire is hung from a rigid support and a load is suspended from the free end. If the increase in length of the composite wire is 1.35 mm, find the ratio of the i) stress in the two wires and ii) strains in the two wires.

($Y_{Al}=7\times10^{10}Nm^{-2}$, $Y_{steel}=2\times10^{11}Nm^{-2}$)

1) 1 : 1, 20 : 7 2) 1 : 1, 10 : 7
3) 20 : 7, 1 : 1 4) 10 : 7, 1 : 1

7. An Aluminium wire and a steel wire of the same length and cross section are joined end to end. The composite wire is hung from a rigid support and a load is suspended from the free end. The young's modulus of steel 20/7 times that of the Aluminium. The ratio of increase of length of steel to Aluminium is ....

1) 20/7 2) 400/49 3) 7/20 4) 49/400

8. The stress versus strain graphs for wires of two materials A and B are as shown in the figure. If Y_A and Y_B are the Young's modulii of the materials, then

1) $Y_B = 2Y_A$
2) $Y_A = Y_B$
3) $Y_B = 3Y_A$
4) $Y_A = 3Y_B$

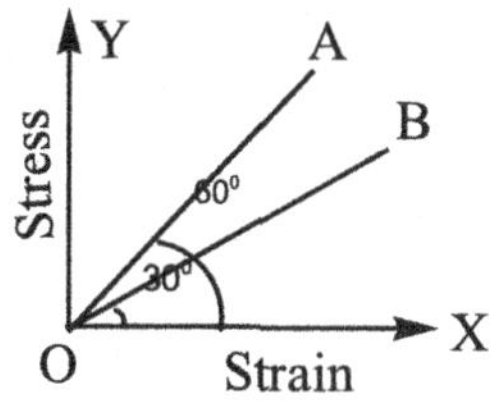

9. One end of a uniform wire of length ℓ and weight W is attached to a point in the roof and weight W/2 is suspended from the lower end. If A is the area of cross section of the wire then stress in the wire at its mid point is

1) W/2A 2) 2W/A 3) 3W/2A 4) W/A

10. When a mass is suspended from the end of a wire the top end of which is attached to the roof of the lift the extension is 'e' when the lift is stationary. If the lift moves up with a constant acceleration g/2 the extension of the wire would be ....

1) 2e/3 2) 3e/2 3) 2e 4) 3e

11. A steel wire of diameter 0.8mm and length 1m is clamped firmly at two points A and B which are 1m apart and in the same horizontal plane. A body is hung from the middle point of the wire such that the middle point sags 1cm lower from the original position. Calculate the mass of the body. Given Y = 2 x 10^{12} dyne/cm^2

1) 60 gr 2) 82 gr 3) 100gr 4) 112 gr

12. A 20kg load is suspended from the lower end of a wire 20cm and 1mm^2 in cross sectional area. The upper half of the wire is iron and the lower half is aluminium

Find the elongation in the wire

(Y_{iron}= 20x10^{10} Nm^{-2} Y_{Al}=7x10^{10} Nm^{-2})

1) 18.9x10^{-3}m 2) 18.9x10^{-5}m
3) 1.78x10^{-3}m 4) None

13. A mass m kg is whirled in a vertical plane by tying it at the end of a flexible wire of length l and area of cross section 'A'. When the mass is at its lowest position the strain produced in the wire is, if Youngs modulus of the wire is 'Y'

(if V= $\sqrt{5g\ell}$)

1) $\frac{AY}{6mg}$ 2) $\frac{6mg}{AY}$ 3) $\frac{5mg}{AY}$ 4) $\frac{AY}{5mg}$

14. A block of mass 2 kg is attached to one end of a wire of cross sectional area 1 mm^2 and is whirled in a vertical circle of radius 40 cm. The speed of the block at the bottom of the circle 5ms^{-1}. Find the elongation of the wire when the block is at the top of the circle. (Y=2x10^{11}Nm^{-2})

1) 0.2892 mm 2) 0.054 mm
3) 0.1446 mm 4) 0.0017 mm

15. The body of mass 5kg is attached to a wire 0.2m long its breaking strength is 5x 10^7 Nm^{-2} It can be rotated in a horizontal circle with maximum angular velocity 5rads^{-1}, then the area of cross-section of the wire is

1) 5 x10^{-7}m^2 2) 5 x 10^{-6}m^2
2) 5 x 10^{-4}m^2 4) 25 x10^{-6}m^2

16. A rectangular block of size 10 cm x 8 cm x 5 cm is kept in three different positions

(I) 10 cm x 8 cm on ground.

(II) 5 cm x 8 cm on ground.

(III) 10 cm x 5 cm on ground.

In each case, the shadded area is rigidly fixed and a definite force F is applied tangentially to the opposite face to deform the block. The displacement of the upper face will be

1) Same in all the three cases
2) Maximum in (I) position
3) Maximum in (II) position
4) Maximum in (III) position

17. A metal rope of density 5000kgm^{-3} has breaking stress 19.6x10^8 Nm^{-2}. This rope is used to measure the depth of the sea. Then the depth of the sea that can be measured without breaking is -----

1) 50x10^3 m 2) 20x10^3 m
3) 30x10^3 m 4) 40x10^3 m

18. A 2 m long rod of radius 1 cm which is fixed from one end is given a twist of 0.8 radians at the other end. The shear strain developed will be

1) 0.002 2) 0.004 3) 0.008 4) 0.016

19. A steel wire of mass 3.16Kg is stretched to a tensile strain of 1×10^{-3}. What is the elastic deformation energy if density ρ=7.9g/cc and Y=2×10^{11}N/m^2 ? [ROORKEE 1998]

1) 4KJ 2) 0.4KJ
3) 0.04KJ 4) 4J

20. A spring of force constant 'K' is cut into two equal parts. The force constant of each part is

1) 2K 2) K/2 3) 4K 4) K/4

21. The rubber cord of a catapult has a cross sectional area $1mm^2$ and total unstretched length 10cm. It is stretched to 12cm and then released to project a missile of mass 5gm. Taking Y for rubber $5x10^8$ Nm^{-2} calculate the velocity of projection ...

1) 10 ms^{-1} 2) 15 ms^{-1}
3) 20 ms^{-1} 4) 25 ms^{-1}

22. Two springs A and B have their force constants K_1 and K_2 ($K_1 > K_2$) when they are stretched by the same force work done is

1) More in case of A 2) More in case of B
3) Same in both the cases 4) None

23. The limiting stress of a typical human bone is $0.9x10^8$ Nm^{-2} while young's modulus is $1.4x10^{10}$ Nm^{-2}. How much energy can be absorbed by two leg bones without breaking if each has a typical length of 50cm and an average cross sectional area of $5cm^2$?

1) 145 J 2) 290 J 3) 490 J 4) 435 J

ANSWERS

1) 2	2) 1	3) 1	4) 1	5) 1
6) 1	7) 3	8) 4	9) 4	10) 2
11) 2	12) 1	13) 2	14) 2	15) 1
16) 4	17) 1	18) 2	19) 3	20) 1
21) 3	22) 2	23) 1		

COMPREHENSION QUESTIONS

Comprehension - I (1 - 3) :

A wire of cross-section A is stretched horizontally between two clamps located $2l$ metre apart. A weight W kg is suspended from the mid point of the wire. If the mid point sags vertically through a distance x<1 then

1. The strain produced in the wire is

1) $\frac{2x^2}{l^2}$ 2) $\frac{x^2}{l^2}$ 3) $\frac{x^2}{2l^2}$ 4) $\frac{x}{l}$

2. The value of stress is

1) $\frac{Wl}{2xA}$ 2) $\frac{Wl}{4xA}$ 3) $\frac{2Wx}{lA}$ 4) $\frac{4xW}{lA}$

3. If the Young's modulus of the material is Y, the value of extension x is

1) $\left(\frac{Wl}{YA}\right)^{1/3}$ 2) $\left(\frac{YA}{Wl}\right)^{1/3}$ 3) $\frac{1}{l}\left[\frac{WA}{Y}\right]^{1/3}$ 4) $l\left[\frac{W}{YA}\right]^{1/3}$

Comprehension - II (4 - 5):

A rod of uniform cross-sectional area A and length L has a weight W. It is supspended verticallyk from a fixed support. If the material of the rod is homogeneous and its modulus of elasticity is Y, then

4. The total elongation produced in the rod due to its own weight is

1) $\frac{WL}{YA}$ 2) $\frac{WL}{2YA}$ 3) $\frac{2WL}{YA}$ 4) $\frac{3WL}{2YA}$

5. If a force F is also applied at the free end of the rod, then the total elongation of the rod is

1) $\frac{(W+F)L}{YA}$ 2) $\frac{(W+F)L}{2YA}$

3) $\frac{2(W+F)L}{YA}$ 4) $\frac{[(W/2)+F]L}{YA}$

Comprehension - III (6 - 7) :

A uniform of length L aqnd mass M is pulled horozontally on a smooth surfce with a force F.

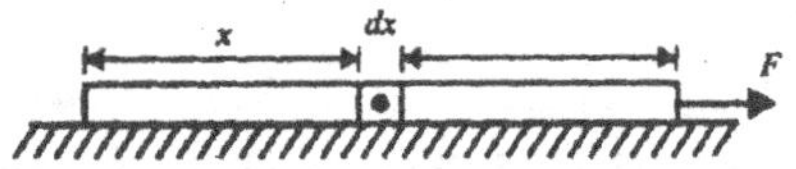

6. The elongation of the rod if the Young's modulus of the material is Y.

1) $\frac{3FL}{2YA}$ 2) $\frac{5FL}{6YA}$ 3) $\frac{2FL}{YA}$ 4) $\frac{FL}{2YA}$

7. If the rod is pulled with two equal and opposite forces each of magnitude F then the elongation produced in the rod is

1) $\frac{3FL}{2YA}$ 2) $\frac{FL}{YA}$ 3) $\frac{2FL}{YA}$ 4) $\frac{FL}{2YA}$

Comprehension - IV (8 - 9):

A light rod of length L=2 m is uspended horizontally from the ceiling by two wires A and B of equal lengths. The wire A is made of steel with the area of cross-section $A_S=1x10^{-5}$ m^2, while the wire B is made of brass of cross-section $A_b=2x10^{-5}$ m^2. A weight W is suspended at a distance x from the wire A as shown in the figure. Take, Young's modulus of steel and brass as $Y_S=2x10^{11}$ N/m^2 and $Y_b=1x10^{11}$ N/m^2.

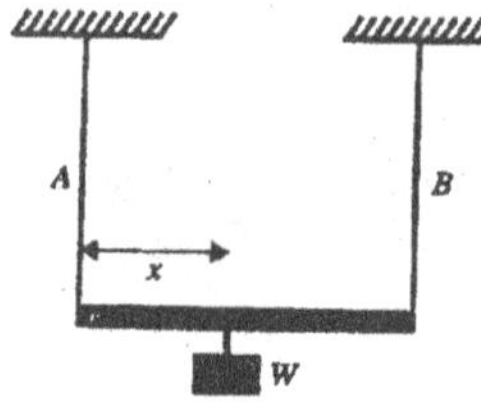

8. **The value of x so that equal stresses are produced in each wire is**

1) 1.33 m 2) 1.5 m 3) 1.6 m 4) 1.75 m

9. **The value of x so that equal strains are produced in each wire is**

1) 1 m 2) 1.33 m 3) 1.5 m 4) 1.75 m

ANSWERS

1) 3	2) 1	3) 4	4) 2	5) 4
6) 4	7) 2	8) 1	9) 1	

ADDITIONAL MATERIAL FOR AIEEE

1. **Relation among elastic contants Y, n, k, σ**

1) $\frac{9}{y}=\frac{1}{k}+\frac{3}{n}$ 2) $y = 2n(1+\sigma)$

3) $y = 3k(1-2\sigma)$ 4) $\sigma = \frac{3k-2n}{2(n+3k)}$

EXERCISE – IV

1. **The Young's modulus and rigidity modulus of a metal are 40 GPa and 20 GPa respectively. Then the bulk modulus is**

1) 13.33GPa 2) 26.67GPa

3) 6.67GPa 4) 120GPa

2. **For a given material, young's modulus is 2.6 times the rigidity modulus. The poisson's ratio is---**

1) 0.5 2) 0.3 3) 0.2 4) 0.6

3. **If the Poissions ratio of a substance is 0.25 then the ratio of its Youngs modulus and rigidity modulus is**

1) 2/5 2) 5/2 3) 4/5 4) 5/4

4. **The Young's modulus of steal is 210 GPa. Its poissions ratio is 0.3. Then its rigidity modulus is**

1) 8.8GPa 2) 80.8GPa

3) 19.6GPa 4) 6.6GPa

5. **For a given material the Youngs modulus is 2.4 times that of its rigidity modulus its poission's ratio is**

1) 2.4 2) 1.2 3) 0.4 4) 0.2

6. **If the Poission's ratio of a solid is 2/5. Find the ratio of its Young's modulus to the rigidity modulus**

1) 5 / 4 2) 7 / 15

3) 14 / 9 4) 14 / 5

7. **If the shear modulus and bulk modulus of a material are $0.42\text{x}10^{11}\text{Nm}^{-2}$ and $0.56 \text{ x } 10^{11} \text{Nm}^{-2}$ respectively then young's modulus of that material is ---**

1) $10^{10}\,\text{Nm}^{-2}$ 2) $1.33\text{x}10^{11}\,\text{Nm}^{-2}$

3) $0.75\text{x}10^{11}\,\text{Nm}^{-2}$ 4) $10^{11}\,\text{Nm}^{-2}$

8. **Y,K,n represent respectively the young's modulus, bulk modulus and rigidity modulus of a body. If rigidity modulus is twice the bulk modulus, then [1999 M]**

1) $y=\frac{5}{18}K$ 2) $y=\frac{9}{5}K$

3) $y=\frac{9}{5}\eta$ 4) $y=\frac{18}{5}K$

9. **If for material Young's modulus is $6.6\times10^{10}\text{N/m}^2$ and bulk modulus is $11\times10^{10}\text{N/m}^2$ then the Possion's ratio is ...**

1) 0.4 2) 0.6 **[1993 E]**

2) 1 4) infinity

10. **For a given material the Young's modulus is 2.4times that of rigidity modulus. Its Possion's radio is [1990 M]**

1) 2.4 2) 1.2 3) 0.4 4) 0.2

ANSWERS

1) 1	2) 2	3) 2	4) 2	5) 4
6) 4	7) 4	8) 4	9) 1	10) 4

Chapter - 2

MECHANICAL PROPERTIES OF FLUIDS

FLUID MECHANICS

SYNOPSIS

Hydrostatics

The study of the fluids at rest is called hydrostatics.

Density

The density of a body is defined as the ratio of mass of the body to the volume occupied by the body

$\therefore$ density $= \dfrac{\text{mass}}{\text{volume}} \Rightarrow \rho = \dfrac{M}{V}$

units : kgm^{-3} (in SI) ; $gm\ cm^{-3}$ in (CGS)

$1 gm/cc = 10^3 kg/m^3$

Dimensional formula $= ML^{-3}$

Density is a scalar quantity

Variation of density with temperature:

In general, with increase in temperature due to the expansion, the volume increases while mass remaining the same, thus the denstity will decrease

$$\rho_0 = \rho(1+\gamma\ \Delta t) \Rightarrow \rho = \frac{\rho_0}{(1+\gamma\Delta t)} \cong \rho_0(1-\gamma\Delta t)$$

where ρ_0 = density at 0^0C

ρ = density at desired temp

γ = coefficient of volume expansion

Δt = rise in temperature

Pascals law

If an external pressure is applied to an enclosed fluid it is transmitted undiminished to every point of the fluid and to the walls of the container.

Applications of Pascals law are, hydraulic lift, hydraulic brakes...... $\dfrac{F_1}{A_1} = \dfrac{F_2}{A_2}$.

Variation of density with pressure

With increase in pressure, volume drecreases density incrases:

$$\rho_0 = \rho\left[1-\frac{\Delta p}{K}\right] \Rightarrow \rho = \rho_0\left(1-\frac{\Delta p}{k}\right)^{-1} = \rho_0\left(1+\frac{\Delta p}{k}\right)$$

where ρ_0 = density at NTP

ρ = density at desired pressure

Δp = increase in pressure

K = Bulk Modulus

Archimedes's Principle

When a body is immersed partly or wholly in a fluid it loses some weight, which is equal to the weight of the fluid displaced by the body

Buoyancy

When a body is partly or wholly immersed in a fluid the fluid exerts force on the body due to hydrostatic pressure normal to the surface and equal to the pressure at that point multiplied by area.

Buoyancy

Weight of fluid displaced = ΔF_B =mg =V ρ g

M=mass of liquid displaced

V=Volume of body immersed or volume of fluid displaced

ρ = density of fluid

g=acceleration due to gravity

- ***Apparent loss of weight of the body = Actual weight–Apparent weight of the body in the fluid***

= [Mg -(Mg-V ρ g) = V ρ g

= weight of fluid displaced

Where M = Mass of the body immersed,

V = volume of immersed part of body

Upthrust or buoyancy is independent of mass, size, density, shape etc of the body. It depends only on the volume of the body immersed inside the fluid, nature (density) of the fluid and accleration due to gravity

Note -(i) :If a body moves down with accleration 'a' (a<g) then buoyancy force = $V_{in}\rho$ (g-a)

(ii) : The buoyant force on an object immersed in a fluid is zero when the container of the fluid falls freely.

- **Relative density (Specific gravity)**

$$RD = \frac{\text{density of the body}}{\text{density of water at } 4^0 c}$$

$$= \frac{\text{Mass of the body}}{\text{Mass of equal volume of water at } 4^0 c}$$

$$RD = \frac{\text{weight of the body}}{\text{weight of equal volume of water at } 4^0 c}$$

$$RD = \frac{\text{weight of the body}}{\text{upthrust exerted by water}}$$

$$= \frac{\text{weight of the body}}{\text{Apparent loss of weight in water}}$$

$$RD = \frac{\text{weight of the body in air}}{\text{weight of body in air} - \text{weight of the body is water}}$$

- R.D. of the body $= \frac{w_1}{w_1 - w_2}$

If loss of weight of a body in water is 'a' and in liquid is 'b' then,

$V\rho_W g = a$; $V\rho_L g = b$

- RD of liquid $= \frac{\rho_L}{\rho_w}$

$$\frac{\text{loss of weight in liquid}}{\text{loss of weight in water}} = \frac{b}{a} = \frac{w_{air} - w_{liquid}}{w_{air} - w_{water}}$$

Flotation

When a body of density ρ_b and volume V is immersed in a liquid of density ρ then the forces acting are

i) The weight of body acting vertically downwards through the centre of gravity of the body

$W = mg = V\rho_b g$

ii) The upthrust acting upwards through the centre of gravity of displaced liquid called centre of buoyancy

$Th = V\rho g$

Fraction of volume of the body outside the liquid:

A body of volume V and density ρ_b is floating in a liquid of density ρ with volume V_{in} immersed in the liquid

$$\therefore V\rho_b = V_{in}\rho$$

$$\Rightarrow V_{in} = V\left[\frac{\rho_b}{\rho}\right]$$

$$\therefore V_{out} = V - V_{in} = V\left[1 - \frac{\rho_b}{\rho}\right]$$

$$f_{out} = \frac{V_{out}}{V} = \left[1 - \frac{\rho_b}{\rho}\right]$$

Floating ice:

1. When ice block floating in water melts, the level of water in the vessel remains unchanged
2. When ice block with cork floating in water melts, the level of water in the vessel remains unchanged
3. When ice block with lead shot or metal ball floating in water melts, the level of water in the vessel decreases
4. If a block of ice, floating in a liquid of density less than that of water, melts, then the liquid level falls
5. If a block of ice, floating in a liquid of density greater than that of water, melts, then the liquid level rises.

Hydrodynamics

The study of fluids in motion is called hydrodynamics.

1. As liquids and gases flow, they are called fluids.
2. An ideal fluid is non-viscous and incompressible.
3. The intermolecular forces of attraction are very weak in liquids So, they have no definite shape. They take the shape of the container.
4. Though the forces of attraction are weak, as they exist, liquids have a definite volume only.
5. As there are no forces of attraction at all between the molecules of a gas, it has no definite shape and no definite volume.
6. The gas expands and fills the whole available volume of the vessel in which it is kept.
7. Cohesive forces are the prime cause for the viscosity of liquids.
8. The fluids encounter resistance to flow.
9. The pressure difference determines the direction of flow of a fluid as the potential difference determines the direction of flow of electric current.
10. When the liquid flows in different layers, the upper layer tries to drag forward or increase the velocity of the lower layer along with it. While the lower layer tries to drag or slow down the upper layer.
11. The relative motion between different layers of the liquid in motion, stops due to the viscous forces that act between different layers and the liquid finally comes to rest.
12. The viscous forces depend upon the nature of the fluid and act tangential to the layers.
13. The internal resistance or friction in a fluid is termed as viscosity.
14. Viscosity is the property of a fluid by virtue of which the relative motion between its different layers is opposed.
15. To maintain the flow of a liquid an external force is to be applied to overcome the viscous force.
16. On account of viscous force a liquid after being stirred comes to rest.
17. When there is no wind, the waves on the sea subside by themselves due to viscous forces.
18. The rain drops strike the earth with terminal velocity and not with acceleration due to viscosity of air.
19. If churning is stopped, the liquid comes to rest itself after some time due to viscosity.

STREAM LINE FLOW

1. The velocity of flow of a liquid depends upon its external pressure.
2. The external pressure drives the liquid and makes it to overcome the viscous drag.
3. When a liquid flows steadily and slowly with a velocity less than a certain velocity called 'critical velocity (Vc)', the flow is called stream line flow.
4. The line along which the particles of the liquid move, one after another, with their velocities constant, but not necessarily be equal at different points on it, is called a stream line.

In stream line flow :

a) every particle follows the same path of its preceding particle.
b) the path followed by a particle is known as stream line.
c) the tangent drawn at any point to the stream line gives the direction of velocity at that point.
d) the velocity at any point remains same but may change from point to point.
e) no radial flow
f) two stream lines never intersect
g) rate of flow of the liquid at any section of the tube is constant
h) the pressure at any point in a given cross-section of the tube is constant
i) the energy is spent in overcoming the viscous drag only
j) the stream lines may be parallel straight lines or curves.

5. The stream lines are crowded where the velocity is more.

Turbulent flow

1. If velocity of the fluid exceeds the critical velocity, the fluid motion becomes disorderly, then the flow of liquid is called turbulent flow.

   In turbulent flow

   a) the liquid flow is zig-zag
   b) velocity of a particle crossing a particular point is not constant with time in magnitude and direction.
   c) eddies (whirlpools) are formed
   d) energy is wasted or dissipated in producing eddy currents

Critical velocity (V_C)

It is that maximum velocity of a fluid above which a stream line flow changes into a turbulent flow.

1. Reynold's formula for critical velocity :

$$\boxed{V_C = \frac{K\eta}{\rho r}}$$ where K= Reynold's number,

η = coeff of viscosity, ρ = density of the fluid, r=radius of the pipe.

2. For velocities well below the critical velocity, the rate of flow is independent of the density of the liquid.
3. For velocities greater than critical velocity, rate of flow depends mainly on the density and is practically independent of viscosity. [Ex. Rapid flow of viscous lava during volcanic eruptions]
4. Smaller the radius of the tube, greater be the chances for the streamline flow.
5. The liquid is incompressible means that the density of the liquid is independent of the variations in pressure.
6. The liquid is non-viscous means that there will be no tangential forces (friction) between different layers of the liquid.
7. The motion of an ideal liquid is irrotational
8. ***Generally liquids of high viscosity have streamline flow and liquids of low viscosity and high density have turbulent flow.***
9. Honey is more viscous than water.

Coefficient of viscosity (η)

1. According to Newton, the viscous force F acting tangentially between any two layer of liquid is
   a) directly proportional to the surface area (A) of the layer,
   b) directly proportional to the difference in velocity (dv) between the two layers, and
   c) inversely proportional to the perpendicular distance (dx) between the two layers.

$$F \alpha A \frac{dv}{dx} \text{ or } F = \eta A \frac{dv}{dx}$$

where dx is called velocity gradient and η is called coefficient of viscosity.

η : It is the ratio of tangential stress to the velocity gradient. (or)

It is the tangential viscous force acting per unit area between two layers moving with unit velocity gradient.

Unit of η : Nsm^{-2} (or) Pa–s

Poise is the C.G.S. unit of η.

1 Pa–s = 10 poise $[\eta] = ML^{-1}T^{-1}$

unit of velocity gradient s^{-1}

2. In a river, at the start, the water flow is highly unsteady. After travelling a long distance, the flow becomes steady due to viscosity.
3. Viscosity of water is used in seperating gold particles from sand particles.
4. Depending upon viscosity of liquid, certain diseases can be diagnised.
5. To reduce friction between moving parts of a machine highly viscous oils, grease etc. (lubricants) are used.
6. A parachutist, due to viscous drag of air, reaches the ground with a small terminal or constant velocity.
7. Due to viscosity of air, clouds float in air.

Coefficient of dynamic viscosity (or) kinematic viscosity (μ): It is defined as the ratio of coefficient of viscosity to density of the fluid.

$$\mu = \frac{\eta}{\rho}$$

$$\therefore unit\ \ Pa - s/m^3$$

$$D.F: ML^{-4}T^{-1}$$

Effect of temperature on viscosity

1. The viscosity of liquids decreases with the increase in temperature because cohesive forces between molecules decreases.
2. The graph between T - η is a rectangular hyperbola for liquids.
3. For gases the viscosity increases with the increase of temperature. because number of collisions between the molecules of the gas increases.

Effect of pressure on viscosity

1. For liquids the viscosity increases with increase in pressure.
2. For gases at normal pressure, viscosity is independent of pressure or density.

Poiseulli's formula

1. According to Poiseullis, the volume of liquid flowing through a capillary tube per second is given by

$$V = \frac{\pi P r^4}{8\eta l},$$

where r = radius of the capillary tube,

P = pressure difference across the ends of the tube

l = length of the tube

2. Fluid resistance : From poiscuille's equation the volume of liquid coming out per sec

$$V = \frac{\pi p r^4}{8 l \eta}, = \frac{P}{\left(\frac{8 l \eta}{\pi r^4}\right)} = \frac{p}{R}$$, where $R = \frac{8 l \eta}{\pi r^4}$ is called fluid resistance (Similar to that i = V/R in electricity)

Capillary tubes in series

When two capillary tubes are connected in series across constant pressure difference 'P' the fluid resistance $R` = R_1 + R_2$

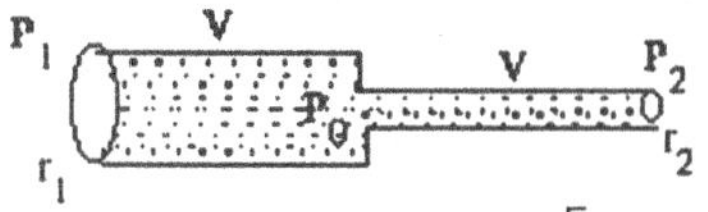

$$R = \frac{8\ell_1\eta}{\pi r_1^4} + \frac{8\ell_2\eta}{\pi r_2^4} = \frac{8\eta}{\pi}\left[\frac{\ell_1}{r_1^4} + \frac{\ell_2}{r_2^4}\right]$$

volume of liquid flowing per second is same through both capillaries,

$$V = \frac{p}{R} = \frac{p}{R_1 + R_2} = \frac{\pi p}{8\eta\left[\frac{\ell_1}{r_1^4} + \frac{\ell_2}{r_2^4}\right]}$$

Here $P = P_1 - P_2 = (P_1 - P_0) + (P_0 - P_2)$

Capillary tubes in parallel

When two capillary tubes are connected in parallel across constant pressure difference 'P' the fluid resistance is given by

$$\frac{1}{R} = \frac{1}{R_1} + \frac{1}{R_2} \Rightarrow R = \frac{R_1 R_2}{R_1 + R_2}$$

$$\frac{1}{R} = \frac{\pi}{8\eta}\left[\frac{r_1^4}{\ell_1} + \frac{r_2^4}{\ell_2}\right]$$

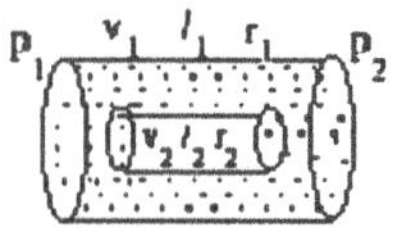

In 'this case the volume of fluid flowing per second in each tube is different, but pressure difference $P = P_1 \sim P_2$ is same

Total volume flowing per second $V = V_1 + V_2$

$$V = \frac{\pi p}{8\eta}\left(\frac{r_1^4}{\ell_1} + \frac{r_2^4}{\ell_2}\right)$$

of same liquid flows through two indentical cappillary tubes of different presssures.

$$\frac{V_1}{V_2} = \frac{p_1}{p_2}$$ (When l, r are constant)

If the fluid through two different tubes of same pressure difference,

$$\frac{V_1}{V_2} = \left(\frac{r_1}{r_2}\right)^4.$$

Equation of continuity

1. For an ideal liquid passing through a tube of non-uniform bore, the product of area of cross-section and velocity of the liquid is constant.

   $A_1V_1 = A_2V_2$

   Where V_1 and V_2 are the velocities of the liquid at areas of cross-section A_1 and $A_2$of the tube.
2. greater the radius of the tube, smaller is the velocity of the liquid.
3. when we try to close a water tap with our fingers, fast jets of water gush out through the small gaps between the fingers.
4. Equation of continuity is the consequence of law of conservation of mass.

Bernoullis theorem :

1. A fluid possesses three types of energy.
   a) Pressure energy = work done in displacing the liquid = PAl.

      Pressure energy per unit volume of the liquid =

      $\frac{PAl}{Al}$ = P, pressure

   b) Kinetic energy per unit volume of the liquid

      $= \frac{1}{2}\rho V^2$

      where ρ = density, V = velocity

   c) Potential energy per unit volume of the liquid = ρgh.
2. The sum of the pressure energy, kinetic energy and gravitational potential energy per unit volume at any point in a stream line flow of a fluid is constant.

   is, $P + \frac{1}{2}\rho V^2 + \rho gh$ = constant :

   or $\frac{p}{\rho g} + \frac{V^2}{2g} + h$ = constant

   Here $\frac{p}{\rho g}$ is called pressure head, h is called gravitational head and $\frac{v^2}{2g}$ is called velocity head.
3. For a horizontal tube, h = 0, $p + \frac{\rho V^2}{2}$ = constant.
4. Pressure is large when the velocity flow is less and vice versa.
5. The theorem can be treated as a form of work-energy theorem or conservation of energy.
6. Pressure in a fluid at a depth h is $P = P_o + hdg$. This is called static pressure [P_o = atmospheric pressure]
7. $\frac{1}{2}\rho V^2$ is dynamic pressure

Applications of Bernoullis theorem

a) A body at rest in a liquid experiences an upward thrust called 'buoyancy'. The fluid is then said to experience a 'static lift'.

b) A body in motion in a fluid experiences a lift known as 'dynamic lift'.

c) Aeroplanes get the dynamic lift to rise.

d) magnus effect of a spinning ball.

e) working of a bunsen burner.

f) Race cars are aerodynamically designed at the lower surface.

g) Flag flutters, when the wind is blowing.

h) A lorry when moves fast, pieces of paper, straw etc. are pulled towards the lorry.

i) When the wind blows at a very high speed at the time of cyclone, the roof of a house is blown off.

j) When the ships move in water parallel to each other in the same direction, they are dragged to come closer and may collide.

k) When air is blown fast under one of the pans of a counter poised balance, it moves down.

l) When air is blown fast between two freely suspended ping pong balls, they come closer.

m) When air is blown fast over a paper, it rises.

n) A man standing close to the track gets a push towards the train, when the train is moving fast.

o) Working of an atomizer, - deep waters remain still, working of a paint gun are also few more examples for Bernoullis theorem.

Torricelli theorem

According to Bernoulli's theorem,

$$P + O + \rho gH = P + \frac{1}{2}\rho\upsilon^2 + \rho g(H - h)$$

$$\frac{1}{2}\rho\upsilon^2 = \rho gh$$

$$\therefore \upsilon = \sqrt{2gh}$$

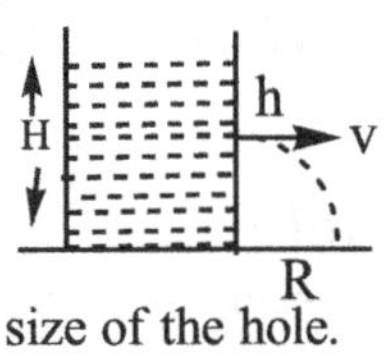

V is independent of the size of the hole.

The formula was derived by Torricelli and is called Toricellis theorem.

V is called velocity of efflux.

Range, $R = V\sqrt{\dfrac{2(H-h)}{g}}$

As the size of the hole increases, the rate of flow increases.

Range is maximum when, $h = H/2$

Stoke's law - Terminal velocity :

1. A spherical body when dropped in a fluid, it sets up relative motion between different layers of the fluid. The viscous force developed increases with increase in velocity and opposes the fall of the body.
2. When the sum of the viscous force and up thrust of the liquid on the body is equal to its weight, the body then begins to fall with a constant velocity, known as terminal velocity(V_T)
3. According to stoke, the viscous force = $6\pi\eta r V_T$. Where η = coeff. of viscosity of the fluid, r = radius of the body and V_T terminal velocity.
4. Upward thrust on the body = $\frac{4}{3}\pi r^3\sigma g$

   where σ = density of fluid

   weight of the body = $\frac{4}{3}\pi r^3\rho g$ where ρ = density of the body.

At stoke's law : for terminal velocity

$$6\pi\eta r V_t = \frac{4}{3}\pi r^3(\rho-\sigma)g$$

$$\boxed{\Rightarrow V_t = \frac{2}{9}\frac{r^2 g(\rho-\sigma)}{\eta}}$$

1. $V_t \propto r^2 \Rightarrow \dfrac{V_{t_1}}{V_{t_2}} = \left(\dfrac{r_1}{r_2}\right)^2$
2. Terminal velocity is directly proportional to the square of the radius of the sphere.
3. When 'n' droplets are falling down with their terminal velocities 'V' are combined and formed a big drop, then terminal velocity of big drop is $\boxed{V_{big} = n^{2/3}\,V}$

   Volume of big drop = n× volume of small drop

   V_{big} = n. V

   $\frac{4}{3}\pi.R^3 = n.\frac{4}{3}\pi r^3$ $\quad R = n^{1/3}.\,r.$
4. $V_T \propto (\rho-\sigma)$
5. If $\rho<\sigma$, body will rise instead of falling. [Ex. air bubble in water]
6. $V_T \propto \frac{1}{\eta}$,greater the viscosity, smaller is the terminal velocity. [Ex. a lead shot quickly attains terminal velocity in glycerine than in water].

EXERCISE - I

1. The buoyant force is equal to

1) weight of the body
2) weight of the liquid displaced by the body
3) difference of the weight of the body and liquid displaced
4) none of the above

2. The ice is floating on water. When it is completely melted, the level of water

1) rises 2) decreases
3) remains same 4) none

3. A boat with bricks is floating on the water. When the bricks are thrown into water; the level of water in the pond :

1) rises 2) decreases
3) remains same 4) may rise or fall

4. A boat in the river enter into the sea water, then it

1) sinks 2) rises 3) remains same 4) none

5. A block of ice containing a lead shot is floating in the water. When ice is completely melted and lead shot sinks to the bottom of the beaker, the level of water

1) decreases 2) increases
3) remains same 4) can not say

6. A boat is containing leaves. When they are thrown into the water, the level of water in the lake

1) remains same 2) falls
3) rises 4) none

7. A body weights w_1 in air and w_2 in water. The specific gravity of the material of the body is

1) $\frac{W_1}{W_2}$ 2) $\frac{W_1-W_2}{W_1}$ 3) $\frac{W_1-W_2}{W_2}$ 4) $\frac{W_1}{W_1-W_2}$

8. A body weighs w_1 in air, w_2 in water and w_3 in a liquid. The density of the liquid is given by

1) $\frac{W_1}{W_2-W_3}$ 2) $\frac{W_2-W_3}{W_1-W_2}$

3) $\frac{W_1-W_3}{W_1-W_2}$ 4) $\frac{W_2+W_3}{2W_1}$

9. If the weight of the body and upthrust on the body in a liquid are equal, then the body

1) Sinks 2) Floats
3) Just completely immerses in the liquid
4) Completely lies out side the liquid

10. A body is floating in a liquid. Weight of the body is

1) More than the weight of the body
2) Equal to weight of the liquid displaced by the body
3) Less than the weight of the body
4) Zero

11. A wooden block with a coin plced on its , top floats in water as shown. After some time the coin falls into water . Then [IIT - 2002]

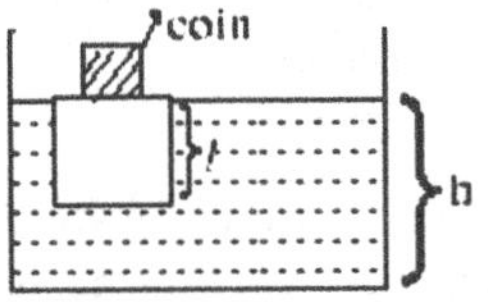

1) l decreases and h increases
2) l increases and h decreases
3) both l and h increase 4) both l and h decrease

12. A body weighs more in

1) Air 2) Hydrogen 3) Water 4) Vaccum

13. Viscosity is the property of liquids by virtue of which :

1) Liquid molecules attract each other
2) Liquid become conducting
3) Liquids oppose relative motion of its parts
4) None

14. The force acting tangential to the layers of the liquid is called :

1) Buoyant force 2) Viscous drag
3) Frictional force 4) Gravitational force

15. The property of viscosity in gas is due to :

1) Cohesive forces between the molecules
2) Collisions between the molecules
3) Not having a definite volume
4) Not having a definite size

16. The correct statement of the following is

1) Layer of water with the ground is at rest
2) Deep water remains more quiet
3) Velocity of efflux does not depend upon density of the liquid
4) All of the above

17. A bird resting on the floor of an air tight box is being carried by a boy. If the bird starts flying, he feels the box as

1) Heavier 2) Lighter
3) Of same weight as before
4) Lighter in the beginning and heavier afterwards

18. Among water and honey, viscosity is more for

1) Water 2) Honey
3) Same for both 4) None

19. The clouds float in air due to

1) Viscosity of air 2) Clouds are heavier
3) At higher altitudes density of air is more
4) None

20. As temperature of liquid increases, its viscosity

1) Increases 2) Decreases
3) Does not change 4) Gases do not have viscosity

21. For liquids, the graph between temperature and coefficient of viscosity is

1) Straight line 2) Parabola
3) Rectangular hyperbola 4) Ellipse

22. At normal pressure, for gases as pressure increases, the viscosity

1) Remains same 2) Increases
3) Decreases 4) None

23. The viscosity of water is

1) More in deep waters 2) More in shallow water
3) Same throughout 4) None

24. Unit of viscosity

1) Pa–s 2) $Nm^{-2}s$ 3) Poise 4) All of the above

25. Unit of velocity gradient

1) s 2) ms^{-1} 3) s^{-1} 4) $m^{-1}s^{1}$

26. The dimensional formula for rate of flow is

1) LT^{-1} 2) $ML^{3}T^{-1}$ 3) $L^{3}T^{-1}$ 4) $L^{-3}T^{1}$

27. The dimensional formula of coefficient of viscosity is

1) $ML^{-1}T^{-1}$ 2) $ML^{-2}T^{-1}$ 3) $M^{-1}L^{1}T^{-1}$ 4) $ML^{2}T^{-3}$

28. Ratio of S.I. unit of Coefficient of viscosity to its C.G.S. unit is

1) 10 2) 0.1 3) 0.01 4) 100

29. When the wind does not blow, the waves subside themselves due to

1) Bernoulli's theorem 2) Flow of continuity
3) Viscosity 4) All of the above

30. Atomizer works on the

1) principle of continuity 2) Bernoulliís principle
3) Stokeís law 4) Archimedi's principle

31. When the wind blows at a high speed, the roof of a house is blown off because :

1) the pressure under the roof increases
2) Pressure above the roof increases
3) Pressure above the roof decreases
4) The wind pushes off the roof

32. When water flows from a narrow pipe into a broader pipe,
1) pressure decreases, rate of flow increases
2) pressure increases, velocity increases
3) rate of flow does not change but velocity increases
4) pressure increases, velocity decreases, rate of flow remains same

33. For a body falling with terminal velocity, the net force on it is :
1) buoyant force 2) weight of the body
3) difference of viscous drag and weight of the body
4) zero

34. As temperature increases, the viscosity of gases:
1) increases 2) decreases
3) does not change 4) may increase or decrease

35. v^2 / 2g is called :
1) pressure head 2) gravitational head
3) both 4) none

36. Bernoulli's theorem is a consequence of
1) Law of conservation of energy
2) Law of conservation of mass
3) Law of conservation of linear momentium
4) None

37. As temperature increases; the viscosity of
1) Gases and liquids increase
2) Liquids and gases decrease
3) Gases increase but liquids decrease
4) Liquids increase but gases decrease

38. A physical balance is counterpoised. The air is blown with a straw under one of the pan. Then that pan :
1) Rises up 2) Remains in the same position
3) Lowers down 4) None

39. Liquid is flowing through a non-uniform tube. The rate of flow is more at
1) Narow region 2) Wider region
3) Same through out 4) None

40. A vessel is containing a liquid. A hole is made at a depth to the side wall of the vessel. The rate of flow of liquid is V. If size (area) of the hole is doubled
1) Velocity of efflux is doubled but rate of flow is halved
2) Both velocity of efflux and rate of flow are doubled
3) Velocity of efflux remains same but rate of flow is doubled
4) Velocity of efflux is halved but rate of flow is quadrupled

41. To reduce friction between the moving parts of a machine, the liquid used must have
1) High viscosity 2) Low viscosity
3) Low density 4) Low boiling point

42. Rain drops of different sizes strike the earth with
1) Same acceleration
2) Same constant velocity
3) Different constant velocities
4) Different accelerations

43. Viscous force is similar to friction in solids, but viscous force
a) is independent of area but friction depends on area
b) is temparature dependent while friction force between solids depends upon normal reaction
c) is velocity dependent while friction is not
d) is velocity independent while friction is velocity dependent
1) a, b, c are correct 2) a, b, c, d are correct
3) b, c are correct 4) c, d are correct

44. Which of the following is the unit of coefficient of viscosity
a) pascal x second **b) poise**
c) N/m^2 **d) stoke**
1) a is correct 2) d is correct
3) a, b are correct 4) a, b, d are correct

45. When a body is weighed in a liquid the apparent loss in it's weight is equal to
a) the upthrust of liquid on the body
b) weight of the liquid displaced by the body
c) the difference in weights of the body in air and liquid
d) weight of water displaced by the body
1) a, b, c are correct 2) a, b are correct
3) a, b, d are correct 4) b, d are correct

46. Which of the following represents conservation laws?
a) equation of continuity
b) Bernoullie's Theorem
c) magnus effect **d) Toricelli's Theoum**
1) c, d are correct 2) a, c are correct
3) a, b are correct 4) a, b, c are correct

47. In the case of motion of a fluid in a tube where area of cross section is maximum

a) velocity is maximum b) pressure is maximum

c) velocity is minimum d) presure is minimum

1) b, c are correct 2) a, d are correct

3) a, b, c are correct 4) c, d are correct

48. When an air bubble moves up from the bottom of a lake

a) velocity decreases and becomes zero

b) acceleration increases and becomes zero

c) velocity increases and becomes constant

d) acceleration decreases and becomes zero

1) a, d are correct 2) a, b are correct

3) c, d are correct 4) c is correct

49. Four small lead balls of volumes $1cm^3$, $2cm^3$, $3cm^3$ and $4cm^3$ are descending down through the same liquid

a) viscous force is more on $4(cm)^3$ ball

b) terminal velocity is more for $4(cm)^3$ ball

c) viscous force is more in $1(cm)^3$ ball

d) terminal velocity is more in $1(cm)^3$ ball

1) a, b are correct 2) c, d are correct

3) a, d are correct 4) b, c are correct

50. Four lead spheres a, b, c & d whose radii 1cm, 2cm, 3cm and 4cm respectively are dropped into different liquids of co-efficients of viscosity $1x10^{-3}$ Pa–s, $2x10^{-5}$ Pa–s, $6.5x10^{-3}$ Pa–s, $15.4x10^{-3}$ Pa–s. Arrange the spheres with their terminal velocity in descending order

1) a, b, c, d 2) b, c, d, a

3) c, d, a, b 4) a, c, d, b

51. An ideal fluid is flowing through four tubes a, b, c, d of radii in the ratio 2 : 7 : 3 : 1 with velocities in the ratio 1 : 2 : 5 : 15 when maintained at different pressures. The ascending order of the amount of fluid flowing through the tubes persecond is

1) a, d, c, b 2) b, c, d, a 3) c, p, b, a 4) b, c, d, a

52. Four spheres, a, b, c, d of same material and equal volume are dropped into four liquids of densities d_1, d_2, d_3 and d_4 ($d_1 > d_2 > d_3 > d_4$). The ascending order of the terminal velocities (η is same)

1) a, b, c, d 2) d, c, b, a 3) d, c, a, b 4) b, c, d, a

53.

Section - A	Section - B
a) Equation of continuity	e) Less than critical velocity
b) Bernoullie's theorem	f) Formation of eddies & vortices
c) Turbulent flow	g) Law of conservation of mass
d) Stream line flow	h) Law of conservation of energy

1) a-h; b-f; c-e; d-g 2) a-g; b-e; c-g; d-f 3) a-f; b-g; c-h; d-e 4) a-g; b-h; c-f; d-e

54.

Section - A	Section - B
a) Incompressible liquid	e) Density constant
b) Turbulent flow	f) Stream lines
c) Tube of flow	g) Constant
d) Fluid flux rate in laminar flow	h) Reynold's no. >2000

1) a-f; b-e; c-g; d-h 2) a-e; b-h; c-f; d-g 3) a-g; b-f; c-e; d-h 4) a-h; b-g; c-e; d-f

55. Match the physical quantities given here under having same units and dimensions

Section - A	Section - B
a) Kinametic viscosity	e) $\frac{1}{2}\rho\left(v_2^2 - v_1^2\right)A$
b) Dynamic lift	f) $\frac{\eta}{\rho}$
c) Bernoullie's theorem	g) a v= constant
d) Equation of continuity	h) $p+\frac{1}{2}\rho v^2+\rho gh=\text{const.}$

1) a-f; b-e; c-h; d-g 2) a-f; b-e; c-g; d-h 3) a-g; b-f; c-e; d-h 4) a-h; b-g; c-f; d-e

INSTRUCTION FOR QUESTIONS :

In each of the following questions, a statement of Assertion (A) is given followed by a corresponding statement of reason (R) just below it. Of the statement mark the correct answer.

1) Both 'A' and 'R' are true and 'R' is the correct explanation of 'A'
2) Both 'A' and 'R' are true and 'R' is not the correct explanation of 'A'
3)'A' is true and 'R' is false
4)'A' is false and 'R' is true

56. A) : Pascal law is working principle of a hydraulic lift.

R) : Pressure is equal to thrust per unit area.

57. A) : Hydrostatic pressure is a vector quantity.

R) : Pressure is force divided by area, and force is a vector quantity.

58. A) : A man is sitting in a boat which is floating on a pond. If the man drinks some water from the pond, the level of the water in the pond decreases.

R) : According to Archimede's principle the weight displaced by body is equal to the weight of the body.

59. A) : A piece of ice floats in water. The level of water remains unchanged when the ice melts completely.

R) : According to Archimede's principle, the loss in weight of the body in the liquid is equal to the weight of the liquid displaced by the immersed part of the body.

60. A) : The shape of an automobile is so designed that its front resembles the stream line pattern of the fluid through which it moves.

R) : The resistance offered by the fluid is maximum.

61. A) : A needle placed carefully on the surface of water may float, whereas a ball of the same material will always sinks.

R) : The buoyancy of an object depends both on the material and shape of the object.

62. A) : A hydrogen filled balloon stops raising after it has attained a certian height in the sky.

R) : The buoyancy of an object depends both on the material and shape of the object.

63. A) : A block of wood floats on the surface of water

R) : Appareut weight of a body floating in a liquid is zero

64. A) : Apparent weight of 1 kg of iron is equal to that of 1 kg of cotton in air

R) : Buoyant force on 1 kg of iron is lesser than Buoyant force on 1 kg of cotton

65. A) : Buoyancy force on a body immersed in a liquid is equal to weight of displaced liquid.

R) : Displaced liquid pushes the body upwards

66. A) : An ice block with lead shot is floating in water. When ice melt completely water level slightly drops.

R) : When ice block with lead shot is placed in water, volume of displaced water is more than that of the ice block.

67. A) : The size of the needle of a syringe controls flow rate better than the thumb pressure exerted by a doctor while administering an injection.

R) : Flow rate is independent of pressure exerted by the thumb of the doctor.

68. A) : For a floating body to be in stable equilibrium, its centre of buoyancy must be located above the centre of gravity.

R) : The torque required by the weight of the body and the upthrust will restore body back to its normal position , after the body is disturbed.

69. A) : All the raindrops hit the surface of the earth with the same constant velocity

R) : An object falling through a viscous medium eventually attainns a terminal velocity.

70. A) : Water flows faster than honey.

R) : The coefficient of viscosity of water is less than honey.

71. A) : The viscosity of liquid increases rapidly with rise of temperature.

R) : Viscosity of a liquid is the property of the liquid by virtue of whih c it opposes the relative motion amongst its differents layers.

72. A) : Machine parts are jammed in winter.

R) : The viscosity of lubricant used in machine parts increase at low temperature.

73. A) : A spinning cricket ball moving through air is not deflected from its normal trajectory

R) : Magnus effect is an application of Bernoulli's principle

74. A) : During takeoff of an aeroplane, velocity of air above the wing is greater than velocity below the wing

R) : Bernoulli's principle is based on law of conservation of energy

75. A) : When air is blown between two ping pong balls which are suspended freely they move closer

R) : When air is blown between two ping pong balls the pressure between the balls decreases

76. A) : Coefficient of viscosity for gasses increases with increase in temperature.

R) : As temperature of gases is increased the collission between molecules increases

77. A) : Raindrops falling from a large height, accelerate initially and then fall with constant velocity

R) : For a spherical body falling with constant velocity, Buoyant force is equal and opposite to the weight of the drop

78. When temperature is increased [2004 M]

a) Viscosity of the gas increases

b) Viscosity of the gas decreases

c) Viscosity of the liquid decreases

d) Viscosity of the liquid increases

1) (a) and (c) are true 2) (b) and (c) are true
3) (b) and (d) are true 4) (a) and (d) are true

79. Consider the follwing two statements A and B and identify the correct choice in the following

A) Viscosity of liquids decreases with decrease in temperature

B) surface tension of a liquid decreases with increase in temperature

1) both A and B are true 2) A is true but B is false
3) A is false but B is true 4) both A and B are flase

80. Scent sprayer is based on [AIIMS 2002]

1) Charle's law 2) Boyle's law
3) Archimedes principle 4) Bermoulli's therorem

81. The speeds of air-flow on the upper and lower surfaces of a wing of an aeroplane are v_1 and v_2 respectively. It A is the cross-sectional area of the wing and 'ρ' is the density of air, then the upward lift is [2006 M]

1) $1/2\rho A(V_1-V_2)$ 2) $1/2\rho A(V_1+V_2)$
3) $1/2\rho A(V^2{}_1-V^2{}_2)$ 4) $1/2\rho A(V^2{}_1+V^2{}_2)$

82. Water flows through a non-uniform tube. Areas of cross -section of parts A,B and C are 25, 5 and 35cm² respectively. Which part has the highest velocity? [IIIT Allahabad 2001]

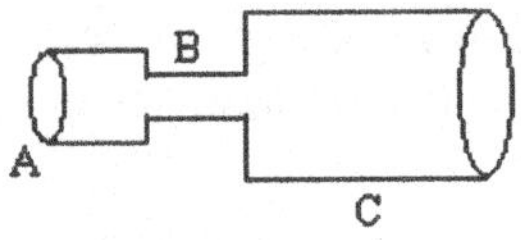

1) A 2) B 3) C 4) all have same velocity

83. A body of density d' is left free in a liquid of density d (where d'>d), what is the downward acceleration of the body while sinking in the liquid ? (A.T.B)

1) $\left(1-\frac{d}{d'}\right)g$ 2) $\left(1-\frac{d'}{d}\right)g$

3) $\left(\frac{d'}{d}-1\right)g$ 4) $\left(\frac{d}{d'}-1\right)g$

ANSWERS

1) 2	2) 3	3) 2	4) 2	5) 1
6) 1	7) 4	8) 3	9) 3	10) 2
11) 4	12) 4	13) 3	14) 2	15) 2
16) 4	17) 3	18) 2	19) 1	20) 2
21) 3	22) 1	23) 1	24) 4	25) 3
26) 3	27) 1	28) 1	29) 3	30) 2
31) 3	32) 4	33) 4	34) 1	35) 4
36) 1	37) 3	38) 3	39) 3	40) 3
41) 1	42) 3	43) 3	44) 3	45) 1
46) 3	47) 1	48) 3	49) 1	50) 4
51) 1	52) 1	53) 4	54) 2	55) 1
56) 2	57) 4	58) 4	59) 1	60) 3
61) 3	62) 3	63) 2	64) 4	65) 2
66) 2	67) 3	68) 1	69) 4	70) 1
71) 4	72) 1	73) 4	74) 2	75) 1
76) 1	77) 3	78) 1	79) 3	80) 4
81) 3	82) 2	83) 1		

EXERCISE - II(A)

BUOYANCY

1. A body is floating in the water such that 6 out of 10 parts of its volume under the water. Its density is

1) 0.6 gm/cc 2) 600 kg/m³
3) both 4) none

2. Two solids A and B float in water. It is observed that A floats with half of its volume immersed and B floats with 2/3 of its volume immersed. The ratio of their densities is

1) 4 : 3 2) 3 : 4 3) 2 : 3 4) 3 : 2

3. A block of ice of density 0.9 gm cc^{-1} and of volume 100 cc floats in water. Then volume of ice inside the water

1) 10 cc 2) 90 cc 3) 50 cc 4) 9 cc

4. **A body weighs 200 gm in air 180 gm in a liquid and 175 gm in water. The density of the material of the body is**

1) 8000 kg/m^3 2) 800 kg/m^3
3) 8 kg/m^3 4) none

5. **A block of solid insoluble in water weighs 24gm in air and 21gm when completely immersed in water. Its weight when completely immersed in liquid of specific gravity 1.1 is**

1) 20.7gm 2) 27.3gm 3) 24gm 4) 3.3gm

6. **An inverted u-tube has its two limbs in water and kerosene contained in two beakers. If water rises to a height of 10cm, to what height does kerosene (density = 0.8 gm/cc) rise in the other limb ?**

1) 10 cm 2) 12.5 cm 3) 15 cm 4) 20 cm

7. **A metal sphere weighing 60gm in air is suspended by a thread in an oil of specific gravity 0.6 . If the tension in the thread is 55gmwt, the specific gravity of the metal sphere is**

1) 9.4 2) 4.5 3) 6.3 4) 7.2

8. **What mass of lead will weigh as much as 8gm of iron, when both are immersed in water ? (Given specific gravities of iron and lead are 8 and 11 respectively). (A.T.B.)**

1) 1.1gm 2) 2.2 gm 3) 5.5 gm 4) 7.7 gm

9. **The base area of a boat is $2m^2$. A man weighing 76kg weight steps in to the boat. Calicutate the depth in to which the boat sinks further in to water. (A.T.B.)**

1) 1.2 cm 2) 2.5 cm 3) 3.8 cm 4) 4.2 cm

10. **If the atmospheric pressure is 76cm of mercury, at what depth of water the pressure will becomes 4 atmospheres ? (A.T.B.)**

1) 31m 2) 21m 3) 11m 4) 9

VISCOUS FORCE

11. **A piston of cross-sectional area 100 cm^2 is used in a hydraulic press to exert a force of 10^7 dyne on water. The cross-sectional area of the other piston which supports a truck of mass 200 kg is,**

1) 1.96 cm^2 2) 0.196m^2 3) 0.98 m^2 4) 98 m^2

12. **A glass plate of surface area 1 m^2 placed on a layer of turpentine of thickness 2 mm. If the coefficient of viscosity of turpentine 0.00149 Nsm^{-2}. The force to be applied to move the plate horizontally with a velocity of 0.1 ms^{-1}. is**

1) 745 N 2) 74.5 N 3) 7.45 N 4) 0.0745 N

13. **A square plate of 0.1 m side moves parallel to a second plate with a velocity of 0.1 m/s, both plates being immersed in water. If the viscous force is 0.002 N and the coefficient of viscosity is 0.01 poise, distance between the plates in m is** (2003 M)

1) 0.1 2) 0.05 3) 0.005 4) 0.0005

14. **In the flow of a viscous liquid over a flat horizontal surface, the velocity is 0.5 cm/s at 1 mm distance above the flat surface at rest. What is the velocity gradient ? (A.T.B.)**

1) 0.05 m/s/cm 2) 0.06 m/s/cm
3) 0.07 m/s/cm 4) 0.08 m/s/cm

POISEULLI'S EQUATION

15. **How should the pressure difference be altered so that the rate of flow of liquid be doubled when radius and length are both doubled? (A.T.B)**

1) 4 times the previous value
2) 2 times the previous value
3) 1/2 times the previous value
4) 1/4 times the previous value

16. **Water is allowed to flow through a capillary tube of length 10cm and diameter 2 mm under a constant pressure difference of 6.5 cm of water level. If 0.16 litres of water flows in 1 minute, its coefficient of viscosity is,**

1) 0.00937 poise 2) 0.0937 poise
3) 0.000937 poise 4) 0.937 poise

EQUATION OF CONTINUITY AND BERNOULI'S THEOREM

17. **A liquid enters a tube of variable cross section with a velocity 3 ms^{-1} through the wider end and leaves from the narrower end whose radius is half of the wider one. The velocity with which it leaves the tube is**

1) 1.2 ms^{-1} 2) 12 ms^{-1} 3) 10ms^{-1} 4) 20 ms^{-1}

18. **The work done by pressure in forcing $2m^3$ of water through a pipe of radius 1 cm, if the pressure difference across a pipe is 10^5Pa is**

1) $3x10^6$J 2) $2x10^5$J 3) $4x10^6$J 4) 0.2 x 10^3J

19. **In a horizontal pipe line pressure falls by 10 Pa between two points separated by a distance of 1 km. The change in kinetic energy per kg of oil when the oil flows from one point to the other is (Density of oil = 800 Kg/m^3)**

1) $\frac{1}{40}J$ 2) $\frac{1}{80}J$ 3) $\frac{3J}{40}$ 4) $\frac{9}{50}J$

20. Volume of the liquid flowing per second through a pipe of diameter d is V. At a point where the radius is d, the velocity of flow is

1) $\frac{V}{d^2}$ 2) $\frac{V}{\pi d^2}$ 3) $\frac{4V}{\pi d^2}$ 4) $\frac{V}{4\pi d^2}$

21. Water is flowing through a horizontal pipe of variable cross - section . The rate of volume flow is 0.2 lit/sec. The pressure is 30,000 Pa. at a point where the area of cross - section is $5x10^{-5}$ m^2. The pressure at the point where the area of cross-section is 10^{-4} is

1) 6000 Pa 2) 36000 Pa 3) 24000 Pa 4) 30000 Pa

22. Due to disease the main artery enlarged and the ara of cross - section became 1.6 times the previous value. The speed of blood in the unenlarged artery is 0.4 ms^{-1} . The excess pressure on the walls of artery due to enlargement is [density of blood is 1060Kg/m^3]

1) 51.675 Pa 2) 21.275 Pa
3) 41.675 Pa 4) 31.275 Pa

23. Air streams horizontally across an aeroplane wing of area 3m^2 weighing 250 kg. The air speed is 60 m/s and 45 m/s over the top surface and under the bottom surface respectively. What is the lift on the wing ? (Density of air 1.293 g/l)

1) 3000 N 2) 3500N 3) 4000 N 4) 3054 N

24. The velocity of air over the upper surface of the wing of an aeroplane is 40m/s and that on the lower surface is 30m/s. If the area of the wing is 5m^2 and the mass of the wing is 300kg, the net force acting on the wing is (Density of air = 1.3kg/m^3 and g=10m/s^2)

1) 725N upward 2) 725N downward
3) 2275 N upward 4) 2275 N downward

25. A horizontal pipe of non - uniform cross - section allows water to flow through it with a velocity 1 ms^{-1} when pressure is 50 kPa at a point. If the velocity of flow has to be2 ms^{-1} at some other point, the pressure at that point should be

1) 51.5 kPa 2) 100 kPa
3) 48. 5 kPa 4) 24.25 kPa

26. A water hose 2 cm in diameter is used to fill a 20 litre bucket. If it takes 1 minute to fill the bucket, then the speed v at which the water leaves the hose is,

1) 150 cm/s 2) 70 cm/s
3) 100 cm/s 4) 106 cm/s

27. Water contained in a tank flows through an orifice of diameter 2 cm, under a constant pressure difference of 10cm of water column. The rate of flow of water through the orifice

1) 44 cc/s 2) 4.4 cc/s
3) 440 cc/s 4) 4400 cc/s

28. An orifice of area 10 sq.mm is made at the bottom of a vessel containing water to a height of 10m above the orifice. What is the force on the vessel

1) 9.98N 2) 9.8N 3) 1.96N 4) 19.6N

29. A large water tank has a hole in its wall near the bottom. the water level above the hole is 4.9m. The horizontal distance from the hole at which the water touches the ground, if the bottom of water tank is 4.9m above the ground

1) 9.8 cm 2) 98 m 3) 9.8 m 4) 980 m

30. A water tank which is on ground has an arrangement to maintain a constant water level of depth 60 cm. Through a hole on its vertical wall at a depth of 20cm from the free surface water comes out and reaches the ground at a certain distance. To have the same horizontal range another hole can be made at a depth of

1) 30 cm 2) 10 cm 3) 40 cm 4) 50 cm

31. A tank full of water has a small hole at its bottom. If one fourth of the tank is emptied in t_1 seconds and the remaining three-fourths of the tank is emptied in t_2 seconds. Then the ratio (t_1/t_2) is [2004 M]

1) $\sqrt{3}$ 2) $\sqrt{2}$ 3) $\frac{1}{\sqrt{2}}$ 4) $\left[\frac{2}{\sqrt{3}}-1\right]$

32. A large tank filled with water to a height "h" is to be emptied through a small hole at the bottom.The ratio of the time taken for the level to fall from h to h/2 and that taken for the level to fall from h/2 to 0 is

1) $\sqrt{2}-1$ 2) $\frac{1}{\sqrt{2}}$ 3) $\sqrt{2}$ 4) $\frac{1}{\sqrt{2}-1}$

33. A large open tank has two holes in the wall. One is a square hole of side L at a depth Y from the top and the other one is a circular hole of radius R at a depth 4Y from the top. When the tank is completely filled with water, the quantities of water flowing out per second from both holes are the same. Then R is equal to [IIT Screening 2001]

1) $\frac{L}{\sqrt{2\pi}}$ 2) $2\pi L$ 3) L 4) $\frac{L}{2\pi}$

34. How much work is done by an agent in forcing $3m^3$ of water through a pipe of radius 2cm, if the difference in pressue at two ends of the pipe is 10^6 N/m² ? (A.T.B.)

1) 3×10^6 J 2) 2×10^6 J 3) 4×10^6 J 4) 1×10^6 J

TERMINAL VELOCITY

35. A metal ball of radius 'r' and density 'd' travels with a terminal velocity 'v' in a liquid of density d/4. The terminal velocity of another ball of radius '2r' and density '3d' in the same liquid is

1) $\frac{44v}{3}$ 2) $\frac{22v}{3}$ 3) $\frac{11v}{3}$ 4) $\frac{3v}{44}$

36. A steel ball of mass 'm' falls in a viscous liquid with a terminal velocity 4 ms^{-1} . If another steel ball of mass '8 m' falls through the same liquid then its terminal velocity is

1) 4 ms^{-1} 2) 8 ms^{-1} 3) 16 ms^{-1} 4) 2 ms^{-1}

37. Two equal drops are falling through air with a steady velocity of 5 cm/s. If the drops coalesce, the new terminal velocity will be

1) 5 x 2 cm/s 2) 5 x $\sqrt{2}$ cm/s

3) 5 x $(4)^{1/3}$ cm/s 4) $5/\sqrt{2}$ cm/s

38. Two metal spheres of densities in the ratio 3 : 2 and diameter in the ratio 1 : 2 are released from rest in two vertical liquid columns of coefficients of viscosity in the ratio 4 : 3. If the viscous force on them is same , then the ratio of their instantaneous velocities is

1) 1 : 2 2) 3 : 2 3) 4 : 3 4) 8 : 3

39. The terminal velocity of a small ball falling in a viscous liquid depends upon i) its mass m ii) its radius r iii) the coefficient of viscosity of the liquid η and iv) acceleration due to gravity. Which of the following relations is dimensionally true for the terminal velocity V =

1) $\frac{Kmg}{\eta r}$ 2) $\frac{Kmgr}{\eta}$ 3) $\frac{Kmg\eta}{r}$ 4) $\frac{Kr\eta}{mg}$

ANSWERS

1) 2	2) 3	3) 2	4) 2	5) 1
6) 1	7) 4	8) 4	9) 3	10) 1
11) 2	12) 4	13) 4	14) 1	15) 4
16) 3	17) 2	18) 2	19) 2	20) 2
21) 2	22) 1	23) 4	24) 2	25) 3
26) 4	27) 3	28) 3	29) 3	30) 3
31) 4	32) 1	33) 1	34) 1	35) 1
36) 3	37)3	38) 2	39)1	

EXERCISE - II(B)

BUOYANCY

1. A block of wood floats in water with 2/3 of its volume submerged.The density of wood is

1) 3/2 $gmcc^{-1}$ 2) 1/3 $gmcc^{-1}$

3) 1/2 $gmcc^{-1}$ 4) 2/3 $gmcc^{-1}$

2. Two blocks A and B float in water. If A floats with1/3 of its volume immersed and B floats with 2/5 of its volume immersed, then the ratio of their densities is

1) 2/15 2) 1/5 3) 5/6 4) 2/3

3. A brass cylinder weighs 90 gm in air. If the density of brass is 9 gm cc^{-1}, the weight of the cylinder in water is

1) 80gm 2) 81gm 3) 10gm 4) 90gm

4. A block of wood floats with 4/5 of its volume inside water but just floats in an oil. The density of the oil is

1) 400 kg/m³ 2) 800 kg/m³

3) 200 kg/m³ 4) 600 kg/m³

5. A solid sphere weighs 10N in air, 6N in water and 8N in a liquid. The specific gravity of the liquid is

1) 0.5 2) 0.25 3) 1 4) 0.8

6. An open U-tube of uniform cross -section contains mercury. When 27.2 cm of water is poured into one limb of the tube, then the difference in levels of liquids of the two sides is (ρ_{water} = 1gm/cm³ and ρ_{Hg} = 13.6gm/cm³)

1) 23.2cm 2) 25.2cm 3) 21.2cm 4) None

7. A sphere weighs 0.2kg in air, 0.1 kg in water and 0.12 kg in a liquid. The specific gravity of the liquid is

1) 0.8 2) 8 3) 1.25 4) 1

8. What mass of lead will weigh as much as 6g of iron, when both are immersed in water. (Given specific gravities of iron and lead are 8 and 11 respectively)

1) 5.5g 2) 6.6g 3) 7.7g 4) 8.8g

9. A piston of cross - section area 100 cm² is used in a hydraulic press to exert a force of 10^7 dynes on the water. The crosssectional area of the other piston which supports an object having a mass 2000 kg is (g = 10m/s²)

1) 100 cm² 2) 10^9 cm²

3) 2 x 10^4 cm² 4) 2 x 10^{10} cm²

10. If the atmospheric pressure is 76cm of mercury at what depth of water the pressure will becomes 2 atmospheres nearly

1)862cm 2)932cm 3) 982cm 4)1033cm

VISCOUS FORCE

11. A plate of area 100 cm^2 is placed on the upper surface of castor oil 2mm thick. Taking the coefficient of viscosity to be 15.5 poise. The horizontal force necessary to move the plate with a velocity of 3 cms^{-1} is

1) 46.5 N 2) 23.25 N 3) 0.2325N 4) 2.325 N

12. A glass plate of dimensions 30 x 10 cm is placed on a table on which there is a layer of machine oil of thickness 0.01mm. The Co-efficient of viscosity of machine oil is 0.5 Ns/m^2. The force required to draw the plate horizontally with a velocity of $5ms^{-1}$.

1) 3000 N 2) 3750 N 3) 7500 N 4) 15000 N

13. Water in a river 20 m deep is flowing at a speed of $10ms^{-1.}$ The shearing stress between the horizontal layers of water in the river in Nm^{-2} is (Coefficient of viscosity of water = 10^{-3} SI units)

1) 1×10^{-2} 2) 0.5×10^{-2} 3) 1×10^{-3} 4) 0.5×10^{-3}

14. In the flow of a viscous liquid over a flat horizonta surface, the velocity is 0.25cm/sec at 2m.m distance above the flat surface at rest. What is the velosicity gradient

1) 0.0625/sec 2) 0.125/sec
3) 0.25/sec 4) 0375/sec

POISEULLI'S EQUATION

15. The coefficients of viscosity of two liquids are in the ratio 2 : 3. What is the ratio of the volumes of liquid collected in the same time using the same vessel completely filled with liquids of densities in the ratio 4 : 5 ? (The capillary tubes used have the same length and same bore diameter)

1) 8 : 15 2) 6 : 5 3) 5 : 6 4) 15 : 8

16. When water (η=0.01 poise) and benzene (η = 0.0065 poise) are allowed to flow through capillary tube, it was found that the same amount of liquids are collected in the same time. But the pressures that caused the flow are different. The pressure on water is 0.015 atmosphere, the corresponding pressure for benzene in atmosphere is

1) 0.0975 2) 0.00975
3) 0.975 4) 9.75

EQUATION OF CONTINUITY AND BERNOULI'S THEOREM

17. If liquid enters a section of tube of radius 5 cm with a velocity 10 m/s, with what velocity does the liquid leave another section of the tube of radius 4 cm ? The tube is horizontal

1) 0.625 m/s 2) 5.625 m/s
3) 15. 625 m/s 4) 24 m/s

18. The workdone in raising a stone of mass 4kg, specific gravity 2, immersed in water, from a depth of 3m to 1m below the surface is,

1) 98.J 2) 19.6 J 3) 29.4 J 4) 39.2 J

19. In a horizontal pipe line of uniform area of cross-section, the pressure falls by 8 Nm^{-2} between two points separated by distance of 1 km. The change in KE per kg of the oil flowing at these points is (density of oil is 800 kg/m^3)

1) $10^{-2}J\,kg^{-1}$ 2) $10^{-5}J\,kg^{-1}$
3) $10^{2}J\,kg^{-1}$ 4) $10^{-3}J\,kg^{-1}$

20. In a horizontal pipe a liquid of density d enters with velocity v and leaves with velocity 2v. The pressure difference between the ends of the pipe is

1) $3dV^2$ 2) $\frac{3}{2}dv^2$ 3) $\frac{1}{4}dv^2$ 4) $\frac{1}{2}dv^2$

21. A horizontal pipe of non -uniform cross - section has water flow through it such that the velocity is 1 ms^{-1} at a point where the pressure is 50 KPa. The pressure at a point where the velocity flow is $2ms^{-1}$ is

1) 1500 Pa 2) 51500 Pa 3) 48500 Pa 4) 3000 Pa

22. A pipe is running full of water. At a point 'A; it tapers from 200 cm^2 cross - sectional area to 100 cm^2 at other point 'B' . If the pressure difference between A and B is 15 cm of water colum; the rate of volume flow of water through the pipe is [g = 10 ms^{-2}]

1) 20 lit/ sec 2) 30 lit / sec
3) 15 lit / sec 4) 5lit / sec

23. The lift force acting on the wing of an aeroplane is 2275N. If the area of the wing is$5m^2$ and velocity of air on the lower surface of the wing is 30m/s, velocity of air on the upper surface of the wing is (density of air is 1.3kg/m^3)

1) 20m/s 2) 25m/s 3) 35m/s 4) 40m/s

24. The pressure on the top surface of an aeroplane wing was 0.90×10^5Pa and at the bottom surface is 0.91×10^5Pa. If the area of each surface is $40m^2$, the net lift force on the wing is

1) 2×10^4N 2) 72×10^5N
3) 6×10^7N 4) 4×10^4N

25. The pressure on the top surface of an aeroplane wing is 0.7×10^3 Pa and the pressure on the bottom surface is 0.75×10^3 Pa. If the area of each surface is 50 m^2, the dynamic lift on the wing is. (2007 - M)

1) 25×10^4 N 2) 0.5×10^4 N
3) 5×10^4 N 4) 0.25×10^4 N

26. A tank containing water has an orifice in one vertical side. It is 10m below the surface level of water in tank. Then velocity of efflux from that small orifice is

1) $7ms^{-1}$ 2) $3.5ms^{-1}$ 3) $9.8ms^{-1}$ 4) $14ms^{-1}$

27. A water tank has a small orifice of area $1mm^2$ in its vertical side. It is 5m below the surface level of water. If $g=10m/s^2$, the volume of water escaping from the orifice in one second

1) $10^{-3}m^3$ 2) $10^{-4}m^3$ 3) $10^{-5}m^3$ 4) $10^{-6}m^3$

28. A large vessel open at the top contains a non viscous liquid of specific gravity 0.8. If a hole is made at a depth of 5m from the free surface, the velocity of efflux is ($g=10m/s^2$)

1) 5m/s 2) 10m/s 3) 20m/s 4) $5\sqrt{2}$ m/s

29. A vessel filled with water stands on a table. In its side, the vessel has a small orifice arranged at a distance 25cm from the bottom of the vessel and at a distance 16cm from the level of water which is kept constant. The horizontal distance from the orifice at which the jet of water fall on the table is

1) 0.2 m 2) 0.4 m 3) 0.6 m 4) 0.8 m

30. A tank with vertical wall is mounted so that its base is at a height H above the horizontal ground. The tank is filled with water to a depth 'h'. A hole is punched in the side wall of the tank at a depth 'X' below the water surface. To have maximum range of the emerging stream, the value of X is [2001E]

1) $\frac{H+h}{4}$ 2) $\frac{H+h}{2}$

3) $\frac{H+h}{3}$ 4) $\frac{3(H+h)}{4}$

31. There is a hole at the bottom of a large open vessel. If water is filled upto a height h, it flows out in time "t". If water is filled to a height 4h, it will flow out in time

1) 4t 2) t/4 3) t/2 4) 2t

32. The level of water in a vessel falls from a height 64 cm to 36 cm in 5 minutes as water flows out through a hole in its bottom. What further time is required for the water level to fall from 36 cm to 16 cm ?

1) 5 minutes 2) 2 minutes
3) 8 minutes 4) 10 minutes

33. A water barrel having water upto a depth 'd' is placed on a table of height 'h'. A small hole is made on the wall of the barrel at its bottom. If the stream of water coming out of the hole falls on the ground at a horizontal distance 'R' from the barrel, then the value of 'd' is [2002E]

1) $\frac{4h}{R^2}$ 2) $4hR^2$ 3) $\frac{R^2}{4h}$ 4) $\frac{h}{4R^2}$

34. A hose shoots water straight up for a distance of 2.5mt. The end opening on the hose has an area of 0.75 cm^2. How much water will comes out in one minute ?

1) 31.3 lit. 2) 3.13 lit.
3) 313 lit. 4) 3.13 lit.

TERMINAL VELOCITY

35. Two metal balls of densities 'd' and '3d' and radii 'r_1' and 'r_2' are travelling in the same liquid. The density of the liquid is 'd/2'. The ratio of their terminal velocities is 4 : 5 respectively. Then the ratio of their radii is

1) 3 : 4 2) 2 : 1 3) 4 : 5 4) 5 : 12

36. The viscous force acting on a lead ball travelling down in a liquid column with a velocity 'v' is F. If another lead ball of double the mass travels down in the same liquid with a velocity 2v then the viscous force acting on it is

1) 16 F 2) 4 F 3) $2^{1/4}$ F 4) $2^{4/3}$ F

37. Two rain drops reach the earth with their terminal velocities in the ratio 4:9. The ratio of their radii is [2005 M]

1) 4:9 2) 2:3 3) 3:2 4) 9:4

38. The ratio of radii of two spheres is 1 : 2, when they are dropped in a viscous liquid the ratio of their terminal speeds is

1) 1 : 1 2) 2 : 1 3) 1 : 2 4) 1 : 4

39. A solid rubber ball of density d and radius R falls through air. Assume that the air resistance acting on the ball is KRV, where V is velocity of the ball and K is a constant. The terminal velocity attained by the ball is

1) $4\pi R^2 dg/3K$ 2) $4\pi Rdg/3K$ [83]

3) $4\pi Rd^2 g/3K$ 4) None

ANSWERS

1) 4	2) 3	3) 1	4) 2	5) 1
6) 2	7) 1	8) 2	9) 3	10) 4
11) 3	12) 3	13)4	14) 2	15) 2
16) 2	17) 3	18) 4	19) 1	20) 2
21) 3	22) 1	23) 4	24) 4	25) 4
26) 4	27) 3	28) 2	29) 2	30) 2
31) 4	32) 1	33) 3	34) 1	35) 2
36) 4	37) 2	38) 4	39) 1	

EXERCISE - III(A)

BUOYANCY

1. A cork is submerged in water by a spring attached to the bottom of a bowl. Then the bowl is kept in an elevator moving with acceleration downwards, the length of spring

1) Increases 2) Decreases

3) Remains unchanged 4) None of these

2. Two metals are mixed together in equal volumes to form an alloy of specific gravity 4. When equal masses of the same two metals are mixed together, the specific gravity of metal is 3, the specific gravity of each metal is

1) 4 and 6 2) 2 and 6 3) 2 and 4 4) 4 and 8

3. A ballon contains 400m^3 of hydrogen. If the mass of the ballon without hydrogen is 34kg. Its net lifting ability is

(Density of hydrogen = 0.09 kg/m^3)

Density of air = 1.3 kg/m^3)

1) 590kg.wt 2) 450kg.wt

3) 70kg.wt 4) 36kg.wt

4. Density of ice is ρ and that of water is σ. What will be the decrease in volume when a mass M of ice melts.

1) $\frac{M}{\sigma-\rho}$ 2) $\frac{\sigma-\rho}{M}$ 3) $M\left[\frac{1}{\rho}-\frac{1}{\sigma}\right]$ 4) $\frac{1}{M}\left[\frac{1}{\rho}-\frac{1}{\sigma}\right]$

5. A large block of ice 5 m thick has a vertical hole drilled in it and is floating in a lake. the minimum length of the rope required to draw a bucketfull of water through the hole is (density of ice = 900 kg/m^3)

1) 0.5 m 2) 1m 4) 4 m 4) 4.5 m

6. A cubical block of wood of side 10 cm floats at the interface between an oil and water with its lower surface 2 cm below the interface. the heights of the oil and water columns are 10 cm each. Density of oil is 0.8g/c.c. The mass of the block is

1) 720 gram 2) 900 gram

3) 960 gram 4) 840 gram

7. An ice berg of density 900 kg/m^3 is floating in water of density 1000 kg/m^3. The percentage of volume of ice cube out side the water is

1) 20% 2) 35 % 3) 10% 4) 25%

*8. A piece of copper having an internal cavity weighs 264 gm in air and 221gm in water. The volume of the cavity is (Density of copper = 8.8gm/cc)

1) 43cc 2) 30cc 3) 13cc 4) 70cc

9. A hemispherical portion of radius R is removed from the bottom of a cylinder of radius R. The volume of the remaining cylinder is V and its mass is M. It is suspended by a string in a liquid of density ρ where it stays vertical. The upper surface of the cylinder is at a depth h below the liquid surface. The force on the bottom of the cylinder by the liquid is [IIT Scren.-2001]

1) Mg

2) Mg - Vρg

3) Mg+ π R^{2}h ρ g

4) ρ g(V+ π R^{2}h)

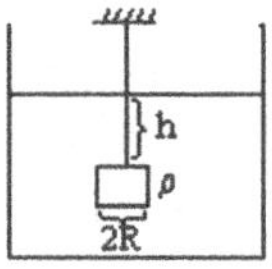

VISCOUS FORCE

10. Water flows through a horizontal tube of diameter 8cm and length 1km at the rate of 40 litres/s. Find the pressure difference required if $\eta = 0.01$ poise for water. (A.T.B.)

1) 1.98 × 10^5 dyne/cm^2

2) 2.98 × 10^5 dyne/cm^2

3) 3.98 × 10^5 dyne/cm^2

4) 4.98 × 10^5 dyne/cm^2

POISEULLI'S EQUATION

11. Two capillary tubes of length 15 cm and 5 cm and radii 0.06 cm and 0.02 cm respectively are connected in series. If the pressure difference across the end faces is equal to pressure of 14 cm high water column then find the pressure difference across the (i) first tube and (ii) second tube.

1) 0.5, 13.5 cm of water column

2) 1.5, 12.5 cm of water column

3) 2.5, 11.5 cm of water column

4) 3.5, 10.5 cm of water column

12. Two capillary tubes of equal length and inner radii 2r and 4r respectively are added in series and a liquid flows through it. If the pressure difference between the ends of the whole system is 8.5 cm of mercury, find the pressure difference between the two ends of the first capillary tube.

1) 8. 0 cm of Hg 2) 9.0cm of Hg

3) 10.0 cm of Hg 4) 11.0 cm of Hg

13. In a planet a sucrose solution of coefficient of viscocity 0.0015N-Sm^{-2} is driven at a velocity of 10^{-3}ms^{-1} through xylem vessels of radius 2 μ m and length 5 μ m. The hydrostatic pressure difference across the length of xylem vessels in Nm^{-2} is [2002M]

1) 5 2) 8 3) 10 4) 15

14. Two capillary tubes of radius r but of lengths ℓ_1 and ℓ_2 are fitted in parallel to the bottom of a vessel. The pressure head is P. What should be the length of a single tube that can replace the two tubes so that the rate of flow is same as before

1) $\ell_1 + \ell_2$ 2) $\dfrac{1}{\ell_1} + \dfrac{1}{\ell_2}$

3) $\dfrac{\ell_1 \ell_2}{\ell_1 + \ell_2}$ 4) $\dfrac{1}{\ell_1 + \ell_2}$

15. A capilary tube is attached horizontally to a constant head arrangement. If the radius capilary tube is increased by 10% then the rate of flow of liquid will change nearly by

1) + 10% 2) + 46% 3) – 10% 4) – 40%

16. The rate of steady volume flow of water through a capilary tube of length 'ℓ' and radius 'r' under a pressure difference of P is V. This tube is connected with another tube of the same length but half the radius in series. Then the rate of steady volume flow through them is (The pressure difference across the combination is P) (2003 E)

1) V/16 2) V/17 3) 16V/17 4) 17V/16

EQUATION OF CONTINUITY AND BERNOULI'S THEOREM

17. Blood flows through a main artery of area of cross section 2cm^2 with a speed of 0.25 ms^{-1}. If it branches into several capillaries of total area of cross section 0.4 m^2, the average speed of blood through each capillary is

1) 0.25 ms^{-1} 2) 1.25 X 10^{-4} ms^{-1}

3) 12.5 X 10^{-2} ms^{-1} 4) 0.5 X 10^{-4} ms^{-1}

18. Water flows at the rate of 2ms^{-1} through a pipe of radius 2 cm and emerges from a nozzle of radius 1 cm. If the nozzle is directed vertically upwards, water rises to a maximum height of (g = 10 ms^{-2})

1) 3.2 m 2) 6.4 m 3) 1.6 m 4) 0.8 m

19. A large vessel with a small hole at the bottom is filled with water and kerosene. The height of the water column is 20 cm and that of the kerosene is 25 cm. The velocity with with which water flows out of the hole is (density of kerosene = 0.8 gm/cc) (Neglect viscous force)

1) 2 ms^{-1} 2) 4 ms^{-1} 3) 2.6 ms^{-1} 4) 3 ms^{-1}

20. A cylindrical drum open at the top contains 30 liters of water. It drains out through a small opening at the bottom. 10 liters of water comes out in time t, the next 10 liters in further time t_2 and the last 10litres in further time t_3. Then

1) $t_1 = t_2 = t_3$ 2) $t_1 > t_2 > t_3$ 3) $t_1 < t_2 < t_3$ 4) $t_2 > t_1 = t_3$

21. A vessel of large uniform cross - sectional area resting on a horizontal surface holds two immiscible, non - viscous and incompressible liquids of densitles d and 2d each of height H/2. If a small hole is punched on the vertical side of the container at a height h$\left(h < \dfrac{H}{2}\right)$, the efflux speed of the liquid at the hole is

1) $\sqrt{2g(H-h)}$ 2) $\sqrt{2g(H-h)}$

3) $\sqrt{\dfrac{g}{2}(3H-4h)}$ 4) $\sqrt{\dfrac{g}{2}(4H-3h)}$

22. Water from a tap emerges vertically down with an initial speed of $1.0 ms^{-1}$. The cross sectional area of tap is $10 x\ 10^{-5}\ m^2$. Assume that the pressure is constant throughout the stream of water, and that the flow is a steady, the cross sectional area of the steam 0.15m below the tap is [UP SET 2000]

1) $5.0 \times 10^{-4} m^2$ 2) $1.0 \times 10^{-5} m^2$
3) $5.0 \times 10^{-5} m^2$ 4) $2.0 \times 10^{-5} m^2$

23. A large open top container of negligible mass and uniform cross sectional area A has a small hole of cross sectional area a in its side wall near the bottom. The container is kept over a smooth horizontal floor and contains a liquid of density ρ and mass m_0. Assuming that the liquid starts flowing through the hole A, the acceleration of the container will be [IIT 97]

1) $\frac{2ag}{A}$ 2) $\frac{ag}{A}$ 3) $\frac{2Ag}{a}$ 4) $\frac{Ag}{a}$

24. A small hole is made at a height of $\left(1/\sqrt{2}\right)m$ from the bottom of a cylindrical water tank. The length of the water column is $\sqrt{2}m$. Find the distance where the water emerging from the hole strikes the ground.

1) 4m 2) 2m 3) 3m 4) $\sqrt{2}$ m

25. There are two holes one each along the oppsite sides of a wide rectangular tank. The cross section of each hole is $0.01\ m^2$ and the vertical distance between the holes is one meter. The tank is filled with water. The net force on the tank in newton when the water flows out of the holes is: (density of water is $1000\ kg/m^3$)

1) 100 2) 200 3) 300 4) 400

26. Water in a storage tank stands 2.5m above the level of a valve in the side of the tank. With what speed the water will rush out neglecting friction (A.T.B.)

1) 5 m/s 2) 7 m/s 3) 9 m/s 4) 11 m/s

27. The average speed of blood in the artery (diameter =2cm) during the resting part of heart's cycle is about 30 cm s^{-1}. Is the flow laminar or turbulent? Density of blood $1.05 x 10^3$ Kg m^{-3}. Explain your answer.

1) laminar 2) turbulent
3) both 4) none

28. The vein of healthy person is horizontal and its length is 0.04m and has internal radius 1 mm. At 37^0C, the average velocity of blood through the vein is 4 cm s^{-1}. Find a) the rate of flow of the blood b) the velocity at the mid point of the vein and c) the decrease in pressure at the other end of the vein.

Take $\eta_{blood} = 4 \times 10^{-1}$ Pas $P_a s$

1) $1.25 \times 10^{-7}\ m^3 s^{-1}$, $0.08\ m\ s^{-1}$, 51.2 $P_a s$
2) $0.08 \times 10^{-7}\ m^3 s^{-1}$, $0.08\ m\ s^{-1}$, 51.2 $P_a s$
3) $1.25 \times 10^{-7}\ m^3 s^{-1}$, $0.08\ m\ s^{-1}$, 5.12 $P_a s$
4) $1.25 \times 10^{-7}\ m^3 s^{-1}$, $0.8\ m\ s^{-1}$, 51.2 $P_a s$

TERMINAL VELOCITY

29. An air bubble of radius 1 mm moves up with uniform velocity of $0.109 ms^{-1}$ in a liquid column of density $14.7 \times 10^3 kg/m^3$, then coefficient of viscosity will be (g = 10 ms^{-2})

1) 0.3 Pa - s 2) 300 Pa - s **[Raj PET 97]**
3) 15 Pa - s 4) 150 Pa-s

30. Two balls of same density falls in a viscous medium. The radius of first ball being double than the radius of second ball, then how many times is the terminal velocity of second ball to that of first ball [IIIT Allahabad 2001]

1) 1 2) 2 3) 4 4) 1/4

31. A ball of radius r and density ρ falls freely under gravity through a distance h before entering water. Velocity of ball does not change even on entering water. If viscosity of water is η, the value of h is given by

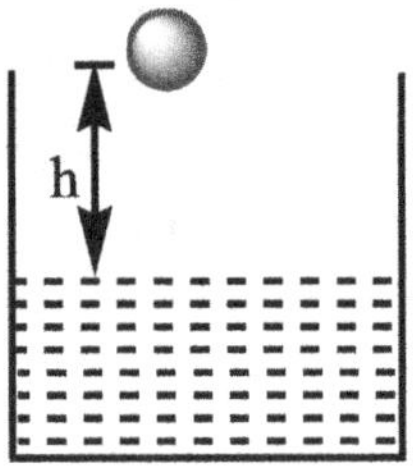

1) $\frac{2}{9}r^2\left(\frac{1-\rho}{\eta}\right)g$

2) $\frac{2}{81}r^2\left(\frac{\rho-1}{\eta}\right)g$

3) $\frac{2}{81}r^4\left(\frac{\rho-1}{\eta}\right)^2 g$

4) $\frac{2}{9}r^4\left(\frac{\rho-1}{\eta}\right)^2 g$

32. What is the radius of a small oil drop while falling through air acquires a terminal velocity of 1mm/ s. The density of oil and air are $0.91 g/cm^3$ and 1.3g/liter respectively and $\eta = 1.8 \times 10^{-4}$ poise for air. (A.T.B.)

1) 2×10^{-4}cm 2) 4×10^{-4}cm
3) 5×10^{-4}cm 4) 3×10^{-4}cm

33. An air bubble of radius 1cm rises through a liquid of density 1.5g/c.c., at a constant speed of 0.25cm/s. What is coefficient of viscosity of liquid, neglecting the density of air? (A.T.B.)

1) 1.31×10^3 poise 2) 2×10^3 poise
3) 3×10^3 poise 4) 4×10^3 poise

34. A sphere floats with exactly half its volume immersed in the liquid. If the mass of the sphere is 25.6g and density of liquid is 1.6gr/cm^3. Calculate the radius of the sphere approximately

1) 1cm 2) 2cm 3) 3cm 4) 4cm

ANSWERS

1) 2	2) 2	3) 2	4) 3	5) 1
6) 4	7) 3	8) 3	9) 4	10) 3
11) 1	12) 1	13) 4	14) 3	15) 2
16) 2	17) 2	18) 1	19) 3	20) 3
21) 3	22) 3	23) 1	24) 4	25) 2
26) 2	27)1	28) 1	29) 1	30) 4
31) 3	32) 4	33) 1	34) 2	

EXERCISE - III(B)

BUOYANCY

1. A sphere of density d is allowed to fall in a liquid of density d/4. The accleration of the body will be

1) g/4 2) 3g/4 3) g/2 4) g

2. Three liquids of densities d, 2d and 3d are mixed in equal proportions of weights. The relative density of mixture is

1) $\frac{11d}{7}$ 2) $\frac{18d}{11}$ 3) $\frac{13d}{9}$ 4) $\frac{23d}{18}$

3. A balloon contains 500m^3 of hydrogen. If the mass of the balloon without hydrogen is 55 kg , its net lifting ability is (Density of hydrogen = 0.09 kg / m^3, density of air = 1.3 kg/m^3)

1) 590 kg.wt 2) 450 kg.wt
3) 70 kg.wt 4) 550 kg.wt

4. 90gm of sulphuric acid of specific gravity 1.8 are mixed with 90gm of water of density 1gm/cc. If the specific gravity of mixture is 1.5, the contraction in volume when liquids are mixed is

1) 140cc 2) 120cc 3) 40cc 4) 20cc

5. A log of wood of mass 120 kg floats in water. The weight that can be on the raft to make it just sink, should be(density of wood = 600 kg/m^3)

1) 80 kg 2) 50 kg 3) 60 kg 4) 30 kg

6. Two bodies re in equilibrium when suspended in water from the arms of a balance. The mass of one body is 36 g and its density is 9 g / cm^3. If the mass of the other is 48 g, is density in g / cm^3 is

1) 4/3 2) 3/2 3) 3 4) 5

7. The fraction of a floating object of volume V_0 and density d_0 above the surface of a liquid of density d will be

1) $\frac{d_0}{d}$ 2) $\frac{dd_0}{d+d_0}$ 3) $\frac{d-d_0}{d}$ 4) $\frac{d-d_0}{d_0}$

8. A balloon filled with hydrogen has a volume of 1000 *lt* and a mass of 1kg. What would be the volume of a block of a very light material which it can just lift? One litre of the material has a mass of 91.3g (Density of air - 1.3 g/l)

1) 13.33 *lt* 2) 3.33 *lt*
3) 1 *lt* 4) none

9. A ball whose density is 0.4×10^3 kg/m^3 falls into water from a height of 9 cm. To what depth does the ball sink

1) 9 cm 2) 6 cm 3) 4.5 cm 4) 2.25 cm

VISCOUS FORCE

10. The relative velocity of two consecutive layers is 8cm/s. If the perpendicular distance between the layers is 0.1 cm, then the velocity gradient will be

1) 8 sec^{-1} 2) 80 sec^{-1} 3) 0.8 sec^{-1} 4) 0.08 sec^{-1}

POISEULLI'S EQUATION

11. Three capillaries of lengths l, 2l and l/2 are connected in series. Their radii are r, r/2 and r/3 respectively. If stream line flow is maintained and the pressure difference across the first capillary tube is p_1 and the pressure difference across (i) the second and (ii) the third capillary tube.

1) 32 p_1 ; 40.5 p_1 2) 35 p_1; 34.6 p_1
3) 13 p_1 ; 10.0 p_1 4) 21 p_1 ; 13.1 p_1

12. An open tank cotnain full of water column of length 60 cm. Three capilary tubes each of length 'l' and same radius are attached to the tank in horizontal postition at a depth 5 cm, 15 cm, 20 cm from top. What is the length of equivelent capillary tube of same radius that should be attaced at the bottom of the tank.

1) 3l/2 2) 2l/3 3) 3l/4 4) 2l/5

13. **Water is conveyed through a uniform tube of 8cm in diameter and 3140m in length at rate of 2×10^{-3} m^3 per second. The pressure required to maintain the flow is (viscosity of water=10^{-3} SI units)** [2001E]

1) 6.25Nm^{-2} 2) 0.625Nm^{-2}
3) 6250Nm^{-2} 4) 0.000625Nm^{-2}

14. **Two capillary tubes of same length but different radii r_1 and r_2 are fitted in parallel to the bottom of a vessel. The pressure head is P. What should be the radius of a single tube that can replace the two tubes so that the rate of flow is same as before**

1) $r_1^4 + r_2^4$ 2) $(r_1^2 + r_2^2)^{\frac{1}{4}}$
3) $(r_1^4 + r_2^4)^{\frac{1}{4}}$ 4) $r_1 + r_2$

15. **We have two capillary tubes T_1 and T_2. Their lengths are ℓ_1 and ℓ_2 and radii of cross - sectin are r_1 and r_2 respectively. The rate of flow of water under a pressure difference P through tube T_1 is 8 cm^3/sec. If $\ell_1 = 2\ell_2$ and $r_1 = r_2$ what will be the rate of flow when the two tubes are connected in series and pressure difference across the combination is same as before (in cm^3/sec)**

1) 4 2) $\frac{16}{3}$ 3) $\frac{8}{17}$ 4) 8

16. **Two capillary of length L and 2L and of radius R and 2R are connected in series. The net rate of flow of fluid through them will be (given $X = \pi PR^4 / 8\eta L$)**

1) $\frac{8}{9}X$ 2) $\frac{9}{8}X$ 3) $\frac{5}{7}X$ 4) $\frac{7}{5}X$

17. **A vessel fitted with a horizontal capillary tube at its base is first filled with a liquid A, and then with a liquid B. The specific gravities of the two liquids are 0.6 and 0.8 respectively. If the time taken to empty the vessel be 1 hr and 2 hr respectively, then determine the ratio of their coefficient of viscosities.**

1) 3 : 8 2) 4 : 5 3) 8 : 3 4) 5 : 4

EQUATION OF CONTINUITY AND BERNOULI'S THEOREM

18. **The velocity of flow of a liquid through a pipe at a point of radius r is v. At a point where the radius is 2r, the mass of the liquid flowing out per second is (d = density of the liquid)**

1) $2\pi r^2 vd$ 2) $4\pi r^2 vd$
3) $\pi r^2 vd$ 4) $\frac{\pi r^2 vd}{4}$

19. **An open tank filled with water has a hole at its bottom. If the total pressure at the bottom of the tank is 2 atmospheres , the velocity of efflux is (nearly)**
(1 Atmosphere = 10^5 Nm^{-2})

1) 30 ms^{-1} 2) 20 ms^{-1}
3) 14 ms^{-1} 4) 9.8 ms^{-1}

20. **A cylinder of weight 20m is completely filled with water. The velocity of efflux of water (in ms^{-1}) through a small hole on the side wall of the cylinder near its bottom is**

1) 10 ms^{-1} 2) 20 ms^{-1} [AIEEE2002]
3) 25.5 ms^{-1} 4) 5 ms^{-1}

21. **The level of water in a vessel falls from a height 36 cm to 16 cm in 6 minutes as water flows out through a hole in its bottom. At the end of next 3 minutes, the height of water level in the vessel will be**

1) 6 cm 2) 3 cm 3) 9 cm 4) 2 cm

22. **A manometer shown below contains a liquid of density ρ. The difference in the levels when the manometer rotates at a constant angular velocity ω about one of its vertical limbs.**
(Let l be the length of horizontal tube)

1) $\frac{\omega^2\ell^2}{2g}$ 2) $\frac{\omega^2\ell^2}{g}$ 3) $\frac{2\omega^2\ell^2}{g}$ 4) $\frac{3\omega^2\ell^2}{g}$

23. **One end of a horizontal oil pipe is closed using a valve and the reading of the barometer attached to the pipe is 2 x 10^5 Pa. When the valve in the pipe is opened the reading of the barometer falls to 10^5 Pa. The velocity of oil flowing through the pipe will be (density of oil = 800 kg m^{-3}) in ms^{-1}**

1) 5 2) 10 3) 2.5 4) $5\sqrt{10}$

24. A horizontal pipe line carries water in a streamline flow. At a point along the pipe where cross sectional area is $10cm^2$, the water velocity is 1 ms^{-1} and the pressure is 2000Pa. The pressure of water at another point where the cross sectional area is $5cm^2$ is (in pa)[IIT 94]

1) 1000 2) 500 3) 3500 4) 1025

25. Tanks A and B open at the top contain two different liquids upto certain height in them. A hole is made to the wall of each tank at a depth 'h' from the surface of the liquid. The area of the hole in A is twice that of in B. If the liquid mass flux through each hole is equal, then the ratio of the densities of the liquids respectively, is [2002E]

1) $\frac{2}{1}$ 2) $\frac{3}{2}$ 3) $\frac{2}{3}$ 4) $\frac{1}{2}$

26. A wind-powered generator converts wind energy into electrical energy. Assume that the generator converts a fixed fraction of the wind energy intercepted by its blades into electrical energy. For wind speed V, the electrical power output will be proportional to [IIT Screening 2001]

1) V 2) V^2 3) V^3 4) V^4

27. A large cylindrical vessel filled with a liquid upto a certain height is empited by a horizontal capillary tube, fitted to its base. Find the ratio of time taken to empty the tank by one half, to the next one fourth.

1) $\sqrt{2}:1$ 2) $1:\sqrt{3}$ 3) $\sqrt{3}:1$ 4) 1 : 1

28. Water is falling down at the rate of 1.8 Kg per minute from a vetical tube. The radius at the bottom of the tube is 0.004 m and pressure around it is 0.76m of Hg. The diameter of the tube at a height of 0.5 m from the bottom is 0.005 m. find the pressure at that point.

1) 0.724 m of Hg 2) 0.824 m of Hg
3) 0.924 m of Hg 4) 0.624 m of Hg

TERMINAL VELOCITY

29. An air bubble of radius 1cm rises from the bottom portion through a liquid of density 1.5g/cc at constant speed of 0.25cm/se. If the density of air is neglected the coefficient of viscosity of the liquid is approximately (In pa. s) [2006 E]

1) 13,000 2) 11300 3) 130 4) 13

30. An iron sphere of mass $20x10^{-3}$kg falls through a viscous liquid with terminal velocity 0.5 m/se. The terminal velocity (in m/se) of another iron sphere of mass $54x10^{-2}$kg is [2005 E]

1) 4.5 2) 3.5 3) 2.5 4) 1.5

31. A drop of liquid having radius 2mm has a terminal velocity $20cms^{-1}$, the terminal velocity of a drop 1 mm radius will be [AFMC 2000]

1) 40 cms $^{-1}$ 2) 20 cms $^{-1}$ 3) 10 cms $^{-1}$ 4) 5 cms $^{-1}$

32. A spherical ball of radius 1×10^{-4} m and density 10^4 kg m^{-3} falls freely under gravity through a distance h before entering a tank of water. If, after entering the water, the velocity of the ball does not change, find h. The coefficient of viscosity of water is 9.8×10^{-6} Pa s.

1) 10. 4 m 2) 12.4 m
3) 25.5 m 4) 20.4 m

33. An air babul of radius 0.5cm rises through a liquid of denstiy 1.5gr/c.c at a constant speed of 0.5cm/sec. What is coefficient of veiscosity of liquid. Negleting the density of air

1) 1.63×10^2 poise 2) 3.2×10^2 poise
3) 4.8×10^2 poise 4) 6.4×10^2 poise

34. A sphere floats with exactly 3/4 its volume immersed in the liquid. If the mass of the sphere is 25.6g and density of liquid is $1.4gr/cm^3$. Calculate the radius of the sphere approximately

1) $(2.8)^{1/3}$ 2) $(3.8)^{1/3}$ 3) $(4.8)^{1/3}$ 4) $(5.8)^{1/3}$

ANSWERS

1) 2	2) 2	3) 4	4) 4	5) 1
6) 3	7) 3	8) 2	9) 2	10) 2
11) 1	12) 1	13) 3	14)3	15)2
16) 2	17) 1	18) 3	19) 3	20) 2
21)3	22)1	23) 4	24) 2	25)4
26) 3	27) 1	28) 1	29) 3	30) 2
31) 4	32) 4	33) 1	34) 4	

COMPREHENSION QUESTIONS

Comprehension - I (1 - 3) :

A cylindrical tank is completely filled with a liquid. At the bottom of the tank, three capillary tubes A, B and C of lengths l, $2l$ and $l/2$ are connected in series. Their radii are, $\frac{r}{2}$, r and 2r respectively. The pressure difference across the entire combination is P.

1. The ratio of rates of flow of liquid in the pipes A, B and C is

1) 1:1:1 2) 2:4:1 3) 1:2:4 4) 1:16:256

2. **The ratio of pressure difference across the pipes A, B and C is**

1) 1:1:1 2) 1:16:256
3) 512:64:1 4) 64:1:16

3. **The pressure difference across the pipe 'B' is**

1) P 2) $\frac{P}{64}$ 3) $\frac{P}{256}$ 4) $\frac{64}{577}P$

Comprehension - II (4 - 6) :

Three metal spheres P, Q and R of densities d, $\frac{d}{2}$ and 2d are falling down in a liquid of density $\frac{d}{4}$ with terminal velocities. The radii of the spheres are $\frac{r}{2}$, r and 2r respectively.

4. **The ratio of the terminal velocities of the spheres P, Q and R is**

1) 3:4:112 2) 1:4:16 3) 3:1:7 4) 1:1:1

5. **The ratio of their momenta is**

1) 128:8:112 2) $\frac{3}{112}:\frac{1}{128}:8$
3) $\frac{3}{128}:\frac{1}{8}:112$ 4) $\frac{3}{256}:\frac{1}{64}:8$

6. **The ratio of their Kinetic energies is**

1) $\frac{3}{448}:\frac{1}{512}:8$ 2) $\frac{3}{512}:\frac{1}{448}:8$
3) $64:\frac{1}{8}:512$ 4) $\frac{3}{512}:\frac{1}{8}:448$

Comprehension - III (7 - 11) :

In one experiment two beakers containing water and a liquid of specific gravity 0.8 are taken. A body 'P' is found to be floating with 40% of its volume above water surface. In another experiment another body 'Q' which weighs 100gm in air found to be weighing 80gm in water.

7. **The density of body 'P' is**

1) 0.4 gm/c.c 2) 0.6 gm/c.c
3) 1.4 gm/c.c 4) 1.6 gm/c.c

8. **The fraction of volume of body 'P' submerged in the liquid is**

1) $\frac{1}{4}$ 2) $\frac{1}{2}$ 3) $\frac{2}{3}$ 4) $\frac{3}{4}$

9. **The ratio of volumes of the body 'P' submerged in water to the liquid is**

1) 4:5 2) 5:4 3) 3:4 4) 4:3

10. **The apparent weight of body 'Q' in the liquid is**

1) 70 gm 2) 84 gm 3) 96 gm 4) 80 gm

11. **The density of body 'Q' is**

1) 4 gm/c.c 2) 2 gm/c.c
3) 5 gm/c.c 4) 8 gm/c.c

Comprehension - IV (12 - 14) :

Water is flowing through a horizontal pipe of variable cross section. The areas of cross section at A, B and C are 200 cm^2, 100 cm^2 and 50 cm^2 respectively. Water is flowing from A to B to C. Velocity water at A is 2 m/s and pressure at A is 50 KPa.

12. **Velocity of water at 'C' is**

1) 4 m/s 2) 8 m/s
3) 1 m/s 4) 0.5 m/s

13. **Pressure at B is**

1) 94 KPa 2) 106 KPa
3) 100 KPa 4) 90 KPa

14. **Rate of flow of water at 'C' is**

1) 40 lit/s 2) 80 lit/s
3) 20 lit/s 4) 10 lit/s

ANSWERS

1) 1	2) 3	3) 4	4) 1	5) 3
6) 4	7) 2	8) 4	9) 1	10) 2
11) 3	12) 2	13) 1	14) 1	

PROBLEMS RELATED TO AIEEE

EXERCISE - IV(A)

1. **Figure shows a U-tube of uniform cross-sectional area A accelerated with acceleration a as shown. If d is the separation between the limbs, then what is the difference in the level of the liquid in the U-tube ?**

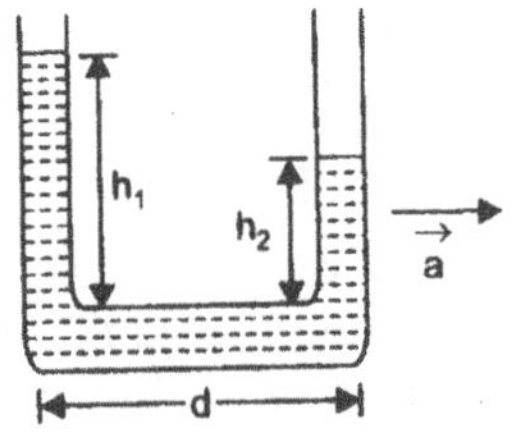

1) $\frac{ad}{g}$ 2) $\frac{d}{ag}$ 3) $\frac{g}{ad}$ 4) $\frac{a}{gd}$

2. **A tube 1 cm^2 in cross-section is attached to the top of a vessel 1 cm high and of cross section 100 cm^2. Water is poured into the system filling it to a depth of 100 cm above the bottom of the vessel as shown in figure then find force exerted by water against the bottom of the vessel ?**

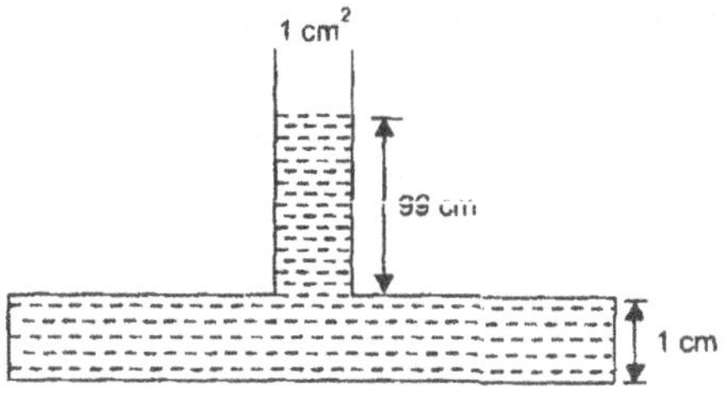

1) 100 N 2) 200 N 3) 300 N 4) 400 N

3. **In a car lift, compressed air exerts a force F_1 on a small piston having a radius of 5 cm. This pressure is transmitted to a second pistion of radius 15 cm. If the mass of the car to be lifted is 1350 kg, what is F_1 ?**

1) 1.5×10^3 N 2) 2.5×10^3 N
3) 3.5×10^3 N 4) 4.5×10^3 N

4. **A 700 g solid cube having an edge of length 10 cm floats in water. What volume of the cube is outside water ?**

1) 300 cm^3 2) 600 cm^3
3) 150 cm^3 4) 450 cm^3

5. **A spring balance reads 10 kg when a bucket of water is suspended from it. What is the reading on the spring balance when an ice cube of mass 1.5 kg is put into the bucket**

1) 22.5 kg 2) 2.25 kg 3) 1.15 kg 4) 11.5 kg

6. **A concrete sphere of radius R has a cavity of radius r which is packed with saw-dust. The specific gravities of concrete and sa-dust are 2.4 and 0.3 respectively. The sphere floats with its entire volume sub-merged under water. Calculate the ratio of mass of concrete and the mass of saw-dust**

1) 1 2) 2 3) 3 4) 4

7. **A piece of brass (Cu and Zn) weighs 12.9 g in air. When completely immersed in water, it weights 11.3 g. The relative densities of Cu and Zn are 8.9 and 7.1 respectively. Calculate the mass of copper in the alloy**

1) 6. 6 gm 2) 7. 6 gm 3) 8.6 gm 4) 5.6 gm

8. **Calculate the fraction of a floating object of density d_0 above the surface of a liquid of density d**

1) $\frac{d-d_0}{d}$ 2) $\frac{d_0-d}{d_0}$

3) $\frac{d_0}{d}$ 4) $\frac{d}{d_0}$

9. **A string is holding a solid block below the surface of a liquid. The density of liquid is greater than that of solid. The tension in the string is T_0. When the system is at rest. What will be the tension in the string if the system has upward acceleration a ?**

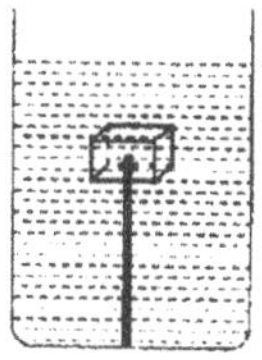

1) $\left[1+\frac{a}{g}\right]$ 2) $T_0\left[1+\frac{g}{a}\right]$

3) $T_0\left[1+\frac{a}{g}\right]$ 4) $T_0\left[1-\frac{g}{a}\right]$

10. **To what height should a cylindrical vessel be filled with a homogeneous liquid to make the force with which the liquid presses on the side of the vessel equal to the force exerted by the liquid on the bottom of the vessel ?**

1) equal to the radius 2) less than raidus
3) more than radius 4) four times of radius

11. **A cube of wood supporting a 200 g just floats in water. When the mass is removed, the cube rises by 2 cm. What is the side of the cube ?**

1) 10 cm 2) 20 cm 3) 30 cm 4) 40 cm

12. **A liquid flows through a pipe of 10^{-3} m radius and 0.1 m length under a pressure of 10^3 Nm^{-2}. Calculate the speed of the liquid coming out of the pipe. The coefficient of viscosity of the liquid of 1.25×10^{-3} decapoise.**

1) 1 $m\,s^{-1}$ 2) 2 $m\,s^{-1}$ 3) 3 $m\,s^{-1}$ 4) 4 $m\,s^{-1}$

13. Figure shows how the stream of water emerging from a faucet "necks down" as it falls. Figure shows two levels separated by a vertical distance h. The cross-sectional areas A_0 and A are marked in the figure. At what rate does water flow from the tap?

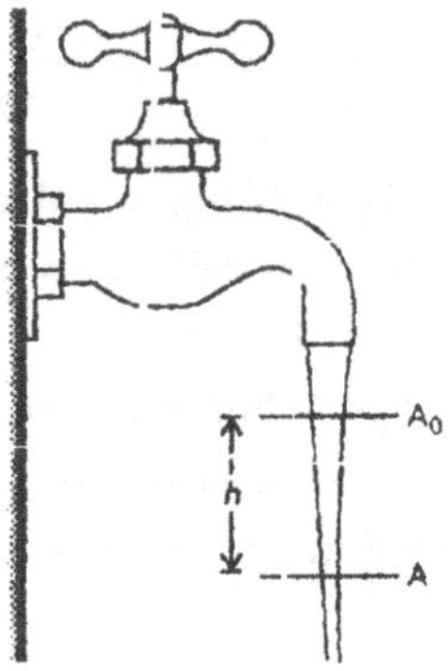

1) $A_0A\sqrt{\frac{2gh}{A^2-A_0^2}}$ 2) $A_0\sqrt{\frac{2ghA}{A^2-A_0^2}}$

3) $A\sqrt{\frac{2ghA_0}{A^2-A_0^2}}$ 4) $A_0A\sqrt{\frac{2gh}{A_0^2-A^2}}$

14. What is the largest average velocity of blood flow in an artery of radius 2×10^{-3} m if the flow must remain laminar ? (Given : viscosity of blood 2.084×10^{-3} Pa s and density of blood = 1060 kg m^{-3}).

1) 0.886 m s^{-1} 2) 0.786 m s^{-1}

3) 0.983 m s^{-1} 4) 0.686 m s^{-1}

15. A horizontal tube has different cross-section at two points A and B. The diameters of the tube at points A and B are 4 cm and 2 cm respectively. The two manometers arms are fixed at points A and B. When a liquid of density 800 kg m^{-3} flows through the tube, the difference of pressure between the arms of manometer is 8 cm. Calculate the rate of flow of liquid nearly, deriving the necessary formula (g = 9.8 m s^{-2})

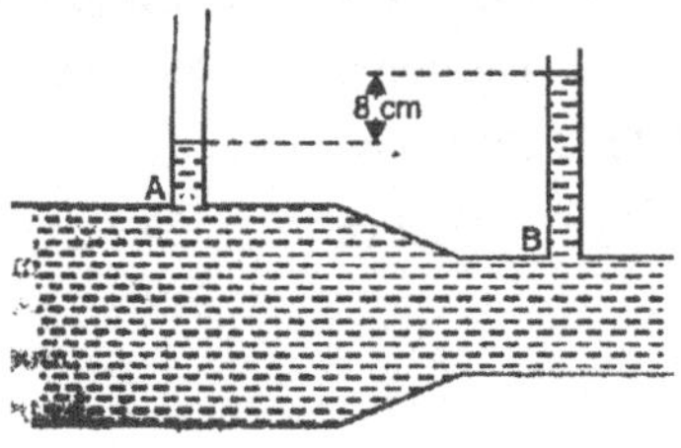

1) 3×10^{-4} m s^{-1} 2) 4×10^{-4} m s^{-1}

3) 2×10^{-4} m s^{-1} 4) 5×10^{-4} m s^{-1}

16. A body of density ρ floats with a volume V_1 of its total volume V immersed in one liquid of density ρ_1 and with the remainder of volume V_2 immersed in another liquid of density ρ_2, where $\rho_1 > \rho_2$. Find the relative volume immersed in two liquids

1) $V_1 = V\left(\frac{\rho-\rho_2}{\rho_1-\rho_2}\right)$, $V_2 = V\left(\frac{\rho_1-\rho}{\rho_1-\rho_2}\right)$

2) $V_1 = V\left(\frac{\rho_1-\rho_2}{\rho_1-\rho}\right)$, $V_2 = V\left(\frac{\rho_1-\rho}{\rho_1-\rho_2}\right)$

3) $V_1 = V\left(\frac{\rho-\rho_2}{\rho_1-\rho}\right)$, $V_2 = V\left(\frac{\rho_1-\rho_2}{\rho-\rho_2}\right)$

4) $V_1 = V\left(\frac{\rho-\rho_2}{\rho_1-\rho}\right)$, $V_2 = V\left(\frac{\rho-\rho_2}{\rho-\rho_1}\right)$

17. A beaker containing water of weight W is placed on a spring balance. A stone weight W is hung and lowered into the water without touching the sides and bottom of the beaker. Then the tension in the string is ....

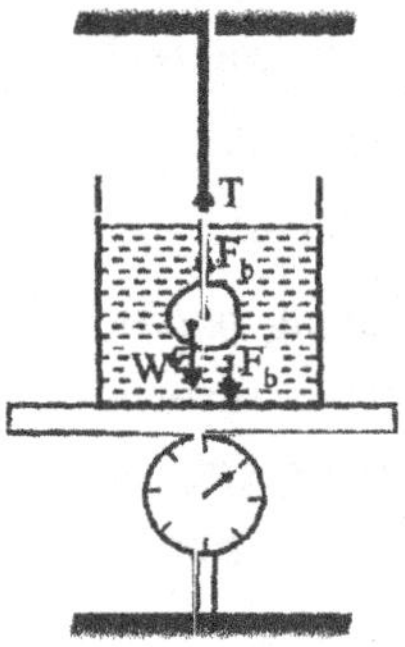

1) $T = W + F_b$

2) $T = W$

3) $T = W - F_b$

4) $T = F_b$

18. A cylinderical tank has a hole of 1 cm^2 at its bottom. If the water is allowed to flow into the tank from a tube above it at the rate of 70 cm^3/s, then find the maximum height upto which water can rise in the tank

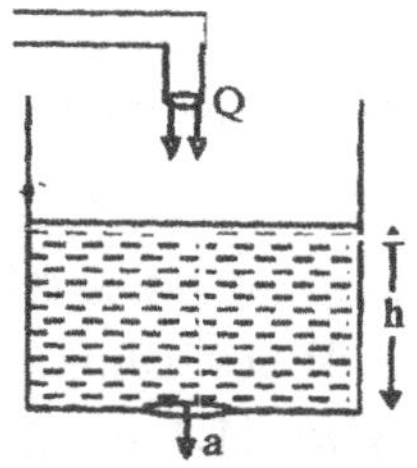

1) 2.5×10^{-2} m

2) 1.25×10^{-2} m

3) 5.5×10^{-2} m

4) 3.5×10^{-2} m

19. A tank filled with water (density $\rho_w = 1000$ kg/m^3) and oil of (density $\rho_{oil} = 900$kg/m^3). The hight of water is 1.0m and oil is 4.0m. Find the velocity of efflux through a hole at the bottom of the tank

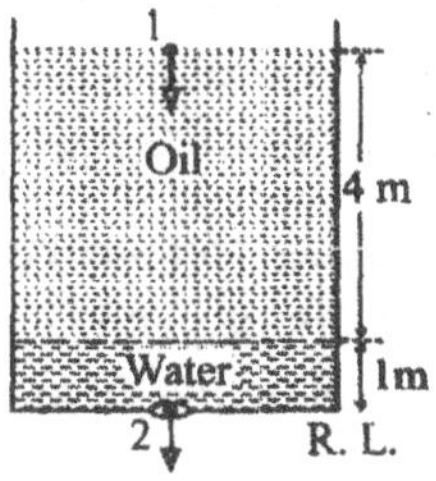

1) 5.9 m/s

2) 6.5 m/s

3) 9.5 m/s

4) 4.5 m/s

20. A large open top container of negligible mass and uniform cross-sectional area A has a small hole of cross-sectional area A/100 in its side wall near the bottom. The container is kept on a smooth horizontal floor and contains a liquid of density ρ and mass m_0. Assuming that the liquid starts flowing out horizontally through the hole at t = 0, calculate the acceleration of the container.

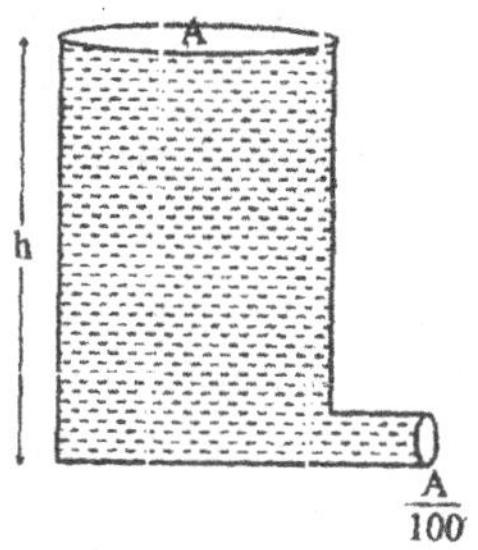

1) $\frac{1}{2}$ m/s^2

2) $\frac{1}{3}$ m/s^2

3) $\frac{1}{4}$ m/s^2

4) $\frac{1}{5}$ m/s^2

21. A closed rectangular tank is completely filled with water and is accelerated horizontally with an acceleration a towards right. Pressure is maximum at ....

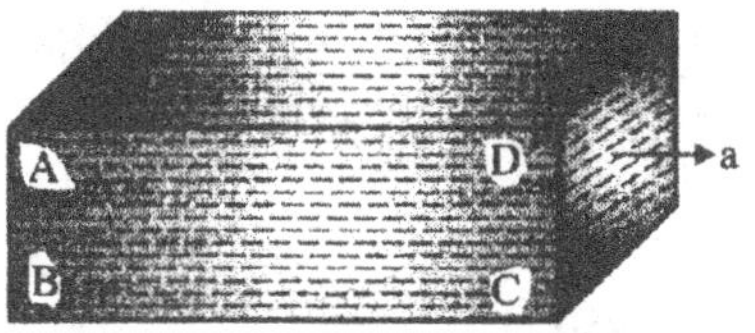

1) B 2) C 3) D 4) A

22. In this figure, an ideal liquid flows through the tube, which is of uniform cross-section. The liquid has velocities v_A and v_B, and pressure P_A and P_B at points A and B respectively

1) $v_A = v_B$ 2) $v_B > v_A$
3) $P_A = P_B$ 4) $P_B < P_A$

23. A homogeneous solid cylinder of length L (L<H/2). Cross-sectional area A/5 is immersed such that it floats with its axis vertical at the liquid-liquid interface with length L/4 in the denser liquid as shown in the figure. The lower density liquid is open to atmosphere having pressure P_0. The density D of solid is given by

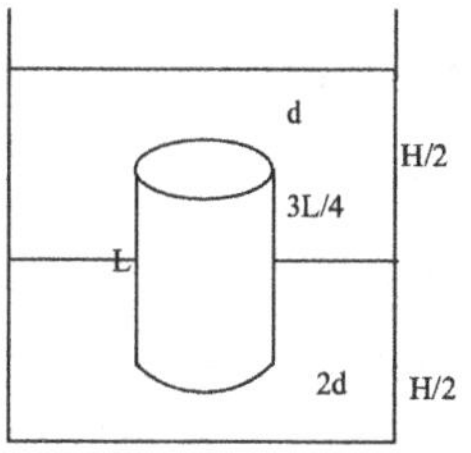

1) $\frac{5}{4}$ d 2) $\frac{4}{5}$ d 3) d 4) $\frac{d}{5}$

24. Two communicating vessels contain mercury. The diameter of one vessel is n times larger than the diameter of the other. A column of water of height h is poured into the left vessel. The mercury level will rise in the right-hand vessel (s=relative density of mercury and ρ = density of water) by

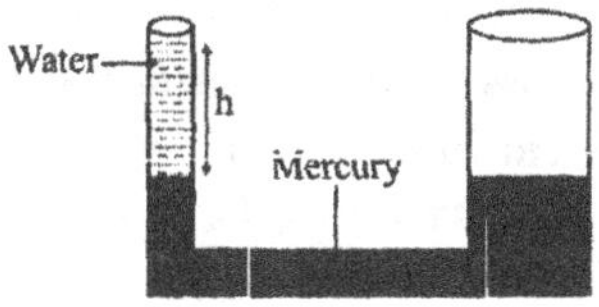

1) $\frac{n^2h}{(n+1)^2s}$ 2) $\frac{h}{(n^2+1)s}$

3) $\frac{h}{(n+1)^2s}$ 4) $\frac{h}{n^2s}$

25. A cylinderical vessel of 90 cm height is kept filled upto the brim. It has four holes 1, 2, 3, 4 which are respectively at heights of 20 cm, 30 cm 45 cm and 50 cm from the horizontal floor PQ. The water filling at the maximum horizontal distance from the vessel comes from

1) Hole number 4

2) Hole number 3

3) Hole number 2

4) Hole number 1

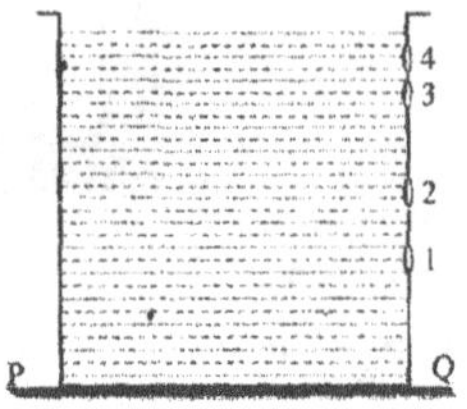

ANSWERS

1) 1	2) 1	3) 1	4) 1	5) 4
6) 4	7) 2	8) 1	9) 3	10) 1
11) 1	12) 1	13) 4	14) 3	15) 2
16) 1	17) 3	18) 1	19) 3	20) 4
21) 1	22) 1	23) 1	24) 2	25) 2

EXERCISE - IV(B)

1. Figure shows a U-tube of uniform cross-sectional area A accelerated with acceleration $4.9 m/s^2$ as shown. If 2m is the separation between the limbs, then the difference in the level of the liquid in the U-tube is

1) 2m

2) 3m

3) 4m

4) 1m

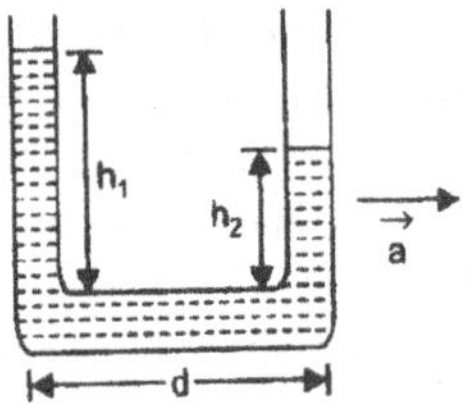

2. A tube 1 cm^2 in cross-section is attached to the top of a vessel 1 cm high and of cross section 100 cm^2. Water is poured into the system filling it to a depth of 100 cm above the bottom of the vessel as shown in figure. Then the weight of water in the system is

1) 19. 9 N

2) 2.99 N

3) 3.99 N 4) 1.99 N

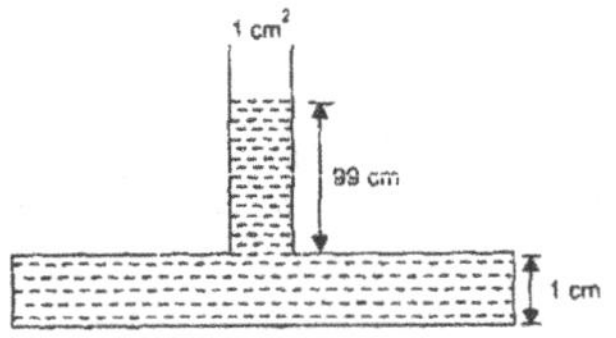

3. In a car lift, compressed air exerts a force F_1 on a msall piston having a radius of 5 cm. This pressure is transmitted to a second pistion of radius 15 cm. If the mass of the car to be lifted is 1350 kg, the pressure necessary to accomplish this task is

1) 1.9×10^5 Pa 2) 19×10^5 Pa
3) 0.19×10^5 Pa 4) 2.9×10^5 Pa

4. A 600 g solid cube having an edge of length 9 cm floats in water. The volume of the cube outside water is

1) 400 cm^3 2) 300 cm^3 3) 200 cm^3 4) 100 cm^3

5. A spring balance reads 10 kg when a bucket of water is suspended from it. What is the reading on the spring balance when an iron piece of mass 7.8 kg suspended by an other string is immersed with half its volume inside the water in the bucket. (Relative density of iron = 7.8) ?

1) 1.05 kg 2) 10.5 kg 3) 105 kg 4) 0.105 kg

6. A hallow spherical iron shell floats almost completely submerged in water. The outer diameter is 58.7 cm and the density of iron is 7.87 g cm^{-3}. What is the approximate inner diameter of the shell ?

1) 56 cm 2) 65 cm 3) 23 cm 4) 32 cm

7. A piece of brass (Cu and Zn) weighs 12.9 g in air. When completely immersed in water, it weights 11.3 g. The relative densities of Cu and Zn are 9 and 3 respectively. the mass of copper in the alloy is

1) 10.2 gm 2) 12.75 gm 3) 15.5 gm 4) 9.8 gm

8. Calculate the fraction of a floating object of density 300 kg m^{-3} above the surface of a liquid of density 800 kg m^{-3}

1) 8/5 2) 5/8 3) 4/5 4) 5/4

9. A string is holding a solid block below the surface of a liquid. The density of liquid is greater than that of solid. The tension in the string is 400 N. When the system is at rest. What will be the tension in the string if the system has upward acceleration 2.45 ms^{-2}?

1) 125 N 2) 250 N

3) 500 N 4) 750 N

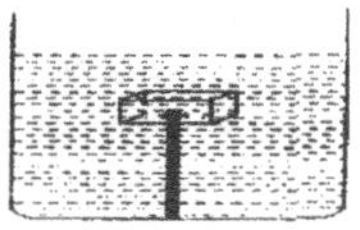

10. To what height should a cylindrical vessel be filled with a homogeneous liquid to make the force with which the liquid presses on the side of the vessel equal to the force exerted by the liquid on the bottom of the vessel of radius 10cm?

1) 20 cm 2) 15 cm 3) 12.5 cm 4) 10 cm

11. **A cube of wood supporting a 400 g just floats in water. When the mass is removed, the cube rises by 2 cm. What is the side of the cube nearly ?**

1) 1.414 cm 2) 14.14 cm 3) 24.24 cm 4) 28.28 cm

12. **A liquid flows through a pipe of 10^{-3} m radius and 0.1 m length under a pressure of 10^3 Nm^{-2}. Calculate the speed of the liquid coming out of the pipe. The coefficient of viscosity of the liquid of 2.5×10^{-3} decapoise.**

1) 0.3 ms^{-1} 2) 0.5 ms^{-1} 3) 0.7 ms^{-1} 4) 0.9 ms^{-1}

13. **Figure shows how the stream of water emerging from a faucet "necks down" as it falls. Figure shows two levels separated by a vertical distance h. The cross-sectional areas A and A/2 are marked in the figure. At what rate does water flow from the tap?**

1) $\sqrt{\frac{2ghA}{3}}$

2) $A\sqrt{\frac{2gh}{5}}$

3) $A\sqrt{\frac{2gh}{3}}$

4) $\frac{A}{3}\sqrt{2gh}$

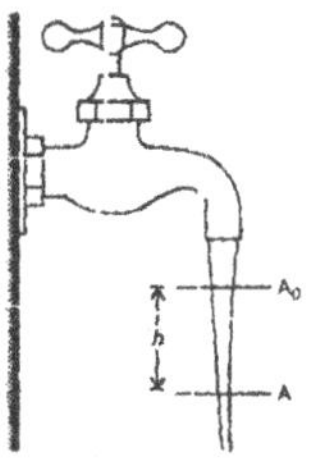

14. **Calculate the minimum pressure required to force the blood from the heart to the top of the head nearly (vertical distance 0.5 m). Assume the density of blood to be 1040 kg m^{-3}. Friction is to be neglected.**

1) 7.6 x 10^3 Nm^{-2} 2) 3.6 x10^3 Nm^{-2}

3) 4.5 x 10^3 Nm^{-2} 4) 2.3 x 10^3 Nm^{-2}

15. **A Pitot's tube is fixed in a main pipe of diameter 15 cm and difference of pressure indicated by the gauge is 4 cm of water column. Calculate the volume of water passing through the main pipe per minute nearly ?**

1) 2.4 x 10^4 cm^3s^{-1} 2) 1.2 x 10^4 cm^3s^{-1}

3) 4.8 x 10^4 cm^3s^{-1} 4) 5.6 x 10^4 cm^3s^{-1}

16. **What is the flow rate of blood flow in an artery of radius 2×10^{-3} m if the flow must remain laminar ? (Given : viscosity of blood 2.084 × 10^{-3} Pa s and density of blood = 1060 kg m^{-3}).**

1) 12.36 $\times10^{-5}$ m^3 s^{-1} 2) 1.236 $\times10^{-5}$ m^3 s^{-1}

3) 22.36 $\times10^{-5}$ m^3 s^{-1} 4) 2.26 $\times10^{-5}$ m^3 s^{-1}

17. **A horizontal tube has different cross-section at two points A and B. The areas of the tube at points A and B are a_1 and a_2 respectively. The two manometers arms are fixed at points A and B. When a liquid of density flows through the tube, the difference of pressure between the arms of manometer is h. The rate of flow of liquid is ,**

1) $a_1a_2\sqrt{\frac{2gh}{a_1^2-a_2^2}}$ 2) $a_1a_2\sqrt{\frac{2gh}{a_1^2a_2^2}}$

3) $a_1a_2\sqrt{\frac{2gh}{a_1^2+a_2^2}}$ 4) $a_1a_2\sqrt{\frac{2g}{a_1^2+a_2^2}}$

18. **A body of density ρ floats with a volume V_1 of its total volume V immersed in one liquid of density 2ρ and with the remainder of volume V_2 immersed in another liquid of density ρ /2. Find the relative volume immersed in two liquids**

1) $\frac{V}{2},\frac{V}{3}$ 2) $\frac{V}{3},\frac{3V}{2}$ 3) $\frac{V}{2},\frac{2V}{3}$ 4) $\frac{4V}{3},\frac{V}{2}$

19. **A beaker containing water of weight W is placed on a spring balance. A stone weight W is hung and lowered into the water without touching the sides and bottom of the beaker. Then the reading of the balance is**

1) $W + F_b$

2) $W - F_b$

3) W

4) F_b

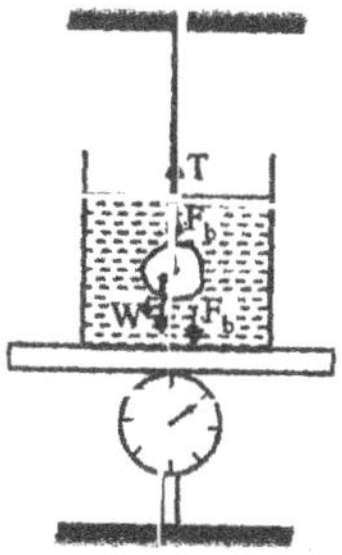

Fig. 3.50

20. **A cylinderical tank has a hole of 2 cm^2 at its bottom. If the water is allowed to flow into the tank from a tube above it at the rate of 1 0 0 cm^3/s, then find the maximum height upto which water can rise in the tank (Take g = 10 m s^{-2})**

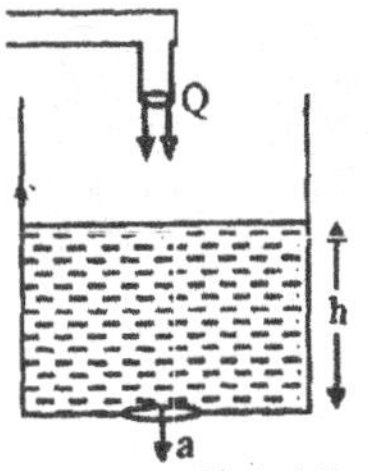

1) 2.5× 10^{-2} m 2) 1.25× 10^{-2} m

3) 5.5× 10^{-2} m 4) 3.5× 10^{-2} m

21. **A bent is lowered into a water stream as shown in the figure . The velocity of stream relative to the tube is equal to v = 2.5 m/s. The closed upper end of the tube located at the height h_0 = 12 cm as a small oriffice. To what height h will the water jet spurt ?**

1) 0.2 m

2) 0.1 m

3) 0.3 m 4) 0.4 m

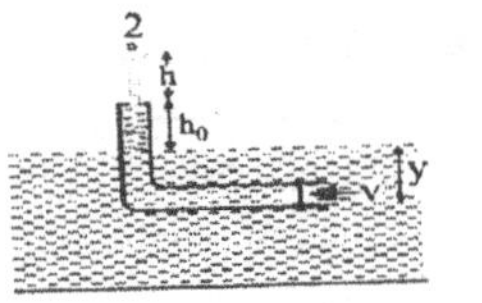

22. **A large open top container of negligible mass and uniform cross-sectional area A has a small hole of cross-sectional area A/100 in its side wall near the bottom. The container is kept on a smooth horizontal floor and contains a liquid of density ρ and mass m_0. Assuming that the liquid starts flowing out horizontally through the hole at t = 0, calculate its velocity when 75% of the liquid has drained out.**

1) $\frac{1}{5}\times\frac{100}{\sqrt{2g}}\sqrt{\frac{m_0}{\rho A}}$

2) $\frac{1}{5}\times\frac{200}{\sqrt{2g}}\sqrt{\frac{m_0}{\rho A}}$

3) $\frac{1}{3}\times\frac{200}{\sqrt{2g}}\sqrt{\frac{m_0}{\rho A}}$

4) $\frac{1}{4}\times\frac{200}{\sqrt{2g}}\sqrt{\frac{m_0}{\rho A}}$

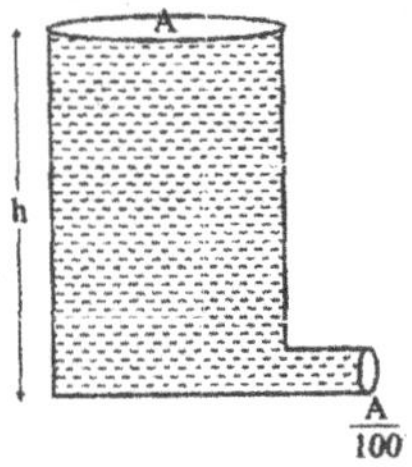

23. **A closed rectangular tank is completely filled with water and is accelerated horizontally with an acceleration a towards right. Pressure is minimum at ....**

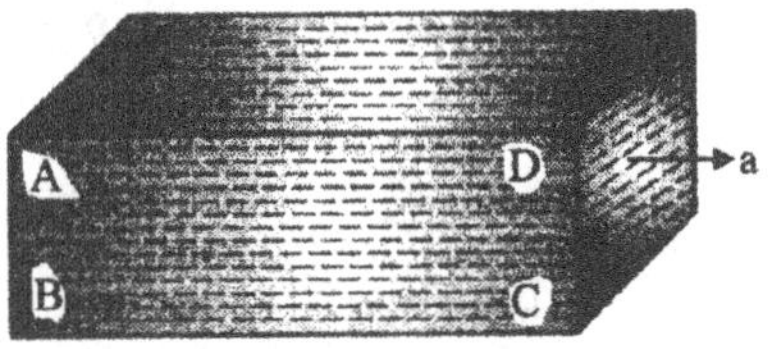

1) A 2) B 3) C 4) D

24. **In this figure, an ideal liquid flows through the tube, which is of uniform cross-section. The liquid has velocities v_A and v_B, and pressure P_A and P_B at points A and B respectively, then**

1) $v_A < v_B$

2) $v_B > v_A$

3) $P_A = P_B$

4) $P_B > P_A$

25. **An incompressible liquid flows through a horizontally tube shown in the following figure. Then the velocity v of the fluid is**

1) 3.0 m/s 2) 1.5 m/s

3) 1.0 m/s 4) 2.25 m/s

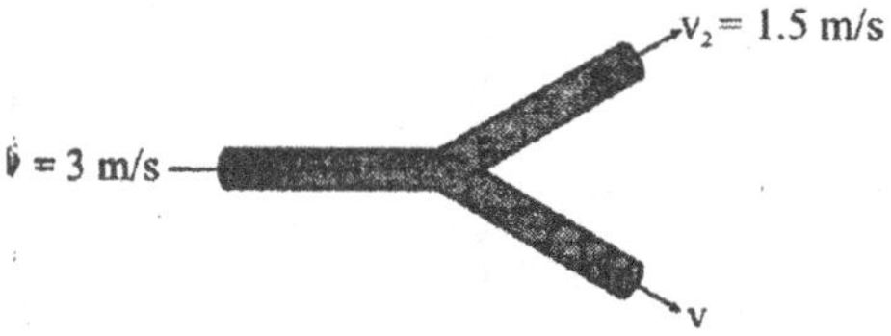

ANSWERS

1) 4	2) 4	3) 1	4) 1	5) 2
6) 1	7) 2	8) 2	9) 3	10) 4
11) 2	12) 2	13) 3	14) 1	15) 1
16) 2	17) 1	18) 3	19) 1	20) 2
21) 1	22) 1	23) 4	24) 4	25) 3

Chapter - 3

MECHANICAL PROPERTIES OF FLUIDS

SURFACE TENSION

SYNOPSIS

- The property of liquid by virtue of which its free surface behaves like streched elastic membrane is called surface tension.
- It does not obey Hooke's law.
- It is the normal force acting per unit length of an imaginary line drawn on the free surface of the liquid.
- $T = F/l$ D.F. - MT^{-2} unit N/m.
- Surface tension arises due to cohesive forces. Surface tension belongs to electro magnetic force.
- Surface tension depends upon.
  - i) Nature of liquid
  - ii) Temperature of liquid
  - iii) Impurities
  - iv) Medium over the liquid
  - v) Charge on the liquid drop
- Water spreads over the glass more than mercury because surface tension of water is less

  S.T. of water < S.T. of Hg.
- Antiseptic ointments should have less surface tension so that they can spread easily.
- Temperature : When temperature is more, surface tension is less (except for molten cadmium and molten copper)
- Surface tension is zero at critical temperature, boiling point, triplepoint.
- Surface tension is molecular phenomenon
- Impurities : partially soluble impurities decreases surface tension.

  Eg : Oil, grease, wax

  Completly soluble impurities increases it

  Eg : salt, proteins. added to water, increases S.T.

• ***Force in addition to the weight of the body required to pull it from the surface of liquid on which is floating $F = Tl$***

1) Needle/wire $F = T \times 2l$

2) A ring $F = 2 \times 2\pi r \times T$

(or) hollow cylinder $F = 4\pi rT$

3) A small disc or solid cylinder $F = 2\pi rT$

4) Annular ring $F = 2\pi(R+r)T$

5) The perimeter of ellipse = $\pi(a+b)$

Area of the ellipse = πab

Where a = length of semi major axis

b = length of semi minor axis

Cohesive forces : Forces of attraction between identical molecules.

- Cohesive forces do not obey inverse square law
- Cohesive forces are not gravitational forces.

Adhesive forces : Forces of attraction between dissimilar molecules.

- **Molecular range :** The maximum distance upto which a molecule can attract another molecule is effectively molecular range (10^{-9}m or 1nm)
- **Sphere of influence** : A sphere imagined with molecule at the centre and molecular range as radius is known as sphere of influence.
- A molecule attracts and is attracted in turn by only those molecules within the sphere of influence
- The resultant force acting on a molecule well inside the liquid is zero.
- The molecules on and near the surface of a liquid experience resultant downward force.
- If a molecule is to be brought from bottom of a liquid to the top, work must be done against these forces. This work is stored in the form of P.E.
- Free surface of a liquid possess P.E.
- A system is stable if it has low P.E.
- The liquid surface tries to have a stable surface by having least P.E. To have least PE, it tries to have least surface area.
- Surface tension is a molecular phenomenon.
- Because of surface tension, liquid surface tries to have least surface area ie., a spherical shape for a given volume.

Applications :

1) Rain drops are spherical
2) Small drops of mercury are spherical (forces due to S.T. are predominant than gravitational forces)
3) Large drops of mercury are flat because of predominant gravitational forces.
4) Ends of glass rod becomes round on heating
5) Manufacture of lead shots is based on this concept.

6) Hair sets well when oil is applied.
7) Bristles of paint brush cling together when taken out of water.
8) An ice block (cubical) is suspended in a gravity less region by means of a thread. When all the ice melts, the water remains there itself attaining a spherical shape.

Surface energy : The potential energy per unit surface area of the liquid (or) workdone to increase surface area of liquid by one unit is surface energy.

Surface energy is numerically equal to surface tension under isothermal conditions.

$$T = \frac{W}{\Delta A} \ \frac{J}{m^2} \quad ; \quad \boxed{W = \Delta A \times T}$$

- Work done in blowing a drop of radius (R) is

$$\boxed{W = 4\pi R^2 T}$$

- Work done in increasing the radius of the drop from R_1 to R_2 is $\boxed{W = 4\pi (R_2^2 - R_1^2)\, T}$

- Work done in blowing a soap bubble of radius (R)

$$\boxed{W = 8\pi R^2 T}$$

Work done in increasing the radius of bubble from R_1 to R_2 is $\boxed{W = 8\pi (R_2^2 - R_1^2)\, T}$

- ***When a large drop of radius (R) splits into n number of identical small drops each of radius (r), then***

i) total surface area increases

ii) energy must be supplied or work must be done

iii) If splitting occurs in adiabatic conditions, the drop uses its internal energy for splitting. So the temperature of drop decreases. So droplets cool.

iv) $\Delta A = 4\pi R^2 \, [n^{1/3} - 1] \Rightarrow W = 4\pi R^2 T (n^{1/3} - 1)$

- ***When (n) small drops of radius (r) merge to form a big drop***

i) energy is released during merging

ii) surface area decreases

iii) in adiabatic conditions, temperature of drop increases

iv) Energy released during merging

$$W = 4\pi r^2 n^{2/3} (n^{1/3} - 1) T$$

- Surface energy of a film $\boxed{W = 2AT}$
- Between two glass plates a water drop is squeezed to form a thin film of thickness (d) and surface area (A). The force required to separate the two plates.

$$F.d = 2AT \longrightarrow \boxed{F = \frac{2AT}{d}}$$

- **Angle of contact :** The angle made by the tangent drawn at the point of contact of liquid surface and solid surface with the solid and is measured inside the liquid.

Angle of contact depends on

i) solid - liquid pair

ii) temperature

iii) impurities

- Angle of contact (θ)

i) decreases with increase in temperature

ii) decreases with partially soluble impurities, like detergents & wetting agents

iii) increases with addition of water proof agent

iv) does not depend upon angle of inclination.

The radius of the bore of a capillary tube is r and the angle of contact of the liquid is θ. when the tube is dipped in the liquid, the radius of curvature of the meniscus of liquid rising in the tube is $\boxed{R = \frac{r}{\cos\theta}}$

- $\theta = 0°$ shape is concave hemispherical (rises)

eg : pure water, clean glass

- $\theta < 90°$ concave (rises)

ordinary water, glass

- $\theta = 90°$ horizontal (non rise or fall)

Eg. water and silver

- $\theta > 90°$ convex (falls)

Eg. Mercury and glass

- $\theta = 180°$ Convex hemispherical

Capillarity : The rise or fall of a liquid in a capillary tube due to surface tension is called capillarity.

Capillarity is due to both cohesive and adhesive forces.

$$2\pi r T \cos\theta = \pi r^2 h d g$$

$T = \frac{hrdg}{2\cos\theta} \Rightarrow \boxed{h = \frac{2T\cos\theta}{rdg}}$

$\boxed{h \propto \frac{1}{r} \propto \frac{1}{\sqrt{A}} \propto \frac{1}{V^{1/3}}}$

Where A is area of cross section of capillary tube.

V is the volume of the liquid.

- Graph between h and r is rectangular hyperbola.
- Volume or mass of liquid rises into capillary tube is directly proportional to radius of capillary tube.

$V \alpha\ r^2.h\alpha\ r^2.\ \frac{1}{r}\left[\because h \propto \frac{1}{r}\right]$

$V \alpha\ r$

- ***When capillary tube is dipped in a beaker containing liquid, rise in liquid is h. If the tube is tilted through an angle (α) with the vertical, then***

  i) the liquid rises into the capillary tube remains same

  ii) length of liquid column rise increases

$\boxed{\ell = \frac{h}{\cos\alpha}} \Rightarrow \boxed{\ell_1\cos\alpha_1 = \ell_2\cos\alpha_2}$

When capillary tube is of insufficient length :

i) liquid do not overflow

ii) curvature of liquid surface decreases

iii) radius of curvature increases

iv) angle of contact increases

The liquid rises and then forms a bubble like out side but will not overflow.

- When the arrangement containing a capillary tube dipped in beaker containing water is taken to height (x) above the ground

  $hg = h^1g\left[1 - \frac{2x}{R}\right]$ [Where R is radius of the earth]

  $\frac{h}{h^1} = \frac{R-2x}{R}$

- If the same arrangement is taken to depth (x) below the ground then $\frac{h^1}{h} = \frac{R}{R-x}$
- If height of liquid column rise is (h) on earth, the height on the surface of moon will be 6h.
- If a capillary tube dipped in beaker is placed in artificial satellite then

  i) liquid rises to full length of the tube whatever may be length of the tube

  ii) liquid does not overflow.

Applications :

1) Action of towel

2) Action of fountain pen

3) Kerosene stove, water rising from roots to leaves in plants.

4) Blotting paper

5) Brick walls are plastered to prevent capillarity

- Due to surface tension, the surface behaves as stretched elastic membrane
- Mosquitoes can walk on surface of water.
- A needle placed gently on the surface of water floats.
- Oil spreads on water. Surface tension $\alpha\ \frac{1}{\text{Area}}$.
- partially soluble impurities decrease surface tension soluble impurities increase surface tension.
- Mosquito breed is destroyed by sprinkling kerosene on stagnant water.
- Detergent cleans the clothes effectively
- Plastic boat fitted with small pieces of camphor moves on surface of water.
- Surface tension decreases with increase of temp.
- Hot water is preferred in washing cloths.
- Hot water is preferred for taking a bath as hot water has less surface tension.

EXCESS PRESSURE

- The difference in pressure between the two sides of liquid surface is called excess pressure.
- Excess pressure in liquid drop (or) an air bubble inside liquid $P = \frac{2T}{R}$
- ***Excess pressure*** inside soap bubble $P = \frac{4T}{R}$ (2 surfaces)

$P\ \alpha\ \frac{1}{r}\ \alpha\ \frac{1}{\sqrt{A}}\ \alpha\ \frac{1}{V^{1/3}}$

A is surface area of bubble

V is volume of bubble

- If a smaller bubble and larger bubble are in communication with each other,

  i) air flows from smaller bubble to larger bubble

  ii) the larger bubble expands and smaller bubble shrinks.

  iii) when two liquid drops of radius R_1 and R_2 merge to form single drop under isothermal condition,

  $4\pi R^2 = 4\pi R_1^2 + 4\pi R_2^2$

Radius of larger drop $R = \sqrt{R_1^2 + R_2^2}$

- Small bubble touching big bubble then radius of curvature of common interface,

$$R = \frac{r_1 r_2}{r_2 - r_1}$$

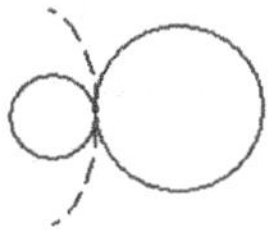

- A bubble of radius (r_1) is inside another bubble of radius (r_2). The radius of a single bubble whose excess pressure is equal to difference in pressures between inside of inner bubble and outside the outer bubble,

$$R = \frac{r_1 r_2}{r_1 + r_2}$$

EXERCISE - I

1. The force of attraction between molecules of same substances are called
1) Cohesive forces 2) Adhesive forces
3) Tensor forces 4) Hardcore forces

2. Forces of adhesion are
1) repulsive forces
2) attraction or repulsion between molecules of different substance
3) attraction or repulsion between of same substance molecules
4) none

3. The range of attraction between molecules is of the order of
1) 10^{-6}m 2) 10^{-9}m 3) 10^{-12}m 4) 10^{-15}m

4. The order of the thickness of surface film of a liquid is
1) 1 $\overset{0}{A}$ 2) 0.1 $\overset{0}{A}$ 3) 5 $\overset{0}{A}$ 4) 0.5 $\overset{0}{A}$

5. Surface tension of a liquid
1) surface phenomenon
2) exists throughout the liquid
3) both 4) none

6. When the surface of a liquid is increased
1) surface tension increases
2) surface energy increases
3) surface tension does not change 4) 2 and 3

7. The main difference between liquid surface and an elastic membrane is
1) Hook's law is not obeyed by liquid surface
2) surface tension increases with increase of surface area
3) both 4) none

8. Unit of surface tension
1) Nm^{-1} 2) Jm^{-2}
3) kgs^{-2} 4) all of the above

9. In the absence of an external force, the shape of liquid drop is determined by
1) surface tension of the liquid
2) density of liquid
3) viscosity of liquid 4) temperature of air only

10. The sphere of influence of a liquid molecule is completely lying inside the liquid then,
1) any two molecules in it can attract
2) molecule at the centre is not attracted by all molecules in the sphere of influence
3) the net force on the molecule at the centre in zero
4) all of the above

11. An iron needle slowly placed on the surface of water floats on it because
1) when inside in water it will displace water more than its weight
2) the density of the material of needle is less than that of water
3) of surface tension 4) of its shape

12. Soap helps in cleaning the clothes because
1) it reduces the surface tension of solution
2) it gives strength to solution
3) it absorbs the dirt 4) none of the above

13. Which of the following has the same dimensional fomula as that of surface tension
1) force constant 2) plancks constant
3) gravitational constant 4) gas constant

14. A drop of kerosene oil placed on water
1) remains as a drop 2) shrinks
3) spreads on the surface of water 4) none

15. The unit of angle of contact is
1) Radian 2) N/m-radian
3) N-radian- meter 4) Sta-radian

16. Two needles are floating on the surface of water. A hot needle when touches water surface between the needles, then they move
1) closer 2) away
3) surface tension 4) all of the above

17. When a soapwater bubble is given a positive charge it expands. If it is given a negative charge, then it
1) expands 2) contracts
3) remains same 4) does not hold negative charge

18. In a gravity free surface, shape of a large drop of liquid is

1) spherical 2) cylindrical
3) neither spherical nor cylindrical
4) nearly spherical

19. When surf (detergent) is added to pure water its surface tension :

1) decreases 2) increases
3) does not change 4) becomes infinite

20. When salt is added to pure water, surface tension of water

1) decreases 2) increases
3) does not change 4) becomes zero

21. A disc of paper of radius R has a hole of radius r. It is floating on a liquid of surface tension T. The force of surface tension of the disc is

1) Tx2pR 2) Tx2p(R-r)
3) Tx2p(R+r) 4) Tx4p(R+r)

22. A metallic wire of density d floats horizontal in water. The maximum radius of the wire so that the wire may not sink is (surface tension of water = T)

1) $\sqrt{\frac{2T}{\pi dg}}$ 2) $\sqrt{\frac{2\pi T}{dg}}$

3) $\sqrt{\frac{2\pi Tg}{d}}$ 4) $\sqrt{2\pi Tgd}$

23. Two pieces of glass plate one upon the other with a little water in between them cannot be separated easily because of [JIMPER 97]

1) inertia 2) pressure
3) surfce tension 4) viscosity

24. A paper disc of radius R from which a hole of radius r is cut out, is floating in a liquid of surface tension, T. The force on the disc due to surfacr tension will be —

1) πRT 2) $\pi(R+r)T$
3) $2\pi T(R+r)$ 4) $4\pi T(R+r)$

25. An ice cube is suspended in vacuum in a gravity free room. As the ice melts it will –

1) fall down on the floor 2) fly up
3) retain its cubical shape
4) change its shape to spherical

26. Two water drops merge with each other to form a large droplet. In this process

1) energy is liberated 2) energy is absorbed
3) energy is neither liberated nor absorbed
4) some mass is converted into energy.

27. The work done to increase unit area of the liquid surface at same temperature is called.

1) surface tension 2) energy of the molecules
3) energy of the liquid 4) none

28. A number of small drops of mercury coalesce adiabatically to form a single drop. The temperature of drop

1) Decreases 2) Increases
3) Remains unchanged
4) May decreases or increase depending upon size

29. N drops of a liquid, each with surface energy E, join to form a single drop, In this process

1) Some energy is absorbed
2) Energy absorbed is $E(n-n^{2/3})$
3) Energy released is $E(n-n^{2/3})$
4) Energy released is $E(2^{2/3}-1)$

30. The amount of work done in blowing a soap bubble such that its diameter increases from d to D (S is the surface tension of solution) [MP PMT 96]

1) $\pi(D^2-d^2)S$ 2) $2\pi(D^2-d^2)S$
3) $4\pi(D^2-d^2)S$ 4) $8\pi(D^2-d^2)S$

31. A drop of water of surface area A and radius R at a temperature T breaks into two drops of area A_1 and A_2, radii r_1 and r_2 and having temperatures T_1 and T_2. Then the correct statement is

1) $A = A_1 + A_2$ 2) $T = T_1 = T_2$
3) $R = r_1 + r_2$ 4) $A < (A_1 + A_2)$

32. A drop of water breaks into two droplets of equal size. In this process which of the following statements is correct?

1) The sum of the temperatures of the two droplets together is equal to the original temperature of the drop
2) The sum of the masses of the droplets is equal to the original mass of the drop
3) The sum of the radii of the two droplets is equal to the radius of the original drop
4) The sum of the surface areas of the two droplets is equal to the surface area of the original drop

33. **If the cohesive and adhesive forces between solid and liquid are equal, shape of the meniscus of the liquid in that capillary tube is**
1) horizontal 2) concave
3) convex 4) none

34. **More liquid rises into a thin capillary tube beause**
1) surface tension decreases
2) surface tension increases
3) rise of liquid is inversely proportional to radius of bore
4) surface tension is proportional to radius of bore

35. **The rise of liquid into capillary tube is h_1. If the apparatus is taken in a lift moving up with acceleration, the height is h_2, then**
1) $h_1 = h_2$ 2) $h_1 < h_2$ 3) $h_1 > h_2$ 4) $h_2 = 0$

36. **The sap in the trees rises up to the branches. This phenomenon is due to**
1) Viscosity 2) Surface tension
3) elasticity 4) all the above

37. **A liquid does not wet the surface of a solid if the angle of contact is**
1) zero 2) less than 90°
3) 45° 4) greater than 90°

38. **Meniscus of mercury in a capillary glass tube is**
1) plane 2) convex up
3) concave 4) cylindrical

39. **If the diameter of a capillary tube is doubled, the rise of water in the tube will be**
1) double 2) half
3) remain the same 4) four times

40. **When two capillary tubes of different diameters are dipped vertically, the rise of the liquid is**
1) same in both the tubes
2) more in tube of large diameter
3) less in tube of smaller diameter
4) more in the tube of smaller diameter

41. **Water rises to a height of 15.0 cm in a capillary tube of height 20 cm above the water level. If the tube is cut at a height of 10cm**
1) Water will come as a fountain from the tube
2) water will stay at a height of 10cm in the tube
3) water will flow down along the walls of the tube
4) none

42. **Capilary tubes with different radii are dipped in water vertically. The relation between the radius 'r' of the tube and rise of water 'h'; in the tube is**
1) hr = constant 2) hr^2 = constant
3) h/r = constant 4) h/r^2 = constant

43. **A capilary tube of radius R is dipped vertically in a liquid with surface tension T. If the rise of the liquid in the tube is 'h' then the weight of the liquid column in the tube is**
1) 2 p RT 2) pR^2T 3) 2p R^2T 4) p RT

44. **At boiling point of a liquid, surface tension is**
1) zero 2) infinite 3) very large 4) none

45. **The angle of contact between pure water and clean glass is**
1) 0° 2) 45° 3) 90° 4) 135°

46. **Liquid neither rises nor falls in a capillary tube when angle of contact is**
1) 0° 2) 90° 3) 120° 4) 135°

47. **The rise of water into a capillary tube placed vertically is h. When it is tilted through an angle, the vertical height of water raised will be :**
1) > h 2) < h 3) h 4) 0

48. **More liquid rises in a thin tube because of**
1) Large value of radius
2) Large value of surface tension
3) Small value of surface tension
4) Small value of radius

49. **Which of the following graphs represent the relation between capillary rise h and bore radius of capillary tube r ?**

1) 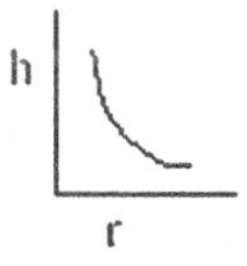

2) 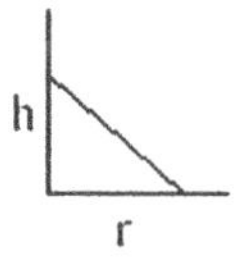

3) 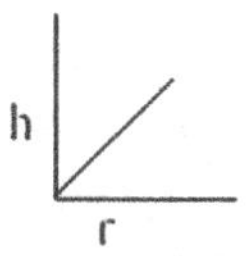

4) 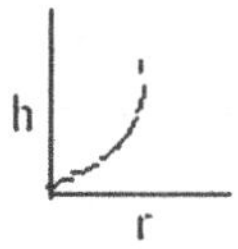

50. **If a wax coated capillary tube is dipped in water, then water in it will**
1) Rise up 2) falls down
3) Some times rise and sometimes fall
4) Rises up and overflows

51. A water proofing agent changes the angle of contact

1) From an obtuse to acute value
2) From an acute to obtuse value
3) From obtuse to p/2   4) From acute to p/2

52. When one end of a glass rod is heated its shape becomes spherical. Its reason is

1) Friction   2) Viscosity
3) Surface tension   4) Gravitation

53. Which, out of the following, is not the unit of surface tension

1) Nm^{-1} 2) $J\text{-}m^{-2}$ 3) $dyne\text{-}cm^{-1}$ 4) $Erg\text{-}cm^{2}$

54. Water rises in a capillary tube upto a height of 4cm. If the length of capillary tube is reduced to 2cm, then

1) Water will come out of the capillary tube
2) Water will over flow in the form of a fountain
3) Capillary tube will break
4) Water will rise upto full length of capillary

55. The angle of contact for silver and pure water is

1) 0°   2) 90°   3) 180°   4) 120°

56. Writing on black board with a piece of chalk is possible by the property of

1) Cohesive force   2) Adhesive force
3) Surface tension   4) Viscosity

57. The number of free liquid surfaces in the air bubble inside a liquid is

1) One   2) Two   3) Four   4) Eight

58. The correct graph between surface tension T and temperature q in low temperature range is

1) 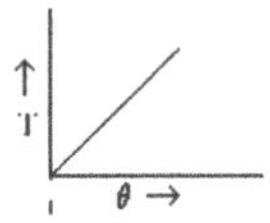

2) 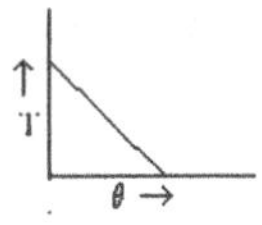

3) 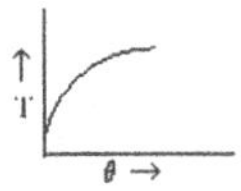

4) 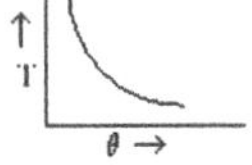

59. The lower end of a capillary tube touches a liquid whose angle of cantct is 90°. The liquid

1) rises into the tube   **[1996 M]**
2) falls in the tube
3) may rise or fall in the tube
4) neither rises nor falls

60. An astronaut tries to fill ink in an ink pen in an artificial satellite. The ink

1) Will be filled into the pen
2) Will not be filled
3) filling will depend upon the quality of ink
4) nothing can be said

61. When a capillary tube is dipped in a liquid, the capillary rise is h_1, when the inner surface is coated with wax, the capillary rise is h_2, then:

1) $h_1 = h_2$ 2) $h_1 < h_2$ 3) $h_1 > h_2$ 4) none

62. For which of the two pairs, the angle of contact is same ?

1) water and glass; glass and mercury
2) pure water and glass; glass and alcohol
3) silver and water; mercury and glass
4) silver and chromium; water and chromium

63. The radius of the bore of a capillary tube is r and the angle of contact of the liquid is θ. when the tube is dipped in the liquid, the radius of curvature of the meniscus of liquid rising in the tube is :

1) $r \sin\theta$ 2) $r/\sin\theta$ 3) $r\cos\theta$ 4) $r/\cos\theta$

64. If two soap bubbles of different radii are in communication with each other

1) air flows from larger bubble into the smaller one until the two bubbles are of equal size
2) the sizes of the bubbles remain unchanged.
3) the air flows from the smaller bubble into the larger one and large bubble grows at the expense of the smaller one.
4) air flows from the larger into the smaller bubble until the radius of the smaller one becomes equal to that of larger one and of the larger one equal to that of the smaller one.

65. Let P_1 and P_2 be the pressures above and below the water meniscus in a capillary tube, then

1) $P_1 = P_2$ 2) $P_1 > P_2$ 3) $P_1 < P_2$ 4) $P_1 = P_2 = 0$

66. The pressure is more on the side of

1) concave side   2) convex side
3) above horizontal surface   4) none

67. Excess pressure inside a soap bubble of radius 'r' and surface tension T is

1) T/2r   2) T/4r   3) 2T/r   4) 4T/r

68. The excess pressure inside a soap bubble is

1) Inversely proportional to the surface tension
2) Inversely proportional to its radius
3) Directly proportional to square of its radius
4) Directly proportional to its radius

69. The radii of two soap bubbles are R_1 and R_2. If they coaleasce, then the radius of curvature of the common surface will be

1) $\dfrac{R_1 + R_2}{2}$ 2) R_1R_2

3) $R_2\text{-}R_1$ 4) $\dfrac{R_1R_2}{R_2 - R_1}$

70. By inserting a capillary tube upto a depth l in water, the rises to a height h. If the lower end of the capillary tube is closed inside water and the capillary is taken out and closed and opened , to what height the water will remain in the tube when $l > h$ [Rajasthan PET 96]

1) zero 2) l+h 3) 2h 4) h

71. When a drop of liquid splits up into number of drops (under isothermal conditions)

a) Volume increases b) Total area increases
c) Energy is absorbed d) Energy is liberated

1) 'a' 'b' are correct 2) 'b' 'c' are correct
3) 'c', 'd' are correct 4) 'a' 'b' 'c' are correct

72. When impurities are added to a liquid it's surface tension

a) always increases b) always decreases
c) If impurities are sparingly soluble surface tension decreases
d) If impurities are highly soluble surface tension increases

1) a, c are correct 2) c, d are correct
3) b, d are correct 4) a, d are correct

73. For a liquid in a vessel if force of Adhesion is more than force of cohesion

a) shape of the meniscus is cancave upwards
b) shape of the meniscus is convex upwards
c) liquid level rises in the capillary tube
d) liquid level drops in the capillary tube

1) a, c are correct 2) c, d are correct
3) b, c are correct 4) b, d are correct

74.

Section - A	Section - B
a) Surface tension decreases	e) Minimum potential energy
b) Capillary rise	f) Domination of adhesive force
c) Spherical shape of rain drops	g) Increase in temperature
d) Tiny droplets of water act as ball bearings	h) Excess of pressure

1) a – e; b – f; c – g; d – h
2) a – h; b – g; c – f; d – e
3) a – f; b – e; c – g; d – h
4) a – g; b – f; c – e; d – h

75. Four mass capilary tubes, a, b, c and d having diameters, 2mm, 1mm, 0.8mm and 0.6mm are dipped in four beakers containing water with 'a' tube vertically, 'b' tube 30° 'c' tube 45° and d tube 60° inclination with the vertical arrange the lengths of water column in the tubes in descending order

1) d, c, b, a 2) d, a, b, c
3) a, c, d, b 4) a, b, c, d

76. Four soap bubbles p, q, r & s have volumes $8cm^3$, $27cm^3$, $64cm^3$ and $125cm^3$ respectively. Then excess of pressures in them in ascending order will be

1) p, s, q and r 2) p, q, r and s
3) s, r, q and p 4) p, s, r, p

77. Four capillary tubes T_1, T_2, T_3 and T_4 having bores of radii 0.1cm, 0.2cm, 0.3 cm and 0.4cm are dipped vertically in water. Descending order of capillary rises is

1) T_2, T_3, T_4, T_1 2) T_4, T_3, T_2, T_1
3) T_1, T_2, T_3, T_4 4) T_4, T_2, T_3, T_1

78. Four capillary tubes p, q, r & s of same length and radii in the ratio 1 : 2 : 6 : 4 respectively are dipped in a beaker containing water. The ascending order of mass of water in the capillary tubes are

1) p, q, r, s 2) p, q, s, r
3) q, p, r, s 4) p, r, s, q

Instruction For Questions

In each of the following questions, a statement of Assertion (A) is given followed by a corresponding statement of reason (R) just below it. Of the statement mark the correct answer.

1) Both 'A' and 'R' are true and 'R' is the correct explanation of 'A'
2) Both 'A' and 'R' are true and 'R' is not the correct explanation of 'A'
3) 'A' is true and 'R' is false
4) 'A' is false and 'R' is true

79. A) : The angle of contact of pure water with glass is acute

R) : The Adhesive force between molecules of water and glass is greater than cohesive force between water molecules

80. A) : A needle placed carefully on the surface of water may float, where as aball of same material sinks

R) : Archimedes principle is applicable for only spherical bodies

81. A) : Kerosene Oil spreads out over water surface

R) : Surface tension is a surface phenomenon

82. A) : Detergents should have small angles of contact.

R) : If angle of contact is small therefore detergents rinse the clothes dust in less time

83. A) : It is better to wash the clothes in cold soap solution

R) : The surface tension of cold solution is more than the surface tension of hot solution

84. A) : When height of a tube is less than liquid rise in the capillary tube, the liquid does not over flow.

R) : Product of radius of meniscus and height of liquid in the capillary tube always remain constant.

85. A) : The impurities always decrease the surface tension of a liquid.

R) : The change in surface tension of the liquid depends upon the degree of contamination of the impurity.

86. A) : The angle of contact of a liquid decrease with increase in temperature.

R) : With increase in temperature , the surface tension of liquid decreases.

87. A) : When n liquid drops combine to form large drop energy is released.

R) : When n liquid drops combine the total surface area decreases.

88. A) : At critical temperature , surface tension of a liquid becomes zero.

R) : At this temperature, intermolecular forces for liquids and gases become equal. Liquid can expand without any restriction.

89. A) : A large soap bubble expands while a small bubble shrinks, when they are connected to each other by a capillary tube.

R) : The excess pressure inside bubble (or drop) is inversely proportional to the radius.

90. A) : When pure water is taken in capillary tube made of glass its surface would be perfectly hemispherical in shape.

R) : Angle of contact for pure water and glass is 0.

91. (A) : At constant temperature, the elongation in a liquid surface does not obey Hooke's law.

(R) : During increase in surface area, the liquid molecules initially present just below the surface will come on to the surface and this is not real elongation

ANSWERS

1) 1	2) 3	3) 2	4) 2	5) 1
6) 4	7) 1	8) 4	9) 1	10) 3
11) 3	12) 1	13) 1	14) 3	15) 1
16) 2	17) 1	18) 1	19) 1	20) 2
21) 3	22) 1	23) 3	24) 3	25) 4
26) 1	27) 1	28) 2	29) 3	30) 2
31) 4	32) 2	33) 1	34) 3	35) 3
36) 2	37) 4	38) 2	39) 2	40) 4
41) 2	42) 1	43) 1	44) 1	45) 1
46) 2	47) 3	48) 4	49) 1	50) 2
51) 2	52) 3	53) 4	54) 4	55) 2
56) 2	57) 1	58) 2	59) 4	60) 2
61) 3	62) 2	63) 4	64) 3	65) 2
66) 1	67) 4	68) 2	69) 4	70) 3
71) 2	72) 2	73) 1	74) 4	75) 1
76) 3	77) 3	78) 2	79) 1	80) 3
81) 2	82) 1	83) 4	84) 1	85) 4
86) 4	87) 2	88) 1	89) 1	90) 1
91) 1				

EXERCISE-II(A)

FORCE DUE TO SURFACE TENSION

1. The surfacetension of water is 72dyne cm^{-1}. Its value in S.I system is

1) 72000 Nm^{-1} 2) 720Nm^{-1}
3) 0.72Nm^{-1} 4) 0.072Nm^{-1}

2. A needle of 12 cm long is on the surface of water. The extra force required, over its weight, to take it out from the surface of water is (surface tension = 72 dyne cm^{-1})

1) 72N 2) 0.864N 3) 0.0173N 4) 7.2N

3. The length of one edge of a glass plate of thickness 0.2 cm is 9.8 cm. If this edge of the glass plate touches the surface of a liquid of surface tension 60 dyne/cm, then it is pulled down with a force of (Assume that angle of contact to be zero)

1) 1100 dyne 2) 2400 dyne
3) 1000 dyne 4) 1200 dyne

4. A uniform wire of 20 cm long is bent into a circle. It is placed gently on the surface of water of surface tension 0.07 Nm^{-1}. The extra force than its weight required to pull it out of the water is

1) 0.014 N 2) 0.028 N
3) zero 4) 0.0035 N

5. **A straight piece of wire 5 cm long is placed horizontally on the surface of water and gently pulled up. It is found that a force of 7×10^{-3} N in addition to weight of wire is required for that purpose, the surface tension of water is**

1) 7×10^{-2} N/m 2) 7×10^{-1} N/m
3) 7×10^{-3} N/m 4) 3.5×10^{-2} N/m

6. **If surface tension of water is 0.07 N/m the weight of water supported in a capillary tube of radius 0.1 mm is**

1) 4.4×10^{-6} N 2) 44×10^{-6} N
3) 22×10^{-6} N 4) 2.2×10^{-6} N

SURFACE ENERGY & WORK DONE

7. **If surface tension of soap solution is equal to $2.5 \times 10^{-2}Nm^{-1}$, the work done to blow a soap bubble of diameter 0.6cm against surface tension forces will be**

1) 5.65×10^{-6}J 2) 11.2×10^{-6}J
3) 5.65×10^{-8}J 4) 11.2×10^{-8}J

8. **A liquid drop of diameter 20mm is split into 1000 small identical droplets. If surface tension of liquid is 72×10^{-3} Nm^{-1}, the amount of workdone in this process in μ J is nearly**

1) 815 2) 07.5 3) 204 4) 81.5

9. **A drop of radius 1mm is sprayed into 1000 droplets of same size. If surface tension of the liquid is 40 dyne (cm^{-1}), the amount of workdone in the process is**

1) 4.53erg 2) 4.53J 3) 4.53×10^{-4}J 4) 9.06×10^{-6}J

10. **A liquid drop of diameter 'D' is split into 27 droplets. If T is surface tension of the liquid the change in energy is**

1) 2pD 2) $2pD^2$ 3) $2pD^2T$ 4) Zero

11. **The workdone in forming a soap film of size 10cmx10cm will be, if the surface tension of soap solution is 3×10^{-2}N/m**

1) 3×10^{-4}J 2) 3×10^{-2}J 3) 6×10^{-4}J 4) 6×10^{-2}J

12. **If the work done in blowing a soap bubble of volume 'v' is w, then the work done in blowing a soap bubble of volume '2v' is**

1) 4 W 2) 8 W 3) $2^{1/3}$ W 4) $4^{1/3}$ W

13. **A soapbubble of radius $\frac{1}{\sqrt{\pi}}$ cm is expanded to double its radius. If the surface tension is 30dynes/cm, the work done is**

1) 700 ergs. 2) 720 ergs
3) 800 ergs 4) 360 ergs

14. **A mercury drop of radius 1 cm is sprayed into 10^6 drops of equal size. The energy expended in joules is (surface tension of mercury is $460 \times 10^{-3}Nm^{-1}$) [EAMCET 2001 E]**

1) 0.057 2) 5.7 3) 5.7×10^{-4} 4) 5.7×10^{-6}

CAPILLARITY

15. **A glass tube of diameter 0.06cm is dipped in methyl alcohol, if angle of contact is 0^0, density of methyl alcohol $0.8gcm^{-3}$ and surface tension $0.023Nm^{-1}$ the height to which it rises in the tube is**

1) 1cm 2) 2cm 3) 4cm 4) 2mm

16. **Water rises to a height of 6cm in a capillary tube if T= 7.2×10^{-2} Nm^{-1} and g = $10ms^{-2}$, the radius of the tube is**

1) 0.24mm 2) 2.4 mm 3) 0.12mm 4) 0.48mm

17. **Water rises to a height of 6.6cm in a a capillary tube of radius 0.2mm. The surface tension of water is (g = $10ms^{-2}$)**

1) 6.5×10^{-3}N/m 2) 6.5×10^{-2}N/m
3) 6.5×10^{-4}N/m 4) 6.5×10^{-1}N/m

18. **A capillary tube of radius 0.2mm is dipped vertically in a beaker containing water if surface tension of water is 7×10^{-2}N/m the rise of water in the tube will be**

1) 7mm 2) 7cm 3) 3.5mm 4) 3.5cm

19. **A capillary tube of diameter 0.4mm is dipped in a beaker containing mercury. If density of mercury is 13.6×10^3 kg/m^3, surface tension $0.49Nm^{-1}$, angle of contact 135^0 the depression of the meniscus in the capillary tube will be**

1) 0.13m 2) 0.013m 3) 0.026m 4) 0.0026m

20. **A liquid rises to a height of 50cm in a capillary tube of diameter 0.04mm. If denisity of liquid is 0.8×10^3 kg m^3 and angle of contact is 20^0, the surface tension of the liquid is (Cos 20^0 = 0.94)**

1) $4.17\frac{N}{m}$ 2) $0.417\frac{N}{m}$
3) $4.17 \times 10^{-2}\frac{N}{m}$ 4) $4.17 \times 10^{-3}\frac{N}{m}$

21. **A capillary tube of radius 0.2mm is dipped vertically in a beaker containing water if surface tension of water is 7×10^{-2}N/m the rise of water in the tube will be**

1) 7mm 2) 7cm 3) 3.5mm 4) 3.5cm

22. **Two capillary tubes of same material dipped are placed vertically in the same liquid. If their radii are in the ratio of 2:3, the ratio of rise of liquid in the tubes is**

1) 2:3 2) 3:2 3) 4:9 4) 9:4

23. **Two similar capillary tubes of sufficient length long are placed in water. One is placed vertically and the other is placed at 60° with the vertical. The length of the liquid rised into tubes is in the ratio**

1) 1 : 3 2) $\sqrt{3}:\sqrt{2}$ 3) 3 : 2 4) 1 : 2

24. **When a capillary tube is dipped in water vertically water rises to a height of 10 mm. The tube is now tilted and makes an angle 60° with vertical. Now length of water column in tube is**

1) 10 mm 2) 5 mm 3) 20 mm 4) 40 mm

25. **A tube of 0.8mm radius is dipped into a liquid with surface tension and density $6x10^{-2}$ N/m and 900kg/m^3 respectively. If the tube is kept vertical, the height of liquid rises in it will be (g = 10 ms^{-2})**

1) 0.017m 2) 0.17m 3) 1.7m 4) 17m

26. **When the capillary tube is lowered into water, the mass of water raised in the tube, above the outside water level is 5gm. If the radius of the tube is doubled, the mass of water that rises in the capillary tube above the outside water level is [1999 M]**

1) 1.25gm 2) 5gm 3) 10gm 4) 20gm

27. **Water rises to a height of 10cm in a capillary tube and mercury falls to a depth of 3.42cm in the same capillary tube. If the density of mercury is 13.6gm/cc and the angle of contact is 135°, ratio of surface tension for water and mercury is**

1) 1 : 0.5 2) 1 : 3

3) 1 : 6.5 4) 1.5 : 1

28. **When a cylindrical tube is dipped vertically into a liquid, the angle of contact is 140^0. When the tube is dipped with an inclination of 40^0 , the angle of contact is [1999 E]**

1) 100^0 2) 140^0 3) 180^0 4) 60^0

29. **A small hole is made at the bottom of a hollow sphere. The water enters into it when it is taken to a depth of 40 cm under water. If the surface tension of water is 0.07 Nm^{-1}, diameter of the hole is**

1) 10/7 mm 2) 10/14 mm

3) 2/15 m m 4) 1/14 mm

EXCESS PRESSURE

30. **The surface tension of soap water is 0.04Nm^{-1}. The excess of pressure inside a soap water bubble of diameter 10 mm is**

1) 64Pa 2) 8Pa 3) 32Pa 4) 16Pa

31. **The pressure inside two soap bubbles is 1.01 and 1.02 atmosphers respectively. The ratio of their volumes is (atmospheric pressure=10^5 Pa)**

1) 1 : 2 2) 2 : 3 3) 8 : 1 4) 3 : 5

32. **Two bubbles of radii 2mm and 3mm when touch each other, the radius of curvature of their interface is**

1) 1.7mm 2) 0.017cm

3) 6mm 4) 0.17cm

33. **A spherical soap bubble of radius 1 cm is formed inside another of radius 3cm. The radius of single soap bubble which maintains the same pressure difference as inside the smaller and outside the larger soap bubble is ....cm [1993 E]**

1) 4/3 2) 3/4 3) 1/2 4) 2

34. **The pressure in air bubble just below the water surface is (Surface tension of liquid is 72 dyne cm^{-1} and atmospheric pressure = 1.0 x 10^{5}Nm^{-2}, Radius of bubble = 0.1mm)**

1) 1.0144x10^{5}Nm^{-2} 2) 2.0274x10^{5}Nm^{-2}

3) Zero 4) 3.0274x10^{5}Nm^{-2}

ANSWERS

1) 4	2) 3	3) 4	4) 2	5) 1
6) 2	7) 1	8) 1	9) 3	10) 3
11) 3	12) 4	13) 2	14) 1	15) 2
16) 1	17) 2	18) 2	19) 3	20) 3
21) 2	22) 2	23) 4	24) 3	25) 1
26) 3	27) 3	28) 2	29) 4	30) 3
31) 3	32) 3	33) 2	34) 1	

EXERCISE-II(B)

FORCE DUE TO SURFACE TENSION

1. **The surface tension of a liquid is 0.04Nm^{-1}. Its value in C.G.S. system is**

1) 4 dyne cm^{-1} 2) 40 dyne cm^{-1}

3) 400 dyne cm^{-1} 4) 0.004 dyne cm^{-1}

2. **A wire ring of diameter 14 cm is gently lowered on a liquid surface so that its plane is parallel to the surface and then pulled up. The extra force applied is 0.0616 N. The surface tension of the liquid is**

1) 7 x 10^{-2} Nm^{-1} 2) 7 x 10^{-3} Nm^{-1}

3) 70 Nm^{-1} 4) 1.7 Nm^{-1}

3. **The additional force required to lift a flat circular disc of radius 5cm from the surface of water with surface tension 75dyne cm^{-1}, will be**

1) 750dyne 2) 750π dyne
3) 30 dyne 4) 60 dyne

4. **A needle of length 5cm is floating on water of suface tension 0.07 Jm^{-2} and is not wetted by water. Then the weight of the needle will be**

1) 7×10^{-3} kgwt 2) 0.7×10^{-3}kgwt
3) 5.2×10^{-2}N 4) 4×10^{-2}N

5. **The maximum force, in addition to the weight required to pull a wire of 5.0cm long from surface of water at temperature 20^0C, is 728 dynes. The surface tension of water is**

1) 7.28 N/cm 2) 7.28 dyne/cm
3) 72.8 dyne/cm 4) 728 dyne/cm

6. **The force of surface tension on a paper disc floating on the surface of water is F. If another paper disc of same outer radius with a hole at its centre is floating force of ST on it if radius of the hole is half of outer radius**

1) F/2 2) 2F/3 3) 3F/2 4) 3F

SURFACE ENERGY & WORK DONE

7. **The work done in blowing a soap bubble of radius 0.2m is(surface tension of water is 0.06N/m)**

1) $192\pi\times10^{-4}J$ 2) $280\pi\times10^{-4}J$
3) $200\pi\times10^{-4}J$ 4) $96\pi\times10^{-4}J$

8. **If n small droplets of water, each of radius r, coalesce to form a single drop of radius R. The amount of heat generated is**

1) $4pT(nr^2-R^2)$ 2) $4pT(nR^2-r^2)$
3) $\frac{4\pi}{T}(nr^2-R^2)$ 4) $\frac{4\pi T}{n}(r^2-R^2)$

9. **The workdone in blowing a soap bubble of radius a under isothermal conditions is (the surface tension of soap film is T)**

1) $4pa^2T$ 2) $8pa^2T$
4) $16pa^2T$ 4) $2pa^2T$

10. **The radius of a soap bubble is increased from $\frac{1}{\sqrt{\pi}}cm$ to $\frac{2}{\sqrt{\pi}}cm$ .If the surface tension of water is 30 dyne/cm, then the work done will be**

1) 180 ergs 2) 360 ergs
3) 720 ergs 4) 960 ergs.

11. **The surface energy of liquid film on a ring of area $0.15m^2$ is (surface tension of liquid is equal to 5 Nm^{-1}) [EAMCET 2000 M]**

1) 0.75J 2) 1.5J 3) 2.25J 4) 3.0J

12. **The work done to blow a bubble is W. The extra work to be done to double its radius is**

1) W 2) 2 W 3) 3W 4) 4 W

13. **The workdone in blowing a spherical soap bubble of diameter 2cm will be (the surface tension of soap solution is $2\times10^{-2}Nm^{-1}$)**

1) 50.2×10^{-6}J 2)50.2J
3) 50.2×10^{-6}erg 4) Zero

14. **The work done in splitting a drop of water of 1mm radius into 10^6 droplets is (S.T of water 72×10^{-3} J/m^2)**

1) 9.98×10^{-5} J 2) 8.95×10^{-5} J
3) 5.89×10^{-5} J 4) 5.98×10^{-5} J

CAPILLARITY

15. **Two capillary tubes of glass of radii 3mm and 2mm are placed vertically in water, ratio of height of water in the tubes is**

1) 3:2 2) 4:9 3) 9:4 4) 2:3

16. **If a capillary tube is tilted to 45° and 60° from the vertical then the ratio of length l_1, l_2 of liquid coloumns in it wil be**

1) $1:\sqrt{2}$ 2) $\sqrt{2}:1$ 3) 2:1 4) 1:2

17. **Two capillaries P and Q are dipped in water. The height of water level in capillary P is 2/3 to the height in Q capillary. The ratio of diameters is**

1) 2:3 2) 3:2 3) 3:4 4) 4:3

18. **Two capillary tubes of same diameter are put vertically one each in two limbs whose relative densities are 0.8 and 0.6 and surface tension are 60 and 50 dyne/cm respectively. Ratio of heights of liquids in the two tubes is**

1) 10/9 2) 3/10 3) 10/3 4) 9/10

19. **In a capillary tube experiment, a vertical 30 cm long capillary tube is dipped in water. The water rises upto a height of 10cm due to capillary action. If this experiment is conducted in freely falling elevator the length of water column becomes**

1) 10cm 2) 20cm 3) 30cm 4) zero.

20. **Water rises to a height h in a capillary tube of certain diameter. This capillary tube is replaced by a similar tube of half the diameter. Now the water will rise to a height of**

1) 4h 2) 3h 3) 2h 4) h

21. The angle of contact between glass and water is 0^0 and it rises in a capillary upto 6cm when its surface tension is 70 dynes/cm. Another liquid of surface tension 140 dyne/cm angle of contact 60^0 and relative density 2 will rise in the same capillary by

1) 12cm 2) 24cm 3) 3cm 4) 6cm

22. A U tube is arranged such that diameter of one limb is 2D and diameter of other is D. Surface tension of water contained in the tube is T. If d is density if water, the difference in the levels of liquid is (angle of contact of liquid is zero)

1) $\frac{2T}{Ddg}$ 2) $\frac{T}{Ddg}$ 3) $\frac{T}{2Ddg}$ 4) $\frac{4T}{Ddg}$

23. Water rises upto a height of 12 cm into a capillary tube when placed vertically. If it is now tilted through 45° from the vertical, length of the water in the capillary tube is

1) 12 cm 2) $6\sqrt{2}$ cm
3) $12\sqrt{2}$ cm 4) 6 cm

24. Water rises in a capillary tube of diameter $0.2\text{x}10^{-2}$m, upto a height of 1.5cm. The surface tension of water is

1) $73.5\text{x}10^{-3}$ Nm^{-1} 2) $73.5\text{x}10^{-3}$$dynecm^{-1}$
3) $35.7\text{x}10^{-3}$ Nm^{-1} 4) $43.5\text{x}10^{-3}$$Nm^{-1}$

25. A capillary tube is immersed in water and the mass of water that rises in the capillary tube is M. If another tube of double the radius is immersed then the mass of water that will rise in the capillary tube is

1) 4 M 2) 2 M 3) M/2 4) M/4

26. Two capillaries of same diameter are dipped into liquids of specific gravity 0.4 and 0.8 respectively. If their surface tensions are in the ratio 6:5 then the ratio of heights of liquids in them will be

1) 12 : 5 2) 5 : 12 3) 4 : 8 4) 8 : 4

27. The meniscus of a liquid in a capillary tube is concave. If F_C is the force of cohesion and F_A is the force of adhesion between molecules of the liquid, then

1) $F_C = F_A$ 2) $F_C > F_A$
3) $F_C < F_A$ 4) All of the above

28. A capillary tube has a uniform bore of radius 1mm. It is placed in water of surface tension 0.07 Nm^{-1}.Mass of water rised into the capillary tube is

1) 44mg 2) 44gm 3) 44kg 4) 4.4kg

29. A narrow capillary tube when dipped in a beaker containing waer, the rise is 20 cm. If the area of cross - section of the bore is reduced to 1/4 th value, water will rise to a height of

1) 10 cm 2) 20 cm 3) 40 cm 4) 80 cm

EXCESS PRESSURE

30. The ratio of radii of two soap bubbles is 1:4, then the ratio of excess of pressures in them will be

1) 4:1 2) 1:4 3) 1:16 4) 16:1

31. The excess of pressure inside the first soap bubble is three times that inside the second bubble. Then ratio of volume of the first to second bubble is

1) 1 : 3 2) 1 : 9 3) 1 : 27 4) 3 : 1

32. Two soap bubbles of radii 4cm and 5cm coalesce to form a common surface. The radius of curvature of this common surface will be

1) 20cm 2) 1cm 3) 1.25cm 4) $\sqrt{41}$ cm

33. A soap bubble of radius 3 cm and another bubble of 4 cm coalesce under isothermal conditions in vacuum. The radius of the new bubble is

1) 3 cm 2) 4 cm 3) 5 cm 4) 7 cm

34. The excess pressure in an air bubble just below the water survface is - (surface tension of lequid is 72 dyne/cm and atmospheric pressure = 1.o13 x 10^5 N/m^2, Radius of bubble = 0.1mm)

1) 1.44 x $10^3 N/m^2$ 2) 1.44 x $10^4 N/m^2$
3) 1.44 x 10^2 N/m^2 4) 1.027 x 10^5 N/m^2

ANSWERS

1) 2	2) 1	3) 2	4) 2	5) 3
6) 3	7) 1	8) 1	9) 2	10) 3
11) 2	12) 3	13) 1	14) 2	15) 4
16) 1	17) 2	18) 4	19) 3	20) 3
21) 3	22) 2	23) 3	24) 1	25) 2
26) 1	27) 3	28) 1	29) 3	30) 1
31) 1	32) 1	33) 2	34) 1	

EXERCISE-III(A)

FORCE DUE TO SURFACE TENSION

1. A wire of length 'l' meters, made of a material of specific gravity 8 is floating horizontally on the surface of water. If it is not wet by water, the maximum diameter of the wire (in millimeters) upto which it can continue to float is (surface tension of water is T = $70\text{x}10^{-3}$ Nm^{-1}) [2006M]

1) 1.5 2) 1.1 3) 0.75 4) 0.55

2. A ring is cut from a platinum tube having 10 cm internal and 11 cm external diameter. It is supported horizontally from a pan of balance, so that it comes in contact with water in a glass vessel. What is the surface tension of water if an extra, 4.752gram weight is required to pull it away from water? ($g = 10\ m/s^2$) :

1) 7.2×10^{-2} N/m 2) 72 N/m
3) 14.4×10^{-2} N/m 4) 144 N/m

3. A soap film is formed in a circular frame loop of thread is lying on the film. If the film inside the loop is broken then the tension in the thread will be –

1) $2\pi r T$ 2) $\pi r T$ 3) rT 4) $\pi r^2 T$

4. A thin metal disc of radius 'r' floats on water surface bends the surface downwards along the perimeter making an angle θ with vertical edge of the disc. If the disc displaces a weight of water W and surface tension of water is T, then the weight of metal disc is

1) $2\pi rT + W$ 2) $2\pi rT\cos\theta - W$
3) $2\pi rT\cos\theta + W$ 4) $W - 2\pi rT\cos\theta$

SURFACE ENERGY & WORK DONE

5. A film of water is formed between two straight parallel wires each 10 cm long and at seperation 5 mm. To increase the distance between them to 6 mm, the work done is 144×10^{-7}J. The surface tension of water is

1) $7 \times 10^{-3}\ Nm^{-1}$ 2) $7.2 \times 10^{-3}\ Nm^{-1}$
3) $0.72 Nm^{-1}$ 4) $0.072\ Nm^{-1}$

6. The area of the surface of water in a vessel is $40 cm^2$. It is transferred into a cylindrical vessel of area of cross section $220 cm^2$. If the surface tension of water is $0.07\ Nm^{-1}$, the workdone in the transfer of water is neglect gravity.

1) 12.6 erg 2) 1.26 erg
3) 12.6×10^{-4}J 4) 1.26×10^{-4}J

7. 8000 identical water drops are combined to form a big drop. Then the ratio of final surface energy to the initial surface energy of all the drops together is [2000 E]

1) 1:10 2) 1:15 3) 1:20 4) 1:25

8. A drop of water fo volume $0.05 cm^3$ is pressed between two glass plates, as a consequence of which it spreads and occupies an area of $40 cm^2$. If the surface tension of water is 70 dyne/cm, then the normal force required to separate out the two glass plates will be in Newton

1) 90 2) 45 3) 22.5 4) 450

9. Water is filled into a container with hexagonal cross section of side 6 cm. If the surface tension of water is $0.075 Nm^{-1}$ then the surface energy of water will be

1) 8.5×10^{-4}J 2) 2.8×10^{-4}J
3) 6.4×10^{-4}J 4) 7×10^{-4}J

10. If a number of small droplets of water each of radius r coalesce to form a single drop of radius R, then the rise of temperature is : (J = mechanical equivalent of heat, d= density of water, T= surface tension of water)

1) $\frac{3Td}{J}\left(\frac{1}{R} - \frac{1}{r}\right)$ 2) $\frac{3J}{TD}\left(\frac{1}{r} - \frac{1}{R}\right)$

3) $\frac{3T}{dJ}\left(\frac{1}{r} - \frac{1}{R}\right)$ 4) none

CAPILLARITY

11. Two parallel glass plates are dipped perpendiculary in a liquid of density p. The separation between the plates is 'd' and the surface tension T. The angle of contact of glass is θ. The capillary rise of the liquid between the plates is :

1) $\left(\frac{T\cos\theta}{pd}\right)$ 2) $\left(\frac{2T\cos\theta}{pgd}\right)$

3) $\left(\frac{2T}{pgd\cos\theta}\right)$ 4) $\left(\frac{T\cos\theta}{pgd}\right)$

12. The lower end of a capillary tube of radius r is placed vertically in water. Then with the rise of water in the capillary, heat evolved is:

1) $+\frac{\pi^2 r^2 h^2}{j}dg$ 2) $+\frac{\pi^2 r^2 h^2}{2j}dg$

3) $-\frac{\pi r^2 h^2}{2j}dg$ 4) $-\frac{\pi r^2 h^2}{j}dg$

13. A U-tube has two vertical limbs of diameters 4mm and 5mm. When water is poured into it, the water level difference between the limbs is ($T = 70$ dyne cm^{-1}) ($g = 10 ms^{-2}$)

1) 0.14cm 2) 0.014m
3) 14mm 4) 0.14mm

14. A capillary tube of radius 0.25mm is submerged vertically in water so that 25mm of its length is outside water. The radius of curvature of the meniscus will be ($T = 75 \times 10^{-3}$ N/m)

1) 0.2 mm 2) 0.4 mm
3) 0.6 mm 4) 0.8 mm

15. **Liquid rises to a height of 2cm in a capillary tube. The angle of contact between the solid and liquid is zero. The tube is depressed more now so that the tip of the capillary tube is only 1 cm above the liquid then the apparent angle of contact betwen solid and liquid is** **[J &K CET 2000]**

1) 0^0 2) 30^0 3) 60^0 4) None

16. **Drops of liquid of density d are floating half immersed in a liquid of density ρ. If the surface tension of liquid is T then the radius of the drop will be (d = density of liquid drop)**

1) $\sqrt{\frac{3T}{g(2d-\rho)}}$ 2) $\sqrt{\frac{6T}{g(2d-\rho)}}$

3) $\sqrt{\frac{2T}{g(2d-\rho)}}$ 4) $\sqrt{\frac{3T}{g(4d-3\rho)}}$

17. **A capillary tube is immersed vertically in water such that the height of liquid column in it is 'x'. This arrangement is taken into a mine of depth 'd' and the height of the liquid column is found to be 'y'. If 'R' is the radius of the earth, then the depth of mine is**

1) $d=R\left(\frac{y-x}{x}\right)$ 2) $d=R\left(\frac{y-x}{y}\right)$

3) $d=R\left(\frac{x}{y-x}\right)$ 4) $d=R\left(\frac{y}{y-x}\right)$

EXCESS PRESSURE

18. **Two soap bubbles combine under isothermal conditions to form a single soap bubble. If in this process, the change in volume is V and change in area is S, then**

1) PV+TS = 0 2) 4PV+3TS=0

3) 3PV+4TS=0 4) 3PV+TS=0

19. **One end of a uniform glass capillary tube of radius r=0.025 cm is immersed vertically in water to a depth h = 1cm. The excess pressure in N/m^2 required to blow an air bubble out of the tube :** **[2004 E]**

Surface tension of water = 7×10^{-2}N/m

Density of water = $10^3 kg/m^3$

Acceleration due to gravity = $10ms^{-2}$

1) 0.0048×10^5 2) 0.0066×10^5

3) 1.0018×10^5 4) 1.0033×10^5

20. **A barometer is made of two identical capillary tubes whose radii are 1.44×10^{-3}m and 7.2×10^{-4} respectively. If the height of a liquid in narrow tube is 0.2m more than height of it in broad tube, then what will be the exact pressure difference ? (Density of liquid $=10^3 kgm^{-3}$; surface tension = 72×10^{-3} and g = $9.8ms^{-2}$)**

1) $1750Nm^{-2}$ 2) $1860Nm^{-1}$

3) $1570Nm^{-2}$ 4) $1680Nm^{-2}$

21. **The lower end a capillary tube of diameter 2.00 mm is dipped in water 8 cm below the surface in a beaker it so that an air bubble is formed at the end of the tube inside water. The excess pressur inside the bubble will be (in Pascal) (T = 73 dyne/ cm)**

1) 146 2) 1.078×10^5

3) 1.0193×10^5 4) 1.02×10^4

22. **The lower end a capillary tube of diameter 2.00 mm is dipped in water 8 cm below the surface in a beaker it so that an air bubble is formed at the end of the tube inside water. The pressure required to blow the bubble will be (in Pascal) (T = 73 dyne/ cm)**

1) 146 2) 1.078×10^5

3) 1.0193×10^5 4) 1.02×10^4

23. **Two soap bubbles A and B are given. rdius of of B is 50% more than radius of A . Soap content (volume) of B as compared to A is**

1) 1.5 2) 2.25 3) 3.375 4) 4

ANSWERS

1) 1	2) 1	3) 3	4) 3	5) 4
6) 3	7) 3	8) 2	9) 4	10) 3
11) 2	12) 2	13) 1	14) 3	15) 3
16) 1	17) 2	18) 3	19) 2	20) 2
21) 1	22) 3	23) 3		

EXERCISE-III(B)

FORCE DUE TO SURFACE TENSION

1. **A thin wire ring of 3cm radius rests flat on surface of a liquid. The pull required to raise the ring before the film breaks is 30.14×10^{-3}N more than it is after. The surface tension of liquid (in Nm^{-1}) is** **[2005E]**

1) 80×10^{-3} 2) 87×10^{-3}

3) 90×10^{-3} 4) 98×10^{-3}

2. The ring of radius 1m is lying on the surface of liquid. It is lifted from the liquid surface by a force of 4 Newton in such a way that the liquid film in it remains intact. The surface tension of liquid will be –

1) $\frac{1}{2\pi}N/m$ 2) $\frac{1}{\pi}N/m$

3) $\frac{1}{3\pi}N/m$ 4) $\frac{1}{4\pi}N/m$

3. A liquid is filled into a semielliptical cross - section with a as semi major axis and b as semi minor axis. The ratio of the surface tension forces on the curved part and the plane part of the tube in vertical postion will be

1) $\frac{\pi(a+b)}{4b}$ 2) $\frac{2\pi a}{b}$ 3) $\frac{\pi a}{4b}$ 4) $\frac{\pi(a-b)}{4b}$

4. A wooden stick 2m long is floating on the surface of water. The surface tension of water 0.07 N/m. By putting soap solution on one side of the surface tension is reduced to 0.06N/m. The net force on the stick will be

1) 0.07 N 2) 0.06 N 3) 0.01 N 4) 0.02 N

SURFACE ENERGY & WORK DONE

5. A film of water is formed between two straight parallel wires of length 10cm each separated by 0.5cm. If their separation is increased by 1 mm, while still maintaining their parallelism, how much work will have to be done? (surface tension of water = 7.2×10^{-2} Nm^{-1}) [MP PET 2001]

1) $7.22\times10^{-6}J$ 2) $1.44\times10^{-5}J$

3) $2.88\times10^{-5}J$ 4) $5.76\times10^{-5}J$

6. If work done in increasing the size of soap film from 10 cm x 6 cm to 10 cm x 11cm is 2×10^{-4} J, then the surface tension is (AIIMS 2000)

1) 2 X 10^{-2} Nm^{-1} 2) 2 X 10^{-4} Nm^{-1}

3) 2 X 10^{-6} Nm^{-1} 4) 2 X 10^{-8} Nm^{-1}

7. What is the ratio of surface energy of 1 small drop and 1 large drop, if 1000 small drops combined to form 1 large drop (CPMT 90)

1) 100 : 1 2) 1000 : 1 3) 10 : 1 4) 1 : 100

8. A drop of liquid pressed between two glass plates spreads to a circle of diameter 10cm. Thickness of liquid film is 0.5mm and coefficient of surface tension is $70\times10^{-3}Nm^{-1}$. The force required to pull them a part is [1994 E]

1)4.4N 20 1.1N 3) 2.2N 4) 2.6N

9. The surface tension of a liquid is 0_2.5 N/m. If a liquid film is formed on a ring of area 0.02 m^2 then its surface energy will be –

1)0.01 Joule 2) 0.02 Joule

3) 0.03 Joule 4) 0.04 Joule

10. Several spherical drops of a liquid of denisty 'd' and radius 'r' coalesec to form a single drop of radius 'R'. If all the energy released is converted into K.E. Then the velocity acquired by the drop is

1) $\sqrt{\frac{6T}{d}\left(\frac{1}{r}-\frac{1}{R}\right)}$ 2) $\sqrt{\frac{3T}{d}\left(\frac{1}{r}-\frac{1}{R}\right)}$

3) $\sqrt{\frac{T}{6d}\left(\frac{1}{r}-\frac{1}{R}\right)}$ 4) $\sqrt{\frac{T}{3d}\left(\frac{1}{r}-\frac{1}{R}\right)}$

CAPILLARITY

11. A capillary tube of radius r is dipped vertically in a liquid of density d, surface tension T and angle of contact θ then the pressure difference just below the two surface, one in the beaker and the other in the capillary tube, is :

1) $\frac{2T}{r}$ 2) $\frac{T}{r\cos\theta}$

3) $\frac{2T\cos\theta}{r}$ 4) $\frac{T\cos\theta}{r}$

12. A liquid of density 'd' and surface Tension 'T' ascends into a capillary tube. Then the potential energy of the liquid is

1) $\frac{2\pi T}{dg}$ 2) $\frac{\pi T^2}{dg}$

3) $\frac{2\pi T^2}{dg}$ 4) $\frac{\pi T^2}{2dg}$

13. When water rises in a capillary tube of radius r to height h, then its potential energy U, If capillary tube of radius 2r is dipped in same water then potential energy of water is U_2. The U_1:U_2 will be

1) 1:1 2) 1:2 3) 2:1 4) 1:4

14. A glass rod of radius 1. 0mm is inserted symmetrically into a vertical capillary tube of radius 2.0 mm such that their lower ends are at the same level. The arrangement is now dipped in water. The height to which the water rises in the capilary tube is (T = 75 x 10^{-3} N/m)

1) 5 mm 2) 10 mm 3) 15mm 4) 20 mm

15. **Water rises to a height of 16.3cm in a capillary of height 18 cm above the water level. If the tube is cut at a height of 12 cm :**
1) water will come as a fountain from the capillary tube
2) water will stay at a height of 12cm in the capillary tube
3) the height of water in the capillary will be 10.3cm
4) water will flow down the sides of capillary tube

16. **Water rises in a capillary tube to a certain height such that the upward force due to surface tension is balanced by 75 x 10^{-4} N force due to the weight of the liquid. If the surface tension of water is 6 x 10^{-2} Nm^{-1}, the inner circumference of the capillary must be**
1) 1.25 x 10^{-2} m 2) 0.50 x 10^{-2} m
3) 6.5 x 10^{-2}m 4) 12.5 x 10^{-2} m

17. **A capillary tube is immersed vertically into water in a vessel kept in a stationary lift, the rise of water in it is h. If the lift moves upwards with an acceleration 'a' then the rise of water in the capillary tube is**
1) $\frac{hg}{a-g}$ 2) $\frac{hg}{a}$ 3) $\frac{hg}{a+g}$ 4) h

EXCESS PRESSURE

18. **Under isothermal conditions two soap bubbles of radii a and b coalesce to form a single bubble of radius c. If external pressure is P_0, the surface tension is equal to :**
1) $\frac{P_o(a^3+b^3-c^3)}{4(a^2+b^2+c^2)}$ 2) $\frac{4(a^3+b^3-c^3)}{(a^2+b^2-c^2)}$
3) $\frac{P_0(c^3-a^3-b^3)}{4(a^2+b^2-c^2)}$ 4) $\frac{P(a^3+b^3-c^3)}{(a^2-b^2-c^2)}$

19. **A capillary tube is dipped in water with is lower end 10 cm below the surface. Water rises in the tube to a height of 5 cm. The pressure required to blow a bubble at the lower end of the tube will be**
1) 1 x 10^5 N/m^2 2) 1.015 x 10^5 N/m^2
3) 2 x 10^5 N/m^2 4) 2.5 x 10^5 N/m^2

20. **A glass capillary tube inner diameter 0.28mm is lowered vertically into water in a vessel. The pressure to be applied on thewater in the capillary tube so that water level in the tube is same as that in the vessel in Nm^{-2} is (surface tension of water = 0.07Nm^{-1}; Atmospheric pressure = $10^5 Nm^{-2}$) [2002 E]**
1) 10^3 2) 99×10^3 3) 100×10^3 4) 101×10^3

21. **The lower end a capillary tube of diameter 2.00 mm is dipped in water 8 cm below the surface in a beaker it so that an air bubble is formed at the end of the tube inside water. The external pressure on the bubble will be (in Pascal) (T = 73 dyne/ cm, atm = 10^5 Pa)**
1) 146 2) 1.078 x 10^5
3) 1.0193 x 10^5 4) 1.02 x 10^4

22. **An air bubble of radius 'r' is formed at a depth 'h' below the surface of water of density 'd' and if 'p_0' is the atm pressure, T is surface tension of water, then the pressure inside the bubble is**
1) $p_0+\frac{2T}{r}$ 2) $p_0+hdg+\frac{2T}{r}$
3) $p_0+hdg+\frac{4T}{r}$ 4) $p_0+\frac{4T}{r}$

23. **An ice block of $\frac{4\pi}{3}$ cm^3 is placed in an artificial satellite. When ice melts into water then its surface area will be, if the densities of ice and water are presumed to be same –**
1) $\frac{4\pi}{3}$ cm^2 2) 4π cm^2
3) $\left(\frac{4\pi}{3}\right)^{1/3}$ cm^2 4) $(36\pi)^{1/3}$ cm^2

ANSWERS

1) 1	2) 2	3) 1	4) 4	5) 2
6) 1	7) 4	8) 3	9) 2	10) 1
11) 3	12) 3	13) 1	14) 3	15) 2
16) 4	17) 3	18) 3	19) 2	20) 4
21) 2	22) 2	23) 2		

COMPREHENSION QUESTIONS

Comprehension - I :

Molecular forces exist between the molecules of a liquid in a container. The molecules on the surface have unequal force leading to a tension on the surface. If this is not compensated by a force, the equilibrium of the liquid will be a difficult task. This leads to an excess pressure on the surface. The nature of the meniscus can inform us of the direction of the excess pressure. The angle of contact of the liquid decided by the forces between the molecules, air and container can make the anlge of contact.

1. **The direction of the excess pressure in the meniscus of a liquid of angle of contact $\frac{2\pi}{3}$ is**

   1) Upward 2) Downward
   3) Horizontal 4) Cannot be determined

2. **If the excess pressure in a soap bubble is p, the excess pressure in an air bubble in soap solution is**

   1) p/2 2) p 3) 2p 4) 4p

3. **In a meniscus of radius r, with excess pressure p and atmospheric pressure p_0, the force experienced is**

   1) $(p-p_0)\pi r^2$ 2) $(p-p_0)2\pi r$
   3) $p\pi r^2$ 4) $p_0 2\pi r$

Comprehension - II :

Two soap bubbles, one of radius 50 mm and the other of radius 80 mm are brought together so that they have a common interface.

4. **Determine the radius of curvature of this interface.**

   1) 114.6 mm 2) 125.6 mm
   3) 133.3 mm 4) 154.6 mm

5. **The radius of curvature of this interface attains which shape from smaller bubble toward larger bubble.**

   1) convex 2) concave
   3) no lconclusion 4) concavo convex

Comprehension - III :

A long capillary tube of radius 0.2 mm is placed vertically inside a beaker of water

6. **If the surface tension of water $7.2 \times 10^{-2} Nm^{-1}$ and the angle of contact between glass and water is zero, then the height of water column in the tube. (nearly)**

   1) 3 cm 2) 9 cm 3) 7 cm 4) 5 cm

7. **If the tube is now pushed into water so that only 5.0 cm of its length is above the surface, then determine the angle of contact between the liquid and glass surface**

   1) $\cos^{-1}\left(\frac{4}{5}\right)$ 2) $\cos^{-1}\left(\frac{5}{7}\right)$
   3) $\cos^{-1}\left(\frac{3}{5}\right)$ 4) $\cos^{-1}\left(\frac{5}{4}\right)$

Comprehension - IV :

A capillary tube of radius r lowered into a liquid of surface tension T and density ρ given the angle of contct = 0^0.

8. **The work done by surface tension**

   1) $\frac{\pi T^2}{\rho g}$ 2) $\frac{4\pi T^2}{\rho g}$ 3) $\frac{T^2}{\rho g}$ 4) $\frac{2T^2}{\rho g}$

9. **The potential energy acquired by the liquid in the capillary**

   1) $\frac{\pi T^2}{2\rho g}$ 2) $\frac{T^2}{2\rho g}$ 3) $\frac{T^2}{\rho g}$ 4) $\frac{2\pi T^2}{\rho g}$

Comprehension - V :

The figure shows a capillary tube of radius r dipped into water. The atmospheric pressure is p_0 and the capillary rise of water is h. S is the surface tension for water-glass

10. **The pressure inside water at the point A (Lowest point of the miniscus) is**

   1) p_0 2) $p_o + \frac{2s}{r}$
   3) $p_o - \frac{2s}{r}$ 4) $p_o - \frac{4s}{r}$

11. **Initially h =10cm. If the capillary tube be now inclined at 45^0, the length of water rised in the tube will be**

   1) 10 cm 2) $10\sqrt{2}$ cm 3) $\frac{10}{\sqrt{2}}$ cm 4) 5cm

12. **Which of the following graphs may represent the relation between capillary rise h and the radius r of the capillary**

1) 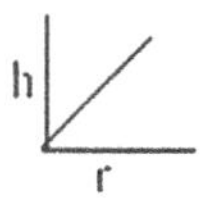

2) 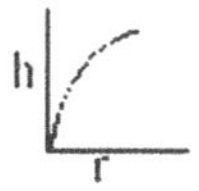

3) 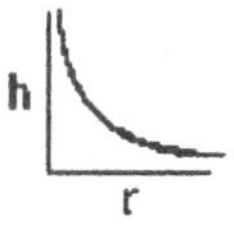

4) 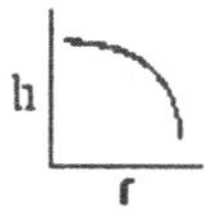

ANSWERS

1) 1	2) 1	3) 3	4) 3	5) 1
6) 3	7) 2	8) 2	9) 4	10) 3
11) 2	12) 3			

Chapter - 4

THERMAL PROPERTIES OF MATTER

THERMOMETRY

SYNOPSIS

1. Heat :

1. The energy transferred between two or more systems by virtue of temperature difference is called heat (or) The energy transferred by a non mechanical method is heat

2. This energy may be transferred between the system and the surroundings.

3. The net flow of energy (Heat flow) is from a body at higher temperature to a body at lower temperature.

4. The unit of heat is Joule.

5. Calorie is the unit of heat in CGS system.

6. 4.186 Joule=1 Calorie, for all calculations it can be taken as 4.2 joule = 1 calorie.

2. Temperature:

1. Temperature is the quantity which determines the direction of flow of heat energy.

2. The concept of temperature is given by Zeroth law of thermodynamics, considering thermal equilibrium.

3. When two systems are in thermal equilibrium, the temperature of the two systems is same.

4. The degree of hotness of a body is its temperature and indicates the relative hotness or coldness.

5. Temperature is a relative measure.

6. Hence a reference point is required for the representation of temperature.

7. The melting point of ice at normal pressure and the boiling point of water at normal pressure are taken as the reference points called, ice point and steam point.

8. Ice point is called lower fixed point and steam point is called upper fixed point.

9. Now a days the triple point of water is taken as fixed point.

10. The tripple point is the temperature at which the three states of matter (ice, water, steam) coexist.

3. Measurement of Temperature:

1. When a substance is heated, its various physical properties change. By measuring the changes of the properties, the temperature of a body is estimated relative to a fixed point.

2. The change in volume of a liquid, the change in pressure of a gas, the change in resistance of a metal wire, the change in vapour pressure, the measurement of thermo emf and the measurement of thermal radiations are some physical properties used in the measurement of temperature.

4. Different scales of temperature:

1. The fundamental interval is same for these scales.

2. The interval between lower fixed point (ice point) and upper fixed point (Steam point) is taken as the fundamental interval.

A. Celsius scale:

1. It is also called as centegrade scale.

2. The fundamental interval is divided into 100 divisions.

3. LFP is represented as 0^0C.

4. UFP is represented as 100^0C.

5. 1 degree difference of celsius is represented as C^0.

B, Fahrenheit:

1. The fundamental interval is divided into 180 divisions.

2. LFP is represented as 32^0F.

3. UFP is represented as 212^0F.

4. 1 degree difference of Fahrenheit scale is F^0.

5. $1F^0 < 1\,C^0$.

6. 180 divisions on Fahrenheit scale = 100 divisions of celsius scale. Hence, $9F^0 = 5C^0$.

7. $0^0F = 255.3K = -17.78^0C$

C. Kelvin Scale (or) Absolute scale:

1. The fundamental interval is divided into 100 divisions.

2. LFP is represented as 273.15K.

3. UFP is represented as 373.15K.

D. Faulty or new scale:

1. The fundamental interval is divided into (Y - X) divisions.

2. LFP is represented as X^0.

3. UFP is represented as Y^0.

4. Measurement is given as Z^0.

5. Relation between different scales:

1. As the fundamental interval is same

$$\frac{\text{Measurement in a scale-LFP of that scale}}{\text{UFP-LFP}} = \text{constant}$$

$$\frac{C-0}{100-0} = \frac{F-32}{212-32} = \frac{K-273}{373-273} = \frac{Z-X}{Y-X}$$

2. Between C and F $\frac{C}{100}=\frac{F-32}{180}$

3. Between F and K $\frac{F-32}{180}=\frac{K-273}{100}$

4. Between Z and C $\frac{C}{100}=\frac{Z-X}{Y-X}$

5. Between Z and F $\frac{F}{180}=\frac{Z-X}{Y-X}$

6. Different scales are introduced depending on the method of estimation of the temperature.

a) **Absolute scale :** Measurement of temperature estimated by the measurement of pressure of a gas at constant volume.

b) **Thermodynamic scale:** Temperature estimated from the results of self acting machine of second law of thermodynamics.

7. Absolute scale and Thermodynamic scales are same.

6. Different types of thermometers :

a) Liquid thermometer :

1. The change in volume of a liquid on heating is used to estimate the temperature.

2. These thermometers are to be kept in contact with the body whose temperature is to be determined.

3. The properties of the liquids used are

i) the thermal capacity must be low, so that it does not absorb large amounts of heat from the body and hence it does not show less value of temperature.

ii) liquid must be a good conductor of heat, so that it attains the temperature of the body quickly.

4. Mercury is used as thermometric liquid. It is opaque and shining in addition to the above required properties.

5. Different liquids are selected for measurement of different ranges.

6. The volume of the bulb of the thermometer is selected depending on the density of liquid and sensitivity. But the surface area must be large so that the exchange of heat is quick.

7. For this reason the volume of bulb used in mercury thermometer is small and cylindrical shape.

8. Ranges of temperature depend on the freezing point and boiling point of thermometric liquid.

9. The range of mercury thermometer is $-35^0C\ to\ 300^0C$ as mercury freezes at -38.8°C, and boils at 356°C, the mercury thermometer can not be used for very low or very high temperature measurements.

10. The range can be increased to 500^0C by filling nitrogen gas at high pressure above mercury column .

11. The alcohol thermometer is $-100^0C\ to\ 60^0C$ alcohol thermometer is suitable for low temperature measurements as it remains a liquid upto -112°C and has large coefficient of expansion.

12. To measure down to -200°C, liquid pentane is used as thermometric liquid.

13. If V_0 is the volume of bulb of thermometer at 0^0C and 'l' is the length of the mercury thread in the capillary tube at t^0C then $l=\frac{V_0t(\gamma_{hg}-\gamma_g)}{A_0(1+2\alpha_g t)}$. Where A_0 area of cross section of capillary bore and $\gamma_g=3\alpha_g$, the volume coefficient of expansion of glass, γ_{Hg} = coefficient of real expansion

14. If the volume of bulb is large when compared to the volume of capillary bore, by best approximate $l=\frac{V_0t(\gamma_{Hg}-\gamma_g)}{A_0}$.

b) Gas thermometers :

1. They are based on Gay Lussac's law (general usage one of the Charles law) given by $P\alpha T$ at constant volume.

2. The relation between pressure and temperature in Celsius at constant volume is given by $P=P_0(1+\beta t)$. where P pressure at t^0C, P_0 is pressure at 0^0C, β is pressure coefficient of a gas and equal to 1/273 and P_{100} is pressure at 100^0C.

$$\frac{P_{100}-P_0}{100P_0}=\frac{P_t-P_0}{tP_0}\quad \therefore t^0C=\left(\frac{P_t-P_0}{P_{100}-P_0}\right)x100.$$

3. The bulb of the gas thermometer must be filled with gas at low pressure and high temperature so that it obeys gas law.

4. All gases at low densities exhibit same expansion behaviour.

5. At absolute zero, the molecular motion of the gas ceases.

6. At absolute zero, the pressure becomes zero.

7. Gas thermometers are more senstive as gases expand more than liquids.

8. These are not portable and slow in action.

9. These are used to standardise the other thermometers and to calibrate them.

10. Constant pressure air thermometer can be used to measure temperature upto 600^0C

11. Constant volume Hydrogen thermometers can be used between -200^0C to 500^0C with Hydrogen in platinum bulb.

12. With Nitrogen gas, the temperatures upto 1500^0C can be measured.

13. With Helium gas the temperatures below -200^0C can be measured.

c) Resistance thermometers:

1. It is based on the variation of resistance of a conducting wire with temperature.

2. It is based on $R_t = R_0(1+\alpha t+\beta t^2)$ where R_t=resistance at t^0C. R_0=resistance at 0^0C. α *and* β are constants.

3. For platinum resistance thermometer $\mathbf{R_{100} = R_0(1+\alpha 100)}$ where α is called the coefficient of resistance between 0^0C and 100^0C. $\alpha = \frac{R_{100} - R_0}{100R_0}$.

If R_t is the resistance of wire at t^0C and R_{100} is the resistance of wire at 100^0C and R_0 is the resistance of wire at 0^0C then $\frac{R_{100}-R_0}{100R_0} = \frac{R_t - R_0}{tR_0}$ $\therefore t = \left(\frac{R_t - R_0}{R_{100} - R_0}\right)100$

4. If R_1 & R_2 are the resistances at temperatures $t_1{}^0C$ & $t_2{}^0C$ then $\alpha = \frac{R_2 - R_1}{R_1 t_2 - R_2 t_1}$. Electrical resistance thermometers can be used over a wide range of -272°C to 1200°C.

5. For different ranges wires of different materials are used.

d) Thermoelectric thermometers:

1. Useful to measure quickly variable temperatures.

2. Useful to measure the temperatures of insects or at a point.

3. Based on the measurement of thermo emf, the temperature is estimated.

4. emf = $At + Bt^2$ where A and B are constants.

5. One junction is kept at 0^0C. Thus t is the temperature of the other junction in ^0C.

6. Maximum emf is produced for Antimany, Bismuth thermocouple, this can be used upto 300^0C only. It is more sensitive than the other thermocouple thermometers.

7. Thermocouples are used to measure temperatures in the range of -250°C to 300°C

e) Radiation pyrometers:

1. They are used to measure the temperature of sun and stars.

2. They are useful to measure very high temperatures.

3. The thermometers need not be kept in contact with the body whose temperature is to be estimated.

4. They are based on stefan's law.

5. The radiation emitted by a black body per unit area per second is proportional to the fourth power of its absolute temperature.

6. The distance between the body and the pyrometer can have any large value.

7. It can be measured, however high the temperature is.

8. The upper limit of temperature can be increased by using appropriate arrangements.

9. Radiation thermometers are used to measure temperature above 3000°C.

f) Vapour pressure thermometers:

1. It is based on the fact that the saturated vapour pressure depends on temperature only.

2. Such thermometers can be used below – 268°C to about -272°C.

EXERCISE - I

1. The equivalance of two systems in thermal equilibrium is represented by the property of

1) heat 2) energy 3) temperature 4) specific heat

2. Temperature can be expressed as a derived quantity in terms of [MP PET 93]

1) length and mass 2) mass and time
3) length, mass and time 4) none of these

3. A Celsius is a unit of [NCERT 73]

1) current 2) heat
3) equivalent to degree centigrade
4) equivalent to degree Kelvin

4. Mercury is used in liquid thermometers because it has

1) high specific heat and high conductivity
2) high specific heat and low conductivity
3) low specific heat and low conductivity
4) low specific heat and high conductivity

5. **The standard fixed point for temperature measurements in use today is**
1) melting point of ice at one atmospheric pressure
2) of pure ice and pure water in equilibrium at one atmosphere pressure
3) at which ice, water and watervapour coexist
4) none of the above

6. **Very low temperatures are measured by**
1) gas thermometers 2) pyrometers
3) thermocouple thermometers
4) vapour pressure thermometers

7. **Thermometers which are not kept in touch with the body to measure temperatures are**
1) pyrometers 2) thermocouples
3) gas thermometers
4) platinum resistance thermometers

8. **Very high temperatures are measured by**
1) pyrometers 2) thermocouples
3) gas thermometers
4) vapour pressure thermometers

9. **The temperature of the sun is measured with [PMT (Punjab) 98; CPMT 98]**
1) platinum thermometer
2) gas thermometer 3) pyrometer
4) vapour pressure thermometer

10. **Of the following thermometers the one which is most useful for the measurement of a rapidly varying temperature is a [CPMT 92]**
1) platinum resistance thermometer
2) gas thermometer
3) thermoelectric thermometer
4) saturation vapour pressure thermometer

11. **Standardisation of thermometers is obtained with [CPMT 96]**
1) Jolly's thermometer
2) platinum resistance thermometer
3) thermocouple thermometer 4) gas thermometer

12. **The gas thermometers are more sensitive than liquid thermometers because**
1) gases expand more than liquids
2) gases are easily obtained 3) gases are much lighter
4) gases do not easily change their states

13. **A constant volume gas thermometer works on [IIT 1980]**
1) Archimede's principle 2) Boyle's law
3) Gay - Lussac's law 4) Charle's law

14. **To measure temperature, most accurately one should use [CPMT 96]**
1) thermocouple thermometer
2) resistance thermometer
3) constant volume gas thermometer
4) mercury thermometer

15. **Of the following gas thermometers, the one which can be used for measuring very low temperatures is**
1) air 2) hydrogen 3) oxygen 4) helium

16. **In a constant volume gas thermometer, the temperature of a bath is measured by**
1) keeping the mass of the gas constant and by noting the changes in volume
2) keeping the mass of the gas constant and by noting the change in pressure
3) keeping the volume of the gas constant and by noting the change in pressure
4) keeping the mass and volume of the gas constant and by noting change in pressure

17. **A temperature T is measured by a constant volume gas thermometer [IIT Screening94]**
1) T is independent of the gas used at all pressures
2) T is independent of the gas used only at high pressure
3) The ideal gas scale agrees with the absolute scale of temperature
4) The ideal gas scale does not agree with the absolute scale of temperature

18. **A constant volume gas thermometer works on [Kerala (PET) 2002]**
1) Archimedes' principle 2) Pascal's law
3) Boyle's law 4) Charles' law

19. **The value of temperature coefficient of resistance thermometer is [Orissa 1991]**
1) low and negative 2) low and positive
3) high and negative 4) high and positive

20. **In thermocouple thermometer which pair of substances is most sensitive**
1) copper 2) silver - constantant
3) bismuth - antimony 4) lead - tin

21. **Two thermometers are constructed in the same way except that one has a spherical bulb and the other a cylindrical bulb; which one will respond quickly to temperature changes**
1) spherical bulb thermometer
2) cylindrical bulb thermometer
3) both equally 4) either

22. Absolute scale of temperature is reproduced in the laboratory by making use of a

1) radiation pyrometer **[SCRA 1998]**
2) platinum resistance thermometer
3) constant volume helium gas thermometer
4) constant pressure ideal gas thermometer

23. Mercury boils at 357^{0}C. However, mercury thermometers are made such that they can measure temperature upto 500^{0}C. This is done by [CPMT 83, AMU 2003]

1) maintaining vacuum above mercury column in the stem of the thermometer
2) filling nitrogen gas at high pressure above the mercury column
3) filling nitrogen gas at low pressure above the mercury column
4) filling oxygen gas at high pressure above the mercury column

24. The temperature of a gas is measured with a

1) platinum resistance thermometer **[CPMT 89]**
2) pyrometer 3) gas thermometer
4) vapour pressure thermometer

25. A clinical thermometer is usually much shorter than a laboratory thermometer, because

1) only a limited range of temperature is required
2) it is marked in Fahrenheit and not in Celsius
3) a long column is inconvenient
4) a long column of mercury can not be introduced in glass tube

26. For a constant volume gas thermometer, one should fill the gas at [EAMCET 96]

1) high temperature and high pressure
2) high temperature and low pressure
3) low temperature and low pressure
4) low temperature and high pressure

27. The resistance R_t of a conductor varies with temperature as shown in fig. If the variation is represented as $R_t = R_0\left(1+\alpha t+\beta t^2\right)$ [CPMT 88]

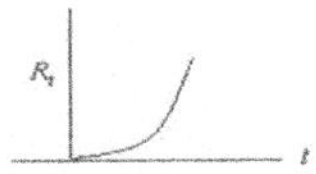

1) α and β are both negative
2) α is positive and β is negative
3) α and β are both positive
4) α is negative and β is positive

28. The resistance R_t of a semi - conductor block varies with temperature is shown in fig. It may be expressed as (α,β being positive constants)

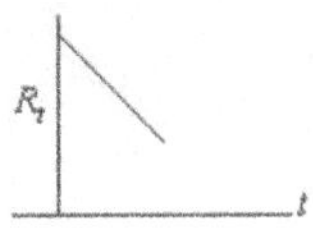

1) $R_t = R_0\left(1+\alpha t+\beta t^2\right)$
2) $R_t = R_0\left(1-\alpha t-\beta t^2\right)$
3) $R_t = R_0\left(1+\alpha t\right)$
4) $R_t = R_0\left(1-\alpha t\right)$

29. Which of the following statements is correct for a bulb thermometer

1) The bulb of the thermometer is made of a conducting material
2) The bulb of the thermometer is made of a poorly conducting material
3) Sole purpose of making the walls of the bulb thin is to provide maximum possible space for liquid
4) Specific heat of liquid used in the bulb must have large value

30. Which of the following statements is true for a thermometer

1) Coefficient of cubical expansion of liquid must be greater than that of bulb material
2) Coefficient of cubical expansion of liquid may be equal to that of bulb material
3) Coefficient of cubical expansion of liquid must be less than that of bulb material
4) Coefficient of cubical expansion of liquid may be greater or less or equal to that of the material of bulb

31. A temperature is measured by a constant volume gas thermometer

1) *T* is independent of the gas used for all pressures
2) *T* is independent of the gas used only at low pressures
3) *T* is independent of the gas used only at high pressure
4) the ideal gas scale does not agree with absoulte scale of temperature

32. Which of the following statements is correct

1) A thermometer should have a small thermal capacity
2) Total heat of the universe is conserved
3) Atmospheric temperature in cities on the sea - coast changes very much
4) Constant volume gas thermometer can be used to measure the temperatures of insects

33. Absolute zero is that temperature at which

1) matter ceases to exist
2) ice melts and water freezes
3) volume and pressure of a gas becomes zero
4) none of the above

34. **On which of the following scales of temperature, the temperature is never negative**
1) Celsius 2) Fahrenheit 3) Reaumur 4) Kelvin

35. **We plot the graph having temperature in ^{0}C on x - axis and in ^{0}F on y - axis. If the graph is straight line, then the correct statement is**
1) the line intercepts the positive x - axis
2) the line intercepts the positive y - axis
3) the line passes through origin
4) the line intercepts the negative axis of both x and y - axes

36. **A thermometric liquid which can be used to measure temperature between - 40^{0}C to + 40^{0}C is [EAMCET 82]**
1) water 2) alcohol 3) mercury 4) phenol

37. **The study of physical phenomenon at low temperatures (below liquid nitrogen temperature) is called [CPMT 92]**
1) refrigeration 2) radiation
3) cryogenics 4) pyrometry

38. **The absolute zero is the temperature at which**
1) water freezes **[AIIMS 1998]**
2) all substances exist in solid state
3) molecular motion ceases 4) none of the above

39. **Given $50^0F, 50^0C$ and $50K$ arrange them in increasing order of temperature**
1) $50^0F, 50^0C \& 50K$
2) $50K, 50^0C \& 50^0F$
3) $50K, 50^0F \& 50^0C$
4) $50^0C, 50^0F \& 50K$

40. **Thermometric substance Range**

List - I	**List - II**
a) air	**e)** -200^0C to 1500^0C
b) Hydrogen	**f) Below** -200^0C
c) Nitrogen	**g)** -200^0C to 500^0C
d) Helium	**h) up to** 600^0C

1) a-g b-h c-e d-f 2) a-h b-g c-f d-e
3) a-h b-g c-e d-f 4) a-e b-f c-g d-h

41. **Thermometer Range**

List - I	**List - II**
a) Vapour pressure	**e)** -250^0C to 3000^0C
b) Thermo couple	**f)** -272^0C to 1200^0C
c) Resistance	**g) above** 3000^0C
d) Radiation	**h) near absolute zero**

1) a-e b-f c-g d-h 2) a-h b-e c-f d-g
3) a-h b-e c-g d-f 4) a-e b-h c-f d-g

42.

List - I	**List - II**
a) Alcohol	**e)** -35^0C to 500^0C
b) Mercury	**f)** -112^0C to 60^0C
c) Liquid pentane	**g)** -35^0C to 300^0C
d) Mercury with a gas at high pressure	**h) Less than –112 up to** -200^0C

1) a-f b-g c-h d-e 2) a-e b-h c-f d-g
3) a-g b-f c-h d-e 4) a-g b-f c-e d-h

ANSWERS

1) 3	2) 4	3) 3	4) 4	5) 3
6) 4	7) 1	8) 1	9) 3	10) 3
11) 4	12) 1	13) 3	14) 3	15) 4
16) 4	17) 3	18) 4	19) 4	20) 3
21) 2	22) 3	23) 2	24) 3	25) 1
26) 2	27) 3	28) 4	29) 1	30) 1
31) 2	32) 1	33) 3	34) 4	35) 2
36) 2	37) 3	38) 3	39) 3	40) 3
41) 2	42) 1			

EXERCISE-II(A)

1. **Express OK on Fahrenheit scale.**
1) -459.67°F 2) 240.6°F 3) -320°F 4) 373°F

2. **On Reaumer scale of temperature the melting point of ice and the boiling point of water are taken as 0°R and 80°R respectively. The freezing point and boiling point of mercury on Celsius scale are 39°C and 367°C respectively. Express these temperature on Reaumer scale**
1) 12.2°R, 127°R 2) 41°R, 290°R
3) 31.2°R, 293.6°R 4) 71°R, 293.6°R

3. **A faulty thermometer has its fixed points marked as 3° and 102°. The temperature of a body as measured by the faulty thermometer is 80°. Find the correct temperature of the body on Celsius scale.**
1) 36.5°C 2) 77.8°C 3) 25.8°C 4) 66.8°C

4. **At what temperature on the Fahrenheit scale will the reading be double of the reading on the Celsius scale?**
1)340°F 2) 240.6°F 3) 330°F 4) 320°F

5. A acurate Celsius thermometer and a Fahrenheit thermometer register 60° and 141° respectively when placed in the same constant temperature enclosure. What is the error in the Fahrenheit thermometer?
1)1°F 2) 3°F 3) -5°F 4) 2°F

6. The reading of centigrade thermometer coincides with that of Fahrenheit thermometer in a liquid. The temperature of the liquid is [BCECE 2003]
1) -40^0C 2) 0^0C 3) 100^0C 4) 300^0C

7. Two thermometers one Celcius and other Fahrenheit are put in a hot bath. The reading on Fahrenheit scale is just three times the reading on Celcius thermometer. The temperature of the bath is
1) 100^0C 2) $(80/3)\,^0$C 3) 80^0C 4) 70^0C

8. The steam point and the ice point of a mercury thermometer are marked as $80^0$and 20^0. What will be the temperature in centigrade mercury scale when this thermometer reads 32^0
1) 20^0C 2) 30^0C 3) 10^0C 4) 0^0C

9. The higher and lower fixed points on a thermometer are separated by 150mm. When the length of the mercury thread above the lower temperature is 30 mm, the temperature reading would be
1) 20^0C 2) 30^0C 3) 10^0C 4) 0^0C

10. A centrigrade and a Fahrenheit thermometer are dipped in boiling water. The water temperature is lowered until the Fahrenheit thermometer registers 176^0F. What is the fall in temperature as registered by the Centigrade thermometer [CBSE 92 ; AIIMS 98]
1) 20^0C 2) 40^0C 3) 60^0C 4) 80^0C

11. A platinum resistance thermometer reads 0^0 C when its resistance is 80 Ω and 100^0 when its resistance is 90 Ω. Find the temperature at the platinum scale at which the resistance is 86 Ω
1) 30^0C 2) 60^0C 3) 20^0C 4) 10^0C

12. The pressure of a gas filled in the bulb of a constant volume gas thermometer at 0^0C and 100^0C are 28.6 cm and 36.6 cm of mercury respectively. The temperature of bulb at which pressure will be 35.0 cm of mercury will be
1) 80^0C 2) 70^0C 3) 55^0C 4) 40^0C

13. A fixed mass of an ideal gas is maintained at constant volume the pressure of the gas at triple point of water is p_{tr} then the thermodynamic temperature of the gas when the pressure is P
1) $273.16\left(\frac{P}{P_{tr}}\right)K$ 2) $273.16\left(\frac{P-P_{tr}}{P_{tr}}\right)K$
3) $273.16\left(\frac{P_{tr}}{P}\right)K$ 4) $273.16\left(\frac{P+P_{tr}}{P}\right)K$

14. The steam point and the ice point of a mercuary thermometer are wrongly marked as 92^0C and 2^0C respectively. The correct temperature read by this thermometer is
1) 2^0C 2) 5^0C 3) 10^0C 4) 20^0C

15. A Fahrenheat thermometer reads 113^0F while a faulty celsius thermometer reads 44^0C. The correction to be applied to the celsius thermometer is
1) -1^0C 2) $+1^0$C 3) $+2^0$C 4) -2^0F

16. On a hypothetical scale X, the ice point is 40^0 and the steam point is 120^0. For another scale Y the ice point and steam points are 30^0
1) -5^0 2) -8^0 3) -10^0 4) -12^0

ANSWERS

1) 1	2) 3	3) 2	4) 4	5) 1
6) 1	7) 2	8) 1	9) 1	10) 1
11) 2	12) 1	13) 1	14) 4	15) 2
16) 3				

EXERCISE- II(B)

THERMOMETRY

1. On the Celsius scale, the absolute zero of temperature is at [CBSE 94]
1) 0^0C 2) -32^0C 3) 100^0C 4) -273.15^0C

2. On Reamuer scale of temperature the melting point of ice and the boiling point of water are taken as 0^0R and 80^0R respectively. The freezing point and boiling point of mercury on Celsius scale are 39°C and 367°C respectively. Express these temperature on Kelvin scale
1) 0K,273K 2) 41K, 290K
3) 312K, 640K 4) 71K, 293.6K

3. The Fahrenheit and Kelvin scales of temperature will give the same reading at
1) - 40 2) 313 3) 574.25 4) 732.75

4. A faulty thermometer has fixed points marked 5 and 95 what is the correct temperature in centigrade when this thermometer reads 59
1) 60^0C 2) 50^0C 3) 30^0C 4) 10^0

5. **At what temperature the reading on Kelvin scale is three times that on Celcius scale**

1) (80 / 3) ^{0}C 2) 136.5^{0}C 3) -68.25^{0}C 4) not possible

6. **If a thermometer reads freezing point of water as 20^{0}C and boiling point as 150^{0}C, how much thermometer read when the actual temperature is 60^{0}C [AFMC 04]**

1) 98^{0}C 2) 110^{0}C 3) 40^{0}C 4) 60^0

7. **Oxygen boils at -183C. This temperature is approximately [CMPT 92]**

1) 215^{0}F 2) -297^{0}F 3) 329^{0}F 4) 361^{0}F

8. **The higher and lower fixed points on a thermometer are separated by 160mm. When the length of the mercury thread above the lower temperature is 40 mm, the temperature reading would be**

1) 40^{0}C 2) 120^{0}C 3) 32^{0}C 4) 25^{0}C

9. **A centrigrade and a Fahrenheit thermometer are dipped in boiling water. The water temperature is lowered until the Fahrenheit thermometer registers 140^0. What is the fall in temperature as registered by the Centigrade thermometer [CBSE 92 ; AIIMS 98]**

1) 30^0 2) 40^0 3) 60^0 4) 80^0

10. **The resistance of a resistance thermometer has values 2.70 Ω and 3.70 Ω at 0^{0}C and 100^{0}C respectively. The temperature at which the resistance is 3.10 Ω is**

1) 30^{0}C 2) 40^{0}C 3) 60^{0}C 4) 70^{0}C

11. **A correct thermometer in Fahrenheit is introduced in a water bath along with a Celsius thermometer. The reading observed are 86^{0}F and 32^{0}C. The correction to be made to the Celsius reading will be**

1) 2.5^{0}C 2) 2^{0}C 3) 1.5^{0}C 4) 3^{0}C

12. **The pressure of hydrogen gas in a constant volume gas thermometer is 80.0 cm at 0^{0}C, 110 cm at 100^{0}C and 95.0 cm at unknown temperature *t*. Then *t* is equal to**

1) 50^{0}C 2) 75^{0}C 3) 95^{0}C 4) 150^{0}C

13. **A constant volume gas thermometer using helium records a pressure of p_1 = 20.0 kPa at triple point of water (273.16K) and a pressure p_2 = 14.3 kPa at the temperature of dry ice (CO_2). Then the temperature of dry ice is**

1) 195.3K 2) 382K 3) 0.11^{0}C 4) 0.014^{0}C

14. **The upper and lower fixed points of a faulty mercury thermometer are 210^{0}F and 34^{0}F respectively. The correct temperature read by this thermometer is**

1) 22^{0}F 2) 80^{0}F 3) 100^{0}F 4) 122^{0}F

15. **When a celsius thermometer reads 90^{0}C, a faulty Fahrenheit thermometer reads 190^{0}F. The correction to be made in the latter scale is**

1)+2^{0}F 2) –2^{0}F 3) –4^{0}F 4) +4^{0}F

16. **On a hypothetical scale A the ice point is 42^0 and the steam point is 182^0. For another scale B. The ice point is –10^0 and steam point is 90^0. If B reads 60^0. The reading of A is**

1) 160^0 2) 140^0 3) 120^0 4) 110^0

ANSWERS

1) 4	2) 3	3) 3	4) 1	5) 2
6) 1	7) 2	8) 4	9) 2	10) 2
11) 2	12) 1	13)1	14) 4	15) 4
16) 2				

Chapter - 5

THERMAL PROPERTIES OF MATTER

EXPANSION OF SOLIDS

SYNOPSIS

1. A solid consists of atoms or molecules packed closely together.
2. Properties of solids depend upon inter atomic forces, attractive or repulsive.
3. Attraction is in between electron cloud of one atom and nuclear charge of the other atom.
4. When they are very close, repulsive force exists between electron clouds of the two atoms.
5. In the equilibrium, these attractive and repulsive forces balance each other.
7. Due to thermal energy, the atom in a solid is always in vibration.

8. **Potential energy curve:**

   a) It is associated with a system of objects where forces exist among them.

   b) A change in their arrangement or configuration causes a change in the potential energy of the system.

   c) As atoms vibrate, due to variation in the seperation, their potential energy changes as 1/2 K x^2. Where K is force constant.

   Variation of P.E. of a pair of atoms with their seperation is a parabola.

   d) The oscillations are simple harmonic.

9. **Harmonic oscillations :**

   a) When vibrations of atoms are simple harmonic, P.E. curve is symmetrical.

   b) On heating, though amplitude increases, mean distance between two consecutive atoms remains same, whatever be the temperature. Hence thermal expansion does not take place.

   c) In simple harmonic oscillations, when P.E. curve is symmetrical, no thermal expansion takes place.

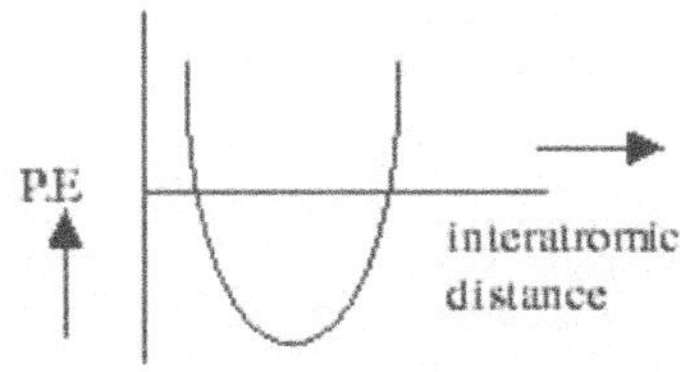

10. Anharmonicity of oscillations :

a) Actually vibrations of atoms are not simple harmonic in real solids

b) Oscillations are anharmonic and phenomenon is called anharmonicity

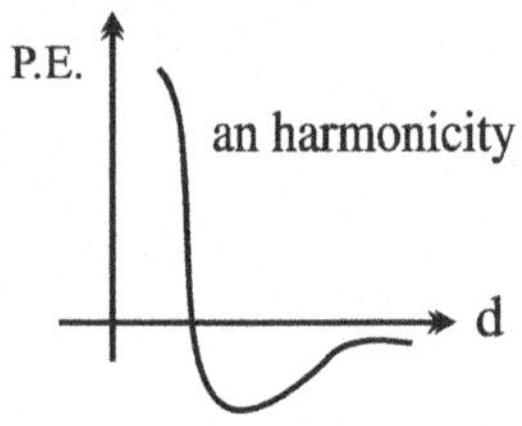

c) The P.E. curve is asymetric.

d) As temperature increases, the equilibrium seperations also increases and body expands.

e) Thermal expansion of solids is due to anharmonicity of vibrations of atoms in the solid.

f) The left branch of the curve is raising more steeply than right branch. Due to this, average seperation increases with temperature.

11. Isotropic solids : Physical properties are same in all directions.

    Ex : glass, rock salt & metals etc. The rocks behave as isotropic with respect to thermal expansion as it is of cubic structure.

12. Anisotropic solids : Physical properties are different in different directions.

    Ex : All crystalline solids, wood etc.

13. In general, solids, liquids and gases expand on heating since distance between the particles increases with increase of temperature.

    Exceptions :

    a) water contracts when heated from 0°C to 4°C.

    b) Silver Iodide, Silica glass, some Nickel Steel alloys, India rubber, castiron, type metal contract on heating

14. Isotropic solids expand equally in all directions. Eg : Metals, glass
15. Anisotropic solids do not expand equally in all directions.

    Eg : Ice land spar [$CaCO_3$], crystals

16. Solids have linear, areal and cubical expansions.

    Hence solids have three coefficients of expansion namely coefficient of lienar expansion α, coefficient of areal expansion β and coefficient of volume expansion γ.

17. Co-efficient of linear expansion of a solid(α)

a) $\alpha = \frac{l_2 - l_1}{l_1(t_2 - t_1)} = \frac{\Delta l}{l_l(\Delta t)}$

S.I. unit : K^{-1}, practical unit : $(°C)^{-1}$

b) change in length $= l_2 - l_1 = \Delta l = \alpha l_1(t_2 - t_1)$

c) Final length $l_2 = l_1\ [1+\alpha(t_2 - t_1)]$

d) $l = l_0\ (1+\alpha t)$, where l_0 is length at 0°C

e) For anisotropic solids like crystals if α_x, α_y, α_z are co-efficients of linear expansions along x, y, z directions then $\alpha = \frac{\alpha_x + \alpha_y + \alpha_z}{3}$

18. Numerical value of co-efficient of linear expansion of a solid

a) depends on nature of material

b) depends on the scale of temperature used

c) is independent of physical dimensions of the body

d) is independent of unit of length chosen

19. Linear expansion or increase in length of rod depends on

a) nature of material b) Initial length of rod

c) rise of temperature

20. Co-efficient of Areal or superficial expansion of a solid (β) :

a) $\beta = \frac{a_2 - a_1}{a_1(t_2 - t_1)}$ b) S.I. unit : K^{-1}

21. Co-efficient of cubical or volume expansion of a solid:

a) $\gamma = \frac{v_2 - v_1}{v_1(t_2 - t_1)}$ b) S.I. unit : K^{-1}

22. Relation among α, β, and γ (for isotopic solids)

a) $\alpha : \beta : \gamma = 1 : 2 : 3$ b) $\beta = 2\alpha = \frac{2}{3}\gamma$

c) $\gamma = 3\alpha = \frac{3}{2}\beta$ d) $\gamma = 3\alpha = \alpha + 2\alpha = \alpha + \beta$.

23. The numerical values of α, β & γ are 5/9 times if Fahrenheit scale is used instead of Celsius scale

24. **Variation of density with Temperature:** The mass of a substance remains same but its volume increases with rise of temperature. The density of the substance varies inversely with temperature.

Density decreases with the increase of temperature.

a) $d_t = \frac{d_0}{1+\gamma t}$ b) $d_2 = \frac{d_1}{1+\gamma\Delta t}$

c) $\gamma = \frac{d_1 - d_2}{d_2(t_2 - t_1)} = \frac{d_o - d_t}{td_t}$ & $\gamma = \frac{d_1 - d_2}{d_2 t_2 - d_1 t_1}$

25. Two rods of different materials will have same difference in lengths at all temperatures, if they have same expansions for equal rise of temperature. Then $l_1\ \alpha_1 = l_2\ \alpha_2$

26. If a hollow pipe and a solid rod of same dimensions made of same material are heated through the same rise of temperature, both expand equally.

27. A thin rod and a thick rod made of same material have same initial lengths. If they are heated to equal temperature difference, both expand equally.

28. A thin rod and thick rod made of same material have same initial lengths. If they are heated by equal quantities of heat, thin rod expands more.

29. A solid sphere and a hollow sphere made of same material have same volume

a) If they are heated to same temperature difference, both expand equally

b) If they are heated by equal quantities of heat, hollow sphere expands more.

30. A solid sphere and another sphere having cavities are made of same material have same outer volume. If they are heated to same temperature difference, both expand equally. Hence increase in volume is independent of cavities.

31. α for pyrex glass is 1/3 of ordinary glass. Hence pyrex glass is used in making thermometers and test tubes used for heating purpose.

32. When a solid is heated, moment of inertia increases since distance of particles from the axis of rotation increases.

33. Invar is an alloy of Steel and Nickel in the ratio 64:36. Invar steel is used in making pendulum, balance wheels of watches, standard metal tapes because it has low α.

34. Pendulum clock:

In summer , the pendulum clock expands

As $T = 2\pi\sqrt{\frac{l}{g}}$

With increase in length l, the time period of pendulum T increases and it makes less number of oscillations than the required oscillations (43,200) per day. Hence the clock loses time and runs slow.

In winter, the pendulum contracts with the decrease in length l, the time period of pendulum T decreases and it makes more oscillations than the required oscillations, (43,200) per day. Hence the clock gains time and runs fast.

change in the period of a pendulum is.

$$\frac{\Delta T}{T}=\frac{1}{2}\alpha(t_2-t_1) \text{ per second}$$

b) Loss or gain in time per second = $\frac{1}{2}\alpha\,\Delta t$

c) Loss or gain in time per day =

$$\frac{1}{2}\alpha\Delta t\times 86400$$

d) Knowing loss or gain in time per second or per day, α of the material of pendulum of the clock can be known.

35. Metal Tapes :

a) Metal tape shows correct reading only at a temperature at which it is calibrated.

b) A metal tape is graduated at t_1°C and is used at t_2°C.

i) If $t_2 > t_1$ and the distance between the divisions increases then

a)observed reading L is less than actual value

b) correction to be applied, $\Delta C = \alpha L(t_2-t_1)$

c) correct reading = L + ΔC

ii) If $t_2 < t_1$ and the distance between the divisions decreases then

a) Observed reading L is greater than actual value.

b) correction to be applied, $\Delta C = \alpha L \Delta t$

c) correct reading = L – ΔC

iii) A scale messures the length of the rod as L cm when both are at which is the $t_1{}^0$C calibration temperature for the scale. If the coefficient linear expansion of the scale and rod are $\alpha_s \,\&\, \alpha_b$. Then the scale reading in centimeter when both are at $t_2{}^0$C is

$$L\left(1+(\alpha_b-\alpha_s)(t_2-t_1)\right)$$

36. Bimetallic strip : a) It is a combination of two metal strips of different linear coefficients and of same length at a particular temperature and rivetted together firmly.

b) When it is heated, it bends such that the metal with higher 'α' lies on the convex side.

Eg : In steel - Brass Bimetallic strip, Brass lies on convex side when it is heated.

c) When it is cooled, it bends the other way such that the metal with higher 'α' lies on concave side.

d) Bimetallic strip is used in Thermostats. Unequal expansion of dissimilar metals is the principle used in Thermostat (Bimetallic strip).

e) Radius of circular arc of bent strip is

$$R=\frac{d}{(\alpha_2-\alpha_1)\Delta t}$$

Here d = Thickness of each strip

$\alpha_2 - \alpha_1$ = Difference in coefficients of linear expansions

Δt = Difference in Temperature.

37. Thermal stress : A rod is rigidly clamped at its ends and prevented without expansion. When it is heated, thermal stress is developed at its ends.

Thermal stress = Y α Δt

Thermal force = (Y α Δt) A

Thermal force depends upon cross-sectional area but not upon length.

Thermal stress is independent of both area and length

38. a) Percentage increase in length= $\frac{\Delta l}{l}x100 = \alpha\Delta t$ (100) of a solid

b) Percentage increase in Area =

$$\frac{\Delta A}{A}x100 = \beta\Delta t\,x\,(100) = 2\alpha\Delta t\,x\,(100) \text{ of a solid}$$

c) Percentage increase in volume =

$$\frac{\Delta V}{V}x100 = \gamma\Delta t\,x\,(100) = 3\alpha\Delta t\,x\,(100) \text{ of a solid}$$

d) Percentage increase is independent of initial dimensions. It depends on nature the of material and rise of temperature only.

EXERCISE - I

1. Substance having very small co-efficient of linear expansion, among the following is

1) Copper 2) Iron 3) Lead 4) Invar

2. The coefficient of volume expansion is [96E]

1) Twice the coefficient of linear expansion

2) Twice the coefficient of areal expansion

3) Thrice the coefficient of areal expansion

4) Thrice the coeffcient of linear expansion

3. **Substance which contracts on heating, among the following is**
   a) Invar **b) Brass**
   c) Silver Iodide **d) Type metal**
   1) a & b 2) b & c 3) c & d 4) d & a
4. **When the temperature of a body increases**
   1) density and moment of inertia increases
   2) density and moment of inertia decreses
   3) density decreases and moment of inertia increases
   4) density increases and moment of inertia decreases
5. **The variation of density of solid with temperature is given by the formula [96 E]**
   1) $d_2 = \frac{d_1}{1+\gamma(t_2-t_1)}$ 2) $d_2 = \frac{d_1}{1-\gamma(t_2-t_1)}$
   3) $d_2 = \frac{d_1}{1-2\gamma(t_2-t_1)}$
   4) $d_2 = \frac{d_1}{1+2\gamma(t_2-t_1)}$
6. **Two Brass rods of same length but with different diameters are heated by equal amounts of heat. The expansion is**
   1) Same in both rods 2) more in thick rod
   3) more in thin rod
   4) depends on material of rods
7. **A Brass stopper snuggly fits in the hole of steel plate. To remove the stopper easily, the system**
   1) should be heated 2) should be cooled
   3) may be heated or cooled
   4) can not be removed by heating or cooling
8. **A ball does not pass through a copper ring at room temperature. The same ball, after heating, passes through the same ring. It shows due to heating, the ball**
   1) expands 2) contracts
   3) does not change 4) 1 or 2
9. **A thick glass tumbler gets cracks when boiling water is poured into it, because of**
   1) poor quality 2) unequal expansions
   3) a of glass is zero 4) a of glass is maximum
10. **Good quality glass tumblers are made of special material whose linear coefficient of expansion is**
    1) large 2) minimum
    3) negative 4) a has no relation
11. **Due to heating, soft rubber**
    1) expands 2) contracts
    3) does not change
    4) 1 or 2 depending on pressure
12. **A solid sphere and a hollow sphere of same material have same mass. When they are heated by 50°C, increase in volume of solid sphere is 5 c.c. The expansion of hollow sphere is**
    1) 5 c.c. 2) more than 5 c.c.
    3) Less than 5 c.c. 4) None
13. **A metal meter scale has two holes at the two ends. when the scale is heated the distance between the two holes**
    1) decreases 2) increases
    3) doesnot change 4) may increase or decrease
14. **In RCC slabs only iron rods are used but not of any other metal because**
    1) it has more ultimate strength
    2) they are cheap
    3) elasticity is more
    4) the coefficient of expansions of concrete and iron are equal.
15. **Among the following solids, the lowest coefficient of expansion is for**
    1) Concrete 2) Rubber
    3) Glass 4) Silicon
16. **If the length of a body is measured in centimeters instead of meters, the coefficient of linear expansion**
    1) increasses 2) decreases
    3) remains same
    4) decreases to 1/100 of the original value
17. **A clock with a brass pendulum keeps correct time at 20°C. If the temperature is 40°C.**
    1) It loses the time
    2) It gains the time
    3) neither loses nor gains the time
    4) All the above three are possible depending on the a of the pendulum
18. **The relation between the coefficients of linear expansion a and volume expansion g is**
    1) a = g 2) g = 3a 3) g = 2 a 4) g = 4a

19. A platinum wire can be sealed through glass, but a brass wire cannot be sealed in glass because

1) Platinum has greater density than brass
2) Platinum and glass have the same value for coefficient volume expansion
3) Platinum is an element while brass is an alloy
4) Platinum does not rust

20. A metal disc having a cirular hole at its centre is heated.If metal expands on heating, the diameter of the hole

1) increases 2) decreases
3) remains unchanged
4) increases or decreases depending on the metal

21. A solid ball of metall has a spherical cavity inside it. The ball is heated. The volume of cavity

1) decreases 2) increases
3) remain unchanged
4) have its shape changed

22. Two spheres of same size are made of same metal, but one is hollow and the other is solid. They are heated to same temperature then

1) Both spheres will expand equally
2) Hollow sphere will expand more than solid one
3) Solid sphere will expand more than Hollow one
4) None of the above

23. **A metal washer has a hole of diameter d_1 and an external diameter d_2 such that $d_2 = 3d_1$. If on heating d_2 increases by 0.3% then d_1**

1) decreases by 0.1% 2) decreases by 0.3%
3) increases by 0.1% 4) increases by 0.3%

24. When two similar rods A and B but of different materials and different lengths are conected end to end and lagged between two strong supports. When the temperature of the rods increases

1) force on A > B but the expansions of the rods are same
2) force on A < B but the expansions of the rods are same
3) forces on A & B are equal but the expansions of the rods differ
4) forces on A & B are equal and the expansions of the rods are same

25. A metal sphere is heated. Maximum percentage increase occurs in its (EAMCET 94 M)

1) Density 2) Volume
3) Surface Area 4) Radius

26. Thermostat works on the principle of

1) on heating metal expands
2) unequal expansions of different metals on heating
3) Both 1 and 2 4) None

27. Two metal strips that constitute a thermostat must necessarily differ in their [IIT 92]

1) mass 2) length
3) resistivity
4) coeffcient of linear expansion

28. A bimetallic strip is heated. It

1) bends towards the metal with lower coefficient of expansion
2) bends towards the metal with higher coefficient of expansion
3) twists itself into helix
4) does not bend

29. A bimetal made of copper and iron strips welded together is straight at room temperature. It is held vertically so that the iron strip is towards the left hand and copper strip is towards right hand. The bimetal strip is then heated. The bimetal strip

1) remains straight 2) bends towards right
3) bends towards left 4) does not change

30. Which of the following statement is wrong

1) Bimetal is used in metal themometer
2) Bimetal is used to generate electricity
3) Bimetal relays are used to open or close electric circuits
4) Bimetal is used in thermostat for regulating heating or cooling of rooms.

MULTIPLE ANSWERS & MATCHING TYPE

31. Pick up the correct statement

a) Cubic crytals exhibit isotropic nature with respect to thermal expansions.
b) Anisotropic solids expand in one direction and contract in perpendicular direction.
c) Many of the crystals are anisotropic solids
d) Rock salt crystal exhibits isotropic nature with respect to thermal expansion.

1) a,b,c are correct 2) b,c,d are correct
3) c,d,a are correct 4) a,b,c,d are correct

32. Pick up the correct statement
a) Sun's temperature is measured by a pyrometer
b) Insects temperature is measured by thermo electric thermometer
c) Temperature of Human body is measured by mercury thermometer.
1) a & b are true 2) b & c are true
3) c & a are true 4) a,b,c are true.

33. When a rod is heated , its linear expansion depends on
(a) Initial length (b) area of cross section
(c) Mass (d) temeprature rise
1) only (a) is correct 2) (a) & (d) are correct
3) (b) & (c) are correct 4) (a) & (c) are correct

34. Th numerical value of coefficient of linear expansion is independent of units of
(a) length (b) temperature
(c) area (d) mass
1) only (a) is correct 2) (a) & (b) are correct
3) (a), (b) & (c) are correct
4) (a), (c) & (d) are correct

35. Expansion during heating
(a) occurs in solids only
(b) causes decrease in weight
(c) is due to increase of interatomic spacing
1) only (a) is wrong
2) (a), (b) & (c) are wrong
3) (a) & (b) are wrong
4) a, b, c are correct

36. When a copper solid sphere is heated, its
(a) moment of inertia increases
(b) Elasticity decreases
(c) density decreases (d) mass increases
1) only (b) is true 2) (a) & (b) are true
3) (a), (b) & (c) are true 4) all are true

37. Due to thermal expansion with rise in temperature
(a) Metallic scale reading becomes lesser than true value
(b) Pendulum clock goes fast
(c) A floating body sinks a little more
(d) The weight of a body in a liquid increases
1) only (a) is correct 2) (a) & (b) are correct
3) (a), (b) & (d) are correct
4) (a), (c) & (d) are correct

38. Which of the following statements are true
(a) Rubber contracts on heating
(b) Water expands on freezing
(c) Water contracts on heating from 0^0C to 4^0C
(d) Water expands on heating from 4^0C to 40^0C
1) (a) is correct 2) (b) and (c) are correct
3) (c) & (d) are correct 4) all are correct

39. When a metal ring having some gap is heated
(a) length of gap increases
(b) radius of the ring decreases
(c) the angle subtended by the gap at the centre remains same
(d) length of gap decreases
1) only (d) is correct 2) (a) and (b) are correct
3) (a) & (c) are correct 4) all are correct

40.

List - I	List - II
a. Isotropic solids	e. Expands on melting
b. Ice	f. Equal expansion in all directions
c. Anisotropic solids	g. contracts on melting
d. Copper	h. unequal expansion in different directions

1) a - f, b - h, c - g, d - e
2) a - f, b - g, c - h, d - e
3) a - g, b - h, c - e, d - f
4) a - e, b - f, c - g, d - h

41.

List - I	List - II
a. Thermal expansion	e. Pendulum clock
b. α,β,γ	f. Depends on Dimensions, Material, Temperature
c. Bimetallic strip	g. Depends on nature of the material only
d. Invar steel	h. to record the continuously changing temperature

1) a - e, b - f, c - g, d - h
2) a - g, b - h, c - e, d - f
3) a - f, b - g, c - h, d - e
4) a - h, b - e, c - f, d - g

42. List - I / List - II

List - I	List - II
a. Thermal stress	e. $3\alpha \ \Delta t \ 100$
b. Loss in time of a pendulum clock per sec.	f. $\frac{d}{(\alpha_2 - \alpha_1)\Delta t}$
c. Percentage increase in volume of a solid	g. $y\alpha\Delta t$
d. Radius of circular arc of a bimetal strip	h. $(1/2)\alpha\Delta t$

1) a - g, b - h, c - e, d - f
2) a - h, b - e, c - g, d - f
3) a - e, b - g, c - f, d - h
4) a - f, b - e, c - h, d - g

ASSERTION & REASON TYPE QUESTIONS

Instruction For Questions **:**

In each of the following questions, a statement of Assertion (A) is given followed by a corresponding statement of reason (R) just below it. Of the statement mark the correct answer.

1) Both 'A' and 'R' are true and 'R' is the correct explanation of 'A'
2) Both 'A' and 'R' are true and 'R' is not the correct explanation of 'A'
3) 'A' is true and 'R' is false
4) 'A' is false and 'R' is true

43 A) :- Metals are isotropic solids
R) : - Metals are composed of large number of minute crystals which are randomly oriented resulting in isotropy

44. A) : Metals expand equally in all directions.
R) : Metals are isotropic

45. A) : If hot liquid is poured in thick glass tumbler it breaks.
R) : Temperature distribution is not uniform in thick bodies leading to differential expansion

46. A) : When a circular disc with a hole is heated, the diameter of the hole increases.
R) : The distance between the molecules at the periphery of the hole increases as solid expands

47. A) : Platinum is used to fuse into glass rods.
R) : Both platinum and glass have almost same values of coefficient of linear expansion.

48. A) : Steel rods are used in R.C.C.
R) : Both steel and R.C.C. have almost same values of coefficient of linear expansion.

49. A) : When a Bimetalic strip made up of Brass and Iron is heated then brass lies on convex side.
R) : Due to lower value of coefficient of linear expansion, brass aquires convex side.

50. A) : When a solid is heated, it expands
R) : The 'PE - interatomic spacing' curve for solids is asymetric

51. A) : Invar is used in metal tapes, balance wheels and pendulum clocks.
R) : The coefficient of linear expansion of invar is very low.

52. A) : A brass disc fits snugly in the hole of the iron plate. To remove the disc the combination must be heated.
R) : A coefficient linear expansion of brass is more than that of iron

53. A) : In summer, the pendulum clock runs slow.
R) : As temperature increases, length of the pendulum decreases and time period decreases.

ANSWERS

1) 4	2) 4	3) 3	4) 3	5) 1
6) 3	7) 2	8) 2	9) 2	10) 2
11) 2	12) 2	13) 2	14) 4	15) 4
16) 3	17) 1	18) 2	19) 2	20) 1
21) 2	22) 1	23) 4	24) 3	25) 2
26) 2	27) 4	28) 1	29) 3	30) 2
31) 4	32)4	33) 2	34) 4	35) 3
36) 3	37)4	38) 4	39) 3	40) 2
41) 3	42) 1	43) 1	44) 1	45) 1
46) 1	47) 1	48) 1	49) 3	50) 1
51) 1	52) 4	53) 3		

EXERCISE- II(A)

1. Coefficient of cubical expansion of a solid is (0.000027/°C). If the temperature is measured on Fahreheit scale, numerical value of coefficient of linear expansion of solid is
1) 0.000009/°F   2) 0.000005/°F
3) 0.000015/°F   4) 0.000018/°F

2. If coefficient of cubical expansion is x times coefficient of superficial expansion, then the value of x is [AFMC 2000]
1) 1.5   2) 2   3) 2.5   4) 3

3. The ratio of the lengths of two rods is 4:3. The ratio of their coefficients of cubical expansion is 2:3. Then the ratio of their linear expansions when they are heated through same temperature difference is
1) 2:1 2) 1:2 3) 8:9 4) 9:8

4. The ratio of lengths of two rods is 1 : 2 and the ratio of coefficient of expansions is 2 : 3. The first rod is heated through 60^0C. Find the temperature through which the second rod is to be heated so that its expansion is twice that of first is
1) 60^0C 2) 40^0C 3) 30^0C 4) 10^0C

5. A wire of length 60 cm is bent into a circle with a gap of 1 cm at its ends. On heating it by 100°C, the length of the gap increases to 1.02 cm. α of material of wire is
1) 2×10^{-4}/°C 2) 4×10^{-4}/°C
3) 6×10^{-4}/°C 4) 1×10^{-4}/°C

6. A metal metre scale gives correct measurement at 0°C. It is generally used at a temperature of 40°C. The correction to be made for every metre is ($\alpha = 10^{-6}/1^0C$)
1) 4×10^{-5} m 2) 4×10^{-5} m to be added
3) 4×10^{-5} m must be deducted
4) None of the above

7. The temperature of a thin uniform rod increases by Δt. If moment of inertia I about an axis perpenducular to its length, then its moment of increases by
1) 0 2) $\alpha I \Delta t$ 3) $2\alpha I \Delta t$ 4) $\alpha^2 I \Delta t$

8. A metal rod has a length of 1m at 30°C. 'α' of metal is 2.5×10^{-5}/°C. The temperature at which it will be shortened by 1mm is
1) –30°C 2) –40°C 3) –10° C 4) 10°C

9. Upon heating, the length of the side of a cube changes by 2%. The volume of the cube changes by
1) 1% 2) 6 % 3) 0.5 % 4) 4 %

10. Radius of a sphere is 100cm at 0°C and 100.1cm at 100°C. Coefficient of cubical expansion of the sphere is
1) 30×10^{-6}/C 2) 10×10^{-6}/°C
3) 60×10^{-6}/°C 4) 20×10^{-6}/°C

11. A brass disc at 0°C has a diameter of 50Cm and a hole of diameter 10cm. For Brass a = 18×10^{-6}/°C. When the disc is heated to 100°C, the diameter of the hole becomes.
1) 10.018 cm 2) 10.180 cm
3) 10.8 cm 4) 10.81 cm

12. Density of a substance at 0°C is 10.6 gm/c.c and at 100°C is 10 gm/c.c. coefficient of linear expansion of solid is
1) 0.0006/°C 2) 0.0004/°C
3) 0.0003/°C 4) 0.0002/°C

13. A steel rod of dimesions 4 x $4cm^2$ is tightly fixed between two supports and is not allowed to expand. It is heated through 2°C. Thermal stress developed is ..... $10^6 N/m^2$ (Y= 20×10^{10} N/m^2 α = 18×10^{-6}/,C)
1) 7.2 2) 2.7 3) 3.6 4) 0.72

14. A length of 2m is measured using a metal tape at 10^0C. It is calibrated at a temperature 30^0C. The actual length is $\left(\alpha = 1 \times 10^{-4}/1C^0\right)$
1) 2.004 m 2) 1.996 m 3) 1.004 m 4) 0.004 cm

15. A metal tape gives correct measurement at 15^0C. It is used to measure a distance of 100m at 45^0C. The error in the measurement, if $\alpha = 12 \times 10^{-6}/1^0C$ is
1) 36 cm 2) 36 m 3) 42 mm 4) 36 mm

16. The ratio of the lengths of an iron rod and an aluminium rod for which the difference in the lengths is independent of temperature is (Coefficient of linear expansion of iron and aluminium are 12×10^{-6}/°C and 24×10^{-6}/°C respectively)
1) 2:1 2) 3:1 3) 1:2 4) 4:1

17. A piece of copper wire has a length of 2m at 10°C. Its length at 100°C is (Coefficient of linear expansion of copper = 17×10^{-6}/°C)
1) 20.00306 m 2) 2.00306 m
3) 4.000306m 4) 15.000306m

18. A piece of steel has a length of 30cm at 15°C. At 90°C its length increases by 0.027 cm. Its coefficient of linear expansion is
1) 15×10^{-6}/°C 2) 12×10^{-6}/°C
3) 15×15^{-6}/°C 4) 20×10^{-6}/°C

19. The density of lead at 0°C is 11.34 g/cm^3. The density of land at 100°C, if the coefficient of linear expansion of lead = 28×10^{-6}/°C is
1) 13.25 gm/cm^3 2) 17.25 gm/cm^3
3) 18.25 gm/cm^3 4) 11.25 gm/cm^3

20. A steel scale is correct at 0°C. The length of a brass tube measured by it at 40°C is 4.5m. The correct length of the tube at 0°C is (Coefficients of linear expansion of steel and brass are 11×10^{-6}/°C and 19×10^{-6}/°C respectively).
1) 4.001m 2) 5.001 m 3) 4.999m 4) 4.501m

21. A brass meter bar is correct at 20°C. The length of a rod as measured by it at 35°C is 50cm. The true length of the rod at 35°C is (coefficient of linear expansion of brass is 19 x 10-6/°C)

1) 40.01 m 2) 25.02 m
3)50.01 m 4) 45.01 m

22. Two thin metal strips, one of brass and the other of iron are fastened together parallel to each other, a distance 2 mm apart to form a bimetallic strip. If the strips are of equal length at 0°C. The radius of the arc formed by the bimetallic strip when heated to 80°C is (Coefficient of linear expansion of brass = 19×10^{-6}/°C and of iron = 12×10^{-6}/°C).

1) 1.50 m 2) 3.57 m
3) 5.57 m 4) 0.57 m

23. An iron ring 1.20 m in diameter is to be put on a wooden cart wheel. The diameter of this wheel is 1.0 cm greater than that of the ring. The temperature of the ring that must be raised before it slips on the wooden wheel is (Coefficient of linear expansion of iron = 12×10^{-6}/°C).

1) 600.4°C 2) 694.4°C
3) 294.4°C 4) 300.4°C

24. A steel bar 0.5 cm square (cross–sectional area) and 25 cm long at 30°C fits into the space between two fixed supports. If the bar is now heated to 230°C, the force it exerts against the supports is (Coefficient of linear expansion forsteel = 11×10^{-6}/°C and Young's modulus for steel 2×10^{11}N/m^2).

1) 11000N 2) 22000N
3) 33000N 4) 44000N

25. A clock, with a brass pendulum, keep correct time at 20°C, but loses 8.212 s per day, when the temperature rises to 30°C. The coefficient of linear expansion of brass is

1) 25×10^{-6}/°C 2) 19×10^{-6}/°C
3) 20×20^{-6}/°C 4) 11×10^{-6}/°C

ANSWERS

1) 2	2) 1	3) 3	4) 2	5) 1
6) 2	7) 3	8) 3	9) 2	10) 1
11) 1	12) 4	13) 1	14) 2	15) 4
16) 1	17) 2	18) 2	19) 4	20) 4
21) 3	22) 2	23) 2	24) 1	25) 2

EXERCISE- II(B)

1. Coefficient of linear expansion of a material is 5×10^{-5}/ ^{0}F. Then coeficient of cubical expansion of that material expressed per ^{0}C is

1) 0.000018/^{0}C 2) 0.00027/ ^{0}C
3) 0.00009/ ^{0}C 4) 0.000015/ ^{0}C

2. If the coefficient of volume expansion of a solid is 0.00027/°C its coefficient of areal expansion is

1) 0.00009/°C 2) 0.00018/°C
3) 0.00027/°C 4) 0.00003/°C

3. A crystal has linear coefficient of expansion 9×10^{-5}, 12×10^{-5}, 7×10^{-5}/k along three mutually perpendicular directions the volume expansion coefficient is

1) $27\times10^{-5}/k$ 2) $26\times10^{-5}/k$
3) $21\times10^{-5}/k$ 4) $28\times10^{-5}/k$

4. Two iron rods have their lengths in the ratio 5 : 3 and diameters in the ratio 2:1. When the rods are heated from 30°C to 100°C, the ratio of their expansions is

1) 5 : 3 2) 3 : 5 3) 5 : 6 4) 6 : 5

5. A wire 100cm long is bent into an arc of a circle with a gap of 2 cm at 0° C. On heating to 80° C, the gap is 2.02 cm .The liner coeffecent of expansion of the material of the wire is

1) 0.0000025/K 2) 0.0000125/K
3) 0.000125/K 4) none of these

6. A metal metre scale gives correct measurement at 40^{0}C. It is generally used at a temperature of 0^{0}C. Find the correction to be made for every metre. $(\alpha = 10^{-6}/1^0C)$

1) 4 X 10^{-5} m 2) 4 X 10^{-5} m to be added
3) 4 X 10^{-5} m must be deducted
4) None of the above.

7. When the temperature of a body increases from t to (t+Δt), its moment of inertia increases from I to (I+ΔI). The coefficient of linear expansion of the body is α ,then $\frac{\Delta I}{I}$

1) $\frac{\Delta t}{t}$ 2) $\frac{2\Delta t}{t}$ 3) $\alpha\Delta t$ 4) $2\alpha\Delta t$

8. A metal rod having a linear coefficient of expansion 2×10^{-5}/°C has a length 1m at 25°C, the temperature at which it is shortened by 1mm is

1) –25°C 2) –125°C 3) 125°C 4) –10°C

9. Coefficient of cubical expansion of a metal cube is γ. Increase in temperature for which the volume of the cube increases by 5% is

1) $0.05\ \gamma$ 2) $0.5\ \gamma$ 3) $0.5/\gamma$ 4) $0.05/\gamma$

10. A copper cube having a length of 10 cm. on each side is heated from 10°C to 110°C. If $\alpha = 12\text{x}10^{-6}/°C$, change in volume in c.c. is

1) 6.3 2) 0.63 3) 3.6 4) 0.36

11. The diameter of a metal ring is D and coefficient of linear expansion is α, if the temperature of the ring is increases by 1^0C, the circumference of the ring will increases by

1) $\frac{\pi D\alpha}{2}$ 2) $2\pi D\alpha$ 3) $\pi D\alpha$ 4) $\pi D^2\alpha$

12. Density of a substance at 0^0C is 10gm/cc, at 100C its density is 9.7gm/cc. The coeffcient of linear expansion of the substance is [98 M]

1) $10^{-4}/^0C$ 2) $3\times10^{-4}/^0C$

3) $10.7\times10^{-4}/^0C$ 4) $10^{-3}/^0C$

13. A wire of cross sectional area $4mm^2$ is fixed between two points at 30°C. Y = $2\text{x}10^{11}$ Pa and $\alpha = 10^{-5}/°C$; when the temperature falls to 20°C, the tension in the string in (newton) is,

1) 600 2) 60 3) 80 4) 800

14. A steel tape is correct at 20^0C. On a day when the temperature is 45^0C a distance of 210 m is measured. The real distance is

$\alpha = 1.2\times10^{-5}/1^0C$

1) 210.063 m 2) 0.063 m

3) 209.237 m 4) 210.63 m

15. An aluminium measuring rod, which is correct at 5^0C measures a certain distance as 88.42 cm at 35^0C. The error in measuring the distance due to the expansion of the rod is $(\alpha = 23\times10^{-6}/1^0C)$

1) 0.02 m 2) 0.03 m

3) 0.12 m 4) 0.06 m

16. The length of the steel rod is to be longer than the copper rod at any temperature by 5 cm. If the α for copper is $1.1 \text{ x } 10^{-5}/^0C$ and α for copper is $1.7 \text{ x } 10^{-5}/^0C$ then at 0^0C length (cm) of the steel rod and that copper rod should be respectively

1) [8.37, 3.37] 2) [10.27, 5.27]

3) [2.57, 7.57] 4) [14.17, 9.17]

17. The length of each rail is 10 m. The linear expansion of steel is $0.000012/^0C$ and the range of variation of temperature at the given place is 15^0C. So the gap to be provided between the rails is

1) 0.0018 m 2) 0.0012 m

3) 0.0022 m 4) 0.02 m

18. A wheel is 80.3 cm in circumference. An iron tyre measures 80.0 cm around its inner face. If the coefficient of linear expansion for iron is $12 \text{ x } 10^{-6}/^0C$, the temperature of the tyre must be raised by

1) 105 ^0C 2) 223 ^0C 3) 312 ^0C 4) 417 ^0C

19. A metal is heated from 0 ^0C to 500 ^0C and its density reduces to $\frac{1}{1.027}$ of its original density. The coefficient of linear expansion for this metal, considering it constant for the given range of temperature is (in $10^{-5}/^0C$)

1) 1.8 2) 2.7 3) 3.8 4) 6.2

20. The length of an iron rod is measured by a brass scale.When both of them are at 20 ^0C, the length is 80 cm. The length of the rod at 100 ^0C as measured by the scale at 100^0C is [The coefficient of linear expansion of brass and iron are $24 \text{ x } 10^{-6}\ (C^0)^{-1}$ and $18 \text{ x } 10^{-6}\ (C^0)^{-1}$ respectively.]

1) 69.9 cm 2) 79.96 cm

3) 89.9 cm 4) 99.9 cm

21. A brass scale of barometre gives correct reading at 2^0C. The coefficient of thermal expansion for brass is 0.00002 $^0C^{-1}$. The barometer reads 75 cm at 27^0C. The atmospheric pressure at 2^0C is

1) 72.9 cm 2) 73.9 cm

3) 74.96 cm 4) 76.9 cm

22. A bimetallic strip of thickness 2 cm consists of zinc and silver riveted together. The approximate radius of curvature of the strip when heated through 50^0C will be : (linear expansivity of zinc and silver are $32 \text{ x } 10^{-6}/^0C$ and $19 \text{ x } 10^{-6}/^0C$ respectively)

1) 30.77 m 2) 61.54 m 3) 15.38 m 4) 7.69 m

23. An iron tyre is to be fitted onto a wooden wheel 1m in diameter. The diameter of the tyre is 5 mm smaller than that of the wheel. The temperature of the tyre that should be increased is [α for iron = $12 \text{ x } 10^{-6}/^0C$.]

1) 418.8^0C 2) 419.8^0C

3) 420.8^0C 4) 421.8^0C

24. A steel ball initially at a pressure of 10^5 Pa is heated from 30^0C to 130^0C keeping its volume constant, The final pressure inside the ball is [coefficient of linear expansion of steel is 1.1×10^{-5} $(C^0)^{-1}$ and Bulk modulus of steel is 1.6 x 10^{11} N/m^2

1) 4.28 x 10^8 Pa 2) 5.28 x 10^8 Pa
3) 6.28 x 10^8 Pa 4) 7.28 x 10^8 Pa

25. A colck with an iron pendulum keeps xorrect time at 15 ^{0}C . If the coefficient of linear expansion of iron is 0.000012/^{0}C and the room temperature is 20^{0}C it loses in a day

1) 1.3 s 2) 2.6 s 3) 3.2 s 4) 3.8 s

ANSWERS

1) 2	2) 2	3) 4	4) 1	5) 3
6) 3	7) 4	8) 1	9) 4	10) 3
11) 3	12) 1	13) 3	14) 1	15) 4
16) 4	17) 1	18) 3	19) 1	20) 2
21) 3	22) 1	23) 1	24) 2	25) 2

EXERCISE- III(A)

1. The steam point and the ice point of a mercuary thermometer are wrongly marked as 92°c $92^0 C$ *and* $2^0 C$ respectively. The correct temperature read by this thermometer is

1) $2^0 C$ 2) $5^0 C$ 3) $10^0 C$ 4) $20^0 C$

2. A Fahrenheat thermometer reads $113^0 F$ while a fualty celsius thermometer reads $44^0 C$. The correction required to be applied to the celsius thermometer is

1) $-1^0 C$ 2) $+1^0 C$ 3) $+2^0 C$ 4) $-2^0 F$

3. On a hypothetical scale X, the ice point is 40^0 and the steam point is 120^0. For another scale Y the ice point and steam points are –30^0 and 130^0 respectively. If X-reads 50^0 The reading of Y is

1) –5^0 2) –8^0 3) –10^0 4) –12^0

4. Distance between two places is 200 km. α a of steel is 12 x 10^{-6}/°C. Total space that must be left between steel rails to allow for a change of temperature from 36°F to 117°F is

1) 1.08km 2) 0.108 km
3) 10.8 km 4) 0.0108 km

5. A steel tape is calibrated at 20°C. When the temperature of the day is –10°C, the percentage error in the measurement with the tape is (α=12 x 10^{-6}/°C)

1) 3.6% 2) 0.36% 3) 0.18% 4) 0.036%

6. A metallic solid sphere is rotating about its diameter as axis of rotation. If the temperature is increased by 200°C, the percentage increase in its moment of inertia is : [Coefficient of linear expansion of the metal = 10^{-5}/°C]

1) 0.1 2) 0.2 3) 0.3 4) 0.4

7. Two marks on a glass rod 10cm apart are found to increase their distance by 0.08mm when the rod is heated from $0^0 C$ *to* $100^0 C$. A flask made of the same glass as that rod measures a volume of 1000 c.c at 0^{0}C. The volume it measures at 100^{0}C in c.c. is [2003 M]

1) 1002.4 2) 1004.2 3) 1006.4 4) 1008.2

8. If a cylinder of diameter 1.0 cm at 30^{0}C is to be slide into a hole of diameter 0.9997 cm in a steel plate at the same temperature, then minimum requried rise in the temperature of the plate is [coefficient of linear expansion of steel=12x10^{-6}/^{0}C] [2001M]

1) $25^0 C$ 2) $35^0 C$
3) $45^0 C$ 4) $55^0 C$

9. A meter rod made of steel is correct at 0°C and other meter rod of same material is correct at 30°C. If α = 12 x 10^{-6}/°C, the difference between their lengths at 20°C is

1) 0.024 cm 2) 0.012 cm
3) 0.036 cm 4) 0.048 cm

10. The ratio of lengths of two rods of different materials is taken as 2 : 3 such that the difference of their lengths is 10 cm at all temperatures. The length of the rod of lesser expansion is

1) 10 cm 2) 20 cm 3) 30 cm 4) 40 cm

11. At 0°C a square steel bar of 1 cm side is rigidly clamped at both ends so that its length cannot increase. Young's modulus of steel is 20x10^{10} N m^{-2} and its coefficient of linear expansion is 11 x 10^{-6}/°C. When the temperature is raised to 10°C, the force exerted on the clamps is

1) 2200 N 2) 3000 N
3) 2400 N 4) 5200 N

12. Two rods of different materials and identical cross sectional areas, are joined face to face at one end and their free ends are fixed to the rigid walls. If the temperature of the surroundings is increased by 30°C, the magnitude of the displacement of the joint of the rod is (length of rods $l_1=l_2=1$unit, ratio of their young's modulii, $y_1/y_2=2$, coefficients of tinear expansion are α_1 and α_2) [2006 E]

1) $5(\alpha_2-\alpha_1)$ 2) $10(\alpha_1-\alpha_2)$

3) $10(\alpha_2-2\alpha_1)$ 4) $5(2\alpha_1-\alpha_2)$

13. Two uniform metal rods of length L_1 and L_2 and their linear coefficients of expansion α_1 and α_2 respectively, are connected to form a single rod of length (L_1+L_2). When the temperature of the combined rod is raised by "t°C", the length of each rod increased by the same amount then $\left[\frac{\alpha_2}{\alpha_1+\alpha_2}\right]$ is [2005 M]

1) $\frac{L_1}{L_1+L_2}$ 2) $\frac{L_1+L_2}{L_1}$ 3) $\frac{L_2}{L_1+L_2}$ 4) $\frac{L_1+L_2}{L_2}$

14. A thin brass sheet at 10°C and a thin steel sheet at 20°C have the same surface area. The common temperature at which both would have the same area is (Coefficient of linear expansion for brass and steel are respectively, 19×10^{-6}/°C are 11×10^{-6}/°C) [2003M]

1) –3.75°C 2) –2.75°C 3) 2.75°C 4) 3.75°C

15. A pendulum clock is 5 seconds fast at a temperature of 15°C and 10 seconds slow at a temperature of 30°C. The temperature at which it gives the correct time is

1) $18^{\circ}C$ 2) $20^{\circ}C$ 3) $22^{\circ}C$ 4) $25^{\circ}C$

16. A pendulum clock gives correct time at 20°C at a place where g= 10m/s². The pendulum consists of a light steel rod connected to a heavy ball. If it is taken to a different place where g = 10.01m/s² at what temperature the pendulum gives correct time (α of steel is 10^{-5}/°C)

1) 30°C 2) 60°C 3) 100°C 4) 120°C

17. A steel meter scale is to be ruled so that millimeter intervals are accurate within about 5×10^{-5} mm at a certain temperature. The maximum temperature variation allowable during the ruling is [coefficient of linear expansion of steel = $10\times10^{-6}k^{-1}$] [2001 E]

1) 2°C 2) 5°C 3) 7°C 4) 10°C

18. An iron ball of diameter 6cm and is 0.01 mm too large to pass through a hole in a brass plate when the ball and plate are at a temperature of 20°C. The temperature at which (both for ball and plate) the ball just pass through the hole is $\left(\alpha_{iron}=12\times10^{-6}/1^{0}C;\ \alpha_{brass}=18\times10^{-6}/1^{0}C\right)$

1) 68°C 2) 48°C 3) 28°C 4) 40°C

19. A glass rod is measured with a zinc scale, both being at 20°c appear to be a metre long. If the scale is correct at 0°c, what is the true length of the glass rod at 0°C ? α of zinc = 0.000029 / °C, α of glass = 0.0000085/°C

1) 80.6 cm 2) 90.4 cm

3) 100.04 cm 4) 121.6 cm

20. Steel rod at 40°C is observed to be 1 m long when measured by a brass scale which is correct at 0°C. correct length of the brass rod at 0° C is ($\alpha_{steel}=12\times10^{-6}/1^{0}C$ and $\alpha_{Brass}=19\times10^{-6}/1^{0}C$)

1) 99.89 cm 2) 98.99 cm

3) 99.97 cm 4) 89.6 cm

ANSWERS

1) 4	2) 2	3) 3	4) 2	5) 4
6) 4	7) 1	8) 1	9) 3	10) 3
11) 1	12) 3	13) 1	14) 1	15) 2
16) 4	17) 2	18) 2	19) 3	20) 3

EXERCISE- III(B)

1. The upper and lower fixed points of a faulty mercury thermometer are 210°F and 34°F respectively. The correct temperature read by this thermometer is

1) 22°F 2) 80°F 3) 100°F 4) 122°F

2. When a celsius thermometer reads 90°C a faulty Fahrenheit thermometer reads 190°F. The correction to be made in the latter scale is

1) +2°F 2) –2°F

3) –4°F 4) +4°F

3. On a hypothetical scale A the ice point is 42° and the steam points is 182° For another scale B. The ice point is –10° and steam point in 90°. If B reads 60°. The reading of A is.

1) 160^{0} 2) 140^{0} 3) 120^{0} 4) 110^{0}

4. The length of each rail is 10m. The linear expansion of steel is 12×10^{-6}/°C and range of variation of temperature at the given place is 15°C. The gap to be left between the rails is

1) 0.0018 m 2) 0.0012 m

3) 0.00022 m 4) 0.05 m

5. A steel tape is calibrated at 20^0C . In summer when the temperature is 40^0C the percentage error in the tape is (α *of steel* $= 11\times10^{-6}/^0C$)
1) 0.22% 2) 0.022% 3) 2.2% 4) 22%

6. The temperature of a thin uniform circular disc, of one metre diameter is increased by 10°C. The percentage increase in moment of inertia of the disc about an axis passing through its centre and perpendicular to the circular face (linear coefficient of expansion = 11×10^{-6}/°C) [2006 E]
1) 0.0055 2) 0.011 3) 0.022 4) 0.044

7. Two marks on a glass rod 10cm apart are found to increase their distance by 0.06mm when the rod is heated from 0^0C to 10^0C. A flask made of the same glass as that rod measures a volume of 1000 c.c at 0^0C. The volume it measures at 100^0C in c.c. is
1) 1018 cc 2) 918 cc
3) 818 cc 4) 718 cc

8. An iron tyre is to be fitted onto a wooden wheel of 1m diameter. The diameter of tyre is 6 mm smaller than that of the wheel. The tyre should be heated so that its temperature increases by a minimum of (coefficient of volume expansion of iron is 3.6×10^{-5}/°C)
1) 500°C 2) 334°C 3) 167°C 4) 1000°C

9. One steel meter bar is correct at 0^0C and another at 25^0 C. the diffennce between their lengths at 20^0C is $\left(\alpha = 12\times10^{-6}/^0C\right)$
1) 0.02 cm 2) 0.03 cm 3) 0.3 cm 4) 0.2 cm

10. The coefficients of linear expansions of steel and brass are 11×10^{-6} / °C and 19×10^{-6} / °C respectively. If their difference in lengths at all temperatures has to be kept constant at 30 cm, their lengths at 0^0C should be
1) 83 cm, 52 cm 2) 71.25cm, 41.25 cm
3) 92 cm, 62 cm 4) 62.25 cm, 32.25 cm

11. A steel rod of length 5m is fixed rigidly between two supports. α of steel = 12×10^{-6}/°C, Y = $2\times10^{11}Nm^{-2}$. With the increase in its temperature by 40° C, the stress developed in the rod is
1) $9.6\times10^5Nm^{-2}$ 2) $9.6 \times 10^6Nm^{-2}$
3) $9.6 \times 10^7Nm^{-2}$ 4) $9.6 \times 10^8 Nm^{-2}$

12. There is some change in length when 33000N tensile force is applied on a steel rod of area of cross section $10^{-3}m^2$. The change in temperature required to produce the same elongation if the steel rod is heated is (The modulus of elasticity is $3 \times 10^{11}N/m^2$ and the coefficient of linear expansion of steel is $1.1 \times 10^{-5}/^0C$) 2008 E
1) 20^0C 2) 15^0C 3) 10^0C 4) 0^0C

13. A steel wire AB of length 100 cm is fixed rigidly at points A and B in an aluminium frame as shown in the figure. If the temperature of the system increases through 100^0C, then the excess stress produced in the steel wire relative to the aluminium?

$\alpha_{Al} = 22\times10^{-6}/1^0C$ and $\alpha_{steel} = 11\times10^{-6}/1^0C$

young's Modulus of steel is 2×10^{11} Nm^{-2}.

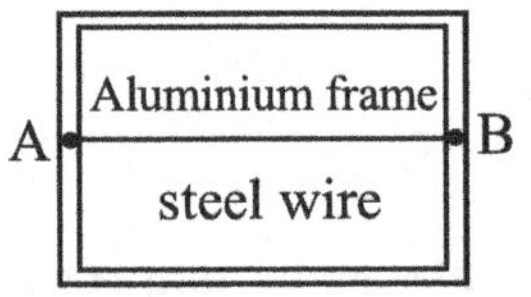

1) 2.2×10^8 Pa 2) 22×10^8 Pa
3) 0.2×10^8 Pa 4) 220×10^8 Pa

14. An iron rod of length 50cm is joined end to end to an aluminium rod of length 100cm. All measurements refer to 20^0C. The coefficients of linear expansion of iron and aluminium are $12 \times 10^{-6}/^0C$ and $24 \times 10^{-6}/^0C$ respectively. The average coefficient of composite rod is
1) $20 \times 10^{-6}/^0C$ 2) $36 \times 10^{-6}/^0C$
3) $12 \times 10^{-6}/^0C$ 4) $48 \times 10^{-6}/^0C$

15. A thin brass sheet at 20°C and a thin steel sheet at 30°C have the same surface area. The common temperature at which both would have the same area is (Coefficient of linear expansion for brass and steel are respectively, 19×10^{-6}/°C are 11×10^{-6}/°C)
1) – 6.25^0C 2) + 6.25^0C
3) – 3.25^0C 4) + 3.25^0C

16. A pendulum of brass has period of 1 second at 20^0C. Coefficient of linear expansion of brass is $1.96 \times 10^{-5}/^0C$. The clock gets delayed in a weak at 30^0C by
1) 8s 2) 59s 3) 224s 4) 504s

17. A metre scale made of steel is calibrated at 20^0C to give correct reading. Find the distance between 50 cm mark and 51 cm mark if the scale is used at 10^0C. Coefficient of linear expansion of steel is $1.1 \times 10^{-5}/^0C$

1) 1.00011 cm
2) 1.0011 cm
3) 1.011 cm
4) 1.000011 cm

18. A steel rod is 4 cm in diameter at 30^0C. A brass ring has an interior diameter of 3.992 cm at 30^0 C. In order that the ring just slides on to the steel rod, the common temperature of the two rods should be
$\left(\alpha_s = 11\times10^{-6}/1^0C;\ \alpha_b = 19\times10^{-6}/1^0C\right)$

1) 200^0C
2) 350^0C
3) 280^0C
4) 300^0C

19. A steel scale measures the length of a copper wire as 80.0cm, When both are at 20^0 C (the calibration temperature for scale). What would be the scale read for the length of the wire when both are at 40^0C (Given $\alpha_{steel} = 11x10^{-6}/^0C$, $\alpha_{copper} = 17x10^{-6}/^0C$)

1) 80.0096 cm
2) 80.0272 cm
3) 1 cm
4) 25.2 cm

20. A zinc rod when measured with brass scale being at 30^0C appears to be 1 m long. If the scale is correct at 0^0C, what is the true length of the zinc rod at 0^0C ? (α of zinc = $0.000029/^0C$ and α_b = $0.000019/1^0C$)

1) 99.97 cm
2) 100.03 cm
3) 91.9 cm
4) 101.2 cm

ANSWERS

1) 4	2) 4	3) 2	4) 1	5) 2
6) 3	7) 1	8) 1	9) 2	10) 2
11) 3	12) 3	13) 1	14) 1	15) 2
16) 2	17) 1	18) 3	19) 1	20) 2

Chapter - 6

THERMAL PROPERTIES OF MATTER

EXPANSION OF LIQUIDS

SYNOPSIS

1. Liquids have definite volume, but do not have definite shape. So, they have only cubical expansion.
2. When a liquid is heated in a container, the container also expands. Hence liquid have two types of cubical expansions. They are apparent and absolute or real cubical expansions.
   Expansion of liquid relative to the container is called apparent expansion.
   Expansion of liquid independent of expansion of container is called real expansion.
   Depending upon these expansions, liquids have two coefficients of expansion,namely coefficient of apparent expansion and coefficient of real expansion.
3. Coefficient of real expansion, $\gamma_R = \frac{v_2 - v_1}{v_1(t_2 - t_1)}$
   Here $v_2 - v_1$ is real increase in volume.
4. Coefficient of apparent expansion
   $$\gamma_a = \frac{v_2 - v_1}{v_1(t_2 - t_1)}$$
   Here $v_2 - v_1$= apparent increase in volume.
5. Relation between the two coefficients is $\gamma_R = \gamma_a + \gamma_g = \gamma_a + 3\alpha$ where γ_g is coefficient of cubical expansion of container and α is coefficient of linear expansion of container.
6. $\gamma_a = \gamma_R - \gamma_g$
   i) If γ_g is +ve and $\gamma_a < \gamma_R$, liquid level raises.
   ii) If $\gamma_g = \gamma_R$, liqiud level remains constant.
   iii) If γ_g is negative then liquid level increases from the begining and increases more rapidly than in case (i).
   iv) If γ_g is +ve and $\gamma_g > \gamma_R$, liquid level falls.
7. If V is the volume of the vessel and V_L is volume of mercury to be taken, for the free space volume to be constant at all temperatures then

(Dilatometer)

$$V(\gamma_g) = V_L(\gamma_R) \qquad \text{ie; } \frac{V}{V_L} = \frac{\gamma_R}{\gamma_g} \approx 7$$

8. Expansion of mercury is nearly 7 times that of glass. If glass vessel is filled with 1/7th of its volume with mercury, the remaining volume of air remains constant at all temperatures.
9. γ_a is determined using specific gravity bottle or pyknometer or weight thermometer using the formula,
   $$\gamma_a = \frac{\text{mass of liquid expelled}}{\text{mass of liquid remained x Temp. difference}}$$
   $$\gamma_a = \frac{(m_2 - m_3)}{(m_3 - m_1)(t_2 - t_1)}$$
   Where m_1 = mass of empty bottle
   m_2 = mass of bottle + full liquid
   m_3 = mass of bottle + remaing liquids
   Cofficient of real expansion $\gamma_R = \gamma_a + \gamma_g$
10. Variation of density of liquid with temperature.
   $$d_2 = \frac{d_1}{1+\gamma_R(t_2 - t_1)} \text{(or)} d_t = \frac{d_0}{1+\gamma_R t}$$
   The density of liquid decreases with increase of temperature.
   $$\gamma_R = \frac{d - d_t}{d_t \cdot t} \quad or \quad \gamma_R = \frac{d_1 - d_2}{d_2(t_2 - t_1)}$$

11. Anomalous expansion of water

a) Liquids expand on heating. But water at 0^0C on heating contracts upto 4^0C. It expands beyond 4^0C like any other liquid . On cooling, water contracts from 100^0C to 4^0C and expands from 4^0C to 0^0C. This is called anomalous expansion of water.
b) When water freezes into ice, it expands. A few other substances such as iron, bismuth, antimony expand when they solidify and such substances are useful for costing.
c) The anomalous expansion of water is demonstrated by Hope's apparatus.
d) This is due to formation of more number of hydrogen bonds.

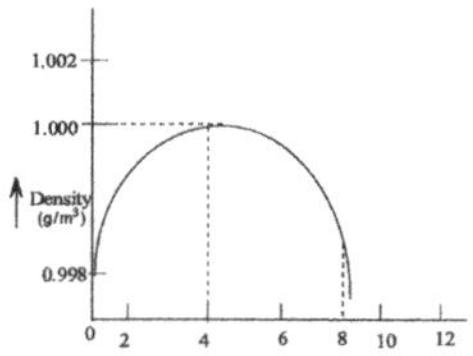

12. If K_1 and K_2 be the fractions of volume of a metal submerged in mercury at t_1^0C and t_2^0C respectively (γ_g and γ_R be the volume coefficient of metal and mercury) then
   $$\Rightarrow \gamma_L = \frac{K_2 - K_1}{K_1 \Delta t} + \frac{K_2}{K_1}\gamma_s$$

EXERCIXE - I

MULTIPLE CHOICE QUESTIONS:

1. **Density of liquid at any temperature 't' is given by $d_t = d_0/(1+\gamma t)$. This equation applies to**

1) all liquids 2) all liquids except water
3) no liquid 4) all liquids except mercury

2. **A block of wood is floating on water at 0°C with certain volume 'p' below the water, when the temperature of water is slowly raised to 10°C, the value of 'p'**

1) Increase upto 4°C and then decreases
2) Increases continuously
3) Decreases continuously
4) Decreases upto 4°C and then increases

3. **A block of wood is floating on water at 0^0C with a certain volume V above water level. The temperature of water is slowly raised from 0^0C to 20^0C. How the volume V change with the rise of temperature**

1) V will remain unchanged
2) V will go decreasing from begining to the end
3) V will decrease till the temperature of water reaches at 4^0C and then it will go on decreasing.
4) V will increase till the temperature of water reaches to 4^0C and then it will go on decreasing

4. **Coefficient of volume expansion of a vessel is greater than the liquid it contains. If the vessel is heated for long time, the level of liquid in the vessel**

1) falls 2) rises
3) remains same
4) first rises and then falls

5. **A liquid with coefficient of volume expansion γ is filled in a container of a material having the coefficient of linear expansion α. If the liquid overflows on heating**

1) $\gamma = 3\alpha$ 2) $\gamma > 3\alpha$
3) $\gamma < 3\alpha$ 4) $\gamma > 3\alpha^3$

6. **The substance having highest coefficient of cubical expansion in the following is**

1) mercury 2) water 3) steel 4) air

7. **The loss of weight of a solid when inmersed in a liquid at 0^0C is m_0 and m_1 at t_1^0C. If γ_S and γ_L are the coefficients of cubical expansions of solid and liquid then m_1 is equal to**

1) $m_0(1-(\gamma_s-\gamma_L)t$ 2) $m_0(1+(\gamma_s-\gamma_L)t$

3) $m_0[(\gamma_S-\gamma_L)]t$ 4) $\dfrac{m_0 t}{(\gamma_S-\gamma_L)}$

8. **When a liquid is heated in a vessel, it is found that γ_R is less than γ_A. This happens when the vessel ----- on heating.**

1) expands 2) contracts
3) does not change 4) none

9. **In an experiment it is observed that for a liquid $\gamma_{Abs} = \gamma_{App}$. Then the vessel ---- on heating.**

1) does not expand 2) does not contract
3) does not expand or contract
4) expands

10. **When a liquid, taken in a long cylindrical vessel of material with linear coeffficient of expansion 'α', is heated, the level of liquid did not change. The volume coeffcient of expansion of liquid is**

1) 3α 2) 2α 3) α 4) 4α

11. **A metal ball immersed in alcohol weighs w_1 at 0°C and w_2 at 50°C. Assuming that the density of the metal is large compared to that of alcohol, then**

1) $w_1 = w_2$ 2) $w_1 > w_2$
3) $w_1 < w_2$ 4) none

12. **A solid is suspended in a liquid. If the liquid is heated, the solid (assume the solid expansion is negligible)**

1) floats 2) sinks
3) no change 4) nothing can be said

13. **A vessel is partly filled with a liquid. Coefficients of cubical expansion of material of the vessel and liquid are γ_g and γ_L respectively. If the system is heated, then the volume unoccupied by the liquid necessarily**

1) decreases if $\gamma_g = \gamma_L$ and $\gamma_g < \gamma_L$
2) increases if $\gamma_g = \gamma_L$ and $\gamma_g < \gamma_L$
3) increases if $\gamma_g = \gamma_L$ and $\gamma_g > \gamma_L$
4) decreases if $\gamma_g = \gamma_L$ and $\gamma_g > \gamma_L$

14. When water is heated from 0°C to 10°C, its volume

1) increases 2) decreases
3) does not change
4) first decreases and then increases

15. By increasing the temperature of a liquid, its

1) volume and density decreases
2) volume and density increase
3) volume increases and density decreases
4) volume decreases and density increases

16. Apparent expansion of liquid does not depend on

1) nature of the liquid 2) colour of the liquid
3) change in temperature 4) material of vessel

17. Volume coefficient of expansion of water is

1) max. at 4°C 2) minimum at 4°C
3) zero at 4°C 4) negative at 4°C

18. Among the following , the liquid having negative coefficient of cubical expansion between 273K and 277K is

1) Ethylene Glycol 2) Mercury
3) Water 4) None

19. In cold countries, water does not freeze at the bottom of a lake in winter, on account of

1) anomalous expansion of water between 4°C and 0°C
2) ice being lighter than water
3) ice being poor conductor of heat
4) all the above three factors

20. The surface water in a lake is going to freeze. Now the temperature of water at the bottom is

1) 274 K 2) 277 K 3) 100 K 4) 0K

21. Water at 4°C is filled to the brim of two beakers A and B of glass of negligible coefficient of expansion. A boy performed experiments by heating the beaker A and cooling the beaker B. He observes

1) Level of water in 'B' went down and water in 'A' flows out
2) Water flows out in both cases
3) Water flows out from B and water level went down in beaker A.
4) In both cases water level went down

MULTIPLE ANSWERS & MATCHING TYPE

22. When a liquid in a vessel is heated its level

(a) will increase
(b) may increase or decreases
(c) will decrease
(d) may remains same

1) only (a) is correct 2) only (b) is correct
3) (c) is correct 4) (b) & (d) are correct

23. certain volume of a liquid is taken in a long glass tube and its temperature is increased at a uniform rate, the rate of increase in the length of the liquid depends on

(a) length of the liquid coloumn
(b) area of cross section of the glass tube
(c) coefficient of expansion of glass

1) only (a) is correct
2) (a) & (b) are correct
3) (b) & (c) are correct
4) (a) (b) & (c) are correct

24. A liquid of coefficient of real expansion γ is partly filled in a vessel of coefficient of linear expansion $\gamma/3$. When the system is heated, then

(a) the volume of space above liquid remains same
(b) the level of liquid relative to vessel remains same
(c) the fraction of volume of liquid in vessel remains same

1) only (a) is correct
2) only (b) & (c) are correct
3) only (c) is true
4) all are true

25. Apparent expansion of a liquid dependsupon

(a) nature of liquid
(b) nature of vessel
(c) temperature rise
(d) scale of temperature

1) only (a) is true
2) (a) & (b) are true
3) (a), (b) & (c) are true
4) (a), (b), (c) & (d) are true

26. A metal ball suspended from a spring balance is immersed in water at 4^0C. If the temperature of water is changed the reading in the balance

(a) may decreases (b) increases

(c) may remains same

1) only (c) is true 2) (b) is true

3) (a) & (c) are true 4) (b) & (c) are true

27. When a liquid in a vessel is heated, its level continuously rises. Then

(a) γ_R is +ve & $\gamma_g = 0$

(b) γ_R is +ve & γ_g is -ve

(c) $\gamma_R = \gamma_A$ (d) $\gamma_R = \gamma_g$

1) (a) is true 2) (a), (b) and (c) are true

3) (c) & (d) are true 4) (b) & (d) are true

28.

List - I	List - II
a. Apparent expansion	e. Nature of vessel & Liquid
b. Real expansion	f. Nature of liquid
c. γ_a	g. Vessel, Liquid & temperature
d. γ_r	h. liquid & temperature

1) a - e, b - f, c - g, d - h

2) a - f, b - g, c - h, d - e

3) a - g, b - h, c - e, d - f

4) a - h, b - e, c - f, d - g

29.

List - I	List - II
a. mass of liquid expelled on heating	e. $\dfrac{m_1}{1+\gamma_a\Delta t}$
b. corrected Barometric height	f. $\dfrac{d_1-d_2}{d_2(t_2-t_1)}$
c. coefficient of real expansion of a liquid	g. $\dfrac{\gamma_a m_1\Delta t}{1+\gamma_a\Delta t}$
d. mass of liquid remaining on heating	h. $h_0[1-(\gamma_r-\alpha)\Delta t]$

1) a - g, b - h, c - f, d - e

2) a - h, b - e, c - g, d - h

3) a - e, b - f, c - g, d - h

4) a - f, b - g, c - h, d - e

30.

List - I	List - II
a. γ_g is +ve & < γ_r	e. liquid level does not change
b. γ_g is -ve	f. liquid level increases continuously
c. $\gamma_g = \gamma_r$	g. liquidlevel decreases
d. $\gamma_g > \gamma_r$	h. liquid level first decreases and then increases

1) a - g, b - e, c - f, d - h

2) a - h, b - f, c - e, d - g

3) a - e, b - f, c - g, d - h

4) a - f, b - g, c - h, d - e

31.

List - I	List - II
a. Temperature of water at the bottom of a lake	e. 273K
b. γ_r of water is -ve between	f. 277K
c. γ_r of water is positive	g. above 277K
d. Temperature of water just below ice layer in a Lake	h. 273K and 277K

1) a - g, b - h, c - e, d - f

2) a - h, b - e, c - f, d - g

3) a - e, b - f, c - g, d - h

4) a - f, b - h, c - g, d - e

ASSERTION & REASON TYPE QUESTIONS

Instruction For Questions :

In each of the following questions, a statement of Assertion (A) is given followed by a corresponding statement of reason (R) just below it. Of the statement mark the correct answer.

1) Both 'A' and 'R' are true and 'R' is the correct explanation of 'A'

2) Both 'A' and 'R' are true and 'R' is not the correct explanation of 'A'

3)'A' is true and 'R' is false

4)'A' is false and 'R' is true

32. A) : Real expansion of liquid does not depend upon material of container

R) : Liquids have no definite shape. They acquire the shape of their containers.

33. A) : 10 litres of Benzene will weigh more in winter than in summer

R) : Due to increase in temperautre, the density of Benzene is more in summer-

34. A) : If w_1, w_2, w_3 are the weights of given body in air, in water at 20^0C and in water at 50^0C respectively, then $w_1 > w_2 > w_3$

R) : Buoyant force due to water decreases in the temperature range of 20^0C to 50^0C

35. A) : A vessel is filled, with water, up to brim at 4^0C. It over flows when the system is cooled are heated

R) : Water has minimum volume at 4^0C.

36. A) : A block is floating on surface of water at 0^0C. As temperature increases from 0^0C to 4^0C, volume of block inside the water decreases.

R) : Density of water is maximum at 4^0C

37. A) : Bottles filled with water, when kept in deep freezer, break.

R) : Water expands on freezing.

38. A) : Aquatic animals, at bottom of water lake, survive in cold countries even when the surface is frozen.

R) : Due to anomalous expansion, the density of water is minimum at 4^0C

39. A) : Water at 4^0C expands both on heating and cooling

R) : H_2O has least volume at 4^0C due to the molecular orientation

40 A) : Anamalous expansion of water is favorable for acquatic animals

R) : The lowest possible temperature at bottom of water lake is 4^0C

ANSWERS

1) 2	2) 4	3) 4	4) 1	5) 2
6) 4	7) 2	8) 2	9) 3	10) 1
11) 3	12) 2	13) 3	14) 4	15) 3
16) 2	17) 3	18) 3	19) 4	20) 2
21) 2	22) 4	23) 4	24) 4	25) 4
26) 4	27) 2	28) 3	29) 1	30) 2
31) 4	32) 2	33) 3	34) 4	35) 1
36) 1	37) 1	38) 3	39) 1	40) 1

EXERCISE– II (A)

1. A vessel is half filled with a liquid at 0°C. When the vessel is heated to 100°C, the liquid occupies 3/4 volume of the vessel. Coefficient of apparent expansion of the liquid is

1) 0.5/°C 2) 0.05/°C
3) 0.005/°C 4) 0.0005/°C

2. A liquid column of height 80cm at 0°C balances the same liquid of height 80.4cm at 100°C. γ_R is

1) 4×10^{-5}/°C 2) 3×10^{-5}/°C
3) 5×10^{-5}/°C 4) 6×10^{-5}/°C

3. γ_A of liquid is 7/8 of γ_R of liquid. α_g of vessel is

1) $\frac{\gamma_R}{8}$ 2) $\frac{\gamma_R}{12}$ 3) $\frac{\gamma_R}{24}$ 4) $\frac{\gamma_R}{36}$

4. Coefficient of real expansion of a liquid is 0.000182/°C. If coefficient of linear expansion of vessel is 0.000009/°C, coefficient of apparent expansion of the liqud is

1) 0.000209/°C 2) 0.000155/°C
3) 0.000173/°C 4) 0.000182/°C

5. Co-efficient of apparent expansions of a liquid in Gold vessel is G and when heated in a silver vessel is S. If coefficient of linear expansion of Gold is A, coefficient of linear expansion of Silver is

1) $\frac{G+S-3A}{3}$ 2) $\frac{G-S+3A}{3}$
3) $\frac{G+S+3A}{3}$ 4) $\frac{G-S+A}{3}$

6. A liquid occupies half of a vessel at a particular temperature. The volume of the unoccupied part remains constant at all temperatures. If α and γ are the coefficients of linear and real expansions of a vessel and liquid, then γ =

1) 3α 2) $3\alpha/2$
3) 6α 4) 9α

7. Densities of a liquid at 30°C and 80°C are 0.6gm/c.c and 0.54 gm/c.c. The coefficient of real expansion of liquid is

1) 0.0002/°C 2) 0.002/°C
3) 0.001/°C 4) 0.0001/°C

8. If on heating a liquid through 80^0C, the mass expelled is $\frac{1}{100}th$ of mass still remaining, the coefficient of apparent expansion of the liquid is [99 E]

1) $12.6\times10^{-4}/^0C$ 2) $0.8\times10^{-4}/^0C$
3) $1.25\times10^{-5}/^0C$ 4) $1.25\times10^{-4}/^0C$

9. When 82 gm of a liquid at 10°C in a specific gravity bottle is heated to 85°C, 2gm of liquid is expelled. Coefficient of apparent expansion of liquid is

1) 0.0033/°C 2) 0.0027/°C
3) 0.000303/°C 4) 0.000333/°C

10. A glass vessel just holds 50gm of a liquid at 0°C. If the coefficient of linear expansion of glass is $8\times10^{-6}/^0C$. The mass of the liquid it holds at 80°C is [coefficient of absolute expansion of liquid $=5\times10^{-4}/^0C$] [96 E]

1) 46 gm 2) 48 gm 3) 51gm 4) 42 gm

11. When a specific gravity bottle filled with mercury at 0° is heated to 100°C, 5.190g of mercury overflows and 300g remains in the specific gravity bottle. Using glycerine in place of mercury, 1.439 of glycerine overflows and 27.333 g remains. The coefficient of real expansion of mercury is 0.000183/°C, then the coefficient of real expansion of glycerine is

1) 5.4×10^{-4}/°C 2) 3.3×10^{-4}/°C
3) 2.4×10^{-4}/°C 4) 10.4×10^{-4}/°C

12. A glass vessel which holds exactly 1000 g of mercury at 15°C is put into boiling water at 100°C. mass of mercury expelled is (Coefficient of absolute expansion of mercury is 0.00018/°C; coefficient of linear expansion of glass is 0.00001/°C).

1)12.60gm 2)24.78gm
3)40.80gm 4)14.78 gm

13. A barometer with a brass scale correct at 0°C reads 70 cm of mercury on a day when the air temperature is 40°C. the correct reading at 0°C is (Coefficient of real expansion of mercury is 0.00018/°C and the coefficient of linear expansion of brass is 0.000018/°C.

1) 60.5cm 2)69.5 cm
3) 20.5cm 4) 50.00cm

14. A specific gravity bottle is completely filled by 271.92 g of mercury at 0°C. the mass of the mercury which would fill in specific gravity bottle at 100°C. is (Coefficient of linear expansion of the material of the specific gravity bottle 8.0 x 10^{-6}/°C and coefficient of real expansion of mercury = 0.000182/°C).

1)267 g 2)277g 3)500g 4)300g

ANSWERS

1) 3	2) 3	3) 3	4) 2	5) 2
6) 3	7) 2	8) 4	9) 4	10) 2
11) 1	12) 1	13) 2	14) 1	

EXERCISE-II (B)

1. A liquid occupy 1/6 of volume at 0°C and 1/5 of volume of vessel at 200°C. γ_A of the liquid is

1) 2×10^{-3}/°C 2) 1×10^{-3}/°C
3) 3×10^{-3}/°C 4) 1.5×10^{-3}/°C

2. In an experiment, to find the coefficient of real expansion of Hg, 70 cm column of Hg at 0°C is found to balance 71.26 cm column of Hg at 100°C. γ_R of mercury is —$\times10^{-6}$/°C

1) 180 2) 1.8 3) 81 4) 8.1

3. For a liquid when heated in a vessel it is found that $\gamma_A=\frac{6}{7}\gamma_R$. Coefficient of linear expansion of the vessel is

1) $\frac{\gamma_R}{21}$ 2) $\frac{\gamma_R}{11}$ 3) $\frac{\gamma_R}{12}$ 4) $\frac{\gamma_R}{14}$

4. When a liquid in a glass vessel is heated, its apparent expansion is $10.3\times10^{-4}/^0C$. Same liquid when heated in a metal, its apparent expansion is $10.06\times10^{-4}/^0C$. The coeffcient of linear expansion of the metal is (α of glass = $9\times10^{-6}/^0C$) [2000 E]

1) $51\times10^{-6}/^0C$ 2) $43\times10^{-6}/^0C$
3) $25\times10^{-6}/^0C$ 4) $17\times10^{-6}/^0C$

5. Coefficient of apparent expansion of a liquid when heated in Brass vessel is B and when heated in Aluminium vessel is A. If coefficient of linear expansion of Brass is x, coefficient of linear expansion of Aluminium is

1) $\frac{B+X-A}{3}$ 2) $\frac{B+3X-A}{3}$
3) $\frac{A+B-3X}{3}$ 4) $\frac{A-B+3X}{3}$

6. Two litre glass flask contains some mercury. It is found that at all temperatures the volume of the air inside the flask remains the same. The volume of the mercury in side the flask is (α for glass = $9\times10^{-6}/^{\circ}C$, γ for mercury = $1.8\times10^{-4}/^{\circ}C$) [2008 M]
1) 1500cc 2) 150cc 3) 3000cc 4) 300cc

7. The coefficinet of real expansion of Hg is $0.18\times10^{-3}/^{0}C$. If the density of Hg at 0°C is 13.6 gn/c.c its density at $200^{0}C$ will be
1) 13.3 gm/c.c 2) 13.13 gm/c.c [85 E]
3) 13.6 gm/c.c 4) 13 gm/c.c

8. If the temperature is increased by 40°C, the mass of liquid expelled is 1/50 of the mass of the remaining liquid. Coefficient of apparent expansion of the liquid is
1) 0.05/°C 2) 0.005/°C
3) 0.0005/°C 4) 0.00005/°C

9. If 52 gm of a liquid is heated in a vessel from 0°C to 100°C . 2 gm of the liquid is expelled. Then the coefficient of apparant expansion of liquid is
1) $4 \times 10^{-4}/ ^{0}C$ 2) $4 \times 10^{-5}/ ^{0}C$
3) $0.25 \times 10^{-4}/ ^{0}C$ 4) $25 \times 10^{-4}/ ^{0}C$

10. A specific gravity bottle contains 100 gm of liquid at 100°C. γ_A of liquid is $6\times10^{-5}/^{0}C$. The mass of the liquid that bottle just holds at 0°C.
1) 100.6 gm 2) 100 gm
3) 106 gm4) 90 gm

11. A container of capacity 700 ml is filled with two different liquids of volume 200 ml and 500 ml with respective volume coefficients as $1.4 \times 10^{-5}/^{0}C$ and $2.1 \times 10^{-5}/^{0}C$ During the heating of the vessel, it is obseved that neither any liquid overflows nor any empty space is created. The volume coefficient of the container is
1) $1.9\times10^{-5}/^{0}C$ 2) $1.9\times10^{-6}/^{0}C$
3) $1.6\times10^{-5}/^{0}C$ 4) $1.6\times10^{-6}/^{0}C$

12. A glass bulb of volume 250cc. is completely filled with mercury at 20°C. The temperature of the system is raised to 100°C. If the coeficient of linear expansiuon of glass is $9 \times 10^{-6}/^{0}C$ and coefficient of absolute expansion of mercury is 1.8×10^{-4}, volume of mercury that overflows is
1) 1.83 cc 2) 2.56 cc 3) 3.06 cc 4) 415 cc

13. The height of the mercury column in a barometer provided with a brass scale corrected at 0°C is observed to be 74.9 cm at 15°C. Find the true height of the column at 15°C. $[\alpha_{brass}=20\times10^{-6}/^{0}C \text{ and } \gamma_{Hg}=175\times10^{-6}/^{0}C]$
1) 74.92 cm 2) 79.92 cm 3) 74.12 cm 4) 72.64 cm

14. A glass vessel just holds 50 gm of toulene at 0°C. If the coefficient of absolute expansion of toulene is $14.76 \times 10^{-4}/^{0}C$ the mass of toulene it holds at 60°C is [α of glass = $9 \times 10^{-6}/^{0}C$]
1) 28 g 2) 35 g 3) 41 g 4) 46 g

ANSWERS

1) 2	2) 1	3) 1	4) 4	5) 2
6) 4	7) 2	8) 3	9) 1	10) 1
11) 1	12) 3	13) 1	14) 4	

EXERCISE– III (A)

1. A mercury thermometer contains 2c.c. of Hg. at 0°C. Distance between 0°C and 100°C marks on the stem is 35cm and diameter of the bore is 0.02cm. γ_A of liquid is
1) 0.000055/°C 2) 0.000066/°C
3) 0.00055/°C 4) 0.000058/°C

2. A cylindrical tube of length 100cm. Contains some mercury inside. If the length of remaining air column remains constant at all temperatures, the height of mercury is (neglect the expansion of cross section)
(γ_R of Hg = $18\times10^{-5}/^{\circ}C$, γ_g of tube = $2.7\times10^{-5}/^{\circ}C$)
1) 10 cm 2) 15 cm 3) 7.5cm 4) 12 cm

3. coefficient of real expansion of mercury is $0.18\times10^{-3}/^{0}C$. If the density of mercury at 0°C is 13.6 gm/c.c., its density at 573K will be [96 E]
1) 12.90 gm/c.c 2) 13.10 gm/c.c
3) 12.95 gm/c.c 4) 12.75 gm/c.c

4. If coefficient of real expansion of a liquid is 1/5500, the temperature at which its density is 1% less than density at 0°C is
1) 55.5°C 2) 100°C 3) 99°C 4) 1°C

5. A pyknometer weights 40gm when empty and 1040 gm when filled with mercury at 0°C. On heating to 100°C, 10 gm of mercury over flows. If the coefficient of real expansion of mercury is 0.0002/°C. The coefficient of cubical expansion of glass is [94 E]
1) $0.00001/^{0}C$ 2) $0.0003/^{0}C$
3) $0.0002/^{0}C$ 4) $0.0001/^{0}C$

6. A glass flask of volume 200cm³ is completely filled with mercury at 20°C. When the flask is heated to 80°C, the amount of mercury that spilt over is (γ_g of glass is 27x10⁻⁶/°C, γ_R of Hg=0.18x10⁻³/°C)
1) 2.16 cm³ 2) 0.32 cm³
3) 1.83 cm³ 4) 2.9 cm³

7. A specific gravity bottle contains 101.6gm. of liquid at 30°C. γ_A of liquid is 16x10⁻⁵/°C. When it is heated to 130°C, mass of liquid expelled is
1) 100 gm 2) 1.6 gm
3) 16 gm 4)1.06 gm

8. A glass bulb of volume 250 c.c is completely filled with mercury at 20°C. The temperature of the sysem is raised to 100°C. If the coefficient of linear expansion of glass is 9 x 10⁻⁶/°C and coefficient of absolute expansion of mercury is 1.8 x 10⁻⁴/°C, the volume of mercury that overflows is nearly [83 E]
1) 3.06 c.c 2) 30.6 c.c
3) 2.5 c.c 4) 25 c.c

9. When a block of iron floats in mercury at 0^0C, a fraction K_1 of its volume is submerged, while at the temperature 60^0C, a fraction K_2 is seen to be submerged. If the coefficient of volume expansion of iron is γ_{Fe} and that of mercury is γ_{Hg} then the ratio $\frac{K_1}{K_2}$ can be expressed as

1) $\frac{1+60\gamma_{Fe}}{1+60\gamma_{Hg}}$ 2) $\frac{1-60\gamma_{Fe}}{1+60\gamma_{Hg}}$

3) $\frac{1+60\gamma_{Fe}}{1-60\gamma_{Hg}}$ 4) $\frac{1+60\gamma_{Hg}}{1+60\gamma_{Fe}}$

10. A piece of metal weighs 46 gm in air. When immerssed in a liquid of specific gravity 1.24, at 27°C it weighs 30 gm. When the temperature of the liquid is raised to 42°C the metal piece weighs 30.5 gm. Specific gravity of liquid at 42°C is 1.20, the coeffcient of linear expansion of solid
1) 2.32 x 10⁻⁵ / °C 2) 6.94 x 10⁻⁵ / °C
3) 4.46 x 10⁻⁵ / °C 4) 5.2 x 10⁻⁵ / °C

ANSWERS

1) 1	2) 2	3) 1	4) 1	5) 4
6) 3	7) 2	8) 1	9) 1	10) 2

EXERCISE– III (B)

1. Volume of mercury in the bulb of a thermometer is 10⁻⁶m³. Area of cross section of capillary tube is 2x10⁻⁷m². If the temperature is raised by 100°C, the length of mercury column is (γ_{Hg}=18x10⁻⁵/°C) [97E]
1) 9 cm 2) 9 mm
3) 18 cm 4) 5 cm

2. A vessel contains a liquid filled with 1/10th of its volume. Another vessel contains same liquid upto 1/8th of its volume. In both case the volume of empty space remains constant at all temperatures. Then the ratio of coefficient of linear expansions of the two vessels is
1) 2 : 5 2) 5 : 2 3) 4 : 5 4) 5 : 4

3. Coefficient of real expansion of turpentine is $9.6\times10^{-4}/^0C$. If its density at 0°C is 0.87 gm/c.c., its density at 373K is
1) 0.79 gm/c.c. 2) 0.74 gm/c.c.
3) 0.7 gm/c.c. 4) 0.84 gm/c.c.

4. Density of a liquid at 0°C is 13.6 gm/c.c., and its coefficient of real expansion is18x10⁻⁵/°C. If the temperature is increased to 200°C, percentage change in its density is
1) 0.36%2) 3.6% 3) 36% 4) 24.4%

5. A glass bottle weighs 50gm and 160gm when filled with a liquid. When it is heated by 100°C, 10gm of liquid overflows. If γ_g is 2x10⁻³/°C, coefficient of cubical expansion of container is
1) (1/3)x10⁻⁴/°C 2) 3x10⁻⁴/°C
3) 6x10⁻⁴/°C 4) 1x10⁻³/°C

6. A specific gravity bottle is filled to the brim with mercury of 400 gm at 0°C. When heated to 90°C, the mass of the mercury that overflows from the specific gravity bottle is [Coefficient of apparent expansion of mercury in glass $=\frac{1}{6500}/^0C$ [2001 M]
1) 5.46 gm 2) 6.54 gm
3) 10.92 gm 4) 13.08 gm

7. 50gm of alcohol is needed to completely fill up a weight thermometer at 20°C. Find the mass of the alcohol which will overflow when it is heated to 30°C temperature. $(\gamma_R = 10.5\times10^{-4}/^0C)$
1) 5.2 gm 2) 0.1 gm
3)0.52 gm 4) 0.05gm

8. **A glass bulb of volume 250cc is completely filled with mercury at 20⁰C. Temperature of the system is raised to 110⁰C . If the coefficient of linear expansion of glass is $9 \times 10^{-6}/^0C$ and the coefficient of absolute expansion of mercury is $1.8 \times 10^{-4}/^0C$. Volume of mercury that overflows is**

1) 3.44cc　　2) 4.34cc
3) 7.88 cc　　4) 6.88 cc

9. **The densities of wood and benzene at 0⁰C are 880 kg/m³ and 900 kg/m³ respectively. The coefficients of volume expansion are 1.2×10^{-3} / ⁰C for wood and 1.5×10^{-3} / ⁰C for benzene. At what temperature will a piece of wood just sink in benzene ?**

1) 83⁰C　2) 73⁰C　3) 63⁰C　4) 53⁰C

10. **A sphere of diameter 7.0 cm and mass 275 g floats in a bath of liquid. As the temperature is raised, the sphere begins to sink at a temperature of 40⁰C. If the density of the liquid is 1.5 g/cm³ at 0⁰C, find the coefficient of cubical expansion of the liquid. Neglect the expansion of the sphere.**

1) 5×10^{-4} / ⁰C　　2) 5×10^{-5} / ⁰C
3) 5×10^{-6} / ⁰C　　4) 5×10^{-3} / ⁰C

ANSWERS

1) 1	2) 3	3) 1	4) 2	5) 4
6) 1	7) 3	8) 1	9) 1	10) 1

COMPREHENSION QUESTIONS

Comprehension - III (8 - 10) :

The anamalous behaviour of water has been explained on the assumption that there are three types of water molecules H_2O,$(H_2O)_2$ and $(H_2O)_3$. These molecules have different volumes per unit mass and are mixed in different proportions at different temperatures so that the maximum density is at 4⁰C due to greater proportion of higher types.

1. **The volume of water at 4⁰C is minimum because the presence of**

1) more number of water molecules of type H_2O
2) more number of Hydrogen bonds
3) Less number of water molecules of type $(H_2O)_2$ and $(H_2O)_3$
4) Less number of Hydrogen bonds between the water molecules

2. **The coefficient real expansion of water at 0⁰C**

1) positive　　2) negative
3) zero　　4) may take any value

3. **Water is taken at 4^0C in a long necked flask made up of silica of negligible expansion. The level of water**

1) first rises and then falls on heating
2) first falls and then rises on cooling
3) rises either on heating or cooling
4) falls either on heating or cooling

ANSWERS

1) 2　　2) 3　　3) 3

Chapter - 7

THERMO DYNAMICS

THERMO DYNAMICS

SYNOPSIS

CALORIEMETRY:

HEAT:

1. Heat is a form of energy that transfers from one body to the other due to temperature difference between them.

Calorie: It is defined as the amount of heat required to rise the temperature of 1g of water through 1^0C (more specifically from 14.5 to 15.5^0C)

* In solids thermal energy is present in the form of kinetic energy, in liquids in the form of translational energy of molecules. In gases it is due to the randum motion of molecules.

Specific heat:

* The amount of heat energy required to rise the temperature of unit mass of a substance through 1^0C(or K) is called specific heat of the substance.
* Specific heat of a substance also depends upon its state i.e solid, liquid and gas.

  In case of water :

  Specific heat of Ice (solid) =0.5 cal/g-c^0(or) 2100 $\frac{J}{kg-k}$

  Specific heat of water (liquid) = 1 cal/g-c^0 (or) 4200 $\frac{J}{kg-k}$

  Specific heat of steam (gas) = 0.47 cal/g-c^0 (or) 1970 $\frac{J}{kg-k}$

* Specific heat of substance at its melting point or boiling point is infinite $\left(as\,\Delta T = 0\right)$
* Specific heat of a substance can also be negative
* Among solids, liquids and gases, specific heat is maximum for H_2 gas

Thermal capcity:

* The heat required to raise the temperature of a given substance by 1^0C is called its thermal capacity.
* Thermal capacity of a substance depends on its mass and nature.

Water - equivalent:

* Water - equivalent of a substance is the mass of water which when given same amount of heat as given to the substance, changes the temperature of water through same range as that of the substance.
* Water equivalent of a substance is expressed in gm of water

Latent heat: The amount of heat required to change the state of unit mass of a substance at a constant temperature is called the latent heat.

* If 'Q' heat changes the state of mass 'm' of a substance at constant temperature, then latent heat

  $$\boxed{L = \frac{Q}{m}} \Rightarrow \boxed{Q = mL}$$

* Latent heat depends on nature and state of substance
* The latent heat during melting or freezing is called the latent heat of fusion while during boiling or condensation is called latent heat of vaporisation

  Note(1): If a solid such as camphor directly converts into gaseous phase at all temperatures the process is called sublimation.

  Note(2): If vapours by releasing heat directly changes into solid state without passing through liquid phase, the process is called

  "Hoar forst"

* In melting or boiling heat supplied is used to increase the internal potential energy of the substance while internal kinetic energy remains constant.
* Clausius and clapeyron has shown the effect of pressure on boiling point of liquids and melting points of solids is related to latent heat heat L(cal/g) thorugh

  $\frac{dP}{dT} = \frac{JL}{T_K\left(V_F - V_I\right)}$ where V_F and V_I are final and initial specific volumes (cc/g).

* In case of boiling as volume of vapours V_F is always greater than volume of liquid V_I. hence $\frac{dP}{dT}$ is positive. i.e boiling point of every liquid rises with increase in pressure
* In case of melting for substances like wax and sulphur which expands on melting,

  $V_F > V_I$ and hence melting point rises with increase in pressure.

* In case of substances like ice and rubber which contracts on melting $V_F < V_I$ and hence $V_F - V_I$ negative i.e melting point is lowered with increase in pressure.
* Addition of imputities changes melting point, and boiling point

Triple point of water

* Triple point is a point in the phase diagram where a solid, liquid and vapour can co -exit (or will be in equilibrium)
* For water the values of pressure and temperature corresponding to triple point are 4.58 mm of Hg (0.006 atm) and 273.16K respectively.

Note: The negative slope of Ice line shows that melting point of Ice decreases with the increase of pressure.

Principle of method of mixtures:

* When two substances at different temperatures are mixed, heat will be transferred from substance at higher temperature to a substance at lower temperature till both acquire same temperature. Heat lost = Heat gained
* Principle of calorimetry represents the law of conservation of heat energy.
* The temperature of mixture can never be lesser than lower temperature and can never be greater than higher temperature $\boxed{T_L \le T_{mix} \le T_H}$
* If 'm' g of steam at 100°C is mixed with 'm' g of ice at 0°C then

  a) Resultant temperature of mixture is 100°C

  b) Mass of steam condensed = $\frac{m}{3}g$

  c) Mass of steam left uncondensed = $\frac{2m}{3}g$

  d) The finial mixture contains $\frac{4m}{3}g$ of water and $\frac{2m}{3}g$ of steam both at 100°C

THERMODNAMICS

* Thermodynamics is a branch of physics which deals with transformation of heat energy into other forms of energy or vice - versa
* Thermodynamic system may be three types

  a) **Open sytem:** It exchanges both energy and matter with the surrounding

  b) **Closed system:** It exchanges only energy but not matter with the system

  c) **Isolated system:** It exchange neither energy nor matter with the surroundings.
* **Thermodynamical variables:** These are macroscopic physical quantities like pressure (P), volume (V) and temperature (T) etc, which are used to describe the state of the system.
* The relation between the thermodynamical variables is called equation of state.
* **Thermal equilibrium:** It two systems are at same temperature they are said to be in thermal equilibrium.
* **Thermodynamic equilibrium:** A system is said to be in thermodynamic equilibrium when it is in a state of thermal, mechanical and chemical equilibrium

Thermodynamic process: If the state of a system changes in such a way that any of P. V, T etc changes, the process is called thermodynamic process.

Cyclic process: It is a process in which the system after passing through a series of changes comes back to its initial state.

Quasi - static process: A quasi-static process can be defined as an infinitesimally slow process in which the system remains in thermal and mechanical (thermodynamic) equilibrium with the surroundings at each and every intermediate stage

Reversible process : A process that can be retraced back in the opposite direction in such a way that the system passes through the same states as in the direct process, and finally the system and the surroundings return to their original states, with no other change anywhere else is called a reversible process.

The conditions required for a process to be reversible are

i) The changes must take place at an infinitesimally slow rate.

ii) The system must always be in thermal and mechanical equilibrium with the surroundings.

Irreversible Process : 'A process that can not be retraced back in the opposite direction' is called an irreversible process. In an irreversible process the system does not pass through the same intermediate states as in the direct process, even if the same initial state is reached.

Zeroth law of thermodynamics :

* "If systems 'A' and 'B' are each in thermal equilibrium with a third system 'C', then A and B are in thermal equilibrium with each other"
* This law leads to the concept of temperature
* It defines temperature as the property of the system which equals that of another system when the two systems are in thermal equilibrium

Joule's law and mechanical equivalent of heat :

* Joule showed that when mechanical energy (work) is converted into heat, the ratio of work done (W) to heat produced (Q) always remains a constant.

  $\frac{W}{Q}$ = constant = J (or) W = JQ

The constant J is called mechanical equivalent of heat

* If a body of mass 'm' falls from height 'h' and its P.E. is converted into heat and is completely taken by the body then

$$mgh = J(ms\,\Delta\theta),\ \Delta\theta = \frac{gh}{JS}$$

* If that body falls from height h_1 and rebounds to height h_2, then $mg(h_1 - h_2) = J(ms\,\Delta\theta)$

$$\Delta\theta = \frac{g(h_1 - h_2)}{JS}$$

* If a block of ice is dropped from a height h which completely melts on reaching the ground. then mgh $= JmL. \Rightarrow h = \frac{JL}{g}$

If only $\frac{1}{n}^{th}$ of mass melts then

$$mgh = J\frac{m}{n}L \Rightarrow h = \frac{JL}{ng}$$

* When a body of mass 'm' moving with velocity V is stopped and all its K.E. is converted into heat energy and is taken by body, then

$$\frac{1}{2}mV^2 = J(ms\theta) \Rightarrow \Delta\theta = \frac{V^2}{2JS}$$

* If a bullet just melts when stopped by an obstacle and if all the heat produced is absorbed by the bullet, then

$$\frac{1}{2}mV^2 = J(ms\Delta\theta + mL) \quad \therefore V^2 = 2J(S\Delta\theta + L)$$

* If an ice block of mass M is dragged on a rough horizontal surface with a constant velocity v, through a distance 'd' and workdone against friction is used as heat which melts ice of mass m then

$$\mu Mgd = JmL$$

$$\Rightarrow m = \frac{\mu Mgd}{JL}$$

Heat, Internal Energy and work in Thermodynamics:

Heat (dQ) : It is the energy that is transferred between a system and its surrounding (or) between two systems because of the temperature difference.

* Heat is a path dependent quantity

Internal energy (U) :

* Internal energy of a system is the energy possessed by the system due to molecular motion and molecular configuration.
* The energy due to molecular motion is internal kinetic energy U_k and that due to molecular configaration is internal potential energy U_p

$U = U_k + U_p$

* In case of ideal gas as there is no molecular attraction, Up = 0, i.e. internal energy of an ideal gas is totally kinetic
* Change in internal energy (dU) is path independent and depends only on the initial and final states of the system. i.e $dU = U_f - U_i$

Work (dw) :

* Work can be defined as the energy that is transferred from one body to the other owing to a force that acts between them
* The amount of external work done by a system as it expands (or) contracts is given by

$$W = \int dW = \int_{V_1}^{V_2} PdV$$

* If 'P' is constant then $W = P(v_2 - v_1)$
* Work is path dependent
* Work is positive during expansion and negative during compression.
* In a cyclic process work done is equal to the area under the cycle.

In this case W is + ve if the cycle is clockwise and W is – ve if the cycle is anticlockwise.

First law of thermodynamics :

* First law of thermodynamics is a statement of conservation of energy in thermodynamical process.
* According to this law, heat given to a system (dQ) is equal to the sum of increase in its internal energy (dU) and the work done (dW) by the system against surroundings.

$$\boxed{dQ = dU + dW}$$

* The first law enables us to define a property of a system called internal energy.
* dQ and dW are path dependent but dU(=dQ–dW) is path independent.

Sign convection :

dQ is +ve when heat is supplied

dQ is –ve when heat is rejected

dU is +ve when temperature increases

dU is –ve when temperature decreases

dW is +ve when work is done by system

dW is –ve when work is done on system

Specific heats of gas :

* The specific heat of a gas at constant volume is defined as the quantity of heat required to raise the temperature of unit mass of gas through 1 K when its volume is kept constant i.e.,

$$c_V = \frac{(\Delta Q)_V}{m\Delta T}$$

* The specific heat of a gas at constant pressure is defined as the quantity of heat required to raise the temperature of unit mass of gas through 1 K when its pressure is kept constant, i.e.,

$$c_P = \frac{(\Delta Q)_P}{m\Delta T}$$

* If instead of unit mass, 1 mole of a gas is considered the specific heats are called molar specific heats, they are denoted by C_V and C_P

$$C_V = \frac{(\Delta Q)_V}{n\Delta T},\ C_P = \frac{(\Delta Q)_P}{n\Delta T}$$

* $C_v = M\,c_v$, $C_p = M\,c_p$
  for c_v, c_p units are joule/kg/K
  for C_V, C_P units are joule/mole/K
* The heat supplied to the gas at constant volume is used only in increasing the temperature (internal energy of the gas)

$$(dQ)_V = dU = nC_VdT$$

* when heat is supplied to a gas at constant pressure, it is used to increase the temperature of the gas and in doing work due to expansion at constant pressure.

$$(dQ)_P = dU + dW$$
$$(dQ)_P = nC_PdT$$

* The relation between c_v, c_p is

  $c_p - c_v = r$ where r is specific gas constant
* Relation between C_V, C_P is
  $C_P - C_V = R$ where R is universal gas constant.
* If for a gas having F degrees of freedom $\gamma = 1 + \frac{2}{F}$

  Hence $C_V = \left(\frac{F}{2}\right)R$, $C_P = \left(1 + \frac{F}{2}\right)R$
* for monatomic gases, like He, A, as F =3

$$C_V = \frac{3}{2}R\ ;\ C_P = \frac{5}{2}R \text{ and } \gamma = \frac{5}{3} = 1.66$$

* for diatomic gases like H_2, O_2 as F = 5

$$C_V = \frac{5}{2}R\ ;\ C_P = \frac{7}{2}R \text{ and } \gamma = \frac{7}{5} = 1.4$$

* for polyatomic gases like NH_3, CO_2 as F = 6

$$C_V = 3R\ ;\ C_P = 4R \text{ and } \gamma = \frac{4}{3} = 1.33$$

* If n_1 moles of one gas mixed with n_2 moles of another gas then for the mixture

$$C_{V_{mix}} = \frac{n_1(C_V)_1 + n_2(C_V)_2}{n_1 + n_2}$$

$$C_{P_{mix}} = \frac{n_1(C_P)_1 + n_2(C_P)_2}{n_1 + n_2}$$

* When a gas is heated at constant pressure.

$$\frac{dU}{dQ} = \frac{1}{\gamma},\ \frac{dW}{dQ} = 1 - \frac{1}{\gamma}$$

Different thermodynamical processes :

Isobaric process :

* When a thermodynamic system undergoes a physical change in such a way that its pressure remains constant, then the change is known as isobaric process.
* Work done in isobaric process is
  $W = P(V_f - V_i) = nR(T_f - T_i)$
* first law of thermodynamics in this case gives $dQ = dU + P\Delta V$
* In this case $dQ = nC_P\,(dT)$

Isochoric process

* When a thermodynamic process undergoes a physical change in such a way that its volume remains constant, then the change is known as isochoric process.
* Work done during this process is $W = 0$
* First law of thermodynamics when applied in this case, $dQ = dU$ i.e., its heat supplied is used to increase the internal energy only (in this case $dQ = nC_VdT$).
* During isochoric heating pressure increases, temperature increases (dQ is +ve, dU is +ve).
* During isochoric cooling pressure decreases, temperature decreases (dQ is –ve, dU is –ve).

Isothermal process

* It is a thermodynamical process in which the temperature of the system remains constant.
* The equation of state for this process is
  PV = constant.
* System is thermally conducting to the surroundings.
* This process takes place slowly.
* In this process internal energy remains constant, i.e., U = constant, dU = 0.
* PV diagram for the process is shown below :

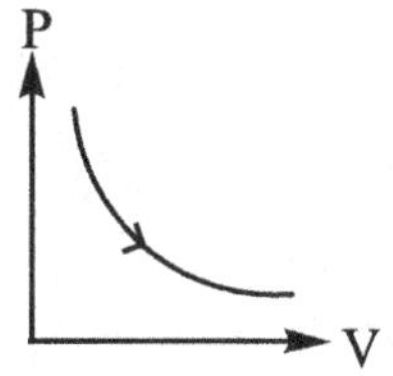

* The slope of isothermal curve is

  Slope $= -\left(\frac{P}{V}\right)$

* Work done during isothermal process is

$$W = nRT \log_e\left(\frac{V_f}{V_i}\right) = 2.303 nRT \log_{10}\left(\frac{V_f}{V_i}\right)$$

$$W = nRT \log_e\left(\frac{P_i}{P_f}\right) = 2.303 nRT \log_{10}\left(\frac{P_i}{P_f}\right)$$

* First law of thermodynamics when applied to this process, dQ = dW i.e., heat supplied is used only to do work.

* Isothermal elasticity $E_\theta = P$

* Specific heat during this process is infinite

Adiabatic Process: If the state of a system changes in such a way that there is no exchange of heat between system and its surroundings, the process is said to be adiabatic.

* System is thermally insulated from the surroundings

* This process takes place quickly

* The equation of state for this process is $PV^\gamma =$ constant

For temperature and volume

$TV^{\gamma-1}$=constant $\Rightarrow T_1V_1^{\gamma-1} = T_2V_2^{\gamma-1}$ or $T \propto V^{1-\gamma}$

For temperature and pressure

$\frac{T^\gamma}{P^{\gamma-1}}$=constant$\Rightarrow T_1^\gamma P_1^{1-\gamma} = T_2^\gamma P_2^{1-\gamma}$ or $T \propto P^{\frac{\gamma-1}{\gamma}}$ or

$P \propto T^{\frac{\gamma}{\gamma-1}}$

* PV diagram for the process is shown below

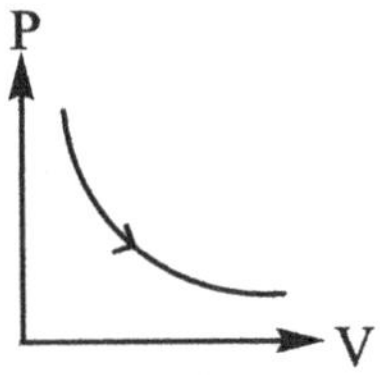

* Slope of adiabatic cure is

$$\text{Slope} = -\gamma\left(\frac{P}{V}\right)$$

* Work done in adiabatic process is

$$W = \frac{P_iV_i - P_fV_f}{\gamma - 1} = \frac{nR(T_i - T_f)}{\gamma - 1} = nC_v(T_i - T_f)$$

* First law of thermodynamics when applied to this process, 0=dU+dW $\Rightarrow$ dU= –dW

* During adiabatic expansion internal energy and temperature decreases

* During adiabatic comperassion internal energy and temperature increases

* Adiabatic elasticity $E_\phi = \gamma P$

* Specific heat of a gas during adabatic change is zero

Comparison between isothermal and adiabatic indicator diagrams:

* Adiabatic curves are more steeper than isothermal curves

i) When a gas expands from some initial volume to final volume, the final pressure is less for adiabatic expansion.

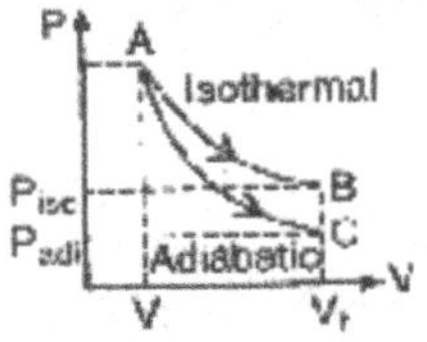

ii) When a gas compresses from some initial volume to a final volume, the final pressure is more in case of adiabatic compression.

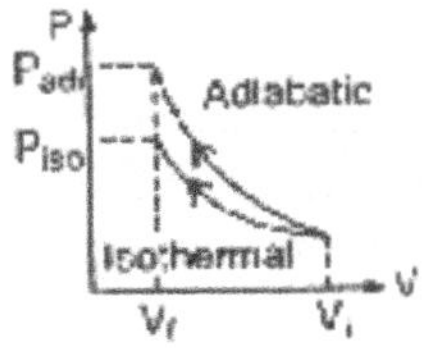

* Graphs of compression process

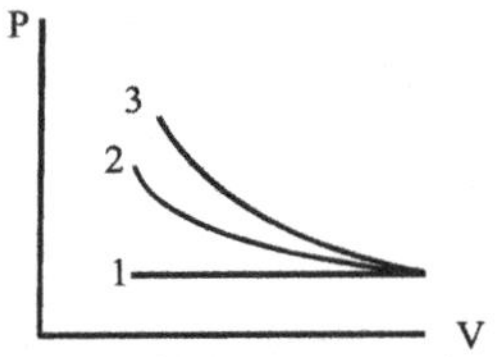

1) isobaric process
2) isothermal process
3) Adiabatic process

a) Work done in adiabatic > isothermal > isobaric process for same change in volume.

* Graphs of expansion process

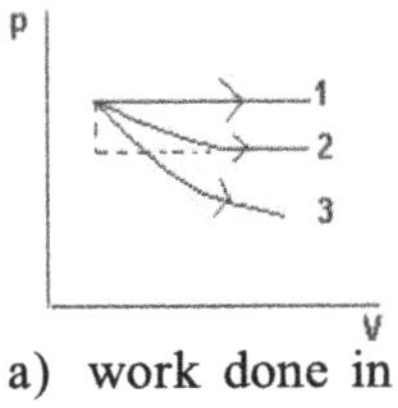

1: isobaric process
2: isothermal process
3: Adiabatic process

a) work done in isobaric process > isothermal >Adiabatic for same change in volume.

Free expansion: A container is divided into two parts.One part is vacuum and the other part contains a gas.If the gas expands into vacuum then it is called free expansion.

here dW=0

If dQ=0 then dU=0

HEAT ENGINE :

A heat engine is a device which converts thermal energy to other useful forms of energy such as mechanical energy, electrical energy.

$\Rightarrow$ Efficiency ,

$$\eta = \frac{W}{Q_1} = \frac{Q_1 - Q_2}{Q_1} = 1 - \frac{Q_2}{Q_1} = 1 - \frac{T_2}{T_1}$$

According to this, efficiency is 100% if $Q_2 = 0$, that is, no heat is rejected to the cold reservoir or sink that is the entire heat absorbed must be converted to mechanical work, which according to Second Law of Thermodynamics is impossible.

In practice, heat engines convert only a fraction of absorbed heat into mechanical work. For example a good automobile engine has an efficiency of about 20% and diesel engines have efficiencies ranging from 35% to 40%. On the basis of this fact the Kelvin-Plank form of Second Law of Thermodynamics states the following.

"no heat engine, operating in a cycle, can absorb thermal energy from a reservoir and perform an equal amount of work. This is equivalent to say that it is impossible to construct a perpetual motion machine."

REFRIGERATOR :

It is a heat Engine running in reverse.

Coefficient of performance,

$$\beta = \frac{Q_2}{W} = \frac{Q_2}{Q_1 - Q_2} = \frac{T_2}{T_1 - T_2}$$

In practice it is desirable to carry out this process with minimum amount of work. If it could be accompanied without doing any work, we would have a "**perfect refrigerator**" which is again in violation with Second Law of Thermodynamics, (Claussius statement).

RELATION BETWEEN η AND β

$$\eta = \frac{1}{1+\beta}$$

CARNOT ENGINE :

Efficiency of Carnot Engine,

$$\eta = 1 - \frac{Q_2}{Q_1} = 1 - \frac{T_2}{T_1} \quad \text{also}$$

In a Carnot cycle $\frac{Q_1}{T_1} = \frac{Q_2}{T_2}$

If an engine consists of a number of sources and sinks, then $\left(\sum \frac{Q}{T}\right)_{source} = \left(\sum \frac{Q}{T}\right)_{sink}$

Second law of Thermodynamics:

Planck statement : It is impossible to construct a heat engine which can completely convert heat energy into mechanical energy without rejecting heat to surroundings. 100% efficiency is not possible.

Kelvin statement :

It is impossible to extract work from a system by cooling it below surrounding temperature.

Classius statement :

It is impossible to transfer heat energy from body at lower temeperature to body at higher temperature unaided by external agency.

Both are identical.

EXERCISE - I

MULTIPLE CHOICE

1. **If temperature scale is changed from centigrade to Fahrenheit, the numerical value of specific heat will**

   1) Increase 2) Decrease
   3) Remains same 4) Doubled

2. **A large ice berg melts at the base and not at the top because**

   1) the base of the iceberg is slightly warm
   2) Ice at the base contains impurities
   3) Lowering of melting point of ice at the bottom due to high pressure
   4) Elevation of melting point of ice at the bottom due to high pressure

3. **During melting process, the heat given to a solid is used in (generally)**

   1) Increasing the temperature
   2) Increasing the density of material
   3) Increasing the average distance between the molecules
   4) Increasing the average K.E. of the molecules

4. **Paraffin wax expands on melting. The melting point of wax will**

   1) Increase with pressure
   2) Decrease with pressure
   3) Not change with pressure
   4) Decrease linearly with pressure

5. **Water is used in car radiators as coolant because of**

   1) Its density is more
   2) High specific heat
   3) High thermal conductivity
   4) Free availability

6. **In the following, the specific heat is minimum for**

   1) water 2) coconut oil
   3) kerosene 4) mercury

7. **A glass of water contains ice cubes floating on it. When all ice melts the level of water in the glass**
1) Increases 2) Decreases
3) Remains same 4) Doubled

8. **When two pieces of ice are pressed together, they combine to form one single piece. This process is called.**
1) Freezing 2) Regelation
3) combination 4) Sublimation

9. **Which of the following pairs of physical quantities may be represented by the same unit?**
1) Heat and temperature
2) Temperature and mole
3) Heat and work 4) specific heat and heat

10. **Which of the following pairs represent units of the same physical quanitity?**
1) kelvin and Joule 2) kelvin and calorie
3) newton and calorie 4) joule and calorie

11. **The temperature of an object is observed to rise in a period. During this period**
1) Heat must be supplied to it
2) heat may have been supplied to it
3) work may have been done on it
4) 2 & 3

12. **Of the following specific heat is maximum for**
1) Mercury 2) Copper
3) Water 4) Silver

13. **Some quantity of heat is given to a substance. The quantity which determines the temperature of the body is**
1) Mass 2) specific heat
3) Water equivalent 4) Thermal conductivity

14. **In a pressure cooker the cooking is fast because**
1) the boiling point of water is raised by the increased pressure inside the cooker
2) The boiling point is lowered
3) More steam is available for cooking
4) more pressure is available at 100°C

15. **A fan produces a feeling of comfort because**
1) A fan supplies cool air
2) Due to evaporation of sweat
3) Fan cools the air
4) Fan increases the humidity in air

16. **Iron, lead and brass spheres of equal masses are heated to a common temperature and are kept on a sheet of wax. The sphere that passes through the plate first is (lead 0.03 cal/g/°C, iron 0.11 cal/g/°C, brass 0.08 cal/g/°C)**
1) lead 2) brass
3) iron 4) all at the same time

17. **A gas is heated first keeping the pressure constant and then keeping the volume constant through the same range of temperature. Heat supplied in the first process to that of supplied in the second process is**
1) More 2) Less
3) Equal 4) More or Less

18. **If a bullet is suddenly stopped the rise in temperature is independent of**
1) Mass of the bullet 2) Material of bullet
3) Velocity of bullet 4) All the above

19. **The triple point for water is**
1) 0^0C at a pressure of 1 atmosphere
2) 273 K at a pressure of 76 mm of Hg
3) -273^0C of a pressure of 4mm. of Hg
4) 273. 16K at a pressure of 613 Pa

20. **Ice, water and steam co-exist at triple point temperature 273.16 K and pressure 4.6 mm Hg In a system in which the triple point conditions of temeprature and pressure exist, the pressure is increased a little while keeping the temperature constant , then the system contains.**
1) ice only 2) water only
3) steam only 4) water and ice

21. **The internal energy U is a unique function of any thermal state, because change in U**
1) Does not depend upon path
2) Depends upon path
3) Corresponds to an adiabatic process
4) Corresponds to an isothermal process.

22. **The internal energy of ideal gas is independent of the gas in bulk but depends only on the**
1) Impurities 2) Pressure
3) Volume 4) Temperature

23. **Gas is taken through a cyclic process completely once. Change in the internal energy of the gas is**
1) infinity 2) zero 3) Small 4) Large

24. **Which of the following does not characterize the thermodynamic state of matter [AIEEE-03]**
1) volume 2) temperature
3) pressure 4) work

25. **What will be the nature of change in internal energy in case of precesses shown below ?**

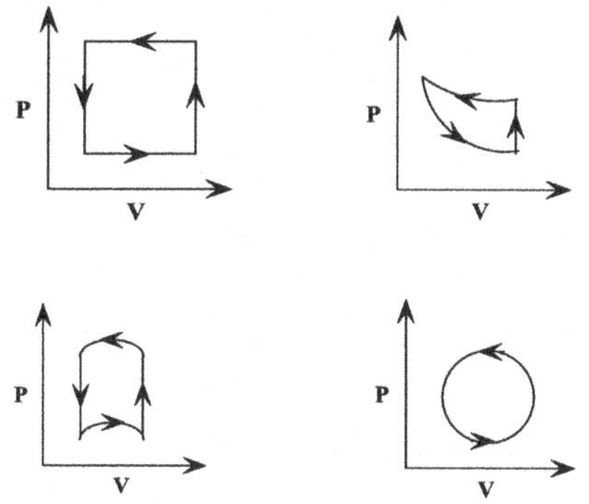

1) + ve in all cases **[AIEEE 2003]**
2) – ve in all cases
3) – ve in 1 and 3 and + ve in 2 and 4
4) zero in all cases

26. **Zeroth law of thermodynamics represents**
a) concept of temperature
b) state of thermal equilibrium of a system
c) that heat is a form of energy
1) a, b are correct 2) b, c are correct
3) a, c are correct 4) all are correct

27. **An ideal gas is heated from 20°C to 40°C under constant pressure. The change in internal energy is (1996M)**
1) zero under constant pressure
2) Double the original value
3) Proportional to change in volume
4) Proportional to change in temperature

28. **First law of thermodynamics is a special case of**
1) Boyle's law 2) Charles' law
3) Law of conservation of energy
4) Law of conservation of momentum

29. **Which of the following is incorrect regarding the first law of thermodynamics ?[AIEEE-05]**
1) It introduces the concept of internal energy
2) It introduces the concept of entropy
3) It is not applicable to any cyclic process
4) It is a restatement of principle of conservation of energy.

30. **When a gas is supplied 'dQ' heat, it performs a work 'dW'. The increase in its internal energy 'dU' is**
1) dU = dQ + dW 2) dU = dQ- dW
3) dU = dW - dQ 4) $dU = \frac{dQ - dW}{2}$

31. **When heat is added to a system, which of the following is not possible ?**
1) Internal energy of the system increases
2) Work is done by the system
3) Neither internal energy increases nor work is done by the system
4) Internal energy increases and also work is done by the system

32. **In a given process on an ideal gas, dW = 0 and dQ < 0. Then for the gas**
1) The temperature will decrease
2) The volume will increase
3) The pressure will remain constant
4) The temperature will increase

33. **Same heat is given to Helium and Oxygen gas of equal masses under constant volume. Then the temperature of the gas which increases more is**
1) Oxygen 2) Helium
3) Rise in temperatures are equal
4) None

34. **Ratio of adiabatic bulk modulus to isothermal bulk modulus is**
1) $C_P - C_V$ 2) C_V/C_P 3) C_P/C_V 4) P

35. **The specific heat capacity of a gas at constant pressure is greater than that at constant volume because**
1) at constant volume , all the heat supplied goes to increase the internal energy of the gas
2) at constant volume there is no change in temperature of the gas when heat is supplied
3) at constant pressure , there is no change in internal energy of the gas when heat is supplied
4) the gas expands when its temperature increases at constant pressure

36. **The temperature of the system decreases in the process of (2007E)**
1) free expansion
2) isothermal expansion.
3) adiabatic expansion
4) isothermal compression

37. **Heat given to Ideal gas in isothermal condition is used to**
1) Increase internal energy
2) To rise its temperature
3) Do external work
4) Do work and to increase temperature

38. **For isothermal change $\frac{\Delta P}{P}$ is equal to**

1) $\frac{1}{2}\gamma\frac{\Delta V}{V}$ 2) $\frac{-\Delta V}{V}$

3) $-\gamma\frac{\Delta V}{V}$ 4) $-\gamma^2\frac{\Delta V}{V}$

39. **For a certain mass of gas Isothermal relations between 'P' and 'V' are shown by graphs at two different temperatures T_1 and T_2 then**

1) $T_1 = T_2$

2) $T_1 > T_2$

3) $T_1 < T_2$

4) $T_1 \geq T_2$

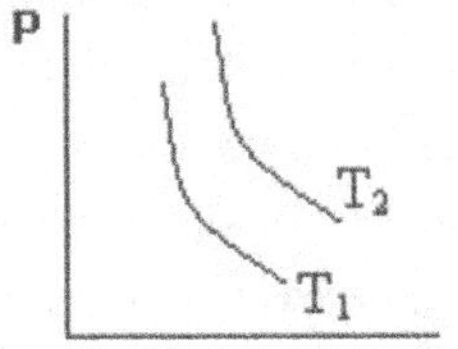

40. **In an isothermal process the specific heat is**

1) Zero 2) Constant

3) Infinity 4) One

41. **Certain amount of heat supplied to an ideal gas under isothermal conditions will result in**

1) a rise in temperature (1997M)

2) doing external work and a change in temperature

3) doing external work

4) an increase in the internal energy of the gas

42. **The isothermal bulk modulus of a perfect gas at a atmospheric pressure is**

1) 1.03×10^5 Newtons/m^2

2) 1.03×10^6 Newtons/m^2

3) 1.0370×10^{10} Newtons/m^2

4) 1.03×10^{11} Newtons/m^2

43. **The change in internal energy of the gas and workdone by the gas in an adiabatic process are**

1) Unequal 2) Equal

3) Only numerically equal

4) No relation.

44. **A liquid in a thermos flask is vigorously shaken. Then the temperature of the liquid**

1) Is not altered 2) Increases

3) Decreases 4) None

45. **When a gas expands adiabatically, its internal energy**

1) Increases 2) Decreases

3) Doesnot change 4) 1 or 2

46. **Which of the following is true in the case of an adiabatic process, where $\gamma = C_P / C_V$?** **(2008 E)**

1) $P^{1-\gamma} T^{\gamma}$ = constant 2) $P^{\gamma} T^{1-\gamma}$ = constant

) PT^{γ} = constant 4) $P^{\gamma} T$ = constant

47. **In the adiabatic compression, the decrease in volume is associated with (2008 E)**

1) increase in temperature and increase in pressure

2) decrease in temperature and increase in pressure

3) decrease in temperature adn decrease in pressure

4) increase in temperature and decrease in pressure

48. **Workdone in compression is least when the process is**

1) Adiabatic 2) Isothermal

3) Isochoric 4) Isobaric

49. **The pressure P and volume V of an ideal gas both increase in a process**

1) It is not possible to have such a process

2) The workdone by the system is positive

3) The temperature of the system increases

4) 2 and 3

50. **Is it possible to change the temperature of a body without supplying heat**

1) Possible in isothermal process

2) Possible in adiabatic process

3) Possible in isochoric processs

4) Possible in isobaric process

51. **For an adiabatic process, the relation between V and T is given by (2007 M)**

1) TV^{γ} = a constant 2) $T^{\gamma}V$ = a constant

3) $TV^{1-\gamma}$ = a constant 4) $TV^{\gamma-1}$ = a constant

52. **In Carnot cycle at the end of the cycle, the temperature of working substance is :**

1) less than initial temperature

2) greater than initial temperature

3) equal to initial temperature

4) none of the above

53. **The efficiency of a carnot heat engine depend upon :**

1) working substance

2) on temperature of sink alone

3) on temperature of source only

4) on temperature of both source and sink

54. **Which statement is incorrect? [AIEEE 2002]**
   1) All reversible cycles have same efficiency.
   2) Reversible cycles have more efficiency than reversible ones
   3) Carnot cycle is a reversible one
   4) Carnot cycle has maximum efficiency

55. **Even Carnot engine cannot give 100% efficiency because we cannot [AIEEE 2002]**
   1) prevent radiation 2) find ideal sources
   3) reach absolute zero temperature
   4) eliminate friction

56. **Heat cannot by itself flow from a body at lower temperature to a body at higher temperature is a statement or consequence of [AIEEE 2003]**
   1) Ist law of thermodynamics
   2) IInd law of thermodynamics
   3) conservation of momentum
   4) conservation of mass

57. **If the temperature of the source is increased, the efficiency of the carnot engine**
   1) increases 2) decreases
   3) remains unchanged
   4) first increases and then becomes constant

58. **Choose the false statement. (ATL)**
   1st Law of thermo dynamics
   1) does not tell us about the efficiency with which heat can be converted in to work.
   2) does not give any idea about the condition under which a body can use its heat energy to produce the work
   3) tells about the direction of heat flow.
   4) defines the internal energy function

59. **From the Graph the curve BC denotes**

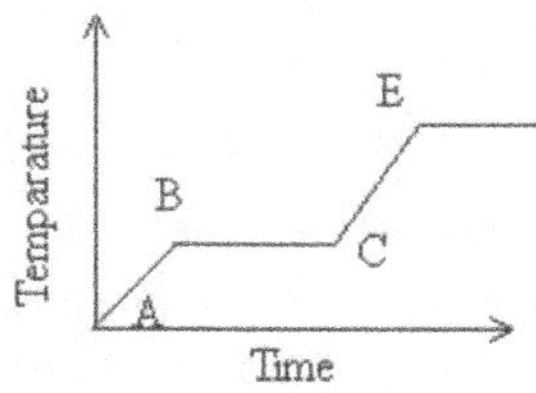

   1) Change of phase (ATL)
   2) Increase of temparature of solid
   3) Decrease of temparature of solid
   4) Increase of temparature of liquid

MORE THAN ONE OPTIONS

60. **Specific heat of a substance depends on**
   a) nature of the substance
   b) mass of substance
   c) heat given to the substance
   1) only a is correct
   2) both a and b are correct
   3) both a, b, c are correct
   4) only a, c are correct

61. **The thermal capacity of a body**
   a) per unit volume is product of density and specific heat
   b) is product of mass and specific heat
   c) is product of volume and specific heat
   1) only a is correct
   2) both a and b are correct
   3) both a, b, c are correct
   4) both b and c are correct

62. **Melting point of a substance**
   a) Decreases on mixing impurities
   b) Increases with increase of pressure for some solids
   c) Decreases with increase of pressure for some solids
   1) both a and b are correct
   2) both b and c are correct
   3) both a and c are correct
   4) a, b, c are correct

63. **In the equation for I law of thermodynamics $dQ = du + dw$**
   a) dQ depends on path
   b)dw depends on path c)du depends on path
   1) only b is correct 2) both a, b are correct
   3) b, c are correct 4) both a, c are correct

64. **The first law of thermodynamics incorporates the concepts of**
   a) equivalance of heat and work
   b) conservation of heat
   c) conservation of energy
   1) both a and b are true 2) both b and c are true
   3) both a and c are true 4) a, b, c are true

65. **Statement A : At absolute zero the substance has internal energy due to the vibrational motion of atoms with in the molecules. This is zero point energy.**

Statement B : First law of thermodynamics does not satisfy the condition under which a body can use its heat energy to produce the work.

1) A is true, B is false 2) A is false, B is true

3) A & B are true 4) A is true, B is false

66. **Molar heat capacity of a gas does not depend on**

a) Its temperature b) Its molecular weight
c) Its atomicity

1) a and b are true 2) a and c are true

3) both b and c are true 4) all a, b, c are true

67. **For a Hydrogen gas**

a) $C_P = 5R/2$ b) $C_V = 5R/2$

c) C_P/C_V is always greater than 1

1) both a and b are true 2) both b and c are true

3) both a and c are true 4) a, b, c are true

68. **The work done by a gas during Isothermal Expansion depends on**

a) temperature of the gas

b) number of moles

c) temperature of surroundings

1) both a and b are true 2) b and c are true

3) both a and c are true 4) a, b, c are true

69. **A gas with $C_P/C_V = \gamma$ goes from an initial state (P_1, V_1, T_1) to a final state (P_2, V_2, T_2) through an adiabatic process. The work done by the gas is**

a) $\frac{nR(T_1 - T_2)}{\gamma - 1}$ b) $\frac{P_1V_1 - P_2v_2}{\gamma - 1}$

c) $nR\gamma(T_1 - T_2)$

1) a and b are true 2) b and c are true

3) both a and c are true 4) a, b, c are true

70. **Melting of ice under atmospheric pressure is**

a) Isothermal process b) Isobaric process
c) Adiabatic process

1) both a and b are correct

2) both b and c are correct

3) both a and c are correct 4) a, b, c are correct

71. **When an isolated gas expands against vaccum, then its**

a) Internal energy increases

b) No work is done

c) Internal energy remains same

1) both a and b are true 2) both b and c are true

3) both a and c are true 4) a, b, c are true

72. **When a gas is compressed to half of its initial volume, if it is comparision between isothermal and adiabatic process, then**

a)Work done is less if the change is Isothermal

b) Final pressure is more if the charge is adiabatic

c) Final pressure is less if the change is Isothermal

1) both a and b are true 2) both b and c are true

3) both a and c are true 4) a, b, c are true

73. **A gas may expand adiabatically or Isothermally. If P, V graphs are drawn and if s_1 and s_2 are the slopes of the process then**

a) $s_1 > s_2$

b) adiabatic curve and Isothermal curve may intersect

c) $s_1 < s_2$

1) both a and b are true 2) both b and c are true

3) both a and c are true 4) a, b, c are true

74. **The following are the P – V diagrams for cyclic process for a gas. Then the heat is absorbed by the gas in the process**

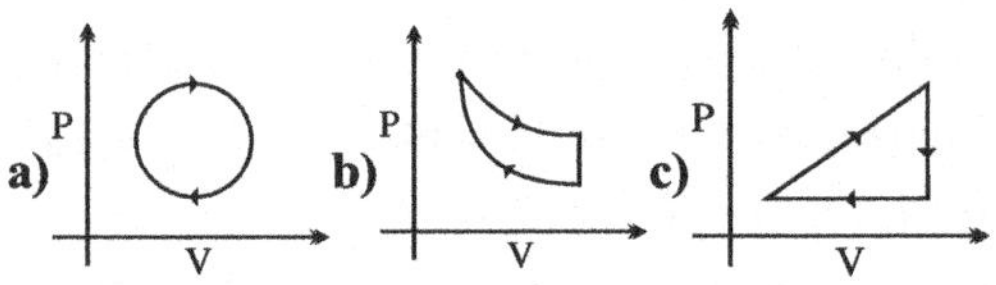

1) in a and b 2) in b and c only

3) in a and c only 4) in all a,b,c

75. **The internal energy of a system remains constant when it undergoes**

a) Cyclic process b) an Isothermal process

c) Any process in which heat given out by the system is equal to work done on the system

1) both a and b are true

2) both b and c are true

3) both a and c are true 4) a, b, c are true

76. **Consider the following statements:**

A) Heat given to an ideal gas under isothermal conditions is used completely to do external work.

B) The change in internal energy in a thermodynamic process is independent of the path.

1) A and B are true 2) Both are false

3) A is true and B is false 4) A is false B is true

77. Consider the following statements and choose the correct answer.

a) if heat is added to a system,its temperature must always increase.

b) if positive work is done by a system in thermodynamic process,its volume must increase. (2007M)

1) both a and b are true

2) a is correct,but b is wrong

3) a is wrong but b is correct

4) both a and b are wrong

78. According to second law of thermodynamics

a) All heat can be converted into work

b) The efficiency of a heat engine is always lesser than unity

c) It is not possible to transfer heat from lower to higher temperature by it self

1) both a and b are true

2) both b and c are true

3) both a and c are true 4) a, b, c are true

79. Consider the statement (A) and (B) and identify the correct answer.

A : First law of thermodynamics specifies the condition underwhich a body can use its heat energy to produce the work.

B : Second law of thermodynamics states that heat always flows from hot body to cold body by itself (2005M)

1) Both (A) and (B are true

2) Both (A) and (B) are false

3) (A) is true , (B) is false

4) (A) is false but (B) is true

80. The second law of thermodynamics is the generalisation of the fact that

I) Heat always flows from hot body to cold body by itself

II) Heat can flow from cold body to hot body itself

III) It is impossible for a self acting machine un aided by any external agency to transfer heat from cold body to hotter body

1) I & II 2) II & III **(ATL)**

3) I & III 4) I, II, III & IV

81. Choose the correct statement (ATL)

A : It is impossible to derive continuous supply of work by cooling a body to a temparature lower than that of the coldest of its surroundings.

B : Heat engine can convert whole of the heat energy supplied to it into useful work.

1) only A 2) only B

3) Both A & B 4) Both A & C

82. Identify the correct statements (ATL)

A : The negative slope of ice line for water shows that melting point of ice decreases with the increase of pressure.

B : Liquids with low latent heat of vaporization are called volatile liquids.

C : Ether and petrol are examples of volatile liquids

1) all are true 2) all are flase

3) A & B are true 4) A&B are wrong

ASSERTION & REASON

These Questions consist of two statements each printed as Assertion and Reason. While answering these questions you are required to choose any one of the following four responses.

1) If both (A) and (R) are true and (R) is the correct explanation of (A)

2) If both (A) and (R) are true and (R) is not the correct explanation of (A)

3) If (A) is true but (R) is false

4) If (A) is false but (R) is true

83. A : One cannot change water into steam by sending steam at 100^{0}C

R : At thermal equilibrium heat exchange cannot take place between two systems

84. A : Specific heat is known as Thermal inertia.

R : The body having higher specific heat cools very slowly and gets heated up very slowly and vice versa (K -1)

85. A : Zeroth law of thermodynamics defines thermal equilibrium

R : At thermal equilibrium of two bodies, they have same internal energies (K -3)

86. A : Workdone by an ideal gas during an expansion is positive

R : Workdone is given by the product of pressure and increase in volume (K-1)

87. A : Workdone by an ideal gas in a cyclic process must be zero

R : During a cyclic process, a gas will come to same state of its parameters. (K -4)

88. A : First law of Thermodynamics explains why entire heat can not be converted into mechanical work.

R : It is impossible to convert heat completely into work without rejecting heat to surroundings. (K -4)

89. A : During an isothermal process, specific heat of a substance is infinity

R : Isothermal process is a constant temperature process (K -1)

90. A : During change of state of a substance internal energy of a substance remains constant

R : During isothermal process of an ideal gas change in internal energy is zero (K -4)

91. A :The internal energy of an ideal gas does not change during an isothermal process

R : The decrease in volume of a gas is compensated by a corresponding increase in pressure when its temperature held constant (K -1)

92. A : During an adiabatic process, an ideal gas expands by decrease in its internal energy only

R : During an adiabatic process, heat cannot be exchanged between a system and its surroundings (K -1)

93. A : During an adiabatic compression temperature of an ideal gas increases

R : Internal energy of a gas can be changed even without addition or withdra of heat but by doing work (K -1)

94. A : Air in a cylinder is suddenly compressed by a piston. If the volume after compression is half the original volume the final pressure is twice the initial pressure.

R : Suddenly compressed process is adiabatic process (K -4)

95. A : During a free expansion of an ideal gas workdone is zero

R : Workdone against vacuum is zero (K -1)

96. A : In isobaric process, during an expansion of a gas both volume and temperature of the gas increase

R : Heat added to a gas partly gets converted into internal energy and remaining to do external work (K -1)

97. A : Heat flows always from a hotter body to a colder body by itself

R : without an external aid heat cannot flow from colder body to a hotter body (K -1)

98. A : It is possible to transfer heat energy from body at lower temperature to body at higher temperature by using external agency

R : External agency extracts larger heat from colder body by doing work rejects smaller heat to hotter body , hence heat transfer is possible. (K -3)

99. A : At absolute zero, the substance has internal energy.

R : At OK the substance posses the vibrational motion of atoms with in the molecules (ATL)(K -1)

100. A : The raise in temparature and internal energy takes place quickly, when the gas heated at constant volume

R :The gas does no work, when it is heated at constant volume. (ATL)(K -1)

ASCENDING (OR) DESCENDING

101. Starting with the same initial conditions an ideal gas expands from volume V_1 to V_2 in three different ways. The work done by the gas is W_1 if the process is purely isothermal, W_2 if purely isobaric and W_3 if purely adiabatic. Then

1) $W_2 > W_1 > W_3$ 2) $W_2 > W_3 > W_1$

3) $W_1 > W_2 > W_3$ 4) $W_1 > W_3 > W_2$

102. In the above case, if P_1, P_2 and P_3 are their final pressure's. Then

1) $P_2 > P_1 > P_3$ 2) $P_2 > P_3 > P_1$

3) $P_1 > P_2 > P_3$ 4) $P_1 > P_3 > P_2$

103. In the above case, if their final temperatures are T_1, T_2 and T_3. Then.

1) $T_2 > T_1 > T_3$ 2) $T_2 > T_3 > T_1$

3) $T_1 > T_2 > T_3$ 4) $T_1 < T_3 < T_2$

Matching

104. Match the following

List – I | **List – II**

a) fraction of total energy that is utilised to increase internal energy of a monoatomic gas — **e) 0.25**

b) fraction of total energy that is involved in expansion for a triatomic gas — **f) infinity**

c) specific heat during isothermal process — **g) zero**

d) specific heat during adiabatic process — **h) 0.6**

1. a-h, b-e, c-g, d-h 2. a-g, b-h, c-f, d-e 3. a-h, b-e, c-f, d-g 4. a-e, b-h, c-f, d-g

105. Match the following :

List – I | **List – II**

a) it is possible to change the temperature of a gas without supplying heat — **e) isothermal process**

b) change in internal energy is zero — **f) adiabatic process**

c) work done is zero — **g) isochoric process**

d) work done by the gas is maximum for a given increment in volume — **h) isobaric process**

1. a-f, b-h, c-g, d-e 2. a-f, b-e, c-g, d-h 3. a-e, b-f, c-g, d-h 4. a-g, b-e, c-h, d-f

106. Match List–I with List–II (the terms have their usual meanings)

List – I | **List – II**

a) Isobaric process — **e) dQ= dw**

b) Isochoric process — **f) d Q= dU+ dw**

c) Isothermal process — **g) dU= – dw**

d) Adiabatic process — **h) dQ= dU**

1) a-e, b-f, c-h, d-g 2) a-f, b-h, c-e, d-g 3) a-f, b-e, c-h, d-g 4) a-g, b-f, c-e, d-h

107. Matching block type

List – I | **List – II**

a) Zeroeth law of thermodynamics — **e) Direction of flow of heat energy**

b) First law of thermodynamics — **f) Zero work done**

c) Free expansion of gas — **g) Temperature**

d) Second law of thermodynamics — **h) Law of conservation of energy**

1) a-f ,b-h,c-g,d-e 2) a-g,b-h,c-f,d-e 3) a-e,b-g,c-f,d-h 4) a-e,b-g,c-h,d-f

108.

List – I | **List – II**

a) Work done on the system — **e) positive**

b) heat energy given to the system — **f) negative**

c) change in internal energy — **g) may be positive (or) negative**

d) During cyclic process change in Internal energy — **h) Zero**

1) a-g,b-e,c-h,d-f 2) a-f,b-g,c-e,d-h 3) a-f,b-e,c-g,d-h 4) a-g,b-e,c-h,d-g

109. Matching block type

List – I	List – II
a) Among solids and liquids specific heat is maximum for	e) Mercury
b) In liquids specific heat is minimum for	f) Water
c) With increase of pressure the boiling point of a liquid	g) Increases
d) With increase of pressure the melting point of a solid substance which contracts on melting	h) Decreases

1) a-f,b-e,c-h,d-g 2) a-e,b-f,c-g,d-h 3) a-e,b-g,c-f,d-h 4) a-f,b-e,c-g,d-h

ANSWERS

1) 2	2) 2	3) 3	4) 1	5) 2
6) 4	7) 3	8) 2	9) 3	10) 4
11) 4	12) 3	13) 3	14) 1	15) 2
16) 1	17) 1	18) 1	19) 4	20) 2
21) 1	22) 4	23) 2	24) 4	25) 4
26) 1	27) 4	28) 3	29) 2	30) 2
31) 3	32) 1	33) 1	34) 3	35) 1
36) 3	37) 3	38) 2	39) 3	40) 3
41) 3	42) 1	43) 3	44) 2	45) 2
46) 1	47) 1	48) 3	49) 3	50) 2
51) 4	52) 3	53) 4	54) 1	55) 3
56) 2	57) 1	58) 3	59) 4	60) 1
61) 2	62) 4	63) 2	64) 3	65) 3
66) 1	67) 2	68) 1	69) 1	70) 1
71) 2	72) 4	73) 1	74) 4	75) 4
76) 1	77) 3	78) 2	79) 4	80) 3
81) 1	82) 1	83) 1	84) 1	85) 3
86) 1	87) 4	88) 4	89) 1	90) 4
91) 1	92) 1	93) 1	94) 4	95) 1
96) 1	97) 1	98) 3	99) 1	100) 1
101) 1	102) 1	103) 3	104) 3	105) 2
106) 2	107) 2	108) 3	109) 4	

EXERCISE – II(A)

Calorimetry :

1. **The densities of two substances are in the ratio 5 : 6 and the specific heats are in the ratio 3 : 5 respectively. The ratio of their thermal capacities per unit volume is**

   1) 2 : 1 2) 1 : 2 3) 4 : 1 4) 1 : 4

2. **The thermal capacity of 10g of a substance is 8 calories. The specific heat is in J/kg-K**

   1) 3360 2) 3630 3) 6330 4) 3306

3. **A piece of metal of mass 112g is heated to 100°C and dropped into a copper calorimeter of mass 40g containing 200g of water at 16°C. Neglecting heat loss, the specific heat of the mettal is nearly, if the equilibrium temperature reached is 24.1°C ,**

   1) 0.294 cal/gm°C 2) 0.394cal/gm°C 3) 0.194 cal/gm°C 4) 0.494cal/gm°C

4. **A copper block of mass 500 gm and specific heat 0.1 cal/gm°C is heated from 30°C to 290°C, the thermal capacity of the block is**

   1) 50cal/°C 2) 50gm 3) 5cal/°C 4) 5gm

5. **Two spheres with radii in the ratio 1 : 2 have specific heats in the ratio x : y and densities in the ratio z : x. The ratio of their thermal capacities is**

   1) Z : 2y 2) Z : 8y 3) Zy : 8 4) xy : 2Z

6. **The specific heat of a substance is 0.09 cal/gm°C. If the temperature is measured on Fahrenheit scale the value of its specific heat in cal/gm/°F is**

   1) 0.09 2) 0.9 3) 0.05 4) 0.5

7. **5 gm of steam at 100°C is passed into calorimeter containing liquid. Temperature of liquid rises from 32°C to 40°C. Then water equivalent of calorimeter and contents is**

   1) 40 gram 2) 375 gram 3) 300 gram 4) 160 gram

8. **The quantity of heat which can rise the temperature of x gm of a substance through t_1°C can rise the temperature of y gm of water through t_2°C is same. The ratio of specific heats of the substances is**

   1) yt_1/xt_2 2) xt_2/yt_1 3) yt_2/xt_1 4) xt_1/yt_2

9. Density of a liquid 'A' is 0.5 g/c.c and that of liquid 'B' is 0.6 g/c.c. Heat capacity of 8 litres of 'A' is equal to that of 10 litres of 'B'. Then their specific heats ratio is

1) 4 : 5 2) 3 : 2 3) 2 : 3 4) 1 : 1

10. Quantity of heat lost in condensation of 10 gm of steam at 100°C is

1) 2.26×10^5J 2) 2.26×10^4J
3) 22.6J 4) 44.52×10^4J

11. Two liquids at temperatures 60°C and 20°C respectively have masses in the ratio 3 : 4 and their specific heats in the ratio 4 : 5. If the two liquids are mixed, the resultant temperature is (2000 E)

1) 70°C 2) 50°C 3) 40°C 4) 35°C

12. Two liquids are at 40°C and 30°C. When they are mixed in equal masses, the temperature of the mixture is 36°C. Ratio of their specific heats is

1) 3 : 2 2) 2 : 3 3) 4 : 3 4) 3 : 4

13. Two liquids of masses M_1 and M_2 and specific heats S_1 and S_2 respectively are mixed. The specific heat of the mixture is

1) $\frac{M_1S_1 + M_2S_2}{M_1 + M_2}$ 2) $\frac{M_1S_1 + M_2S_2}{2(M_1 + M_2)}$

3) $\frac{2(M_1S_1 + M_2S_2)}{M_1 + M_2}$ 4) $\frac{M_1S_1 - M_2S_2}{M_1 + M_2}$

14. 6 gm of steam at 100°C is mixed with 6 gm of ice at 0°C.The mass of steam left uncondensed is

1) 2 gm 2) 4 gm 3) 3 gm 4) 1 gm

15. 10 grams of steam at 100°C is mixed with 50 gm of ice at 0°C then final temperture is

1) 20°C 2) 50°C 3) 40°C 4) 100°C

Joul's law

16. A piece of lead falls from a height of 100m on a fixed non-conducting slab which brings it to rest. If the specific heat of lead is 30.6 cal/kg °C, the increase in temperature of the slab immediately after collision

1) 6.72°C 2) 7.62°C
3) 5.62°C 4) 8.72°C

17. Hailstones fall from a certain height. If only 1% of the hailstones melt on reaching the ground, find the height from which they fall. ($g = 10\ ms^{-2}$. L = 80 calorie/g and J=4.2J/calorie)

1) 336 m 2) 236 m 3) 436 m 4) 536 m

18. A copper block of mass 1 kg slides down on a rough inclined plane of inclination 37° at a constant speed. Find the increase in the temperature of the block as it slides down through 60cm assuming that the loss in mechanical energy goes into the copper block as thermal energy. (Specific heat of copper = 420 J $kg^{-1}K^{-1}$, g = $10ms^{-2}$)

1) 6.6×10^{-3}°C 2) 7.6×10^{-3}°C
3) 8.6×10^{-3}°C 4) 9.6×10^{-3}°C

19. A steel ball of mass 0.1 kg falls freely from a height of 10m an bounces to a height of 5.4m from the ground. If the dissipated energy in this process is absorbed by the ball, the rise in its temperature is (specific heat of steel = 460 $JKg^{-1}K^{-1}$) (g=$10ms^{-2}$). (2000 M)

1) 0.01°C 2) 0.1°C 3) 1°C 4) 1.1°C

20. Two spheres A and B with masses in the ratio 2 : 3 and specific heat 2 : 3 fall freely from rest. If the rise in their temperatures on reaching the ground are in the ratio 1 : 2 the ratio of their heights of fall is

1) 3 : 1 2) 1 : 3 3) 4 : 3 4) 3 : 4

21. From what height a block of ice must fall into a well so that $\frac{1}{100}$th of its mass may be melted (g = 10 m/s²)

1) 300 m 2) 336 m 3) 660 m 4) none

22. Two identical balls 'A' and 'B' are moving with same velocity. If velocity of 'A' is reduced to half and of 'B' to zero, then the rise in temperatures of 'A' to that of 'B' is

1) 3 : 4 2) 4 : 1 3) 2 : 1 4) 1 : 1

23. A lead bullet (specific heat = 0.032 cal/gm °C) is completely stopped when it strikes a target with a velocity of 300 m/s. The heat generated is equally shared by the bullet and the target. The rise in temperature of bullet will be

1) 16.7°C 2) 1.67°C
3) 167.4°C 4) 267.4°C

24. A man of 60 kg gains 1000 cal of heat by eating 5 mangoes. His efficency is 56%. To what height he can jump by using this energy?

1) 4m 2) 20 m 3) 28 m 4) 0.2 m

First law of thermodynamics

25. Find the external work done by the system in kcal, when 20 kcal of heat is supplied to the system the increase in the internal energy is 8400 J)J=4200J/kcal)

1) 16 kcal 2) 18 kcal 3) 20 kcal 4) 19 kcal

26. Heat of 30 kcal is supplied to a system and 4200 J of external work is done on the system so that its volume decreases at constant pressure. What is the change in its internal energy. (J = 4200 J/kcal)

1) 1.302×10^5 J 2) 2.302×10^5 J
3) 3.302×10^5 J 4) 4.302×10^5 J

27. Air expands from 5 litres to 10 litres at 2 atm pressure. External workdone is

1) 10J 2) 1000J 3) 3000 J 4) 300 J

28. Heat given to a system is 35 joules and work done by the system is 15 joules. The change in the internal energy of the system will be

1) – 50 J 2) 20 J 3) 30 J 4) 50 J

29. A gas is compressed at a constant pressure of 50 N/m² from a volume 10m³ to a volume of 4m³. 100J of heat is added to the gas then its internal energy

1) Increases by 400J 2) Increases by 200J
3) Decreases by 400J 4) Decreases by 200J

30. 2 kg of air is heated at constant volume.The temperature of air is increased from 293 K to 313 K.If the specific heat of air at constant volume is 0.718 kJ/kg K, the amount of heat absorbed in kJ and kcal is, (J = 4.2/kcal)

1) 6.838 kJ 2) 7.8 kJ
3) 14.68 kJ 4) 28.72 kJ

31. The Molar heat capacities of nitrogen at constant pressure and constant volume are 29.11kJ/k mole K and 20.81 kJ/kmole K, respectively. When 5 gram of nitrogen is heated from 290 to 310 K i) the decrease in its internal energy and ii) the external work done are, (Molecular weight of nitrogen = 28)

1) 74.32 J, 29.64 J 2) 29.64 J, 74.32 J
3) 64.32 J, 19. 64 J 4) 94.32 J 39.64 J

32. Five kilomoles of oxygen is heated at constant pressure. The temperature of the oxygen gas is increased from 295K to 305K. If the molar heat capacity of oxygen at constant pressure is 6.994 kcal/kmole K. The amount of heat absorbed is in kcal

1) 249.7 2) 44 3) 349.7 4) 539.7

33. Two moles of air, when heated through 10 K expand by an amount of 1.66×10^{-3} m³ under a constant pressure of 10^5 N/m². If C_v = 20.81 J/mole K, then C_p is,

1) 29. 11J/mole K 2) 19.11 J/mole K
3) 32.11 J/mole k 4)19.11J/mole k

34. Four moles of a perfect gas heated to incrcrease its temperature by 2°C absorbes heat of 40 cal at constant volume. If the same gas is heated at constant pressure the amount of heat supplied is, (R = 2 cal/mol K)

1) 28 cal 2) 56 cal
3) 84 cal 4) 94 cal

35. Five moles of hydrogen is heated through 20 K under constant pressure. If R = 8.312J/mole K find the external work done.

1) 831.2 J 2) 83.12J
3) 166.24 J 4) 16.62J

C_P, C_V and their relations

36. Find the change in internal energy in joule. When 10g of air is heated from 30°C to 40°C (Cv = 0.172 kcal/kg K, J = 4200 J/kcal)

1) 62.24 J 2) 72.24 J
3) 52.24 J 4) 82.24 J

37. The temperature of 5 moles of a gas at constant volume is changed from 100°C to 120°C. The change in internal energy is 80J. The total heat capacity of the gas at constant volume will be in joule/kelvin is (2003E)

1) 8 2) 4 3) 0.8 4) 0.4

38. When an ideal diatomic gas is heated at constant pressure, the fraction of heat energy supplied which is used in doing work to maintain pressure constant is

1) 5/7 2) 7/2 3) 2/7 4) 2/5

39. When a monoatomic gas expands at constant pressure, the percentage of heat supplied that increases temperature of the gas and in doing external work in expansion at constant pressure is

1) 100%, 0 2) 60%, 40%
3) 40%, 60% 4) 75%, 25%

40. H calories of heat is required to increase the temperature of one mole of monoatomic gas from 20°C to 30°C at const. volume. The quantity of heat required to increase the temperature of 2 moles of a diatomic gas from 20°C to 25°C is at constant volume is

1) 4H/3 2) 5H/3 3) 2 H 4) 7H/3

41. 1/2mole of Helium gas is contained in a container at S.T.P. The heat energy needed to double the pressure of the gas, keeping the volume constant (heat capacity of the gas = 3 J g^{-1} K^{-1})

1) 3276 J 2) 1638 J
3) 819 J 4) 409.5 J

42. How much heat energy in joules must be supplied to 14 gms of nitrogen at room temperature to raise its temperature by 40^0C at constant pressure ? (Mol. wt. of N_2 = 28 gm, R = constant) (2009 M)

1) 50R 2) 60R 3) 70R 4) 80R

43. For a gas, the difference between the two specific heats is 4150J $Kg^{-1} K^{-1}$ and the ratio of specific heats is 1.4. What is the specific heat of the gas at constant volume in J $Kg^{-1} K^{-1}$?

1)8475 2) 5186 3)1660 4) 10375

44. The specific heat of air at constant pressure is 1.005 kJ/kg K and the specific heat of air at constant volume is 0.718 kJ/kgK. Find the specific gas constant.

1) 0.287 kj/k K 2) 0.21 kJ/kg K
3) 0.34 kJ/kg K 4) .19 kJ/kg K

45. The specific heat of Argon at constant volume is 0.3122kJ/kg K. Find the specific heat of Argon at constant pressure if R = 8.314 kJ/kmole K. (Molecular weight of argon = 39.95)

1) 5203 2) 5302 3) 2305 4) 3025

46. If the ratio of the specific heats of steam is 1.33 = and R = 8312J/kmole K find the molar heat capacities of steam at constant pressure and constant volume.

1) 33.5 kJ/kmole, 25.19 kJ /kg K
2) 25.19 kJ/k mole, 33.5 kJ/kg K
3) 18.82 kJ/k mole, 10.82 kJ/k mole
4) 24.12 kJ /k mole, 16.12 kJ/k mole

47. The volume of 1 kg of hydrogen gas at N.T.P is 11.2 m^3. Specific heat of hydrogen at constant volume is 10046J $kg^{-1}K^{-1}$. Find the specific heat at constant pressure in $Jkg^{-1}k^{-1}$.

1) 120.2 2) 142.2 3) 163.4 4) 182.3

Different thermodynamic process

48. One mole of an ideal gas undergoes an isothermal change at temperature 'T' so that its volume v is doubled. R is the molar gas constant. Work done by the gas during this change is (2008 M)

1) RT log 4 2) RT log 2
3) RT log 1 4) RT log 3

49. One mole of O_2 gas having a volume equal to 22.4 litres at 0^0C and 1 atmospheric pressure in compressed isothermally so that its volume reduces to 11.2 litres. The work done in this process is

1) 1672.5J 2) 1728J 3) –1728J 4) –1572.5J

50. The isothermal Bulk modulus of an ideal gas at pressure P is

1) P 2) γP 3) P/2 4) P/γ

51. Diatomic gas at pressure 'P' and volume 'V' is compressed adiabatically to 1/32 times the orginal volume. Then the final pressure is

1) P/32 2) 32 P 3) 128 P 4) P / 128

52. The pressure and density of a diatomic gas $(\gamma = 7/5)$ change adiabatically from (P, d) to (P^1, d^1). If $\frac{d^1}{d} = 32$, then $\frac{P^1}{P}$ should be

1) 1/128 2) 32
3) 128 4) none of the above

53. An ideal gas at a pressures of 1 atmosphere and temperature of 27^0C is compressed adiabatically until its pressure becomes 8 times the initial pressure, then the final temperature is $(\gamma = 3/2)$

1) 627^0 C 2) 527^0C 3) 427^0C 4) 327^0C

54. The volume of a gas is reduced adiabatically to $\frac{1}{4}$ of its volume at 27^0C, if the value of $\gamma = 1.4$, then the new temperature will be

1) $350 \times 4^{0.4}K$ 2) $300 \times 4^{0.4}K$
3) $150 \times 4^{0.4}K$ 4) None of these

55. Two moles of an ideal monoatomic gas at 27^0C occupies a volume of V.If the gas is expanded adiabatically to the volume 2V, then the work done by the gas will be $(\gamma = 5/3)$

1) –2767.23J 2) 2767.23J
3) 2500J 4) –2500J

56. A container of volume $1m^3$ is divided into two equal compartments, one of which contains an ideal gas at 300 K. The other compartment is vaccum. The whole system is thermally isolated from its surroundings. The partition is removed and the gas expands to occupy the whole volume of the container. Its temperature now would be [Manipal 95]

1) 300 K 2) 250 K
3) 200 K 4) 100 K

57. An ideal gas has volume V_0 at 27^0C. It is heated at constant pressure so that its volume becomes $2V_0$. The final temperature is

1) 54^0C 2) 32.6^0C 3) 327^0C 4) 150 K

58. How much work to be done in decreasing the volume of and ideal gas by an amount of $2.4 \times 10^{-4}\ m^3$ and constant normal pressure of $1 \times 10^5\ N/m^2$

1) 28 joule 2) 27 joule
3) 24 joule 4) 25 joule

59. A gas at 10°C temperature and 1.013×10^5 Pa pressure is compressed adiabatically to half of its volume. If the ration of specific heats of the gas is 1.4, what is its final temperature?

1) 103°C 2) 123°C
3) 93°C 4) 146°C

60. Find the work done by a gas when it expands isothermally at 37°C to four times its initial volume.

1) 3753J 2) 3573J 3) 7533J 4) 5375J

Graphs

61. An Ideal gas is taken around ABCA as shown in the above diagram. The work during a cycle is [KCET 2001]

1) 2PV
2) PV
3) $\frac{1}{2PV}$
4) Zero

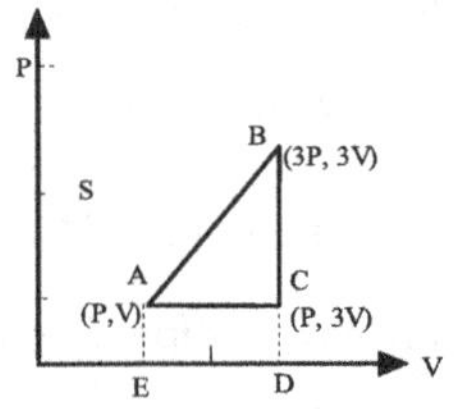

62. Heat energy absorbed by a system in going through a cyclic process shown in figure is [AIIMS 1995; BHU 2002]

1) $10^7 \pi J$
2) $10^4 \pi J$
3) $10^2 \pi J$
4) $10^{-3} \pi J$

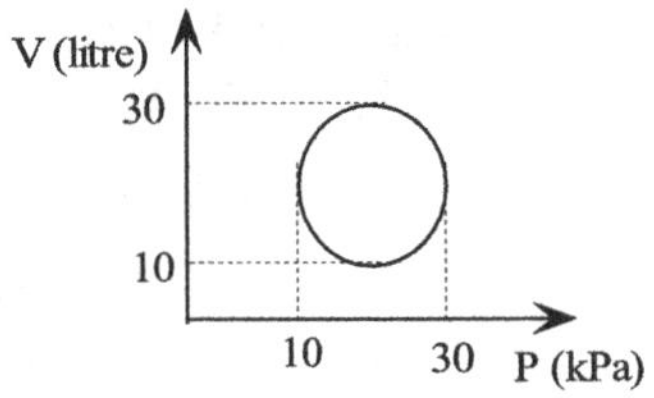

Heat Engine

63. The efficiency of a heat engine if the temperature of the source is 100°C and sink is 27°C is nearly

1) 0.1 2) 0.2 3) 0.3 4) 0.4

64. A Carnot engine working between 300 K and 600 K has work output of 800 J per cycle. What is amount of heat energy supplied to the engine from source per cycle [DMPT 1999; Pb PMT 2002; Kerala PMT 2004]

1) 1800 J/cycle 2) 1000 J/cycle
3) 2000 J/cycle 4) 1600 J/cycle

65. In a mechanical refrigerator, the low temperature coils are at a temperature of – 23°C and the compressed gas in the condenser has a temperature of 27°C. The theoretical coefficient of performance is [UPSEAT 2001]

1) 5 2) 8 3) 6 4) 6.5

66. An ideal heat engine exhausting heat at 77°C is to have a 30% efficiency. It must take heat at [BCECE 2004]

1) 127°C 2) 227°C 3) 327°C 4) 673°

ANSWERS

1) 2	2) 1	3) 3	4) 1	5) 2
6) 3	7) 2	8) 3	9) 2	10) 2
11) 4	12) 1	13) 1	14) 2	15) 3
16) 2	17) 1	18) 3	19) 2	20) 3
21)1	22) 2	23)2	24) 1	25) 2
26) 1	27) 2	28) 3	29) 2	30) 4
31) 1	32) 3	33) 1	34) 2	35) 1
36) 2	37)2	38)1	39) 3	40) 2
41) 2	42) 3	43) 4	44) 1	45) 1
46) 1	47) 2	48) 2	49) 4	50) 1
51) 3	52) 3	53) 4	54) 2	55) 2
56) 1	57) 3	58) 3	59) 1	60) 2
61) 1	62) 3	63) 2	64) 4	65) 1
66) 2				

EXERCISE – II(B)

Calorimetry

1. Two lead spheres at the same temperature have radii in the ratio 1:2. What is the ratio of their heat capacities?

1) 1:8 2) 1:4 3) 8:1 4) 4:1

2. The water equivalent of a substance is 20g. If the mass of the substance is 200gm the specific heat of the substance is (in $J\ kg^{-1}K^{-1}$)

1) 240 2) 360 3) 420 4) 630

3. A piece of metal of mass 100g is heated to 85° and dropped into a copper calorimater of mass 40g containing 246g of water at 20°C Neglecting heat loss the specific heat of the metal is nearly if the equilibrium temperature is 25°C is in Cal/ gm°C

1) 0.21 2) 0.33 3) 0.41 4) 0.51

4. A metal block absorbs 4500 cal of heat when heated from 30°C to 80°C. Its thermal capacity is

1) 90 gm 2) 90 cal / °C 3) 9 gm 4) 9 cal / °C

5. There are two brass spheres at the same temperature. If their radii are in the ratio 2 : 3. The ratio of their thermal capacities is

1) 4 : 9 2) 2 : 3 3) 8 : 27 4) 27 : 8

6. Specific heat of mercury is 0.03 cal/gm°C. Its value in S.I is

1) 116 2) 126 3) 162 4) 216

7. A calorimeter takes 200 cal of heat to rise its temperature through 10°C. Its water equivalent in gm is

1) 2 2) 10 3) 20 4) 40

8. 50 g of copper is heated to increase its temperature by 10°C. If the same quantity of heat is given to 10 gm of water, the rise in temperature is (specific heat of copper = 420 $JKg^{-1}K^{-1}$, specific heat of water = 4200 $Jkg^{-1}K^{-1}$) (2000 M)

1) 5°C 2) 6°C 3) 7 °C 4) 8°C

9. Density of liquid 'A' is 0.2 / C.C. and that of liquid 'B' is 0.4 g/C.C. Heat capacity of 4 litres of 'A' is equal to that of 5 litres of 'B'. Then their specific heats ratio is

1) 5 : 2 2) 2 : 5 3) 3 : 2 4) 5 : 6

10. The amount of heat required to convert 1 gm of Ice at 0^0 to 1 gm of steam at 100^0 C.

1) 720 cal 2) 640 cal 3) 540 cal 4) 80 cal

11. Two liquids A and B are at 25°C and 15°C. Their masses are in 2 : 3 and specific heats in ratio 3 : 4 then resultant temperature when they are mixed is

1) 16.66°C 2) 18.33°C 3) 20°C 4) 10°C

12. 4g of liquid 'A' at 60°C is mixed with 1 gram of liquid 'B' at 50°C. If final temperature is 55°C then their specific heats are in the ratio

1) 4 : 1 2) 1: 1 3) 2 : 1 4) 1: 4

13. A liquid of mass M and specific heat S is heated to a temperature T. Another liquid of mass M/2 and specific heat 2S is heated to a temperature 2T. If these two liquids are mixed, the resultant temperature of the mixture is

1) $\frac{2}{3}T$ 2) $\frac{3}{5}T$ 3) $\frac{1}{3}T$ 4) $\frac{3}{2}T$

14. 5g ice at 0°C is mixed with 5g of steam at 100°C. What is the final temperature ?

1) 0°C 2) 50°C 3) 75°C 4) 100°C

15. Equal masses of ice at 0°C and water at 100°C are mixed. The resultant temperature is

1) 50°C 2) 0°C 3) 10°C 4) 5°C

Joul's Law

16. A piece of aluminium falls from a height of 200m on a fixed non conducting slab which brings it to rest. If the specific heat of aluminium is 210 Cal/kg. the increase in temperature of the slab immediately after collision (assume that there is no loss of heat)

1) 2.2°C 2) 3.3°C 3) 4.4°C 4) 5.5°C

17. Hail stonces fall from certain height. If only 2% of the mass of the hail stone melt on reacting the ground,, the height from which they fall is ($g = 10\ ms^{-2}$, L = 80 cal/gm and J = 4.2J/cal)

1) 33.6 km 2) 67.2 km 3) 672 m 4) 336 m

18. A copper block of mass 5 kg slides down along a rough inclined plane of inclination 30° with a constant speed. The increase in the temperature of the block as it slides down through 100cm assuming that the loss of mach anical energy goes into copper block as thermal energy. (specific heat of copper 420 J kg^{-1} k^{-1} g = 10 ms^{-2})

1) 1.19×10^{-3} °C 2) 2.38×10^{-3}°C

3) 1.19×10^{-2} °C 4) 2.38×10^{-2} °C

19. A brass sphere of mass 0.2 kg falls freely from a height of 20m and bornces to q height pf 8m from the ground. If the dissipated energy in this process is dosorbed by the sphere the rise in its temperature is (specific heat of brass = 360 J kg^{-1} k^{-1} and g = $10ms^{-2}$).

1) 0.44°C 2) 0.66°C 3) 0.77°C 4) 0.88°C

20. Two spheres 'A' and 'B' of masses in the ratio 1 : 2. Specific heats in the ratio 2 : 3 falls from heights 'h_1' and 'h_2'. On reaching the ground rise in temperatures are equal, then h_1/h_2 =

1) 3 : 2 2) 2 : 9 3) 2 : 3 4) 2 : 1

21. 2kg ice block should be dropped from 'x km' height to melt completly. The 8 kg ice should be dropped from a height of

1) 4 x Km 2) x Km 3) 2 x Km 4) x/2 Km

22. Two metal balls having masses 50g and 100g collides with a target with same velocity. Then the ratio of their rise in temperatures is

1) 1 : 2 2) 4 : 1 3) 2 : 1 4) 1 : 1

23. A lead bullet of 10g travelling at 300m/s strikes against a block of wood and comes to rest. Assuming 50% of heat is absorbed by the bullet, the increase in its temperature is(sp-heat of lead is 150J/Kg-K)(2001 E)

1) 100^0C 2) 125^0 3) $150C^0$ 4) $200C^0$

24. A man of 60 kg gains 1000 cal of heat by eating 5 mangoes. His efficency is 28%. To what height he can jump by using this energy?
1) 2m 2) 20 m 3) 28 m 4) 0.2 m

First law of thermodynamics

25. Find the external workdonw by the system in K cal, when 12.5 k cal of heat is supplied to the system and the corresponding increasing in internal energy is 10500 J (J = 4200 J/kcal)
1) 15 K cal 2) 12.5 kcal
3) -10.0 kcal 4) 7.5 kcal

26. Heat of 20 K cal is supplied to the system and 8400 J of external work is done on the system so that its volume decreases at constant pressure. The change in internal enrgy (J = 4200 J /kcal)
1) 9.24×10^5 J 2) 7.56×10^5 J
3) 8.4×10^5 J 4) 10.5×10^5 J.

27. A gas expands from 40 litres to 90 litres at a constant pressure of 8 atmosphers. Work done by the gas during the expansion is
1) 4×10^{-4}J 2) 4×10^4J
3) 4×10^3J 4) 4×10^2J

28. To a system 300 joules of heat is given and it does 60 joules of work. How much does the internal energy of the system change in this process
1) 240 Joule 2) 156. 5 Joule
3) –300 Joule 4) –528. 2 Joule

29. A gas under constant pressure of 4.5×10^5Pa when subjected to 800KJ of heat, changes the volume from 0.5m3 to 2 m3 . The change in the internal energy of the gas is (2002M)
1) 6.75×10^5J 2) 5.25×10^5J
3) 3.25×10^5J 4) 1.25×10^5J

30. 2 moles of hydrogen is heated at constant volume. The temperature of the gas is increased from 280 K to 320 K. If the specific heat of hydrogen at constant volume is 10.183 kJ/kg K, find the amount of heat absorbed in kJ. (Molecular weight of hydrogen = 2.016)
1) 1.642 2) 2.12 3) 1.821 4) 1.12

31. The specific heats of hydrogen at constant pressure and constant volume are 14.307 kJ/kg K and 10.183 kJ/kg K, respectively. When one mole of hydrogen is cooled from 0°C to-20°C find (i) the decrease in its internal energy and (ii) the ratio of the decrease in its internal energy to the total energy lost.
1) 206.8J, 0.421 2) 407.3J, 0.7117
3) 601.6J, 0.812 4) 506.7J, 06881

32. Find the change in internal energy in joule, when 10g of air is heated from 30°C to 40°C.
(C_v = 0.172 kcal/kg K, J = 4200 J/kcal)
1) 72.24J 2) 68.56J
3) 59.64J 4) 42.24J

33. Four moles of a gas when heated through 20°C expand by 6.3×10^{-3} m^3 under a constant pressure of 10^5 N/m^2. If 30 J $mole^{-1}$ k^{-1} then the C_p in J $mole^{-1}$ k^{-1} is,
1) 37.9 2) 38.9 3) 27.9 4) 37.1

34. Two moles of a perfect gas heated to increase its temperature by 4°C absorbs heat of 32 Cal at constant volume. If the same gas is heated at constant pressure the amount of heat supplied is (R = 2 Cal /mol K)
1) 32 Cal 2) 48 cal
3) 56 cal 4) 64 cal.

35. Two moles of hydrogen is heated through 10k under constant pressure. If R = 83 J/mole K the external work done in Joule is
1) 830 J 2) 415 J 3) 249J 4) 166 J

C_P, C_V and their relations

36. Find the change in internal energy in joule when 20gm of a gas is heated from 20°C to 30°C (G = 0.18 kcal/kg K; J = 4200J/kcal)
1) 72.8 J 2) 151.2 J
3) 302 J 4) 450 J

37. When two moles of a gas is heated from O^0 to 10°C at constant volume, its internal energy changes by 420J. The molar specific heat of the gas at constant volume
1) 5.75 J K^{-1} $mole^{-1}$ 2) 10.55 J K^{-1} mole
3) 21 J K^{-1} $mole^{-1}$ 4) 42 J K^{-1} mole

38. When an ideal diatomic gas is heated at constant pressure, the fraction of the heat energy supplied which increases the internal energy of the gas is
1) 2/5 2) 3/5 3) 2/7 4) 5/7

39. When a diatomic gas expands at constant pressure, the percentage of heat supplied that increases temperature of the gas and in doing external work in expansion at constant pressure is
1) 60%, 40% 2) 40%, 60%
3) 28.57%, 71.42% 4) 71.42%, 28.57%

40. 294 joules of heat are required to rise the temperature of 2 mole of an ideal gas at constant pressure from 30^0C to 35^0C. The amount of heat required to rise the temperature of the same gas through the same range of temperature at constant volume (R = 8.3 Joules / mole / K) is

1) 126 J 2) 211 J 3) 294 K 4) 378 J

41. 1672 cal of heat is given to one mole of oxygen at 0^0C keeping the volume constant. Rice in temperature is (C_P = 0.2 cal/gm/^{0}K and R = 2 cal/mole/^{0}K)

1) 33.6 ^{0}C 2) 36.3 ^{0}C
3) 63.3^{0}C 4) 334.4^{0}C

42. 0.5 mole of diatomic gas at 27^0C is heated at constant pressure so that its volume is trebled. If R = 8.3 J $mole^{-1}k^{-1}$ then work done is

1) 4980 J 2) 2490 J 3) 630 J 4) 1245 J

43. The molar specific heat of a gas at constant volume is 20 Joule/gm mol/K. The value of γ for it will be

1) $\frac{11}{10}$ 2) $\frac{7}{5}$ 3) $\frac{5}{3}$ 4) $\frac{3}{2}$

44. The specific heat of air at constant pressure is 1.005 kJ/kg K and the specific heat of air at constant volume is 0.718 kJ/kg K. If the universal gas constant is 8.314 kJ/kmole K find the molecular weight of air.

1) 28.97 2) 24.6 3) 22.8 4) 19.6

45. Calculate the specific heat of a gas at constant volume from the following data. Density of the gas at N.T.P. = 19 × 10^{-2} kg/m^2, (C_p/C_v) = 1.4, J = 4.2 × 10^3 J/kcal; atmospheric pressure = 1.013 × 10^5 N/m^2. in kcal 1 kgk

[Hint : Molar heat capacity at constant volume, $C_v = \frac{R}{\gamma - 1}$; specific heat at constant volume,

$$c_v = \frac{C_v}{M} = \frac{1}{M}\left(\frac{R}{\gamma-1}\right) = \frac{PV}{M(\gamma-1)T}$$

$$= \frac{P}{\rho T(\gamma-1)}\frac{J}{kg\,K} = \frac{P}{J\rho T(\gamma-1)}\frac{kcal}{kg\,K}.$$

1) 2.162 2) 1.612
3) 1.162 4) 2.612

46. If the ratio of specific heats of neon is 1.667 and R = 8312 J/kmole K find the specific heats of neon at constant pressure and constant volume. (Molecular weight of neon = 20.183)

1) 1.029, 0.6174 2) 1.831, 0.921
3) 1.621, 0.421 4) 0.862, 0.246

47. The volume of 1kg of oxygen gas at NTP is 0.7 m^3. Specific heat of oxygen at constant volume is 653 J $kg^{-1}k^{-1}$. The specific heat of oxygen at constant pressure in J kg^{-1} K^{-1} is (atmosphereic pressure 10^5 Pa)

1) 713 2) 813 3) 913 4) 1013

Different thermodynamical process

48. One mole of ideal gas expands isothermally to double its volume at 27°C. Then the work done by the gas is nearly

1) 2760 cal 2) 414 cal
3) 1380 cal 4) 600 cal

49. One mole of an ideal gas expands at a constant temperature of 300 K from an initial volume of 10 litres to a final volume of 20 litres. The work done in expanding the gas is (R = 8.31 J/mole – K) [MP PMT 1995; UPSEAT 2000]

1) 750 joules 2) 1728 joules
3) 1500 joules 4) 3456 joules

50. The isothermal bulk modulus of a perfect gas at normal pressure is

1) $1.013\times10^5 N/m^2$ 2) $1.013\times10^6 N/m^2$
3) $1.013\times10^{-11} N/m^2$ 4) $1.013\times10^{11} N/m^2$

51. A gas for which γ =1.5 is suddenly compressed to 1/4 th of the initial volume. Then the ratio of the final to initial pressure is (2001)

1) 1:16 2) 1:8 3) 1:4 4) 8:1

52. The pressure and density of a monoatomic gas $(\gamma = 5/3)$ change adiabitically from (P_1, d_1) to (P_2, d_2). If $\frac{d_2}{d_1} = 8$ then $\frac{P_2}{P_1}$ should be [2004E]

1) $\frac{1}{32}$ 2) 32 3) 128 4) $\frac{1}{8}$

53. Air is filled in amotor tube at 27^{0}C and at a pressure of 8 atmospheres. The tube suddenly bursts, then temperature of air is [Given γ of air = 1.5]

1) 27.5^{0}C 2) 75^{0}C
3) 150 K 4) 150^{0}C

54. **A mono gas initially at 27°C is compressed adiabatically to one eight of its origianl volume. The temperature after compression will be**

1) 10°C 2) 887°C 3) 927°C 4) 144°C

55. **One gm mol of a diatomic gas $(\gamma = 1.4)$ is compressed adiabatically so that its temperature rises from 27°C to 127°C. The work done will be**

1) 2077.5 joules 2) 207.5 joules
3) 207.5 ergs 4) none of the above

56. **A container of volume $2m^3$ is divided into two equal compartments, one of which contains an ideal gas at 400 K. The other compartment is vaccum. The whole system is thermally isolated from its surroundings. The partition is removed and the gas expands to occupy the whole volume of the container. Its temperature now would be**

1) 400 K 2) 250 K
3) 200 K 4) 100 K

57. **If 300 ml of a gas at 27°C is cooled to 7°C at constant pressure, then its final volume will be [Pb.PET 1999; BHU 2003; CMPT 2004]**

1) 540 ml 2) 350 ml
3) 280 ml 4) 135 ml

58. **Two kg of water is converted into steam by boiling at atmospheric pressure. The volume changes from 2×10^{-3} m^3 to 3.34 m^3. The work done by the system is about**

1) – 340 kJ 2) – 170 kJ
3) 170 kJ 4) 340 kJ

***59** **At 27°C and pressure of 76 cm of Hg the volume of a diatomic gas is 2000 cm^3. If it is compressed adiabatically to a volume of 1000 cm^3, what are its pressure and temperature? (γ =1.4)**

1) 200.5 cm of Hg, 122.9°C
2) 180.4 cm of Hg, 84.2°C
3) 120 cm g kg 80°C
4) 162.4 cm of Hg 92°C

60. **The work done on a gas when it is compressed isothermally at 27°C to half of the initial volume is (nearly)**

1) 3436 J 2) -1718 J
3) +1718 J 4) -3436J

GRAPHS

61. **A thermodynamic system is taken through the cycle PQRSP process. The net work done by the system is [Orissa JEE 2002]**

1) 20 J
2) – 20 J
3) 400 J
4) – 374 J

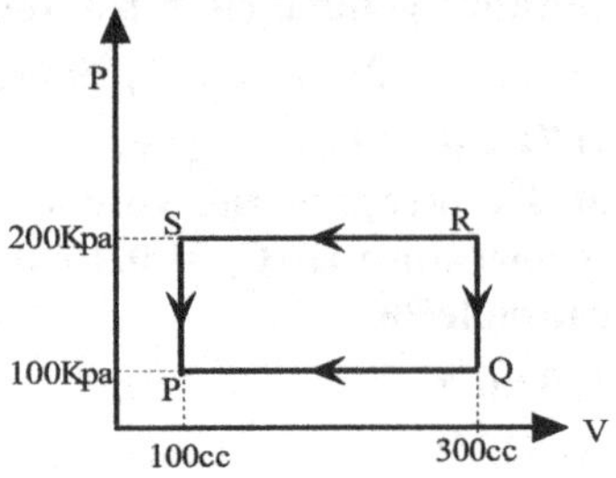

62. **A system is taken through a cyclic process represented by a circle shown. The heat absorbed by the system is**

1) $\pi \times 10^3$ J
2) $\frac{\pi}{2}$ J
3) $4\pi \times 10^2$ J
4) πJ

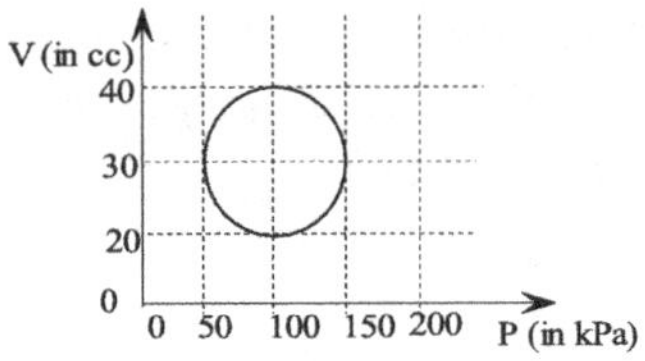

HEAT ENGINE

63. **The efficiency of a heat engine if the temperature of source 227°C and that of sink is 27°C nearly**

1) 0.4 2) 0.5 3) 0.6 4) 0.7

64. **A Carnot engine takes 3×10^6 cal. of heat froma reservoir at 627°C, and gives it to a sink at 27°c. The work done by the engine is [AIEEE 2003]**

1) 4.2×10^6 J 2) 8.4×10^6 J
3) 16.8×10^6 J 4) zero

65. **The coefficient of performance of a Carnot refrigerator working between 30°C and 0° C is**

1) 10 2) 1 3) 9 4) 0

66. **The temperature of sink of Carnot engine is 27°C. Efficiency of engine is 25%. Then temperature of source is [DCE 2002; CPMT 2002]**

1) 227°C 2) 327°C
3) 127°C 4) 27°C

ANSWERS

1) 1	2) 3	3) 1	4) 2	5) 3
6) 2	7) 3	8) 1	9) 1	10) 1
11) 2	12) 4	13) 4	14) 4	15) 3

16) 1	17) 3	18) 3	19) 1	20) 3
21) 2	22) 4	23) 3	24) 1	25) 3
26) 1	27) 2	28) 1	29) 4	30) 1
31) 2	32) 1	33) 1	34) 2	35) 4
36) 2	37) 3	38) 4	39) 4	40) 2
41) 4	42) 2	43) 2	44)1	45) 3
46) 1	47) 3	48) 2	49) 2	50) 1
51) 4	52) 2	53) 3	54) 3	55) 1
56) 1	57) 3	58) 4	59) 1	60) 2
61) 2	62) 2	63) 3	64) 2	65) 3
66) 3				

EXERCISE - III (A)

Calorimetry

1. Steam at 100°C is passed into 22 grams of water at 20°C. When resultant temperature is 90°C, then weight of the water present is

1) 27.33 g 2) 24.8 g 3) 2.8 g 4) 30 g

2. 50 grams of ice at 0°C is mixed with 40 gm of water at 60°C. Then the ratio of ice and water at equilibrium is

1) 2 : 7 2) 7 : 2 3) 2 : 3 4) 2 : 5

3. The temperature of equal masses of three different liquids A,B and C are 12° C, 19°c and 28°C respectively. The temperature when A and B are mixed is 16°C and when B and C are mixed it is 23°C. What should be the temperature when A and C are mixed ?

1) 20.26 °C 2) 15.87 °C
3) 25 °C 4) 30 °C

4. A calorimeter of water equivalent 100 grams contains 200 grams of water at 10°C. A solid of mass 500 grams at 45°C is added to the Calorimeter. If equilibrium temperature is 25°C then, the specific heat of the solid is

1) 0.45 cal g^{-1} $°C^{-1}$ 2) 0.1 cal g^{-1} $°C^{-1}$
3) 4.5 cal g^{-1} $°C^{-1}$ 4) 0.01 cal g^{-1} $°C^{-1}$

5. 30 gms of water at 30°C is in a beaker. Which of the following, when added to water, will have greatest cooling effect? (Specific heat of copper = 0.1 cal/gm°C) (1997)

1) 100 gms of water at 10°C
2) 15 gms of water at 0°C
3) 3 gms of ice at 0°C
4) 18 gms of copper at 0°C

06. 10 gm of ice at – 10° C is mixed with 100 gm of water at 50° C contained in a calorimeter weighing 50gm. (Specific heat of water = 1 cal. gm^{-1}. $°C^{-1}$, Latent heat of ice = 80 cal / gm, specific heat of ice = 0.5 cal/gm / °C and specific heat of copper = 0.09 cal/gm/°C). The final temperature reached by the mixtures is

1) 25.5°C 2) 30.0 °C **(2008 M)**
3) 38.2° C 4) 40.0° C

07. The fraction of ice that melts by mixing equal masses of ice at -10°C and water at 60°C is

1) $\frac{6}{11}$ 2) $\frac{11}{16}$ 3) $\frac{5}{16}$ 4) $\frac{11}{15}$

08. 'n' number of liquids of masses m,2m,3m, 4m, ... having specific heats s, 2s,3s,4s, ... are at temperatures t, 2t, 3t, 4t . . . . are mixed. The resultant temperature of mixture is

1. $\frac{3n}{(2n+1)}t$ 2. $\frac{2n(n+1)}{3(2n+1)}t$

3. $\frac{3n(n+1)}{2(2n+1)}t$ 4. $\frac{3n(n+1)}{(2n+1)}t$

09. A heater melts 0°C ice in a bucket completely into water in 6 minutes and then evaporates all that water into steam in 47 minutes 30 sec. If latent heat of fusion of ice is 80 cal/gram, latent heat of steam will be (specific heat of water is 1 cal / gm / °C)

1) 536 Cal/gram 2) 533.3 Cal/gram
3) 540 Cal/gram 4) 2.268×10^6 J/Kg

10. A tap supplies water at 10°c and another tap at 100°c. How much hot water must be taken so that we get 20kg of water at 35°c.

1) 40/9 kg 2) 50/9 kg
3) 20/9 kg 4) 130/9 kg

***11. 50g of steam at 100°C is passed into 250 g of ice at 0°C. Find the resultant temperature if ltent heat of steam is 540 cal/g, latent heat of ice is 80 cal/g and specific hea of water is 1 cal/g°C.**

1) 40°C 2) 30°C
3) 20°C 4) 10°C

12. Power of a man who can chew 0.3 kg ice in one minute is(in Cal/s)

1) 400 2) 4 3) 24 4) 240

Joul's law

13. A bullet of mass $10x10^{-3}$kg moving with a speed of $20ms^{-1}$ hits an ice block (0°C) of 990g kept at rest on a frictionless floor and gets embedded in it. If ice takes 50% of K.E, the amount of ice melted (in grams) approximately is (J=4.2J/cal) (Latent heat of ice = 80cal/g) (2006M)

1) 6 2) 3
3) $6x10^{-3}$ 4) $3x10^{-3}$

14. A lead bullet, of initial temperature 27° C and speed 'v' kmph pnetrates into a solid object and melts. If 50% of the kinetic energy is used to heat it, the value of v in kmph is (for leads melting point = 600 K, latent heat of fusion = 2.5 x 10^4 J Kg^{-1} ; specific heat = 125 J kg^{-1} K^{-1}) (2008 M)

1) 1800 2) 3600
3) 1200 4) 1000

15. A 10 KW drilling machine is used to drill a bore in a small aluminium block of mass 8 kg. How much is the rise of temperature in 2.5 minutes assuming 50% of the power is used up in heating the machine itself and remaining is lost to the surroundings. (specific heat of aluminium is 0.9$Jgm^{-1}K^{-1}$)

1) 92°C 2) 103°C 3) 77°C 4) 122°C

16. A steel drill is making 180 revolutions per minute, under a constant torque of 5 N-m. If it drills a hole in 7 s. in a steel block of mass 600 gm, rise in temperature of the block is (s = 0.1 cal/gm/°C)

1) 2.6°C 2) 1.3°C 3) 5.2°C 4) 3°C

***17. A metal shpere of radius r and specific heat S is rotated about an axis passing through its centre at a speed of n rotations per second. It is stopped and 50% of its energy is used in increasing its temperature, then the raise in temperature of the shere is (2002E)**

1) $\frac{2}{5}\frac{\pi^2 n^2 r^2}{S}$ 2) $\frac{1}{10}\frac{\pi^2 n^2}{r^2 S}$

3) $\frac{7}{8}\pi r^2 n^2 S$ 4) $\frac{5(\pi rn)}{14S}$

First law of thermodynamics

18. When 20 J of work was done on a gas, 40J of heat energy was released. If the initial internal energy of the gas was 70J, what is the final internal energy?

1) 50 J 2) 60 J 3) 90 J 4) 110 J

19. When 1 gm of water changes from liquid to vapour phase at constant pressure of 1 atmosphere, the volume increases from 1cc to 1671cc. The heat of vaporisation at this pressure is 540 cal/gm. Increase in internal energy of water is (1 atmosphere = 1.01 x 10^6 dyne/cm^2)

1) 4200J 2) 8200J
3) 1200J 4) 2100 J

C_P, C_V and their relations

20. 70 cal of heat is required to rise the temperature of 2 moles of an ideal gas at constant pressure from 30°C to 35°C. What is the amount of heat required to rise the temperature of same gas through the same range at constant voume ? (R = 2cal/mole/K)

1) 28 J 2) 50 Cal
3) 75 J 4) Zero

21. When an ideal diatomic gas is heated at constant pressure, its internal energy increased by 50cal then the work done by the gas is

1) 30 cal 2) 50 cal
3) 70 cal 4) 20 cal

22. If for hydrogen $C_P - C_V$ = m and for nitrogen $C_P - C_V$ = n, Where C_P and C_V refer to specific heats per unit mass respectively at constant pressure and constant volume, the relation between m and n is (molecular weight of hydrogen = 2 and molecular weight or nitrogen = 14) (2002)

1) n = 14m 2) n =7m
3) m = 7n 4) m = 14n

23. One mole of a monoatomic gas is mixed with one mole of a diatomic gas. What will be the γ for the mixture? [AIEEE-2002, 2004]

1) 1.5 2) 1.54 3) 1.4 4) 1.45

24. The ratio of specific heats of a gas is γ. The change in internal energy of one mole of gas when the volume changes from v to 2v at constant pressure "p" is (2005E)

1) $\frac{pv}{\gamma-1}$ 2) pv 3) $\gamma-1$ 4) $\frac{pv}{\gamma}$

DIFFERENT THERMODYNAMICAL PROCESS

25. γ for a gas is 5/3. An ideal gas at 27°C is compressed adiabatically to 8/27 of its original volume. The rise in temperature of the gas is

1) 450°C 2) 375°C 3) 225°C 4) 402°C

26. At N.T.P. one mole of diatomic gas is compressed adiabatically to half of its volume $\gamma = 1.41$. The work doen on gas will be

1) 1280 J 2) 1610 J
3) 1824 J 4) 2025 J

27. 5 moles of hydrogen $(\gamma = 7/5)$ intially at S.T.P. are compressed adiabatically so that its temperature becomes 400°C. The increase in the internal energy of the gas in kilo-J is ($R=8.30 Jmole^{-1}k^{-1}$) (2002E)

1) 21.55 2) 41.50 3) 65.55 4) 80.55

28. During an adiabatic process, the pressure of a gas is proportional to the cube of its absolute tempereature.The value of C_p/C_v for that gas is (AIEEE-2003 & 2003M)

1) 3/5 2) 4/3 3) 5/3 4) 3/2

29. During adiabatic expansion of gas, volume increases by 5%. If $\gamma = 3/2$ find the percentage change in pressure?

1) 7.5 % 2) –2.5 % 3) –15.5 % 4) –7.5 %

30. The equation of a certain gas can be written as $\frac{T^{7/5}}{P^{2/5}} = const.$. Its specific heat at constant volume will be

1) $\frac{3}{2}R$ 2) $\frac{5}{2}R$ 3) $\frac{7}{2}R$ 4) 2R

31. The tyre of motor car contains air at 15°C. If the temperature increases to 35°C the approximate percentage increase in pressure is (ignor expansion of tyre) (2005 E)

1) 7 2) 9 3) 11 4) 13

32. A monoatomic ideal gas, initially at temperature T_1, is enclosed in a cylinder fitted with a frictionless piston. The gas is allowed to expand adiabatically to a temperature. T_2 by releasing the piston suddenly. IF L_1 and L_2 are teh lengths of the gas column before and after expansion respectively, then T_1/T_2 is given by [IIT - JEE (Screening) 2000]

1) $\left(\frac{L_1}{L_2}\right)^{\frac{2}{3}}$ 2) $\frac{L_1}{L_2}$

3) $\frac{L_2}{L_1}$ 4) $\left(\frac{L_2}{L_1}\right)^{\frac{2}{3}}$

33. Three samples of the same gas x , y and z , for which the ratio of specific heats is r=3/2, have initially the same volume. The volumes of each sample is doubled , by iosbaric process in the case of y and by isothermal process in the case of z. If the initial pressures of the samples x,y and z are in the ratio $2\sqrt{2}:1:2$, then the ratio of their final pressures is

1) 2 : 1 : 1 2) 1 : 1 : 1
3) 1 : 2 : 1 4) 1 : 1 : 2

34. Two cylinder A and B fitted with pistons contain equal number of moles of an ideal monoatomic gas at 400K. The piston A is free to move, while that of B is held fixed. The same amount of heat is given to the gas in each cylinder. If the rise in temperature of the gas in A is 42K, then the rise in temperature of the gas in B is (γ=5/3)(2007E)

1) 21 K 2) 35 K 3) 42 K 4) 70 K

35. Two identical vessels A and B with frictionless pistons contain the same ideal gas at teh same temperature and the same Volume V. The masses of gas in A and B are m_A and m_B respectively. The gases are allowed to expand isothermally tothe same final volume 3V. The change in pressures of the gas in A and B are found to be ΔP and 1.5 ΔP respectively. Then [2004 E]

1) $9m_A = 4m_B$ 2) $3m_A = 2m_B$
3) $2m_A = 3m_B$ 4) $4m_A = 9m_B$

36. 'm' grams of a gas of a molecular weight M is flowing in an isolated tube with velocity 2V. IF the gas flow is suddenly stopped the rise in its temperature is; (γ = ratio of specific heats; R = universal gas constant; J = Mechanical equilent of heat) [2006M]

1) $\frac{2MV^2(\gamma-1)}{RJ}$ 2) $\frac{mv^2(\gamma-1)}{M2RJ}$

3) $\frac{mv^2\gamma}{2RJ}$ 4) $\frac{Mv^2\gamma}{2RJ}$

37. Heat is supplied to a diatomic gas at constant pressure. The ratio of $\Delta Q:\Delta U:\Delta W$ is:

1) 5 : 3 : 2 2) 5 : 2 : 3
3) 7 : 5 : 2 4)7 : 2 : 5

Graphs

38. In the figure given two processes A and B are shown by which a thermo-dynamical system goes from initial to final state F. If ΔQ_A and ΔQ_B are respectively the heats supplie to the system then

1) $\Delta Q_A = \Delta Q_B$

2) $\Delta Q_A \geq \Delta Q_B$

3) $\Delta Q_A < \Delta Q_B$

4) $\Delta Q_A > \Delta Q_B$

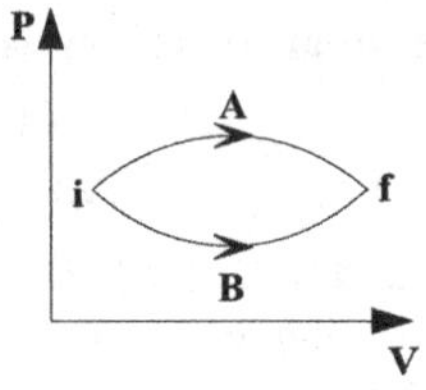

39. The P –V diagram of a system undergoing thermodynamic transformation is shown in figure. The work done on the system in going from $A \to B \to C$ is 50 J and 20 cal heat is given to the system. The change in internal energy between A and C is [UPSEAT 2002]

1) 34 J

2) 70 J

3) 84 J

4) 134 J

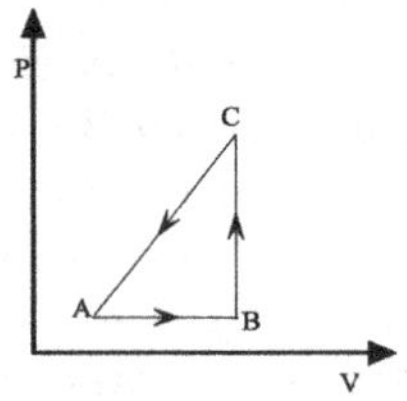

40. A cyclic process ABCD is shown below in the given P - V diagram. In the following answers the one that represents the same process as in P - V diagram : (2006 E)

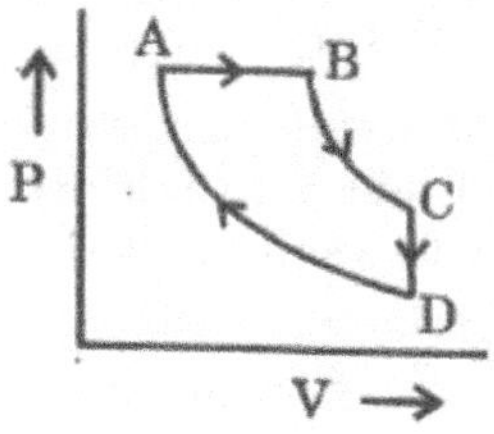

1) 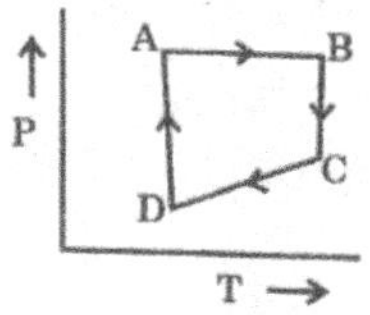

2) 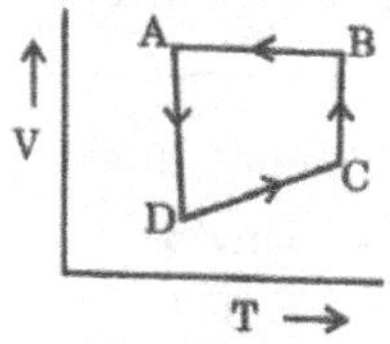

3) 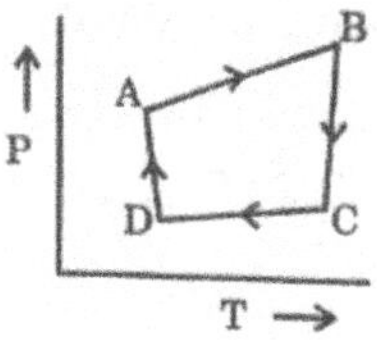

4) 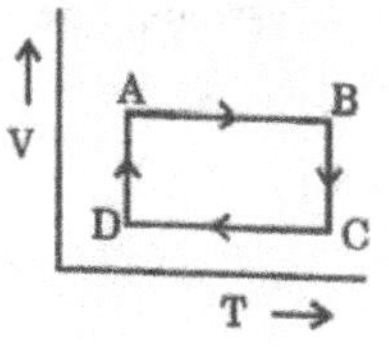

41. An ideal gas is subjected to a cyclic process involving four thermodynamic states; the amounts of heat (Q) and work (W) involved in each of these states are

$Q_1 = 6000$ J, $Q_2 = -5500$ J ; $Q_3 = -3000$ J; $Q_4 = 3500$ J

$W_1 = 2500$ J; $W_2 = -1000$ J; $W_3 = -1200$ J; $W_4 = x$J.

The ratio of the net work done by the gas to the total heat abosrbed by the gas is 'η'. The values of x and η respetively are (2009 E)

1) 500 ; 7.5% 2) 700 ; 10.5%

3) 1000 ; 21% 4) 1500 ; 15%

Heat Engine

42. A Carnot's engine is made to work between 200°C and 0°C first and then between 0°C and –200°C. the ratio of efficiencies of the engine in the two cases is [KCET 2002]

1) 1.73:1 2) 1:1.73 3) 1:1 4) 1 : 2

43. A scientist says that the efficiency of his heat engine which operates at source temperature 127°C and sink temperatue 27°C is 26%, then

1) It is impossible

2) It is possible but less probable

3) It is quite probable

4) Data are incomplete

44. Efficiency of a Carnot engine is 50% when temperature of outlet is 500 K. In order to increase efficiency up to 60% keeping temperature of intake the same what is temperature of outlet [CBSE PMT 2002]

1) 200 K 2) 400 K 3) 600 K 4) 800 K

45. An ideal refrigerator has a freezer at a temperature of –13°C. The coefficient of performance of the engine is 5. The temperature of the air (to which heat is rejected) will be [BHU 2000; CMPT 2002]

1) 325°C 2) 325 K 3) 39°C 4) 320°C

46. A reversible engine converts one-sixth of the heat input into work. When the temperature of the sink is reduced by 62°C, the efficiency of the engine is doubled. The temperature of the source and sink are [CBSE PMT 2000]

1) $80^0C, 37^0C$ 2) $95^0C, 28^0C$

3) $90^0C, 37^0C$ 4) $99^0C, 37^0C$

ANSWERS

1) 2	2) 1	3) 1	4) 1	5) 1
6) 3	7) 2	8) 3	9) 2	10) 2
11) 1	12) 1	13) 4	14) 1	15) 2
16) 1	17) 1	18) 1	19) 4	20) 2
21) 4	22) 3	23) 1	24) 1	25) 2
26) 3	27) 2	28) 4	29) 4	30) 2
31) 4	32) 4	33) 2	34) 4	35) 2
36) 1	37) 3	38) 4	39) 4	40) 1
41) 2	42) 2	43) 1	44) 2	45) 3
46) 4				

EXERCISE - II (B)

Calorimetry :

1. **Steam is passed into a calorimeter with water having total thermal capacity 110 cal/gm and initial temperature 30°C. If the resultant temperature is 90°C, the increase in the mass of the water is**

   1) 12 gm 2) 1.2 gm 3) 5 gm 4) 12.4 gm

2. **10 grams of ice at - 20°C is added to 10 grams of water at 50°C. The amount of ice and water that are present at equilibrium respectively**

   1) 0,20g 2) 5g, 15g
   3) 5g, 10g 4) 10g, 10g

3. **Three liquids A, B and C of masses 400g, 600g and 800g are at 30°C, 40°C and 50°C respectively. When 'A' and 'B' are mixed resultant temperature is 36°C. When B and C are mixed resultant temperature is 44°C. Then ratio of their specific heats are**

   1) 2:1:1 2) 3:2:1
   3) 2:2:1 4) 1:4:9

4. **A copper calorimeter contains 100g of water at 16°C. When 15g of ice is added to it, the resultant temperature of the mixture is 4°C. Water equivalent of the calorimeter is**

   1) 10g 2) 20g 3) 5g 4) 25g

5. **Which one of the following would rise the temperature of 20 gm of water at 30°C most, when added to water ? [Specific heat of copper is 0.1 cal/gm °C].**

   1) 20 gm of water at 40°C
   2) 10 gm of water at 27°C
   3) 1 kg of copper at 25°C
   4) 4 gm of water at 80°C

6. **300g of water at 25°C is added to 100g of ice at 0°C. The final temperature of the mixture is**

   1) $\frac{-5}{3}$ °C 2) $\frac{-5}{2}$ °C
   3) -5 °C 4) 0°C

7. **The temperature of ice is $-20^{o}C$ [specific heat = 0.5 k.cal/(kg- °C)] and that of water at 60°C. They are mixed in equal amounts. What part of the ice will be melted ?**

   1) (5/6) th 2) 3/4 th
   3) whole of it will be melted
   4) (5/11) th

8. **A liquid of mass M and specific heat S is heated to a temperature T. Another liquid of mass M/3 and specific heat 2S is heated to a temperature 2T. If these two liquids are mixed, the resultant temperature of the mixture is**

   1) $\frac{2}{3}$T 2) $\frac{3}{5}$T 3) $\frac{1}{3}$T 4) $\frac{7}{5}T$

9. **An electric kettle can heat certain quantity of water from 19°C to 100°C in 6 minutes, time it takes to boil the same quantity of water into steam at 100°C in (minutes)**
   (L_{steam} = 540 cal/gm)

   1) 40 2) 30 3) 20 4) 50

10. **A Tap supplies water at 20°C and another tap at 80°C. Then amount of hot water that should be mixed with cold water to get 36 litres of water at 45°C is**

    1) 16.36 litres 2) 10 litres
    3) 21 litres 4) 24 litres

11. **1 g of steam at 100°C is passed into a calorimeter of water equivalent 10g containing water of mass90 g. Find the resultant temperature of the mixture if the initial temperature of waer is 27°C.**

    1)36.2°C 2) 38.4°C 3) 39.4°C 4) 33.7°C

12. **The power of a refigerator which freezes 100gm of water at 10°C into ice at 0°C in 100sec is**

    1) 1400 W 2) 1200 W 3) 378 W 4) 1000 W

Joul's law

13. **A 42kg block of ice moving an rough horizontal surface stops due to friction, after some time. If the initial velocity of the decelerating block is 4m/s, the mass of ice in (kg) that has melted due to heat generated by friction is : (latent heat of ice is $3.36 x 10^5$ J/kg) (2005M)**

    1) 10^{-3} 2) $1.5x10^{-3}$ 3) $2x10^{-3}$ 4) $2.5x10^{-3}$

14. A lead bullet just melts when stopped by an obstacle. Assuming that 2/7 th of the heat of absorbed by the obstacle. IF initial temperature is 27^0C, melting point lead is 327^0C, specified heat 0.03 cal/gm^{0}C, latent heat of fusion of lead = 6 cal/gm and J = 4.2 J/cal, velocity of that bullet is

1) 42 ms^{-1} 2) 21 ms^{-1}
3) 210 ms^{-1} 4) 420 ms^{-1}

15. A drilling machine is used to drill a hole in a 10 kg metal block. Power of the drilling machine is 1 KW and it is used for 5 minutes. IF 40% of the work done by the drilling machine is converted into heat in the metal block, the rise in its temperature is (specific heat - 0.115 cal gm^{-1} $^0C^{-1}$ and J = 4.175 J/Cal)

1) 20^0C 2) 25^0C
3) 35^0C 4) 50^0C

16. A steel drill is making 90 revolutions per minute, under a constant torque of 2.5 N-m. If it drills a hole in 14 s. in a steel block of mass 300 gm, rise in temperature of the block is (s = 0.1 cal/gm/°C)

1) 2.6°C 2) 1.3°C
3) 5.2°C 4) 3°C

17. A metal shpere of radius r and specific heat S is rotated about an axis passing through its centre at a speed of n rotations per second. It is stopped and 25% of its energy is used in increasing its temperature, then the raise in temperature of the shere is (2002E)

1) $\frac{1}{5}\frac{\pi^2 n^2 r^2}{S}$ 2) $\frac{1}{10}\frac{\pi^2 n^2}{r^2 S}$

3) $\frac{7}{8}\pi r^2 n^2 S$ 4) $\frac{5(\pi r n)}{14S}$

First Law of Thermodynamics

18. In a thermodynamic process the pressure of a fixed mass of gas is changed. In this process gas releases 20J heat and 8J of work is done on the gas. If initial internal energy of the gas was 30J, then final internal energy is

1) 2J 2) 42J 3) 18J 4) 58 J

19. 1cc of water at 100°C is given 540 cal of heat and the steam formed occupies 1671cc. at 1 atm pressure. Then workdone against atmospheric pressure in this process is nearly(in Cal)

1) 540 2) 500 3) 40 4) 100

C_P, C_V and their relations

20. 3 moles of a monoatomic gas requires 60 cal heat for 5°C rise of temperature at constant volume, then heat required for 5 moles of same gas under constant pressure for 10°C rise of temperature is (R=2 cal/mole/°k)

1) 200 cal 2) 400 cal
3) 100 cal 4) 300 cal

21. A diatomic gas does 100J of work when it is expanded isobarically. The heat given to the gas during this process is

1) 700J 2) 350J 3) 175J 4) 1050J

22. If C_p and C_v denote the specific heats of nitrogen per unit mass at constant pressure and constant volume respectively then [AIEEE 2007]

1) $C_p - C_v = R$ 2) $C_v - C_p = 28R$

3) $C_p - C_v = R/28$ 4) $C_p - C_v = R/14$

23. A gaseous mixture consists of 16g of helium and 16 g of oxygen. The ratio C_p/C_v of the mixture is [AIEEE 2005]

1) 1.4 2) 1.54
3) 1.59 4) 1.62

24. The ratio of specific heats of a gas is γ. The change in internal energy of one mole of gas when the volume changes from v to 3v at constant pressure "p" is (2005E)

1) $\frac{2pv}{\gamma-1}$ 2) 2pv 3) $\gamma-1$ 4) $\frac{2pv}{\gamma}$

Different thermodynamical process

25. A gas is suddenly compressed to 1/4th of its original volume at normal temperature. The increase in its temperature is $(\gamma = 1.5)$

1) 273 K 2) 573 K
3) 373 K 4) 473 K

26. Two moles of an ideal monoatomic gas at 27^0C occupies a volume of V. If the gas is expanded adiabatically to the volume 2V, then the work done by the gas will be

$[\gamma = 5/3, R = 8.31 J/molK]$

1) – 2767.23 J 2) 2767.23 J
3) 2500 J 4) – 2500 J

27. One mole of an ideal gas with $\gamma = 1.4$ is adiabatically compressed so that its temperature rises from 27^0C and 35^0C. The change in the internal energy of the gas is (2001M)

1) -166 J 2) 166J 3) 168 J 4) 168J

28. The relation between P and T for monoatomic gas during adiabatic process is $P \propto T^C$. The value of C is

1) 5/2 2) 2/5 3) 5/3 4) 3/5

29. The volume of air increases by 5% in its adiabatic expansion. The percentage decrease in its pressure will be

1) 5% 2) 6% 3) 7 % 4) 8%

30. Certain perfect gas is found to obey $PV^{3/2}$ = constant during adiabatic process. If such a gas at initial temperature T is adiabatically compressed to half of the initial volume, its final temperature will be

1) $\sqrt{2}T$ 2) 2T 3) $2\sqrt{2}T$ 4) 4T

31. A given mass of a gas is compressed isothermally until its pressure is doubled. It is then allowed to expand adiabatically until its original volume is restored and its pressure is then found to be 0.75 of its initial pressure. The ratio of the specific heats of the gas is approximately : [2006 E]

1) 1.20 2) 1.41 3) 1.67 4) 1.83

32. A monoatomic ideal gas, initially at temperature T_1, is enclosed in a cylinder fitted with a frictionless piston. The gas is allowed to expand adiabatically to a temperature. T_2 by releasing the piston suddenly. IF L_1 and L_2 are the lengths of the gas column before and after expansion respectively, then $\frac{L_1}{L_2}$ is given by

1) $\left(\frac{T_1}{T_2}\right)^{\frac{2}{3}}$ 2) $\frac{T_1}{T_2}$ 3) $\frac{T_2}{T_1}$ 4) $\left(\frac{T_2}{T_1}\right)^{\frac{3}{2}}$

33. Work of 146kJ is performed in order to compress one kilo mole of gas adiabatically and in this process the temperature of gas increases by 7°C, the gas is [AIEEE 2006]

1) monoatomic 2) diatomic 3) triatomic
4) a mixture of monoatomic and diatomic

34. Two cylinders A and B fitted with pistons contain equal amounts of an ideal diatomic gas at 300 K. The piston of A is free to move while that of B s held fixed. The same amount of heat is given to the gas in each cylinder. If the rise in temperature of the gas in A is 30 K, then the rise in temperature of the gas in B is

1) 30 K 2) 18 K 3) 50 K 4) 42 K

35. Two identical containers A and B with frictionless pistons contain the same ideal gas at the same temperature and the same Volume V. The masses of gas in A is m_A adn that in B is m_B respectively. The gas in each cylinder is now allowed to expand isothermally to the same final volume 2 V. The changes in the pressure in A and B are found to be ΔP and 1.5 ΔP respectively.Then

1) $4m_A = 9m_B$ 2) $2m_A = 3m_B$
3) $3m_A = 2m_B$ 4) $9m_A = 4m_B$

36. 'm' grams of a gas of a molecular weight M is flowing in an isolated tube with velocity v. If the gas flow is suddenly stopped the rise in its temperature is; (γ = ratio of specific heats; R = universal gas constant; J = Mechanical equilent of heat)(2006 M)

1) $\frac{Mv^2(\gamma-1)}{2RJ}$ 2) $\frac{mv^2(\gamma-1)}{M2RJ}$

3) $\frac{mv^2\gamma}{2RJ}$ 4) $\frac{Mv^2\gamma}{2RJ}$

37. A mono atomic gas is supplied the heat Q very slowly keeping the pressure constant. The work doen by the gas will be

1) $\frac{2}{3}Q$ 2) $\frac{3}{5}Q$ 3) $\frac{2}{5}Q$ 4) $\frac{1}{5}Q$

<u>**GRAPHS**</u>

38. A system goes from A to B via two processes I and II as shown in figure. If ΔU_1 and ΔU_2 are the changes in internal energies in the processes I and II respectively, then [AIEEE 2005]

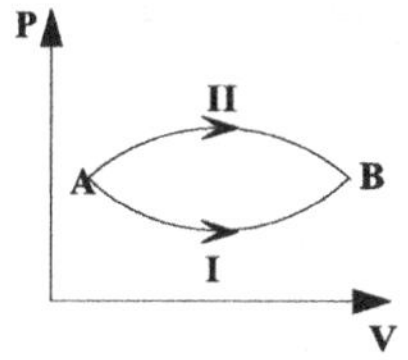

1) $\Delta U_2 > \Delta U_1$
2) $\Delta U_2 < \Delta U_1$
3) $\Delta U_1 = \Delta U_2$
4) Relation between ΔU_1 and ΔU_2 can not be determined

39. **The P – V diagram of a system undergoing thermodynamic transformation is shown in figure. The work done by the system in going from $A \rightarrow B \rightarrow C$ is 30J and 40J heat is given to the system. The change in internal energy between A and C is [BCECE 2005]**

1) 10 J
2) 70 J
3) 84 J
4) 134 J

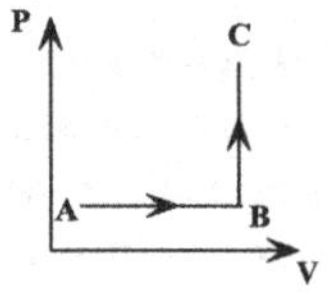

40. **When a system is taken from state i to a state f along path iaf, Q = 50J and W = 20 J. Along path ibf, Q = 35J. If w = – 13J for the return path fi, Q for this path is [AMU (Med.) 2000]**

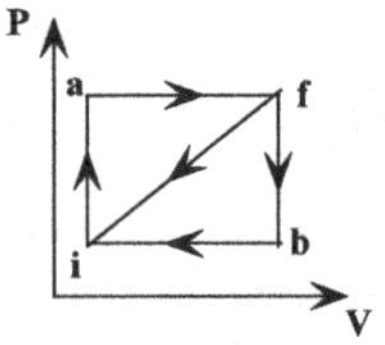

1) 33J 2) 23 J 3) – 7 J 4) – 43 J

40. **An ideal gas is subjected to a cyclic process ABCD as depicted in the P –V diagram given Which of the folloiwing curves represents the equivalent cyclie process? (2009 M & E)**

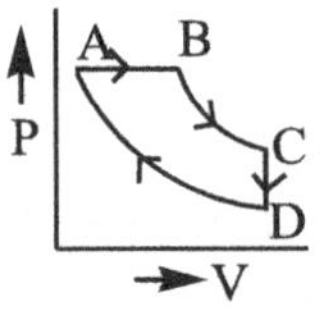

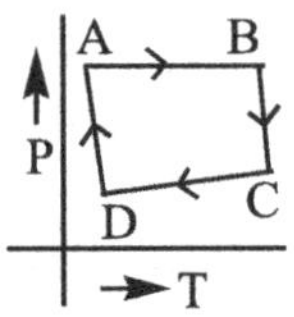

Ⓐ

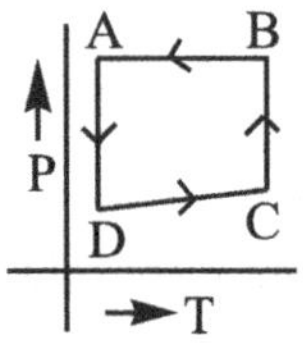

Ⓑ

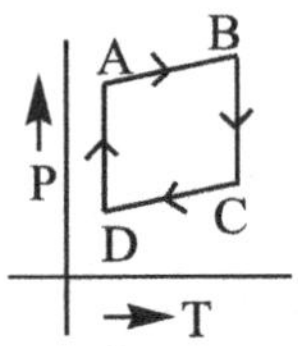

Ⓒ

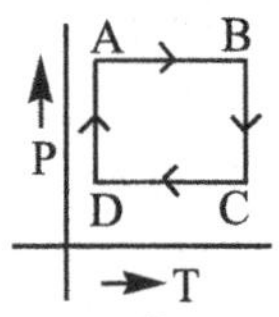

Ⓓ

1) A 2) B 3) C 4) D

Heat Engine

42. **In a Carnot engine, when T_2 = 0°C and T_1 = 200°C, its efficiency is η_1 and when T_1 = 0°C and T_2 = – 200°c, its efficiency is η_2, then what is η_1/η_2**

1) 0.577 2) 0.733
3) 0.638 4) cannot be calculated

43. **For which combination of working temperatures the efficiency of Carnot's engine is highest [KCET 200]**

1) 80 K, 60 K 2) 100 K, 80 K
3) 60 K, 40 K 4) 40K, 20 K

44. **An ideal heat engine working between temperature T_1 and T_2 has an efficiency η, the new efficiency if both the source and sink temperature are doubled, will be [DPMT 2000]**

1) $\frac{\eta}{2}$ 2) η 3) 2η 4) 3η

45. **In a mechanical refrigerator, the low temperature coils are at a temperature of – 23°C and the compressed gas in the condenser has a temperature of 27°C. The theoretical coefficient of performance is [UPSEAT 2001]**

1) 5 2) 8 3) 6 4) 6.5

46. **A Carnot engine whose low temperature reservoir is at 7°C has an efficiency of 50%. It is desired to increase the efficiency to 70%. By how many degrees should the temperature of the high temperature reservoir be increased [UPSEAT 2005]**

1) 840 K 2) 280 K 3) 560K 4) 380 K

ANSWERS

1) 1	2) 2	3) 3	4) 3	5) 4
6) 4	7) 2	8) 4	9) 1	10) 1
11) 4	12) 3	13) 1	14) 4	15) 2
16) 1	17) 1	18) 3	19) 3	20) 4
21) 2	22) 3	23) 4	24) 1	25) 1
26) 2	27) 2	28) 1	29) 3	30) 1
31) 2	32) 4	33) 2	34) 4	35) 3
36) 1	37) 3	38) 3	39) 1	40) 1
41) 4	42) 1	43) 4	44) 2	45) 1
46) 4				

COMPREHENSION QUESTIONS

Comprehension - I (1 - 2) :

The temperature of an ideal gas in a sealed 0.50 m^3 contrainer is reduced from 370K to 270 K. The final pressure of the gas is 80 kPA. The molar heat capacity at constant volume of the gas is 280.0 J/mol. K.

1. The heat absorbed by the gas, in kJ, is closest to

   1) zero 2) +50 3) +65 4) –50

2. The work done by the gas, in kJ, is closest to

   1) zero 2) +50 3) +65 4) –50

Comprehension - II (3- 4):

A compression, at a constant pressure of 100 kPa, is performed on 6.0 moles of an ideal monatomic (C_V = 3/2 R). The compression reduces the volume of the gas from 0.20 m^3 to 0.12 m^3.

3. The change in the internal energy of the gas, in kJ, is closest to

   1) zero 2) +12 3) +20 4) –12

4. The work done by the gas, in kJ, is closest to

   1) zero 2) +8 3) +20 4) –8

Comprehension - III (5 - 6) :

A monatomic ideal gas (C_V = 3/2 R) undergoes an isothermal expansion at 300 K, as the volume increased from 0.08 m^3 to 0.40 m^3. The final pressure is 100 kPa.

5. The change in the internal energy of the gas, in kJ, is closest to

   1) zero 2) +32 3) +64 4) –32

6. The heat transfer to the gas, in kJ, is closed to

   1) zero 2) +32 3) +64 4) –32

Comprehension - IV (7 - 9) :

A cyclinder contains 12 moles of an indeal gas at a temperature of 300 K. The gas is compressed at constant pressure until the final volume equals 0.80 times the initial volume. The molar heat capacity at constant volume of the gas is 24.0 J/mol. K.

7. The heat absorbed by the gas, in kJ, is closest to

   1) +6 2) +17 3) +23 4) –17

8. The change in the internal energy of the gas, in kJ, is closest to

   1) +6 2) +17 3) +23 4) –17

9. The adiabatic constant for the gas is closest to

   1) 1.35 2) 1.40 3) 1.47 4) 1.56

Comprehension - V (10 - 12):

A sealed tank contains 30 moles of an ideal gas, at an initial temperature of 270 K. The pressure of the gas is increased until the final pressure equals 1.6 times the initial pressure. The heat capacity at constant pressure of the gas is 34.0 J/mol. K.

10. The heat absorbed by the gas, in kJ, is closest to

    1) +125 2) +165 3) +205 4) –78

11. The change in the internal energy of the gas, in kJ, is closest to

    1) +125 2) +165 3) +205 4) –78

12. The adiabatic constant for the gas is closest to

    1) 1.29 2) 1.32 3) 1.35 4) 1.38

Comprehension - VI (13 - 15) :

Twenty moles of a monatomic ideal gas ($\gamma = 5/3$) undergo an adiabatic process. The initial pressure is 400 kPa and the initial temperature is 450 K. The final temperature of the gas is 320 K.

13. The final volume of the gas, in SI units, is closest to

    1) 0.19 2) 0.23 3) 0.27 4) 0.31

14. The heat absorbed by the gas, in kJ, is closest to

    1) zero 2) +32 3) +54 4) +32

15. The change in the internal energy of the gas, in kJ, is closest to

    1) zero 2) +32 3) +54 4) +32

Comprehension - VII (16 - 19) :

An adiabatic process is performed on 15 moles of an ideal gas. The initial temperature is 320 K and the initial volume is 0.80 m^3. The final volume is 0.40 m^3. The adiabatic constant for the gas is 1.44.

16. The final temperature of the gas, in SI units, is closest to

    1) 435 2) 450 3) 465 4) 480

17. The heat absorbed by the gas, in kJ, is closest to

    1) zero 2) +14 3) +32 4) –24

18. The work doen by the gas, in kJ, is closest to

    1) zero 2) +14 3) +32 4) –32

19. The heat capacity at constant pressure for the gas, in SI units, is closest to

    1) 23 2) 25 3) 27 4) 29

Comprehension - VIII (20 - 21) :

Two Carnot heat engines operate in tandem as follows: engine A takes in 16.0 kJ per cycle from a heat reservoir at a temperature of 440 K. The heat rejected by engine A is received by engine B, which performs 3.5 kJ of net work per cycle. Engine B, in turn, rejects heat at a temperature of 300 K.

20. The temperature at which engine A rejects heat to engine B, in SI units, is closest to

1) 365 2) 375 3) 385 4) 395

21. The work per cycle performed by engine A, in kJ, is closest to

1) 1.6 2) 1.8 3) 2.0 4) 2.2

Comprehension - IX (22 - 23):

A real (non-Carnot) heat engine, operating between heat reservoirs at temperature of 650 K and 270K, performs 3.5 kJ of net work, and rejects 8.5 kJ of heat, in a single cycle.

22. The thermal efficiency of this heat engine is closest to

1) 0.23 2) 0.26 3) 0.29 4) 0.32

23. The thermal efficiency of a Carnot heat engine, operating between the same heat reservoirs, in percent, is closest to

1) 40 2) 45 3) 50 4) 60

Comprehension - X (24 - 26):

Two moles of helium gas are taken over the cycle ABCDA, as shown in the P-T diagram.

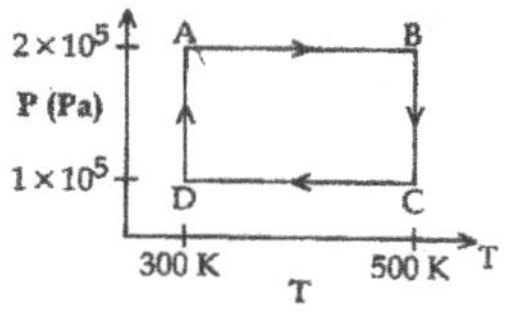

(2009 AIEEE)

24. Assuming the gas to be ideal the work done on the gas in taking it from A to B is:

1) 300R 2) 400R 4) 500R 4) 200R

25. The work done on the gas in taking it from D to A is

1) + 414R 2) – 690 R 3) + 690 R 4) – 414R

26. The net work done on the gas in the cycle ABCDA is:

1) 276 R 2) 1076 R 3) 1904 R 4) zero

ANSWERS

1) 4	2) 1	3) 4	4) 4	5) 1
6) 3	7) 1	8) 4	9) 4	10) 1
11) 1	12) 2	13) 4	14) 1	15) 4
16) 1	17) 1	18) 4	19) 3	20) 4
21) 1	22) 3	23) 2	24) 2	25) 4
26) 1				

Chapter - 8

WAVE MOTION
(Part - A)

WAVE MOTION

SYNOPSIS

Introduction

There are essentially two ways of transporting energy from the place where it is produced to the place where it is desired to be utilized. The first involves the actual transport of matter. For example, a bullet fired from a gun carries its kinetic energy with it which can be used at another location. The second method by which energy can be transported is much more useful and important, it involves what we call a wave process.

A wave is a disturbance that propagates in space, transports energy and momentum from one point to another without the transport of **matter.** Waves are every where whether we recognize or not, we encounter waves on a daily basis. Sound waves, visible light waves, radio waves, ripples on water surface, earthquake waves and waves on a string are just a few examples of waves.

Waves can be one, two or three dimensional according to the number of dimensions in which they propagate energy. Waves moving along strings are one dimensional, ripples on liquid surface are two dimensional, while sound and light waves are three dimensional.

Types of Waves

Waves can be classified in a number of ways based on the following characteristics

On the basis of necessity of medium

i)Mechanical waves: Require medium for their propagation e.g., Waves on string and spring, waves on water surface, sound waves, seismic waves.

ii) Non-mechanical waves: Do not require medium for their propagation are called e.g., Electromagnetic waves like, light, heat (Infrared), radio waves, γ -rays, x-rays etc.

On the basis of vibration of particle:

On the basis of vibration of particle of medium waves can be classified as transverse waves and longitudinal waves.

1) **Transverse waves:** i) Particles of the medium vibrates in a direction perpendicular to the direction propagation of wave

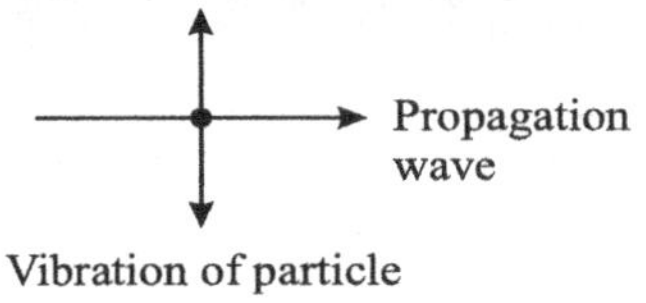

ii) It travels in the form of crests(C) and troughts(T)

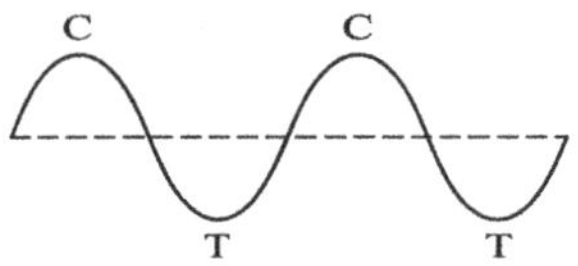

iii) Transverse waves can be transmitted through solids, they can be setup on the surface of liquids. But they cannot be trnasmitted into liquids and gases.

iv) Medium should posses the property of rigidity

v) Transverse waves can be polarised.

vi) Movement of string of a sitar or violin, movement of the membrane of a Tabla or Dholak, movement of kink on a rope waves setup on the surface of water.

2) **Longitudinal waves :** i) Particles of a medium vibrate in the direction of wave motion.

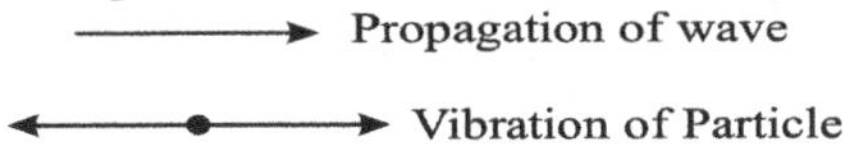

ii) It travels in the form of compression (C) rarefaction (R).

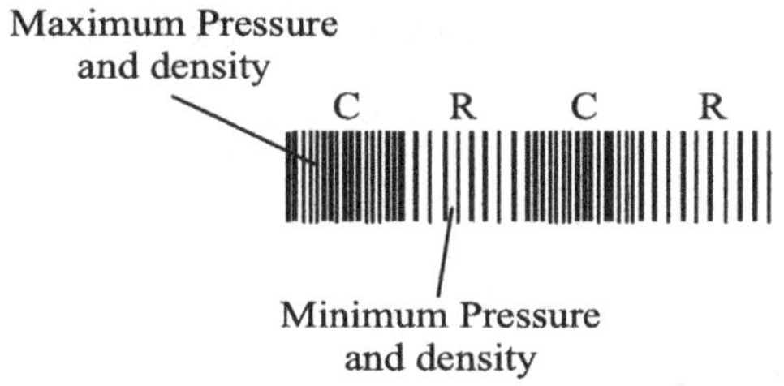

iii) These waves can be transmitted through solids, liquids and gases because for propagation, volume elasticiy is necessary.
iv) Medium should posses the property of elasticity.
v) Longitudinal waves can not be polarized.
vi) Sound waves travel through air, vibration of air column in organ pipes vibration of air column above the surface of water in the tube of resonance apparatus.

On the basis of energy propagation

i) Progressive wave: These waves advances in a medium with definite velocity. These waves propagate energy in the medium. Eg: Sound wave and light waves.

ii) Stationary wave: These waves remains stationary between two boundaries in medium. Energy is not propagated by these waves but it is confined in segments (or loops) e.g., Wave in a string, waves in organ pipes.

Simple Harmonic wave

When a wave passes through a medium, if the particles of the medium execute simple harmonic vibrations, then the wave is called a simple harmonic wave. A graph is drawn (fig.) with the displacement of the particles from their mean positions, at any given instant of time, on the y-axis and their location from origin on x-axis.

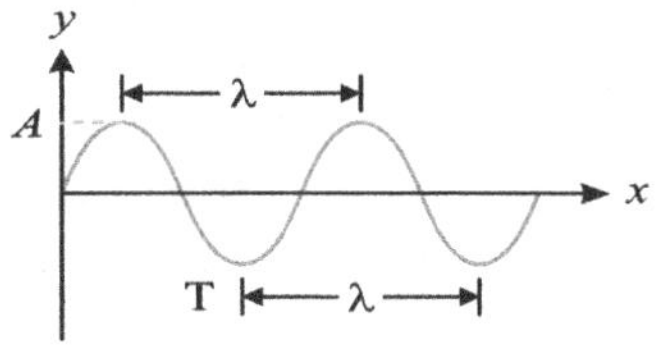

Characteristics of wave:

1. **Amplitude (A):** Maximum displacement of a vibrating particle of medium from its' mean position is called amplitude.
2. **Wavelength (λ) :** It is equal to the distance travelled by the wave during the time in which any one particle of the medium completes one vibration about its mean position.
   Or
   Distance travelled by the wave in one time period is known as wavelength.
   Or
   It is the distance between the two successive points with same phase.

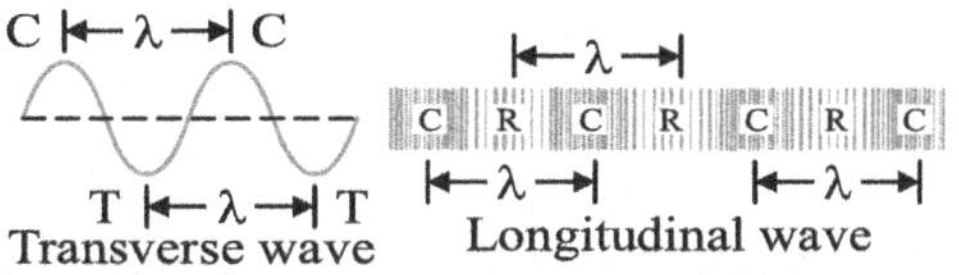

3. **Frequency (n):** Frequency of vibration of a particle is defined as the number of vibrations completed by particle in one second.
   (Or)it is the number of complete wavelengths traversed by the wave in one second.
   Unit of frequency is hertz (Hz) or per second.
4. **Time period (T):** Time period of vibration of particle is defined as the time taken by the particle to complete one vibration about its mean position
   Or it is the time taken by the wave to travel a distance equal to one wavelength.
   Time period = 1/Frequency $\Rightarrow T = 1/n$

5. **Wave pulse:** It is a short wave produced in a medium when the disturbance is created for a short time.

6. **Wave train:** A series of wave pulse is called wave train.

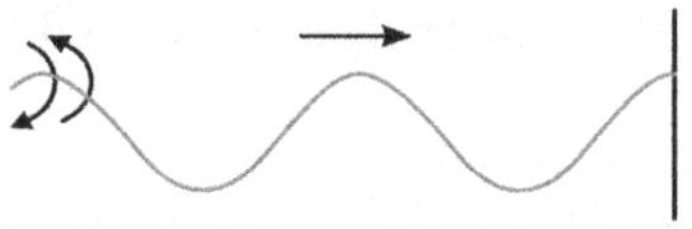

7. **Wave function:** It is a mathematical description of the disturbance created by a wave. For a string, the wave function is a displacement. For sound waves it is a pressure or density fluctuation where as for light waves it is electric or magnetic field.
Now let us consider a one dimensional wave travelling along x-axis. During wave motion, a particle with equilibrium position x is displaced some distance y in the direction perpendicular to the x-axis. In this case y is a function of position (x) and time (t).
i.e., y = f(x, t). This is called wave function.
Let the wave pulse be travelling with a speed v. After a time t, the pulse reaches a distance vt along the +x-axis as shown. Thus the motion of the particle p^1 at distance 'x' at time 't' is same as the motion of the particle p at time $t = 0$ at position $x_0 = x - vt$. Hence the wave functioon now can be represented as $y = f(x - vt)$.

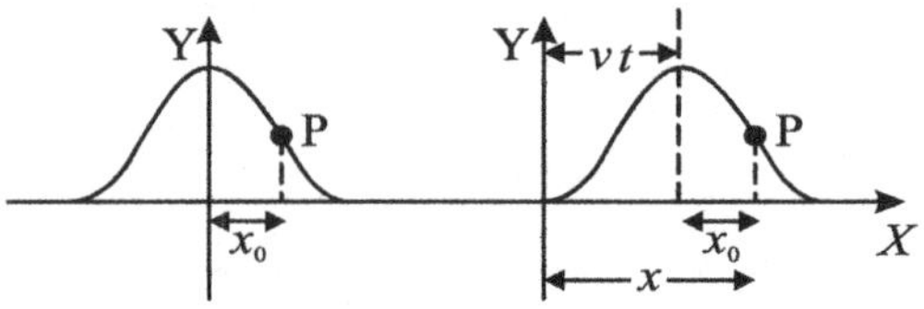

(A) Pulse at time $t=0$ (B) Pulse after time t

In general, then we can represent the transverse position y for all positions and times, measured in stationary frame with the origin at O, as
y(x, t) = f(x – vt) .... (i)
Similarly, if the pulse travels to the left, the transverse position of elements of the string is described by
y(x, t) = f(x + vt) .... (ii)
The function y, sometimes called the wave function, depends on the two variables x and t. For this reason, it is often written y(x, t), which is read "y as a function of x and t".

Note-1: The equation y = f(vt – x) represents the displacement of the particle at x = 0 as time passes

Representation of $y = f\left(t - \frac{x}{v}\right)$ Representation of $y = f(x - vt)$

Note-2: If order of a wave function to represent a wave, the three quantities x, v, t must appear in combinations $(x + vt)$ or $(x - vt)$.

Thus $y = (x - vt)^2, \sqrt{(x - vt)}, Ae^{-B(x-vt)^2}$ etc., represents travelling waves while

$y = (x^2 - v^2t^2), (\sqrt{x} - \sqrt{vt}),\ A\sin(4x^2 - 9t^2)$ etc. do not represent a wave.

8. **Harmonic wave:** If a travelling wave is a sin or cos function of $(x \pm vt)$ the wave is said to be harmonic or plane progressive wave.

9. **The differential form of wave equation:**
All the travelling waves satisfy a differential equation which is called the wave equation. It is given by

$$\frac{\partial^2 y}{\partial t^2} = v^2 \frac{\partial^2 y}{\partial x^2}; \text{ where } v = \frac{\omega}{k}$$

It is satisfied by any equation of the form $y = f(x \pm vt)$

10. **Angular wave number (or) propagation constant (k):** Number of wavelengths in the distance 2π is called the wave number or propagation constant i.e., $k = \frac{2\pi}{\lambda}$
It is unit is rad/m.

11. **Wave velocity (v):** It is the distance travelled by the disturbance in one second. It only depends on the properties of the medium and is independent of time and position.

$$v = n\lambda = \frac{\lambda}{T} = \frac{\omega\lambda}{2\pi} = \frac{\omega}{k}$$

12. **Phase:** Phase gives the state of the vibrating particle at any instant of time as regards to its position and direction of motion.

- Phase is the angular displacement from its mean position. $\theta = (\omega t \pm kx)$
- If phase is constant then the shape of wave remains constant.

Equation of Progressive Wave :

1. If during the propagation of a progressive wave, the particles of the medium perform SHM about their mean position, then the wave is known as a harmonic progressive wave.
2. Suppose a plane simple harmonic wave travels from the origin along the positive direction of x-axis from left to right as shown in the figure

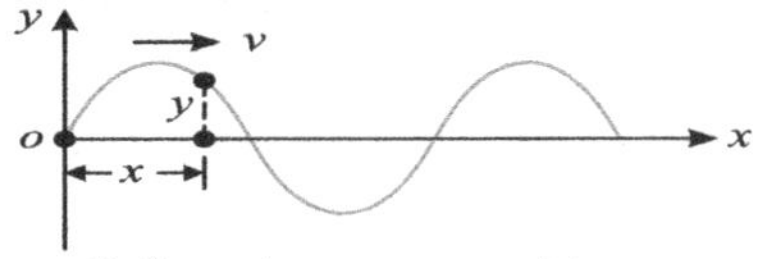

The displacement y of a particle at O from its mean position at any time t is given by
$y = A\sin\omega t$. --- (1)

The wave reaches the particle p after time $t = \frac{x}{v}$.
So that the motion of the particle 'p' which is at a distance 'x' at a time 't' is same as motion of the particle at $x = 0$, at the earliear time $t - \frac{x}{v}$.

Hence the displacement 'y' of the particle 'p' at 'x' at a time 't' in equation (1) by $\left(t - \frac{x}{v}\right)$.

$$y = A\sin\omega\left(t - \frac{x}{v}\right) = A\sin(\omega t - kx) \left(\because k = \frac{\omega}{v}\right)$$

In general along x-axis, y=A Sin (ωt $\pm$ kx)
+ sign for a wave travelling along -ve X direction
- sign for a wave travelling along +ve X direction
where y is displacement of the particle after a time t from mean position, x is displacement of the wave, A is Amplitude.
ω is angular frequency or angular velocity & $\omega = 2\pi / T = 2\pi n$
k is propagation constant & k= $2\pi/\lambda$

- For a given time 't', $y - x$ graph gives the shape of pulse on string.

Various forms of progressive wave function:

(i) $y = A\sin(\omega t \pm kx)$ (or) $y = A\sin(kx \pm \omega t)$

(ii) $y = A\cos(\omega t \pm kx)$ (or) $y = A\cos(kx \pm \omega t)$

(iii) $y = A\sin\left(\omega t \pm \frac{2\pi}{\lambda}x\right)$

(iv) $y = A\sin 2\pi\left[\frac{t}{T} \pm \frac{x}{\lambda}\right]$

(v) $y = A\sin\frac{2\pi}{T}\left(t \pm x\frac{T}{\lambda}\right)$

(vi) $y = A\sin\frac{2\pi}{\lambda}(vt \pm x)$

(vii) $y = A\sin\omega\left(t \pm \frac{x}{v}\right)$

(viii) $y = A\sin 2\pi\left(\frac{t}{T} \pm \frac{x}{\lambda}\right)$

General Expression for a Sinusoidal Wave

$Y = A\sin(kx - \omega t + \phi)$ (or)

$Y = A\sin(\omega t - kx + \phi)$

where ϕ is the phase constant, just as we learned in our study of periodic motion. This constant can be determined from the initial conditions.

Positive and Negative Initial Phase Constants.

In general, the equation of a harmonic wave travelling along the positive x-axis is expressed as $y = A\sin(kx - \omega t \pm \phi)$. Where ϕ is called the initial phase constant. It determines the initial displacement of the particle at x = 0 when t = 0.

i)Positive initial phase constant $y = A\sin(kx - \omega t + \phi)$ .The sine curve starts from the left of the origin.

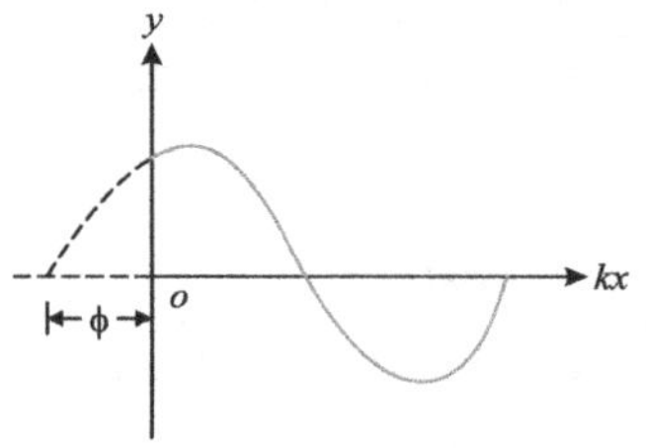

ii) Negative initial phase constant $y = A\sin(kx - \omega t - \phi)$. The sine curve starts from the left of the origin.

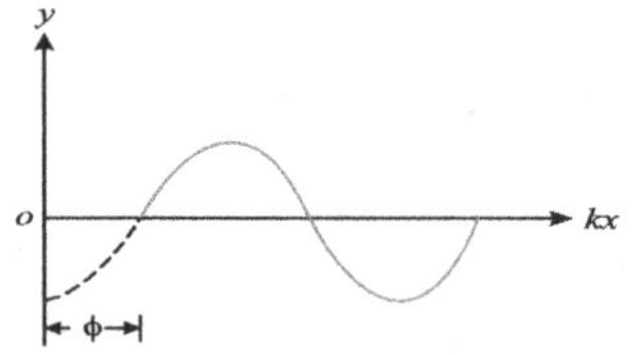

Change in Phase with time for a constant x, i.e., at a fixed point in the medium

$$[\phi]_{t_1} = 2\pi\left(\frac{t_1}{T} - \frac{x}{\lambda}\right) + \phi; [\phi]_{t_2} = 2\pi\left(\frac{t_2}{T} - \frac{x}{\lambda}\right) + \phi$$

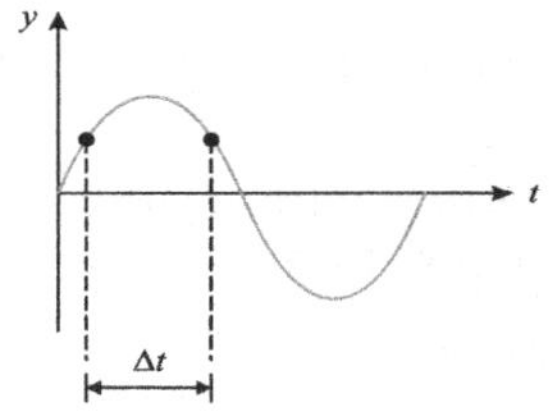

(For the wave travelling in positive x-direction)

$$\Delta\phi = [\phi]_{t_2} - [\phi]_{t_1} = \frac{2\pi}{T}\times(t_2 - t_1) = \frac{2\pi}{T}\times\Delta t$$

$$\Rightarrow \Delta\phi = \frac{2\pi\times\Delta t}{T}$$

Phase difference $= \frac{2\pi}{T}\times \textit{Time difference}$

Variation of Phase with Distance

At a given instant of time t = t, phase at $x = x_1$,

$$[\phi]_{x_1} = 2\pi\left(\frac{t}{T} - \frac{x_1}{\lambda}\right) + \phi$$

(For the wave travelling in positive x-direction and phase at $x = x_2$,

$$[\phi]_{x_2} = 2\pi\left(\frac{t}{T} - \frac{x_2}{\lambda}\right) + \phi$$

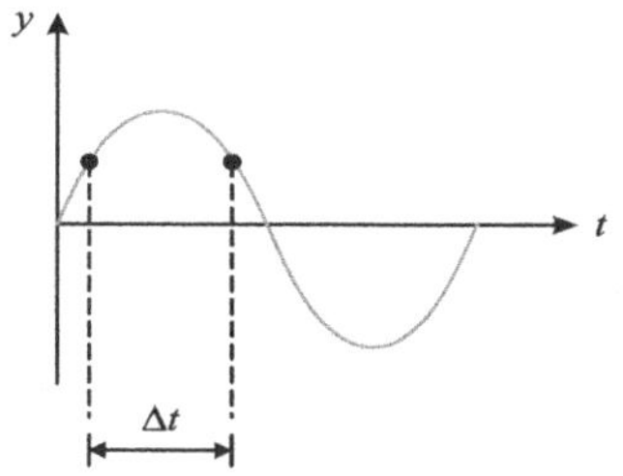

$$\Rightarrow \Delta\phi = [\phi]_{x_2} - [\phi]_{x_1} = \frac{2\pi}{\lambda}(x_2 - x_1) = \frac{2\pi}{\lambda}\Delta x$$

$$\Delta\phi = \frac{2\pi}{\lambda}\Delta x$$

i.e., Phase difference $= \frac{2\pi}{\lambda}\times \textit{Path difference}$

Particle Velocity: The rate of change of displacement y w.r.t time t is known as particle velocity.

Hence from $y = A\sin(\omega t - kx)$

Particle velocity, $v_p = \frac{\partial y}{\partial t} = A\omega\cos(\omega t - kx)$

Maximum particle velocity $(v_p)_{max} = A\omega$

Also $\frac{\partial y}{\partial t} = -\frac{\omega}{k}\times\frac{\partial y}{\partial x}$

- Particle velocity at a given position and time is equal to negative of the product of wave velocity with slope of wave at that point i.e.

$$v_{particle} = -v_{Wave}\left(\frac{\partial y}{\partial x}\right)$$

Particle velocity = –(wave velocity) × slope of wave curve

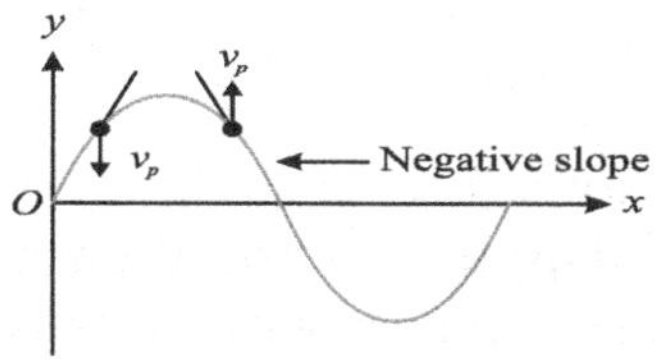

ENERGY, POWER AND INTENSITY OF A WAVE

If a wave given by $y = A\sin(\omega t - kx)$ is propagating through a medium, the particle velocity will be

$$v_p = \frac{\partial y}{\partial t} = A\omega\cos(\omega t - kx)$$

If ρ is the density of the medium, kinetic energy of the wave per unit volume will be

$$= \frac{1}{2}\rho\left[\frac{\partial y}{\partial t}\right]^2 = \frac{1}{2}\rho\omega^2 A^2\cos^2(\omega t - kx)$$

and its maximum value will be equal to energy per unit volume i.e., energy density U.

$$U = \frac{1}{2}\rho A^2\omega^2$$

The energy associated with a volume $\Delta V = S\Delta x$ will be (where 'S' is the area of cross section).

$$\Delta E = U\Delta V = \frac{1}{2}\rho A^2\omega^2 S\Delta x$$

The power (rate of transmission of energy) will be $P = \frac{\Delta E}{\Delta t} = \frac{1}{2}\rho v\omega^2 A^2 S$

$\left[\text{as } \frac{\Delta x}{\Delta t} = v, \text{(Speed of wave)}\right]$

Intensity is defined as power per unit area.

$$I = \frac{\Delta E}{S\Delta t} = \frac{P}{S} = \frac{1}{2}\rho v\omega^2 A^2 = 2\pi^2 f^2 A^2 \rho v$$

If frequency f is constant then $I \propto A^2$

Reflection and Refraction of Waves : When waves are incident on a boundary between two media a part of incident waves returns back into the initial medium (reflection) while the remaining is partly absorbed and partly transmitted into the second medium (refraction)

Boundary conditions: Reflection of a wave pulse from some boundary depends on the nature of the boundary.

Rigid end: When the incident wave reaches a fixed end, it exerts an upward pull on the end, according to Newton's third law the fixed end exerts an equal and opposite downward force on the string. It result as inverted pulse or phase change of π.

Crest (C) reflects as trough (T) and vice-versa. Time changes by $\frac{T}{2}$ and Path changes by $\frac{\lambda}{2}$

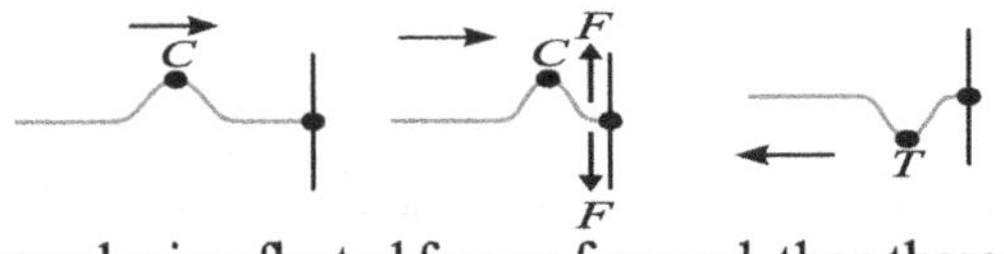

Free end: When a wave or pulse is reflected from a free end, then there is no change of phase (as there is no reaction force).

Crest (C) reflects as crest (C) and trough (T) reflects as trough (T), Time changes by zero and Path changes by zero.

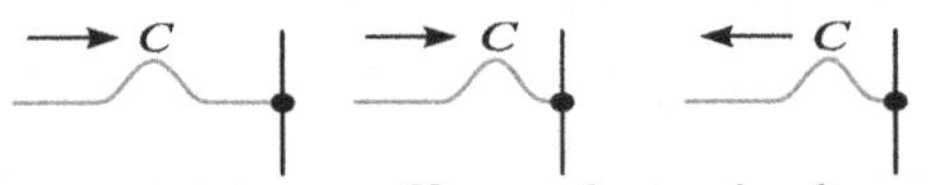

Note: Exception: Longitudinal pressure waves suffer no change in phase from rigid end. i.e., compression pulse reflects as compression pulse. On the other hand if longitudinal pressure wave reflects from free end, it suffer a phase change of π, i.e., compression reflects as rarefaction and vice-versa.

Wave in a combination of string

(i) Wave goes from thin to thick string

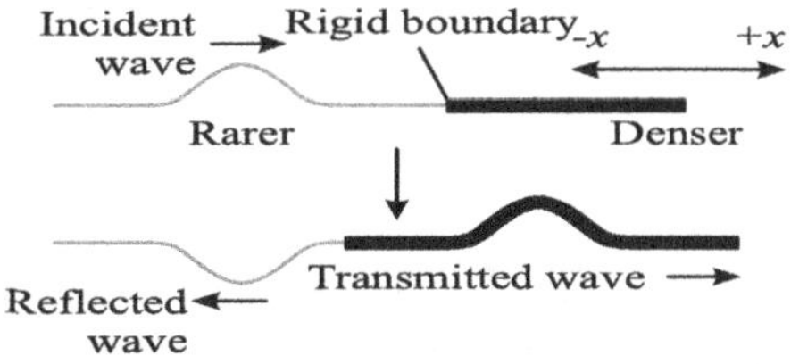

Incident wave $y_i = a_i \sin(\omega t - k_1 x)$

Reflected wave

$$y_r = a_r \sin[\omega t - k_1(-x) + \pi]$$

$$= -a_r \sin(\omega t + k_1 x)$$

Transmitted wave, $y_t = a_t \sin(\omega t - k_2 x)$

(ii) Wave goes from thick to thin string

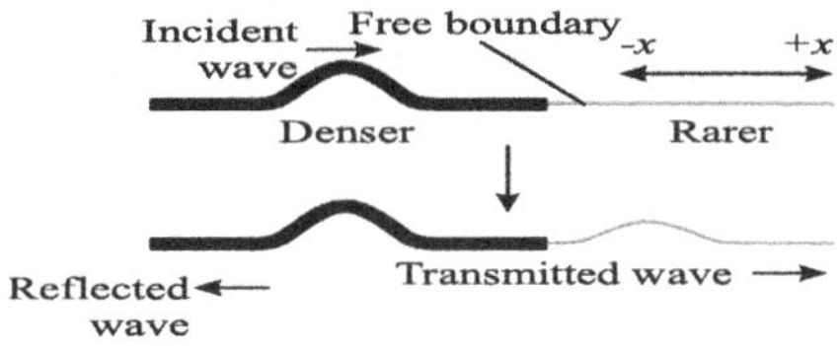

Incident wave $y_i = a_t \sin(\omega t - k_1 x)$

Reflected wave $y_r = a_r \sin[\omega t - k_1(-x) + 0] \quad = a_r \sin(\omega t + k_1 x)$

Transmitted wave $y_t = a_t \sin(\omega t - k_2 x)$

Note: Ratio of amplitudes: It is given as follows

$$\frac{a_r}{a_i} = \frac{k_1 - k_2}{k_1 + k_2} = \frac{v_2 - v_1}{v_2 + v_1} \text{ and } \frac{a_t}{a_i} = \frac{2k_1}{k_1 + k_2} = \frac{2v_2}{v_1 + v_2}$$

The Speed of A Travelling Wave

i) Let a wave moves along the +ve x-axis with velocity 'v' as shown in fig.

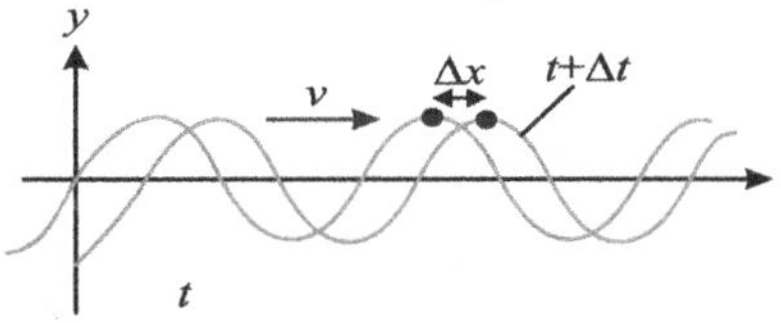

ii) Let a crest shown by a dot (•) moves a distance Δx in time Δt. The speed of the wave is $v = \Delta x / \Delta t$.

iii) We can put the dot (•) on a point with any other phase. It will move with the same speed v (otherwise the wave pattern will not remain fixed).

iv) The motion of a fixed phase point on the wave is given by, $y = \sin(kx - \omega t)$.

v) For the same particle displacement 'y' at two different positions, $kx - \omega t = \text{constant}$ -----(1)

$$\Rightarrow k\Delta x - \omega \Delta t = 0$$

$$\Rightarrow \frac{\Delta x}{\Delta t} = \frac{\omega}{k} \Rightarrow v = \frac{\Delta x}{\Delta t} = \frac{\omega}{k}$$

$$\Rightarrow v = \frac{2\pi n}{2\pi / \lambda} = n\lambda$$

$(\because \omega = 2\pi n \text{ and } k = 2\pi / \lambda)$

Speed of transverse wave in a string

i) Let a transverse pulse is travelling on a stretched string as shown in fig(a).

a) 

b) 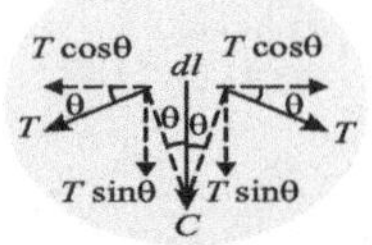

ii) Now consider a small element of length dl on this pulse as shown fig (b). Let this element is forming an arc of radius R and subtending an angle 2θ at center of curvature C.
iii) We can see that two tensions T are acting on the edges of dl along tangential directions as shown.
iv) The horizontal components of these tensions cancel each other, but the vertical components add to form a radial restoring force in downward direction, which is given as

$F_R = 2T\sin\theta \approx 2T\theta$ $\qquad$ $(as \sin\theta \approx \theta)$

$= T\frac{dl}{R}$ .....(1) $\qquad \left[2\theta = \frac{dl}{R}\right]$

v) If 'μ' be the mass per unit length (Linear density) of the string, the mass of this element is given as $dm = \mu dl$. In our reference frame if we look at this element, it appears to be moving toward left with speed v then we can say that the acceleration of this element in our reference frame is

$a = \frac{v^2}{R}$ .....(2)

Now from equations (1) and (2) we have

$$F_R = \frac{dmv^2}{R} \text{ or } T\frac{dl}{R} = \frac{(\mu dl)v^2}{R}$$

or $v = \sqrt{\frac{T}{\mu}}$ .....(3)

Spacial cases:

1. If A is the area of cross-section of the wire then linear density $\mu = M/L = \rho AL/L = \rho A$

   $\Rightarrow v = \sqrt{\frac{T}{\rho A}} = \sqrt{\frac{S}{\rho}}$; where $S = Stress = \frac{T}{A}$

2. If string is stretched by some weight then

   $T = Mg \Rightarrow v = \sqrt{\frac{Mg}{\mu}}$

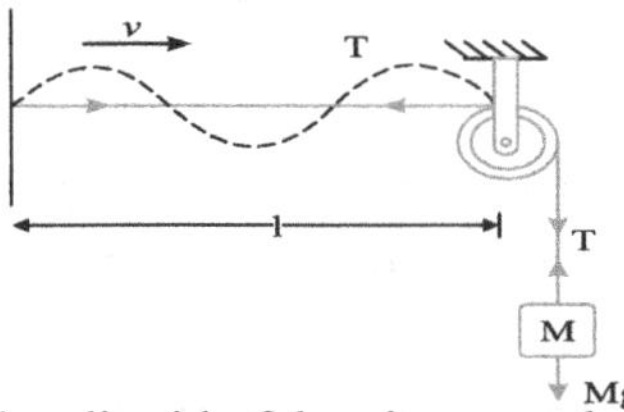

3. If suspended weight is immersed in a liquid of density σ and ρ = density of material of the suspended load then

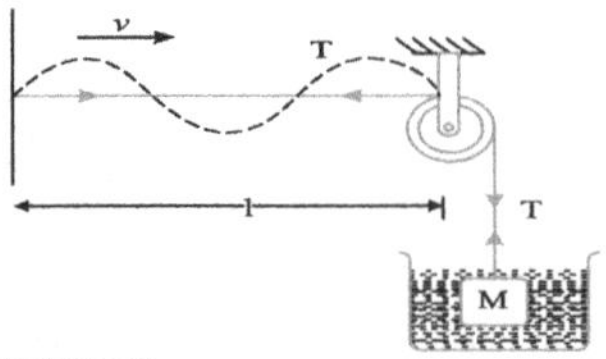

$$T = Mg\left(1 - \frac{\sigma}{\rho}\right) \Rightarrow v = \sqrt{\frac{Mg(1-\sigma/\rho)}{\mu}}$$

4. If v_1, v_2 are the velocities of transverse waves while the load is in air medium and in water medium respectively, the relative density of material of load is $d = \frac{v_1^2}{v_1^2 - v_2^2}$

5. If v_1, v_2 and v_3 are the velocities of transverse waves while the load is in air, in water and in a liquid mediums respectively, the relative density of material of load is $d = \frac{v_1^2 - v_3^2}{v_1^2 - v_2^2}$.

6. If the temperature a string varies through $\Delta\theta$ then the thermal force(tension) developed due to elasticity of string is $T = YA\alpha\Delta\theta$

$\therefore v = \sqrt{\frac{YA\alpha\Delta\theta}{\mu}} = \sqrt{\frac{Y\alpha\Delta\theta}{\rho}}$

where Y = Young's modulus of elasiticyt of string, A = Area of cross section of string,

α = Temperature coefficient of thermal expansion, ρ = Density of wire $= \frac{\mu}{A}$

7. Velocity of wave in vertical strings. If a thick string is suspended vertically then

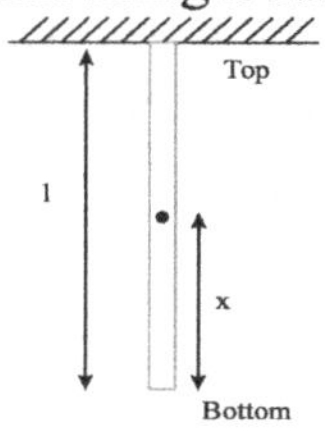

Velocity at the bottom $v_B = 0$

($\because$ tension $T_B = 0$)

Velocity at the top $v_T = \sqrt{\frac{T_T}{\mu}} = \sqrt{\frac{mg}{\mu}} = \sqrt{gl}$

($\because$ tension $T_B = mg = \mu lg$)

The average velocity of wave

$$v_{avg} = \frac{v_T + v_B}{2} = \frac{\sqrt{gl}}{2}$$

$\therefore$ The time taken by the transverse pulse generated at bottom to reach the top is given by $t = \frac{l}{v_{avg}} = 2\sqrt{\frac{l}{g}}$

Note: Velocity at a distance x from bottom $v = \sqrt{gx}$

The time taken to reach the point p from bottom is $v_x = \frac{x}{v_{avg}} = 2\sqrt{\frac{x}{g}}$

W.E-1: ***A longitudinal progressive wave is given by the equation $y = 5 \times 10^{-2} \sin\pi (400t + x)m$. Find (i) amplitude (ii) frequency (iii) wave length and (iv) velocity of the wave. (v) velocity and acceleration of particle at $x = \frac{1}{6}m$ at $t = 0.01$ s (vi) maximum particle velocity and acceleration.***

Sol. Comparing with the general equation of the progressive wave y = Asin(ωt + kx) we find, $\omega = 400\pi$ and $k = \pi$

We find

(i) $A = 5 \times 10^{-2}$ m. (ii) $n = \frac{\omega}{2\pi} = \frac{400\pi}{2\pi} = 200Hz$

(iii) $\lambda = \frac{2\pi}{k} = \frac{2\pi}{\pi} = 2m$ (iv) $v = \frac{\omega}{k} = \frac{400\pi}{\pi} = 400ms^{-1}$

(v) $v_p = A\omega\cos t(\omega t + kx) = 10\sqrt{3}ms^{-1}$ $a_p = -A\omega^2\sin(\omega t + kx) = -4\times10^4 ms^{-2}$

(vi) $v_{max} = A\omega = 20\pi\ ms^{-1}$ $a_{max} = A\omega^2 \Rightarrow 8\times10^4 ms^{-2}$

W.E-2: ***The wave function of a pulse is given by*** $y = \frac{3}{(2x+3t)^2}$ ***where x and y are in metre and t is in second.***

(i) Identify the direction of propagation. ***(ii) Determine the wave velocity of the pulse.***

Sol. (i) Since the given wave function is of the form y = f(x + vt), therefore, the pulse travels along the negative x-axis.

(ii) Since 2x + 3t = constant for the same particle displacement 'y'. Therefore, by differentiating with respect to time, we get $2\frac{dx}{dt} + 3 = 0 \Rightarrow v = \frac{dx}{dt} = \frac{-3}{2} = -1.5m/s$

W.E-3: ***Figure shows a snapshot of a sinusoidal travelling wave taken at t = 0.3s. The wavelength is 7.5 cm and the amplitude is 2 cm. If the crest P was at x = 0 at t = 0, write the equation of travelling wave.***

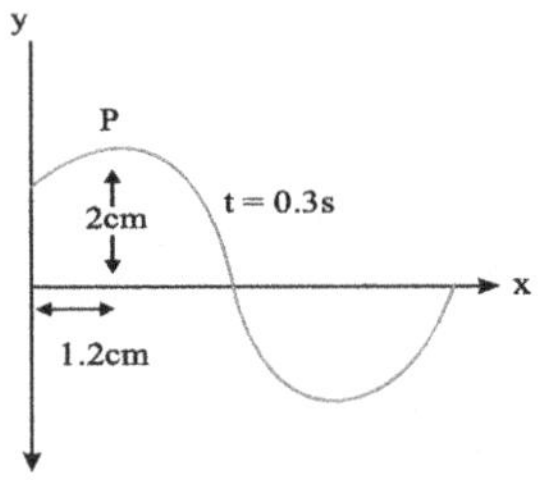

Sol. The wave has travelled a distance of 1.2 cm in 0.3s. Hence, speed of the wave, $v = 1.2/0.3 = 4cm/s$ and $\lambda = 7.5cm$

$\Rightarrow k = \frac{2\pi}{\lambda} = \frac{2\pi}{7.5} = 0.84cm^{-1}$

$\therefore$ Angular frequency $\omega = vk = 4\times0.84 = 3.36 rad/s$

Since the wave is travelling along positive x-direction and crest (maximum displacement) is at x = 0 at t = 0, we can write the wave equation as, $y = A\sin\left(kx - \omega t + \frac{\pi}{2}\right)$

(or) $y(x,t) = A\cos(kx - \omega t)$

Therefore, the desired equation is, $y(x,t) = (2)\cos[(0.84)x - (3.36)t]cm$

W.E-4: ***A copper wire is held at the two ends by rigid supports. At 30°C, the wire is just taut, with negligible tension. Find the speed of transverse waves in this wire at 10°C if Y = 1.3×10¹¹N/m²,*** $\alpha = 1.7\times10^{-5}/°C$ ***and*** $\rho = 9\times10^3 kg/m^3$

Sol. $v = \sqrt{\frac{Y\alpha\Delta\theta}{\rho}} = \sqrt{\frac{1.3\times10^{11}\times1.7\times10^{-5}\times(30-10)}{9\times10^3}} = 70m/s$

W.E-5: ***A 4 kg block is suspended from the ceiling of an elevator through a string having a linear mass density of 19.2 × 10^{-3} kg m^{-1}. Find the speed with which a wave pulse can travel on the string if the elevator accelerates up at 2 ms^{-2}? (g = 10 ms^{-2})***

Sol. $v=\sqrt{\frac{T}{\mu}}=\sqrt{\frac{M(g+a)}{\mu}}=\sqrt{\frac{4(10+2)}{19.2\times10^{-3}}}=50ms^{-1}$.

W.E-6: ***A uniform rope of length 12 m and mass 6 kg hangs vertically from a rigid support. A block of mass 2 kg is attached to the free end of the rope. A transverse pulse of wavelength 0.06 m is produced at the lower end of the rope. What is the wavelength of the pulse when it reaches the top of the rope ?***

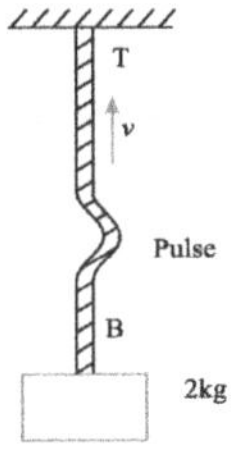

Sol. Now as $v=\sqrt{(T/\mu)}$

$$\frac{v_T}{v_B}=\sqrt{\frac{T_T}{T_B}}=\sqrt{\frac{(6+2)g}{2g}}=2$$

So, $\lambda_T=2\lambda_B=2\times0.06=0.12m$

W.E-7: ***A uniform rope of mass 0.1 kg and length 2.45m hangs from a ceiling. (a) Find the speed of transverse wave in the rope at a point 0.5m distant from the lower end, b) Calculate the time taken by a transverse wave to travel the full length of the rope (g = 9.8 m/s²)***

Sol. a) If M is the mass of string of length L, the mass of length x of the string will be (M/L)x.

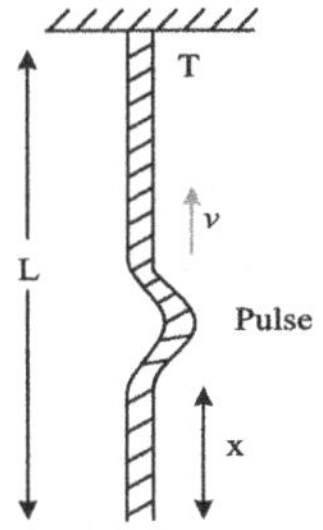

$$\therefore T=\frac{Mx}{L}g$$

So, $v=\sqrt{\frac{T}{\mu}}=\sqrt{\frac{Mgx}{L\left(\frac{M}{L}\right)}}=\sqrt{gx}$.........(1)

Hence x = 0.5m

So, $v=\sqrt{0.5\times9.8}=2.21m/s$

b) $v=\frac{dx}{dt}\Rightarrow\sqrt{gx}=\frac{dx}{dt}\Rightarrow dt=\frac{dx}{\sqrt{gx}}$

$$\Rightarrow\int_0^t dt=\int_0^L\frac{1}{\sqrt{g}}x^{-1/2}dx\Rightarrow t=2\sqrt{\left(L/g\right)}$$

Here, L = 2.45 m, $\therefore t=2\sqrt{(2.45/9.8)}=1s$

W.E-8: ***The strings, shown in figure, are made ofsame material and have same cross-section. The pulleys are light. The wave speed of a transverse wave in the string AB is v_1 and in CD it is v_2. Find v_1/v_2.***

Sol: If T_1 and T_2 are the tensions in strings AB and CD respectively then $T_2=2T_1$.

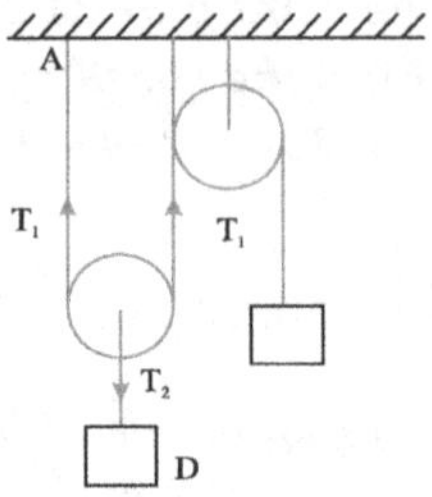

As $v \propto \sqrt{T} \Rightarrow \frac{v_1}{v_2} = \sqrt{\frac{T_1}{T_2}} \Rightarrow \frac{v_1}{v_2} = \frac{1}{\sqrt{2}}$

W.E-9: ***Two blocks each having a mass of 3.2kg are connected by wire CD and the system is suspended from the ceiling by another wire AB. The linear mass density of the wire AB is 10 g/m and that of CD is 80 g/m. Find the speed of a transverse wave pulse produced in AB and CD and ratio of speeds of transverse pulse in AB to that in CD.***

Sol. Tension in string AB is $T_{AB} = 6.4kg = 64N$

Thus speed of transverse wave in string AB is $v_{AB} = \sqrt{\frac{T_{AB}}{\mu_{AB}}} = \sqrt{\frac{64}{10\times10^{-3}}} = \sqrt{6400} = 80m/s$

Tension in string CD is $T = 3.2kg = 32N$

Thus speed of transverse waves in string CD is $v_{CD} = \sqrt{\frac{T_{CD}}{\mu_{DC}}} = \sqrt{\frac{32}{80\times10^{-3}}}$

$= \sqrt{400} = 20m/s \Rightarrow \frac{v_{AB}}{v_{CD}} = \frac{80}{20} = 4:1$

WE-10 ***A progressive wave travels in a medium M_1 and enters into another medium M_2 in which its speed decreases to 75%. What is the ratio of the amplitude and intensity of the***
a. Reflected and the incident waves, and
b. Transmitted and the incident waves?

Sol. let A_i, A_r and A_t be the amplitudes of the incidents, reflected, and transmitted waves. Given that, velocity in the medium refracted is 75% of that in the initial medium.

$v_2 = \frac{3}{4}v_1$

a. $\frac{A_r}{A_i} = \frac{v_2 - v_1}{v_2 + v_1} = \frac{\frac{v_2}{v_1} - 1}{\frac{v_2}{v_1} + 1} = \frac{\frac{3}{4} - 1}{\frac{3}{4} + 1} = -\frac{1}{7}$

i.e., the required ration is $\left|\frac{A_r}{A_i}\right| = 1:7$ and $I \propto A^2 \Rightarrow \frac{I_r}{I_i} = \frac{1}{49}$

b. $\frac{A_t}{A_i} = \frac{2v_2}{v_2 + v_1} = \frac{2v_2/v_1}{\frac{v_2}{v_1} + 1} = \frac{2\left(\frac{3}{4}\right)}{\frac{3}{4} + 1} = \frac{6}{7}$

i.e., the required ratio is $\left|\frac{A_t}{A_i}\right| = 6:7$ and $I \propto A^2 \Rightarrow \frac{I_t}{I_i} = \frac{36}{49}$

W.E-11: ***A long wire PQR is made by joining two wires PQ and QR of equal radii as shown. PQ has length 4.8m and mass 0.06kg. QR has length 2.56 m and mass 0.2kg. The wire PQR is under a tensioon of 80N. A sinusoidal wave pulse of amplitude 3.5cm is sent along the wire PQ from the end P. No power is dissipated during the propagation of the wave pulse.***
a. Find the time taken by the wave pulse to reach the other end R of the wire.
b. The amplitudes of reflected and transmitted wave pulse after incident on the joint Q.

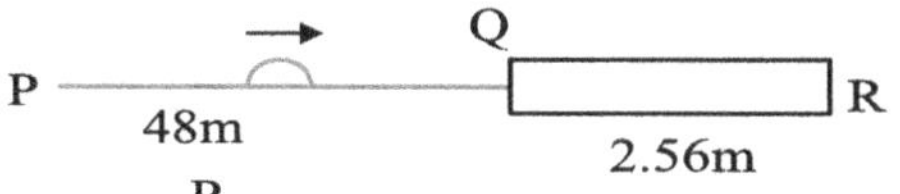

Sol. a.

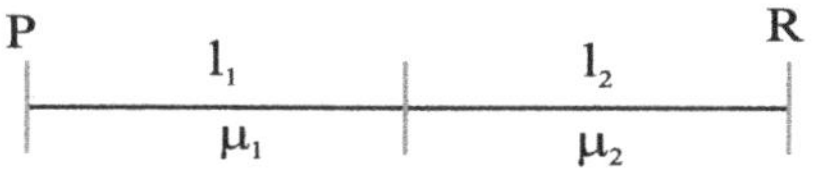

$$\mu_1 = \frac{M_1}{l_1} = \frac{0.06}{4.8} = \frac{1}{80} kg/m$$

$$\mu_2 = \frac{M_2}{l_2} = \frac{0.2}{2.56} = \frac{20}{256} kg/m$$

$$v_1 = \sqrt{\frac{T}{\mu_1}} = \sqrt{\frac{80}{\frac{1}{80}}} = 80m/s$$

$$v_2 = \sqrt{\frac{T}{\mu_2}} = \sqrt{\frac{80}{20/256}} = \sqrt{256\times4} = 32m/s \left(\because \rho = \frac{\mu}{A}\right)$$

$$t = t_1 + t_2 = \frac{l_1}{V_1} + \frac{l_2}{V_2} = \frac{4.8}{80} + \frac{2.56}{32} = 0.06 + 0.08 = 0.14\text{sec}$$

b. $$A_r = \left[\frac{v_2 - v_1}{v_2 + v_1}\right]A_i = \frac{32-80}{32+80}\times3.5 = -1.5cm$$

thus $A_r = 1.5cm$ and -ve sign represents that the reflected pulse suffers a phase difference of π radian.

$$A_t = \left[\frac{2v_2}{v_1+v_2}\right]A_i = \frac{2\times32}{80+32}\times3.5 = 2cm.$$

W.E-12: ***A wave pulse starts propagating in +ve X-direction along a non-uniform wire of length 'L', with mass per unit length given by $\mu = M_o + \alpha x$ and under a tension of TN. Find the time taken by the pulse to travel from the lighter end (x = 0) to the heavier end.***

Sol. $$v = \frac{dx}{dt} = \sqrt{\frac{T}{\mu}} = \sqrt{\frac{T}{M_o + \alpha x}}$$

$$\int_0^L (M_o + ax)^{1/2} dx = \int_0^L \sqrt{T} dt$$

$$\left[\frac{2}{3}\frac{(M_o+ax)^{3/2}}{\alpha}\right]_0^L = \sqrt{T}[t]_0^t$$

$$t = \frac{2}{3\alpha\sqrt{T}}\left[(M_o + \alpha L)^{3/2} - M_o^{3/2}\right]$$

W.E-13: ***A stretched string is forced to tranmit transverse waves by means of an oscillator coupled to one end. The string has a diameter of 4 mm. The amplitude of the oscillation is 10^{-4} m and the frequency is 10 Hz. Tension in the string is 100N and mass density of wire $4.2\times10^3 kgm^{-3}$. Find***

(a) the equation of the waves along the string

(b) the energy per unit volume of the wave

(c) the average energy flow per unit time across any section of the string

Sol.(a) Speed of transverse wave on the string is $v = \sqrt{\frac{T}{\rho A}} (\because \mu = \rho A)$

$$v=\sqrt{\frac{100}{(4.2\times10^3)\left(\frac{\pi}{4}\right)(4.0\times10^{-3})^2}}$$

$=43.53ms^{-1}$

$\omega=2\pi n=20\pi\ rad/s=62.83$ rad/s

$k=\frac{\omega}{v}=1.44m^{-1}$

$\therefore$ Equation of the waves along the string $y(x,t)=A\sin(kx-\omega t)$

$=(10^{-4}m)\sin\left[(1.44m^{-1})x-(62.83\text{rad}s^{-1})t\right]$

(b) Energy per unit volume of the string,

$u=$ energy density $=\frac{1}{2}\rho\omega^2A^2$

$u=\left(\frac{1}{2}\right)(4.2\times10^3)(62.83)^2(10^{-4})^2$

$=8.29\times10^{-2}Jm^{-3}$

(c) Average energy flow per unit time P = power $=\left(\frac{1}{2}\rho\omega^2A^2\right)(Sv)=(u)(Sv)$

$P=(8.29\times10^{-2})\left(\frac{\pi}{4}\right)(4.0\times10^{-3})^2(43.53)=4.53\times10^{-5}Js^{-1}$

Principle of Superposition:

1. The displacement at any time due to a number of waves meeting simultaneously at a point in a medium is the vector sum of the individual displacements due to each one of the waves at that point at the same time.
2. If $\vec{y_1},\vec{y_2},\vec{y_3}$.......... are the displacements at a particular time at a particular position, due to individual waves, then the resultant displacement. $\vec{y}=\vec{y_1}+\vec{y_2}+\vec{y_3}+$.........

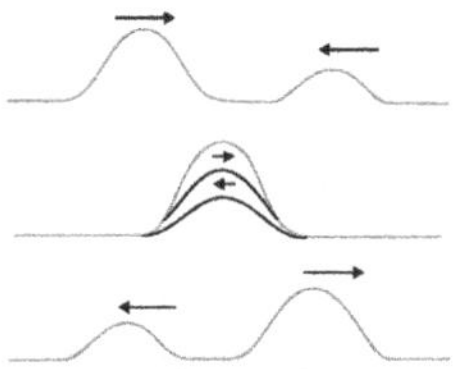

3. Important applications of superposition principle.
   i) Interference of waves: Adding waves that differ in phase.
   ii) Formation of stationary waves: Adding waves that differ in direction.
   iii) Formation of beats: Adding waves that differ in frequency.

Interference of Sound waves

1. When two waves of same frequency, same wavelength, same velocity (nearly equal amplitude) moves in the same direction. Their superimposition results in the interference.
2. Due to interference the resultant intensity of sound at a point is different from the sum of intensities due to each wave separately.
3. Interference is of two type (i) Constructive interference (ii) Destructive interference
4. In interference energy is neither created nor destroyed but is redistributed.
5. For observable interference, the sources (producing interfering waves) must be coherent.

6. Let at a given point two waves arrives with phase difference ϕ and the equation of these waves is given by

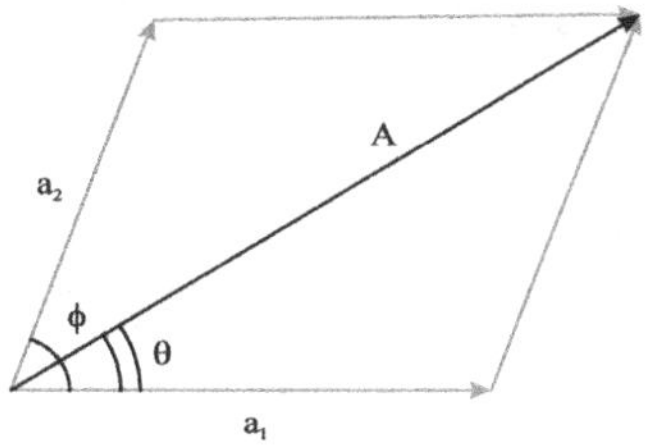

$y_1 = a_1 \sin \omega t$, $y_2 = a_2 \sin(\omega t + \phi)$ then by the principle of superposition $\vec{y} = \vec{y_1} + \vec{y_2}$

$\Rightarrow y = a_1 \sin \omega t + a_2 \sin(\omega t + \phi) = A \sin(\omega t + \phi)$

where $A = \sqrt{a_1^2 + a_2^2 + 2a_1a_2 \cos\phi}$ and $\tan\theta = \dfrac{a_2 \sin\phi}{a_1 + a_2 \cos\phi}$

Since Intensity (I) $\propto (Amplitude\ A)^2$ $\Rightarrow \dfrac{I_1}{I_2} = \left(\dfrac{a_1}{a_2}\right)^2$

Therefore, the resultant intensity is given by $I = I_1 + I_2 + 2\sqrt{I_1 I_2} \cos\phi$

Table: Constructive and destructive interference	
When the waves meet a point with same phase, constructive interference is obtained at that point (i.e., maximum sound)	When the waves meet a point with opposite phase, destructive interference is obtained at that point (i.e., minimum sound)
Phase difference between the waves at the point of observation ϕ=0°(or) $2n\pi$	Phase difference ϕ = 180° (or) $(2n-1)\pi$; n = 1,2,......
Phase difference between the waves at the point of observation $\Delta = n\lambda$ i.e., even multiple of $\lambda/2$)	Phase difference $\Delta = (2n-1)\dfrac{\lambda}{2}$ (i.e., odd multiple of $\lambda/2$)
Resultant amplitude at the point of observation will be maximum $A_{max} = a_1 + a_2$ If $a_1 = a_2 = a_0$ $\Rightarrow A_{max} = 2a_0$	Resultant amplitude at the point of observation will be minimum $A_{max} = a_1 - a_2$ If $a_1 = a_2 \Rightarrow A_{max} = 0$
Resultant intensity at the point of observation will be maximum $I_{max} = I_1 + I_2 + 2\sqrt{I_1 I_2}$ $= \left(\sqrt{I_1} + \sqrt{I_2}\right)^2$ If $I_1 = I_2 = I_0 \Rightarrow I_{max} = 4I_0$	Resultant intensity at the point of observation will be minimum $I_{min} = I_1 + I_2 + 2\sqrt{I_1 I_2}$ $= \left(\sqrt{I_1} - \sqrt{I_2}\right)^2$ If $I_1 = I_2 = I_0 \Rightarrow I_{min} = 4I_0$

7. $\dfrac{I_{max}}{I_{min}} = \left(\dfrac{\sqrt{I_1} + \sqrt{I_2}}{\sqrt{I_1} - \sqrt{I_2}}\right)^2 = \left(\dfrac{a_1 + a_2}{a_1 - a_2}\right)^2 = \left(\dfrac{\dfrac{a_1}{a_2} + 1}{\dfrac{a_1}{a_2} - 1}\right)^2$

W.E-14: ***Two loud speakers L_1 and L_2, driven by a common oscillator and amplifier, are arranged as shown. The frequency of the oscillator is gradually increased from zero and the detector at D records a series of maxima and minima.***

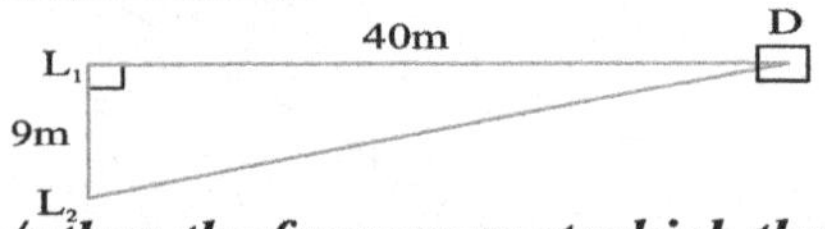

If the speed of sound is 330 m/s then the frequency at which the first maximum is observed is

Sol.

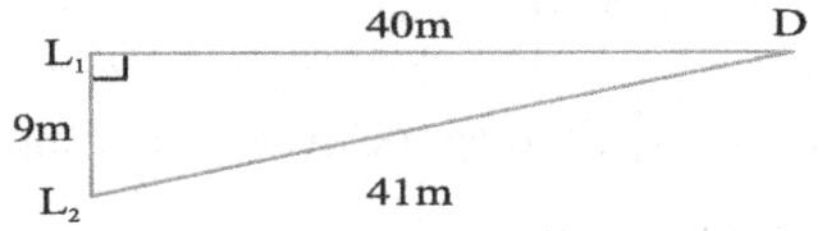

It is clear from figure that the path difference between L_1D and L_2D is $\Delta x = 41 - 40 = 1m$

For maximum $\Delta x = N\lambda$ where N = 1, 2, 3.... for Ist maximum N = 1, $\lambda = \frac{v}{n}$

$\Delta x = 1\times\frac{v}{n} \Rightarrow 1 = 1\times\frac{330}{n} \Rightarrow n = 330Hz$

W.E-15: ***Two coherent narrow slits emitting of wavelength λ in the same phase are placed parallel to each other at a small separation of 2λ. The sound is detected by moving a detector on the screen S at a distance $D(>>\lambda)$ from the slit S_1 as shown in figure.***

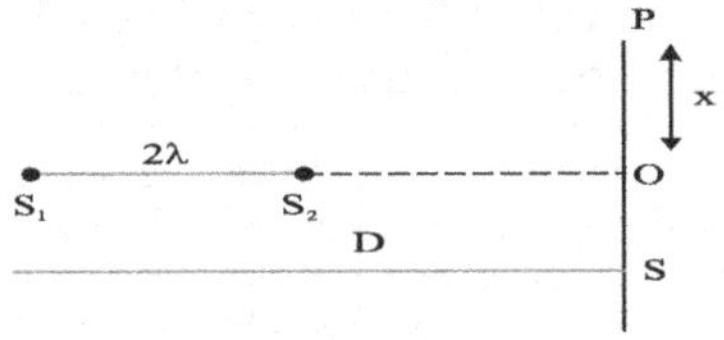

Find the distance x such that the intensity at P is equal to the intensity at O.

Sol.

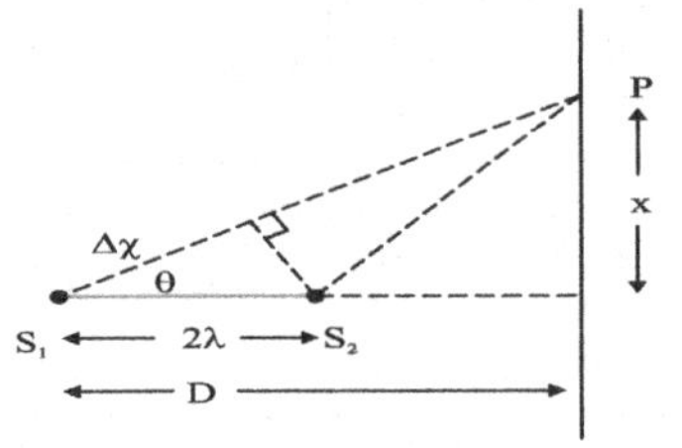

From figure, we get $\cos\theta = \frac{\Delta x}{2\lambda} \Rightarrow \Delta x = 2\lambda\cos\theta$ ------(1)

For maximum intensity path difference

$\Delta x = N\lambda$--------(2)

From equations (1) and (2) we get

$2\lambda\cos\theta = N\lambda \Rightarrow 2\cos\theta = N$

at least p is Ist maxima $\Rightarrow N = 1$

$\therefore \cos\theta = \frac{1}{2} \Rightarrow \theta = 60°$

$\tan\theta = \frac{x}{D} \Rightarrow x = D\tan 60 \Rightarrow x = \sqrt{3}D$

Standing Waves or Stationary Waves:

When two sets of progressive wave trains of same type (both longitudinal or both transverse) having the same amplitude and same time period/frequency/wavelength travelling with same speed along the same straight line in opposite directions superimpose, a new set of waves are formed. These are called stationary waves or standing waves.

These waves are formed only in a bounded medium.

In practice, a stationary wave is formed when a wave train is reflected at a boundary. The incident and reflected waves then interface to produce a stationary wave.

1. Suppose that two super imposing waves are incident wave $y_1 = a\sin(\omega t - kx)$ and reflected wave $y_2 = a\sin(\omega t + kx)$ (As y_2 is the displacement due to reflected wave from a free boundary)

   Then by principle of superposition $y = y_1 + y_2 = a\left[\sin(\omega t - kx) + \sin(\omega t + kx)\right]$

   ($\because \sin C + \sin D = 2\sin\frac{C+D}{2}\cos\frac{C-D}{2}$) $\Rightarrow y = 2a\cos kx \sin\omega t$

   (If reflection takes place from rigid end, then equation of stationary wave will be $y = \pm 2a\sin kx\cos\omega t$)
2. As this equation satisfies the wave equation.

   $\frac{\partial^2 y}{\partial t^2} = v^2\frac{\partial^2 y}{\partial x^2}$. It represents a wave
3. As it is not of the form $f(ax \pm bt)$, the wave is not progressive.
4. Amplitude of the wave $A_{sw} = 2a\cos kx$.
5. **Nodes (N):** The points where amplitude is minimum are called nodes.

   i) Distance between two successive nodes is $\lambda/2$

   ii) Nodes are at permanent rest.

   iii) At nodes air pressure and density both are high.

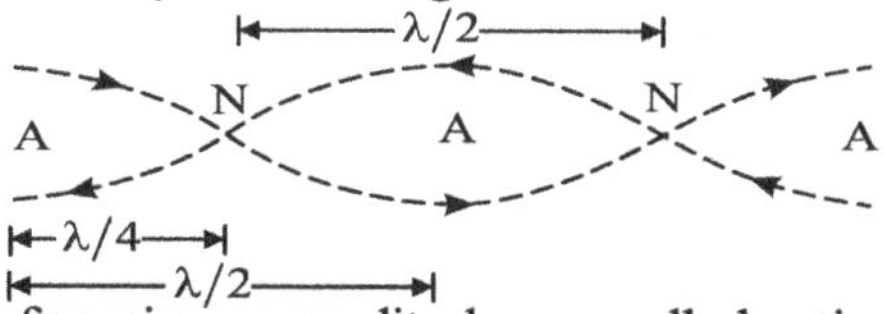

6. **Antinodes (A):** The points of maximum amplitudes are called antinodes.

   (i) The distance between two successive antinodes is $\lambda/2$

   (ii) At antinodes air pressure and density both are low.

   (iii) The distance between a node (N) and adjoining antinode (A) is $\lambda/4$
7. **Amplitude of standing waves in two different cases:**

Table: : Amplitude in two different cases

Reflection at open end or free boundary	**Reflection at closed end or rigid boundary**
$A_{sw} = 2a\cos kx$	$A_{sw} = 2a\sin kx$
Amplitude is maximum when cos kx = ≠ 1 $\Rightarrow kx = 0, 2\pi, ...n\pi$ $\Rightarrow x = 0, \frac{\lambda}{2}, \lambda.......\frac{n\lambda}{2}$ Where $k = \frac{2\pi}{\lambda}$ and n = 0, 1, 2, 3.....	Amplitude is maximum when cos kx = ≠ 1 $\Rightarrow kx = \frac{\pi}{2}, \frac{3\pi}{2}....\frac{(2n-1)\pi}{2}$ $\Rightarrow x = \frac{\lambda}{4}, \frac{3\lambda}{4}..............$ Where $k = \frac{2\pi}{\lambda}$ and n = 1, 2, 3,.......
Amplitude is minimum when cos kx = 0 $\Rightarrow kx = \frac{\pi}{2}, \frac{3\pi}{2}....\frac{(2n-1)\pi}{2}$ $\Rightarrow x = \frac{\lambda}{4}, \frac{3\lambda}{4}.........$	Amplitude is minimum when sin kx = 0 $\Rightarrow kx = \frac{\pi}{2}, \frac{3\pi}{2}....\frac{(2n-1)\pi}{2}$ $\Rightarrow x = 0, \frac{\lambda}{2}, \lambda.......\frac{n\lambda}{2}$

SR-Physics

Terms related to the Application of Stationary wave

1. **Harmonics:** The frequency which are the integral multiple of the fundamental frequency are known as harmonics e.g. if n be the fundamental frequency, then the frequencies n, 2n, 3n ...... are termed as first, second, third .... harmonics.
2. **Overtone:** The harmonics other than the first (fundamental note) which are actually produced by the instrument are called overtones. e.g. the tone with frequency immediately higher than the fundamental is defined as first overtone.
3. **Octave:** The tone whose frequency is doubled the fundamental frequency is defined as Octave.
   i) If $n_2 = 2n_1$ it means n_2 is an octave higher than n_1 or n_1 is an octave lower than n_2.
   ii) If $n_2 = 2^3\ n_1$, it means n_2 is 3-octave higher or n_1 is 3-octave lower.
   iii) Similarly, if $n_2 = 2^n n_1$ it means n_2 is n-octave higher or n_1 is n octave lower.
4. **Unison:** If time period is same i.e., two frequencies are equal then vibrating bodies are said to be in unison.

Standing Waves on a String

1. Consider a string of length l, stretched under tension T between two fixed points.
2. If the string is plucked and then released, a transverse harmonic wave propagates along it's length and is reflected at the end.
3. The incident and reflected waves will superimpose to produce transverse stationary waves in a string.
4. Nodes (N) are formed at rigid end and antinodes (A) are formed in between them.
5. Number of antinodes = Number of nodes – 1
6. Velocity of wave (incident or reflected wave) is given by $v = \sqrt{\frac{T}{\mu}}$.
7. Frequency of vibration (n) = Frequency of wave $= \frac{v}{\lambda} = \frac{1}{\lambda}\sqrt{\frac{T}{\mu}}$
8. For obtaining p loops (p-segments) in string, it has to be plucked at a distance $\frac{l}{2p}$ from one fixed end.
9. **Fundamental mode of vibration**

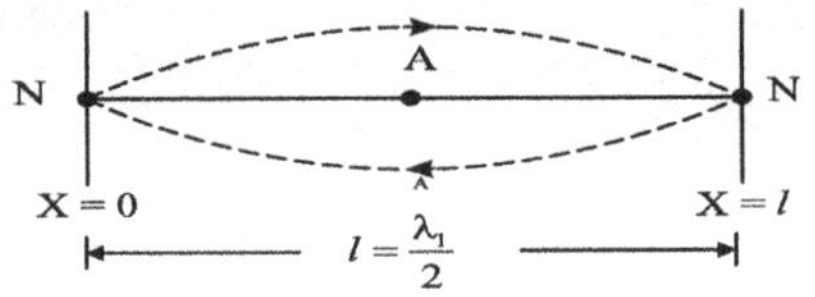

i) Number of loops p = 1

ii) Plucking at $\frac{l}{2}$ (from one fixed end)

iii) $l = \frac{\lambda_1}{2} \Rightarrow \lambda_1 = 2l$

iv) Fundamental frequency or first harmonic

$$n_1 = \frac{1}{\lambda_1}\sqrt{\frac{T}{\mu}} = \frac{1}{2l}\sqrt{\frac{T}{\mu}}$$

10. **Second mode of vibration:**

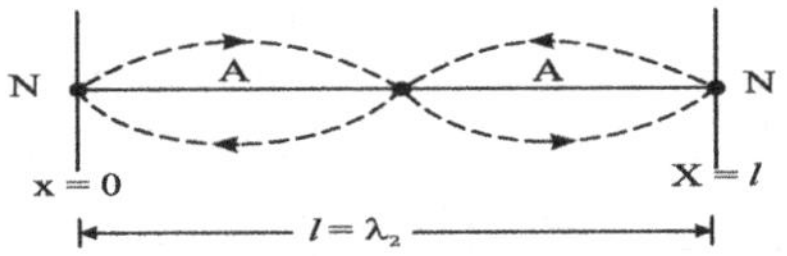

i) Number of loops p = 2 ii) Plucking at $\frac{l}{2\times 2} = \frac{l}{4}$ (from one fixed end) iii) $l = \lambda_2$

iv) Second harmonic or first over tone. $n_2 = \frac{1}{\lambda_2}\sqrt{\frac{T}{\mu}} = \frac{1}{l}\sqrt{\frac{T}{\mu}} = 2n_1$

11. **Third mode of vibration:**

i) Number of loops p = 3

ii) Plucking at $\frac{l}{2\times 3} = \frac{1}{6}$ (from one fixed one)

iii) $l = \frac{3\lambda_3}{2} \Rightarrow \lambda_3 = \frac{2l}{3}$

iv) Third harmonic or second over tone.

$$n_3 = \frac{1}{\lambda_3}\sqrt{\frac{T}{\mu}} = \frac{3}{2l}\sqrt{\frac{T}{\mu}} = 3n_1$$

12. **More about string vibration**

i) In general, if the string is plucked at length $\frac{l}{2p}$, then it vibrates in p segments (loops) and we have the pth harmonic n $n_p = \frac{p}{2l}\sqrt{\frac{T}{\mu}}$

ii) All even and odd harmonics are present. Ratio of harmonic = 1 : 2 : 3......

iii) Ratio of over tones = 2 : 3 : 4 ......

iv) General formula for wavelength $\lambda = \frac{2l}{P}$; where P = 1, 2, 3,..... correspond to 1st, 2nd, 3rd modes of vibratio of the string.

v) General formula for frequency $n = P \times \frac{v}{2l}$

vi) **Positions of nodes:** $x_N = 0, \frac{l}{P}, \frac{2l}{P}, \frac{3l}{P}........l$

vii) **Positions of antinodes:** $x_{AN} = \frac{l}{2P}, \frac{3l}{2P}, \frac{5l}{2P}.....\frac{(2P-1)l}{2P}$

Sonometer

1. It is an apparatus, used to produce resonance (matching frequency) of tuning fork (or any source of sound) with stretched vibrating string.
2. It consists of a hollow rectangular box of light wood. The experimental set up fitted on the box is shown below.

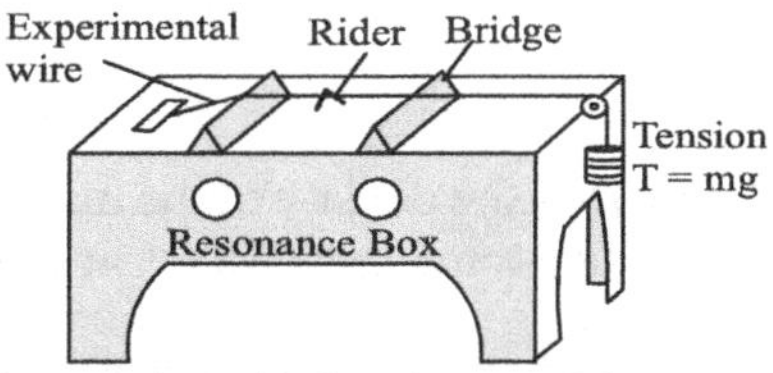

3. The box serves the purpose of increasing the loudness of the sound produced by the vibrating wire.
4. If the length of the wire between the two bridges is l, then the frequency of vibration is $n = \frac{1}{2l}\sqrt{\frac{T}{\mu}} = \sqrt{\frac{T}{\pi r^2 \rho}}$

(r = Radius of the wire, ρ = Density of material of wire) μ = mass per unit length of the wire

5. **Resonance:** When a vibrating tuning fork is placed on the box, and if the length between the bridges is property adjusted then if

$(n)_{Fork} = (n)_{string} \rightarrow$ rider is thrown off the wire.

Laws of string

i) Law of length: If T and μ are constant then $n \propto \frac{1}{l} \Rightarrow nl = \text{constant} \Rightarrow n_1 l_1 = n_2 l_2$

If % change is less than 5% then $\frac{\Delta n}{n} = -\frac{\Delta l}{l}$ or $\frac{\Delta n}{n} \times 100\% = -\frac{\Delta l}{l} \times 100\%$

ii) Law of mass: If T and l are constant then $n \propto \frac{1}{\sqrt{\mu}} \Rightarrow \frac{n_1}{n_2} = \sqrt{\frac{\mu_2}{\mu_1}}$

If % change is less than 5% then $\frac{\Delta n}{n} = -\frac{1}{2}\frac{\Delta \mu}{\mu}$

or $\frac{\Delta n}{n} \times 100\% = -\frac{\Delta \mu}{\mu} \times 100\%$

iii) Law of density: If T, l and r are constant then $n \propto \frac{1}{\sqrt{\rho}} \Rightarrow n\sqrt{\rho} = \text{const} \Rightarrow \frac{n_1}{n_2} = \sqrt{\frac{\rho_2}{\rho_1}}$

If % change is less than 5% then $\frac{\Delta n}{n} = -\frac{1}{2}\frac{\Delta \rho}{\rho}$ or $\frac{\Delta n}{n} \times 100\% = -\frac{\Delta \rho}{\rho} \times 100\%$

iv) Law of tension: If l and μ are constant then $n \propto \sqrt{T}$

$\Rightarrow \frac{n}{\sqrt{T}} = \text{const} \Rightarrow \frac{n_1}{n_2} = \sqrt{\frac{T_1}{T_2}} = \sqrt{\frac{M_1}{M_2}}$

If % change is less than 5% then $\frac{\Delta n}{n} = \frac{1}{2}\frac{\Delta T}{T}$ or $\frac{\Delta n}{n} \times 100\% = -\frac{\Delta T}{T} \times 100\%$

Tuning fork

i) It is a U shaped metal bar made of steel or an alloy with a handle attached at the bend.
ii) When it is struck against a hard rubber pad, its prongs begin to vibrate as shown in figure(a).

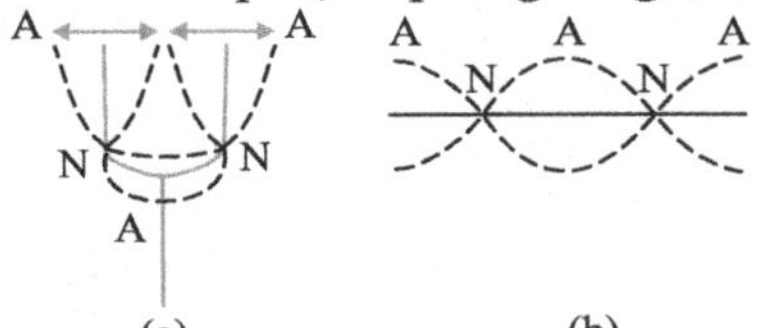

(a) (b)

iii) A tuning fork emits a single frequency note, i.e., a fundamental with no overtones.
iv) A tuning fork may be considered as a vibrating free bar as shown figure(b) that has been bent into U-shape.
v) Two antinodes are formed one at each free end of the bar which are in phase.
vi) The frequency of a tuning fork of arm length 'l' and thickness 'd' in the direction of vibration is given by

$$n = \frac{d}{l^2}v = \frac{d}{l^2}\sqrt{\frac{Y}{\rho}}, \left[\because v = \sqrt{\frac{Y}{\rho}}\right]$$

where Y is the Young's modulus and ρ is the density of the material of the turning fork.
vii) Using the tuning fork we can produce transverse waves in solids and longitudinal waves in solids, liquids and gases.
viii) Transverse vibrations are present in the prongs. Longitudinal vibrations are present in the shank.
ix) Loading or waxing a tuning fork increases its inertia and so decreases its frequency, while filing a tuning fork decreases its inertia and so increases its frequency.
x) When tuning fork is heated its frequency decreases due to decrease in elasticity.

W.E-16: ***The vibrations of a string of length 60 cm fixed at both ends are represented by the equation.*** $y = 4\sin\left[\frac{\pi x}{15}\right]\cos(96\pi t)$ ***Where x and y are in cm and t in sec.***

a) What is the maximum displacement at $x = 5$ cm ?
b) What are the nodes located along the string ?

c) What is the velocity of the particle at $x = 7.5$ cm and $t = 0.25$ s ?

d) Write down the equations of component waves whose superposition gives the above wave.

Sol. a) For x = 5cm, $y = 4\sin(5\pi/15)\cos(96\pi t)$

(or) $y = 2\sqrt{3}\cos(96\pi t)$

So y will be maximum when cos ($96\pi t$) = 1 i.e., $(y_{max})_{x=5} = 2\sqrt{3}cm$

b)At nodes amplitude of wave is zero.

$4\sin\left[\frac{\pi x}{15}\right] = 0 (or) \frac{\pi x}{15} = 0, \pi, 2\pi, 3\pi......$

So x = 0, 15, 30, 45, 60 cm [as length of string = 60cm]

c) As $y = 4\sin(\pi x/15)\cos(96\pi t)$

$\frac{dy}{dx} = -4\sin\left[\frac{\pi x}{15}\right]\sin(96\pi t)\times(96\pi)$

So the velocity of the particle at x = 7.5cm and t = 0.25s,

v_{pa} = -384π sin(7.5π/15) sin (96π × 0.25)

v_{pa} = -384π ×1×0=0

d) $y = y_1 + y_2$ with $y_1 = 2\sin\left[96\pi t + \frac{\pi x}{15}\right]$

$y_2 = -2\sin\left[96\pi t - \frac{\pi x}{15}\right]$

W.E-17: ***A guitar string is 90 cm long and has a fundamental frequency of 124 Hz. Where should it be pressed to produce a fundamental frequency of 186 Hz?***

Sol. Since T is constant we have $n \propto \frac{1}{l}$

$l_2 = \frac{n_1}{n_2} l_1 = \frac{124}{186} \times 90 = 60cm$

Thus, the string should be pressed at 60cm from an end.

W.E-18: ***A wire having a linear mass density 5.0×10^{-3} kg/m is streched between two rigid supports with a tension of 450 N. The wire resonates at a frequency of 420 Hz. The next higher frequency at which the same wire resonates is 490 Hz. Find the length of the wire.***

Sol. Suppose the wire vibrates at 420 Hz in its nth harmonic and at 490 Hz in its (p + 1)th harmonic.

$\frac{490}{420} = \frac{p+1}{p} (or) p = 6$

$420 = \frac{6}{2l}\sqrt{\frac{450}{5.0\times10^{-3}}} \quad \therefore l = \frac{900}{420} = 2.1m$

W.E-19: ***The equation of a standing wave produced on a string fixed at both ends is where 'y' is measured in cm. What could be the smallest length of string?***

Sol. Comparing with $y = 2A\sin kx \cos wt$

We have $k = \frac{\pi}{10} \quad \Rightarrow \lambda = 20cm$

If the string vibrates in 'p' loops then length of string 'l' is $\frac{p\lambda}{2}$. $\Rightarrow \frac{p\lambda}{2} = l$

'ℓ' is minimum if p = 1 $\Rightarrow l = \frac{\lambda}{2} = 10cm$

W.E-20: ***The equation for the vibration of a string fixed at both ends, vibrating in its third harmonic is given by*** $y = 0.4\sin\frac{\pi x}{10}\cos 600\pi t$ ***where x and y are in cm***

1) What is the frequency of vibration? ***2) What are the position of nodes?***
3) What is the length of string?
4) What is the wavelength and speed of transverse waves that can interfer to give this vibration?

Sol. Comparing with

$y = 2A\sin kx\cos\omega t$ we have

1) $\omega = 600\pi$ gives n = 300 Hz

2) To get the position of nodes $\sin\frac{\pi x}{10} = 0$

i.e., $\frac{\pi x}{10} = N\pi$ where N = 0, 1, 2....

Hence nodes occur at x = 0, 10, 20 cm...

3) Since the string is in 3rd harmonic

$l = 3\frac{\lambda}{2}$ gives $l = 30$cm; $\left[\because \lambda = \frac{2\pi}{k} = \frac{2\pi}{\pi/10} = 20cm\right]$

4)Speed of wave $v = n\lambda = 300 \times 20 = 60ms^{-1}$.

W.E-21 ***A sonometer wire has a length of 114 cm between two fixed ends. Where should two bridges be placed to divide the wire into three segments whose fundamental frequencies are in the ratio 1 : 3 : 4 ?***

Sol. In case of a given wire under constant tension, fundamental frequency of vibration $n \propto (1/l)$

$\therefore l_1 : l_2 : l_3 = \frac{1}{1} : \frac{1}{3} : \frac{1}{4} = 12:4:3$

$\therefore l_1 = 72cm; l_2 = 24cm; l_3 = 18cm$

$\therefore$ First bridge is to be placed at 72 cm from one end.

Second bridge is to be placed at 72 + 24 = 96 cm from one end

W.E-22: ***An aluminium wire of cross-sectional area 10^{-6} m^2 is joined to a copper wire of the same cross-section. This compound wire is stretched on a sonometer, pulled by a load of 10 kg. The total length of the compound wire between two bridges is 1.5 m of which the aluminium wire is 0.6 m and the rest is the copper wire. Transverse vibrations are set up in the wire in the lowest frequency of excitation for which standing waves are formed such that the joint in the wire is a node. What is the total number of nodes observed at this frequncy excluding the two at the ends of the wire ? The density of aluminium is 2.6 × 10^4 kg/m^3.***

Sol. As the total length of the wire is 1.5 m and out of which $L_A = 0.6$ m, so the length of copper wire

$L_c = 1.5 - 0.6 = 0.9$ m. The tension in the whole wire is same (=Mg = 10g N) and as fundamental frequency of vibration of string is given by

$$n = \frac{1}{2L}\sqrt{\frac{T}{\mu}} = \frac{1}{2L}\sqrt{\frac{T}{\rho A}} \quad [\because \mu = \rho A]$$

so $n_A = \frac{1}{2L_A}\sqrt{\frac{T}{\rho_A A}}$ and $n_c = \frac{1}{2L_c}\sqrt{\frac{T}{\rho_c A}}$ ------- (1)

Now as in case of composite wire, the whole wire will vibrate with fundamental frequency

$n = p_A n_A = p_C n_C$ -------- (2)

Substituting the values of f_A and f_c from Eqn.(1)in(2)

$$\frac{p_A}{2\times0.6}\sqrt{\frac{T}{A\times2.6\times10^3}} = \frac{p_c}{2\times0.9}\sqrt{\frac{T}{A\times1.0401\times10^4}}$$

i.e., $\dfrac{p_A}{p_c}=\dfrac{2}{3}\sqrt{\dfrac{2.6}{10.4}}=\dfrac{2}{3}\times\dfrac{1}{2}=\dfrac{1}{3}$

So that for fundamental frequency of composite string, $p_A = 1$ and $p_c = 3$, i.e., aluminium string will vibrate in first harmonic and copper wire at second, overtone as shown in figure.

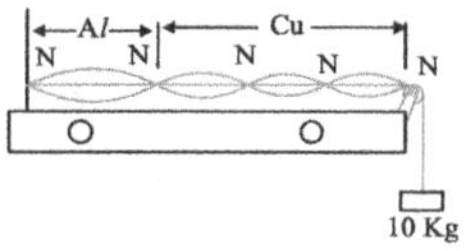

$\therefore n = n_A = 3n_C$

This in turn implies that total number of nodes in the string will be 5 and so number of nodes excluding the nodes at the ends = 5 - 2 = 3

W.E-23: ***A wire of uniform cross-section is stretched between two points 1 m apart. The wire is fixed at one end and a weight of 9 kg is hung over a pulley at the other end produces fundamental frequency of 750 Hz.***

(a) What is the velocity of transverse waves propagating in the wire ?

(b) If now the suspended weight is submerged in a liquid of density (5/9) that of the weight, what will be the velocity and frequency of the waves propagating along the wire ?

Sol. a) In case of fundamental vibrations of string $(\lambda/2)=L$, i.e., $\lambda = 2\times1=2m$

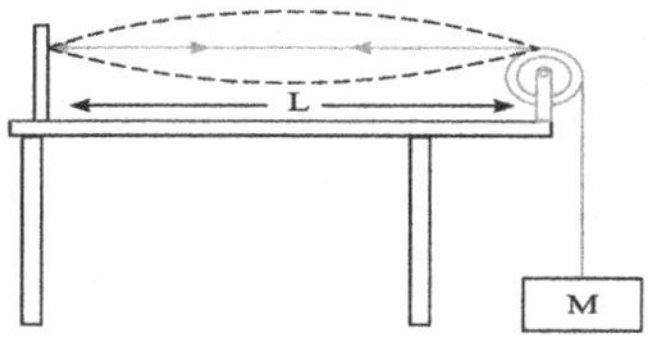

Now as $v=n\lambda$ and n = 750 Hz,

$v_T = 2\times750 = 1500 m/s$

b) Now as in case of a wire under tension $v=\sqrt{\dfrac{T}{\mu}} \Rightarrow v\alpha\sqrt{T} \Rightarrow \dfrac{v_B}{v_T}=\sqrt{\dfrac{T_B}{T_T}}$

$\Rightarrow v_B = 1500\sqrt{\dfrac{T_B}{T_A}} \Rightarrow 1500\sqrt{\dfrac{mg[1-\rho_l/\rho_b]}{mg}} = 1000 m/s$

From $v=n\lambda \Rightarrow n_B = \dfrac{v_B}{l_B}=\dfrac{1000}{2}=500Hz$

WE-24: ***A wire of density 9×10^3 kg/m³ is stretched between two clamps 1 m apart and is subjected to an extension of 4.9×10^{-4} m. What will be the lowest frequency of transverse vibrations in the wire ? ($Y = 9\times10^{10}$ N/m²)***

Sol. In case of fundamental vibrations of a string

$$n=\frac{1}{2L}\sqrt{\frac{Y}{\rho}\frac{\Delta L}{L}}$$

$$=\frac{1}{2\times1}\sqrt{\frac{9\times10^{10}\times4.9\times10^{-4}}{9\times10^{3}\times1}}=35Hz$$

W.E-25: ***A string 120 cm in length sustains a standing wave, with the points of string at which the displacement amplitude is equal to $\sqrt{2}$ mm being separated by 15.0 cm, Find the maximum displacement amplitude.***

Sol. From figure. points A, B, C, D and E are having equal displacement amplitude.

Further, $x_E - x_A = \lambda = 4\times15 = 60cm$

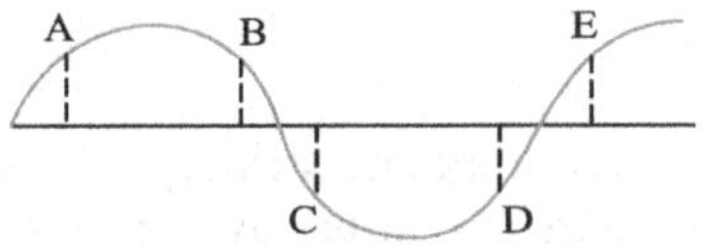

As $\lambda = \frac{2l}{n} = \frac{2\times120}{n} = 60$ $\quad\therefore\quad n = \frac{2\times120}{60} = 4$

So, it corresponds to 4th harmonic.

Also, distance of node from A is 7.5 cm and no node is between them. Taking node at origin, the amplitude of stationary wave can be written as, $A_{sw} = A_{max}\sin kx$

$$A_{sw} = \sqrt{2}mm; k = \frac{2\pi}{\lambda} = \frac{2\pi}{60} \text{ and } x = 7.5 \text{ cm}$$

$$\therefore\ \sqrt{2} = A_{max}\sin\left(\frac{2\pi}{60}\times7.5\right) = A_{max}\sin\frac{\pi}{4}$$

Hence, $A_{max} = 2mm$

Sound Waves :

Sound is a form of energy propagated in the form of longitudinal waves. This energy causes the sensation of hearing on reaching the ear. Any vibrating body could be a source of sound. Longitudinal mechanical waves can be transmitted in all the three states of matter namely, solids, liquids and gases. According to their range of frequencies longitudinal mechanical waves are divided into the three categories.

1) Longitudinal waves having frequencies below 20Hz are called infrasonic waves. These are created by earthquakes, elephants and whales. Infrasonic waves can be heard by snakes.

2) Longitudinal waves having range, of frequencies lying between 20Hz and 20kHz are called audible sound waves. The audible wavelength is 16.5 mm to 16.5m at S.T.P when velocity of sound is 330 m/s. These are generated by tuning forks, streched stings and vocal cords.

The human ear can detect these waves.

3) Longitudinal waves having frequencies greater than 20 kHz are called ultrasonics. The human ear can't detect these waves. These waves can be produced by high frequency vibrations of a quartz crystal under an alternating electric field. These waves can be detected by mosquito, fish and dog etc.

Application of ultrasonic waves :

i) The fine internal cracks in a metal can be detected by ultrasonic waves.

ii) They are used for determining the depth of the sea and used to detect submarine.

iii) They can be used to clean clothes and fine machinery parts

iv) They can be used to kill animals like rats, fish and frogs etc.

Characteristics of Sound

- Hearing of sound is characterised by following three parameters.

1) **Loudness (Refers to Intensity) :**

It is the sensation received by ear due to intensity of sound

Greater the amplitude of vibration, greater will be intensity $(I \alpha A^2)$ and so louder will be sound.

The loudness being the sensation, depends on the sensitivity of listener's ear. Loudness of a sound of a given intensity may be different for different listeners.

The average energy transmitted by a wave per unit normal area per second is called intensity of a wave.

$I = \frac{E}{At}$. Its SI Unit : W/m^2

- It is the average power transmitted by a wave through the given area.

$$I = \frac{P_{avg.}}{\text{area}};\quad I = 2\pi^2 n^2 A^2 \rho v$$

where ρ – density of medium, v – velocity of wave, A - Amplitude, n - Frequency

- Human ear responds to sound intensities over a wide range from 10^{-12} W/m^2 to 1 W/m^2.

- In a spherical wave front (i.e. wave starting from a point source) , the amplitude varies inversely with distance from position of source i.e, $A\alpha\frac{1}{r}\Rightarrow I\propto\frac{1}{r^2}$

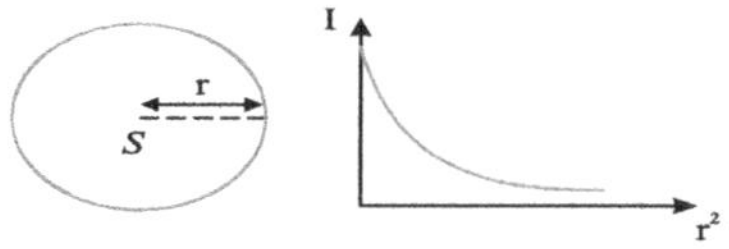

- In a cylindrical wave front (i.e. wave starting from a linear source), the amplitude varies inversely as the square root of distance from the axis of source i.e. , $A\propto\frac{1}{\sqrt{r}}\Rightarrow I\propto\frac{1}{r}$

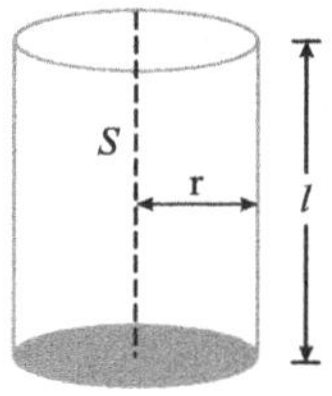

Sound level in decibles is given by $\beta=10\log\left(\frac{I}{I_0}\right)$

If β_1 and β_2 be the sound levels corresponding to sound intensities I_1 and I_2 respectively. Then,

$\beta_1=10\log\frac{I_1}{I_0}$ and $\beta_2=10\log\frac{I_2}{I_0}$

$\therefore\beta_2-\beta_1=10\left(\log\frac{I_2}{I_0}-\log\frac{I_1}{I_0}\right)$

(or) $\beta_2-\beta_1=10\log\left(\frac{I_2}{I_1}\right)$

2) **Pitch (Refers to Frequency):**
The shrillness or harshness of sound is known as pitch. Pitch depends on frequency. Higher the frequency, higher will be the pitch and shriller will be the sound.

3) **Quality or Timber (Refers to Harmonics):**
It is the sensation received by ear due to waveform. Quality of a sound depends on number of overtones. i.e, harmonic present.

Velocity of Sound

- The equation for velocity of sound through a medium is given by $v=\sqrt{\frac{E}{\rho}}$
  where E = modulus of elasticity; ρ=density
- As modulus of elasticity is more for solids and less for gases, so
  $v_{\text{solids}}>v_{\text{liquids}}>v_{\text{gases}}$
- In case of solids $v=\sqrt{\frac{Y}{\rho}}$,
  where Y is Young's modulus,
- In case of fluids (liquids and gases) $v=\sqrt{\frac{B}{\rho}}$
  where B is the Bulk modulus

Velocity of sound in Gases :

- **Newton's formula :**
  Newton assumed that the propagation of sound in a gas takes place under isothermal conditions.
- Isothermal Bulk modulus , $B = P$

$$\therefore v_s = \sqrt{\frac{P}{\rho}}$$

- At S.T.P. $v = \sqrt{\frac{1.013x10^5}{1.29}} \approx 280ms^{-1}$

  Which is less than the experimental value ($332\,m/s$)
- **Laplace's correction:** Laplace assumed that the propagation of sound in a gas takes place under adiabatic conditions.
- Adiabatic Bulk modulus, $B = \gamma P$

$$\therefore v = \sqrt{\frac{\gamma P}{\rho}} = \sqrt{\frac{\gamma PV}{m}} = \sqrt{\frac{\gamma nRT}{m}} = \sqrt{\frac{\gamma RT}{M}}$$

  where V = volume, m is mass, M = molecular weight. T is absolute temperature
- For air $\gamma = 1.4$. Therefore

  At STP $v_0 = 280\sqrt{1.4} \approx 330ms^{-1}$, which agrees with the experimentally calculated value.
- Velocity of sound in a gas is directly proportional to the square root of the absolute temperature

$$\frac{v_t}{v_o} = \sqrt{\frac{T}{T_o}} = \left(\frac{t+273}{273}\right)^{1/2} \left(\because v \propto \sqrt{T}\right)$$

$$\Rightarrow v_t = v_o\left(1+\frac{t}{546}\right)$$

$$\Rightarrow v_t = v_0 + \frac{v_0 t}{546} = v_0 + 0.61t°C$$

Note:

1) When temperature rises by $1°C$ then velocity of sound increases by 0.61 m/s
2) The velocity of sound increases with increase in humidity. Sound travels faster in moist air than in dry air at the same temperature, because density of humidity air is less than that of dry air.

   $\rho_{\text{moist air}} < \rho_{\text{dry air}} \Rightarrow v_{\text{moist air}} > v_{\text{dry air}}$
3) The velocity of sound at constant temperature in a gas does not depend upon the pressure of air.
4) Amplitude, frequency, phase, loudness, pitch, quality donot effect velocity of sound.

W.E-26: ***Find the speed of sound in a mixture of 1 mol of helium and 2 mol of oxygen at 27°C.***

Sol. $\gamma_{mix} = \frac{C_{Pmix}}{C_{Vmix}} = \frac{(19R/6)}{(13R/6)} = \frac{19}{13}$

$$M_{mix} = \frac{n_1M_1 + n_2M_2}{n_1 + n_2} = \frac{1\times4 + 2\times32}{1+2}$$

$$= \frac{68}{3}\times10^{-3}\,kg/mol\ ;$$

$$v = \sqrt{\frac{\gamma_{mix}RT}{M_{mix}}} = \sqrt{\frac{19}{13}\times\frac{8.314\times300}{68\times10^{-3}/3}} \approx 401m/s$$

W.E-27: ***A window whose area is $2m^3$ opens on a street where the street noise result in an intensity level at the window of $60dB$. How much 'acoustic power' enters the window via sound waves. Now if an acoustic absorber is fitted at the window, how much energy from street will it collect in five hours ?***

Sol. Sound level $\beta = 10\log\left(\frac{I}{I_o}\right)$

$\Rightarrow 60 = 10\log\left(\frac{I}{I_o}\right) \Rightarrow \frac{I}{I_o} = 10^6 \Rightarrow I = 10^6 I_o$

$\Rightarrow I = 10^6 \times 10^{-12} = 10^{-6} W/m^2$

but intensity $I = \frac{E}{At} \Rightarrow E = IAt$

$E = 10^{-6} \times 2 \times 5 \times 3600 = 36 \times 10^{-3} J$

Various forms of longitudinal wave:

As we know, during a longitudinal wave propagation the particles of the medium oscillate to produce pressure and density variation along the direction of the wave. These variations result in series of high and low pressure (and density) regions called compression and rarefactions respectively. Hence the longitudinal wave can be in terms of displacement of particles called displacement wave y(x, t) or in terms of change in pressure called pressure wave $\Delta P(x,t)$ or change in density called density wave $\Delta d(x,t)$.

1) **Pressure Wave:**

i)A longitudinal sound wave can be expressed either in terms of the longitudinal displacement of the particles of the medium or in terms of excess pressure produced due to compression or rarefaction. (at compression, the pressure is more than the normal pressure of the medium and at rarefaction the pressure is lesser than the normal).

ii) If the displacement wave is represented by $y = A\sin(\omega t - kx)$ then the corresponding pressure wave will be represented by $\Delta P = -B\frac{dy}{dx}$ (B = Bulk modulus of elasticity of medium)

$\therefore \Delta P = BAk\cos(\omega t - kx) = \Delta P_0 \cos(\omega t - kx)$

where ΔP_0 = pressure amplitude $= BAk$

iii) Pressure wave is $\pi/2$ out of phase(lags) with displacement wave. *i.e.* pressure is maximum when displacement is minimum and vice-versa.

Note1: At the centre of compression and rarefaction particle velocity is maximum and at the boundary of compression and rarefaction particles are momentarly of rest. This is explained as in a harmonic progressive wave v_P = -(slope of y-x)$\times v \Rightarrow \frac{v_p}{v} = -\frac{dy}{dx}$

Since the change in pressure of the medium $\Delta P = -B\left(\frac{dy}{dx}\right) \Rightarrow \Delta P = B\left(\frac{v_p}{v}\right)$

i.e., for a given medium, B and v are constants. Where v_p is maximum, Δp is also maximum, which is true at y = 0

Note 2: As sound sensors (e.g ear or mike) detect pressure changes, description of sound as pressure wave is preferred over displacement wave.

2) **Density wave form :** Let ρ_o be the normal density of the medium and $\Delta\rho$ be the change in density of the medium during the wave propagation.

Then fraction of change in volume of the element $\frac{\Delta v}{v} = -\frac{\Delta\rho}{\rho_0}$ $\left(\because v = \frac{m}{\rho}\right)$

According to definition of Bulk's modulus

$$B = -\Delta P\left(\frac{v}{\Delta v}\right) = \Delta P\left(\frac{\rho_0}{\Delta\rho}\right) \Rightarrow \Delta\rho = \frac{\rho_0}{B}.\Delta p \qquad \Rightarrow \Delta\rho = \frac{\rho_0}{B}(\Delta p)_{max} Cos(kx - \omega t)$$

$$\Rightarrow \Delta\rho = \rho_0 Ak\,Cos(kx - \omega t) \quad \left(\because (\Delta p)_{max} = BAk\right) \Rightarrow \Delta\rho = (\Delta\rho)_{max}\,Cos(kx - \omega t),$$

where $(\Delta\rho)_{max} = \rho_0 Ak$ is called density amplitude. Thus the density wave is in phase with the pressure wave and this is 90^0 out of phase (lags) with the displacement wave as shown in the figure.

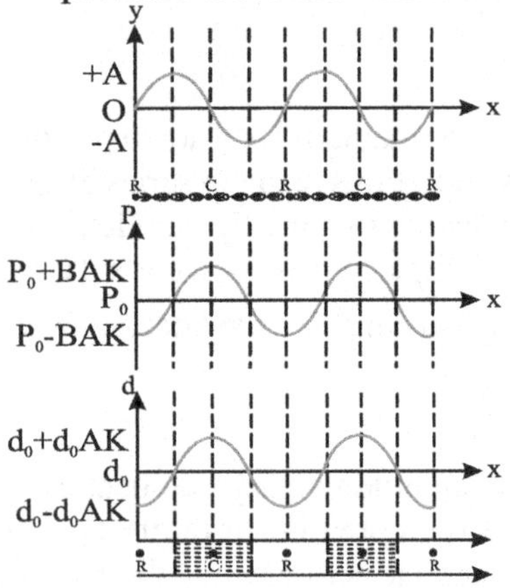

Note 1: The relation between density amplitude and pressure amplitude is $(\Delta\rho)_{max} = (\Delta p)_{max}\left(\frac{\rho}{B}\right)$

Note 2: Average Intensity $I = \frac{P}{S} = \frac{1}{2}\rho\omega^2 A^2 v$

In terms of pressure amplitude, sound intensity

$$I = \frac{1}{2}\rho\omega^2\left(\frac{\Delta p_{max}}{Bk}\right)^2 v = \frac{1}{2}\frac{(\Delta p_{max})^2}{\rho v}$$

$$[\because (\Delta P)_{max} = BAk, k = \frac{\omega}{v} \text{ and } B = \rho v^2]$$

Thus intensity of wave is proportional to square of pressure amplitude or displacement amplitude or density amplitude and is independent of frequency.

W.E-28: ***What is the maximum possible sound level in dB of sound waves in air? Given that density of air*** $= 1.3 kg/m^3, v = 332 m/s$ ***and atmospheric pressure*** $P = 1.01\times10^5\ N/m^2$.

Sol. For maximum possible sound intensity, pressure amplitude of wave will be equal to atmospheric pressure, i.e., $p_0 = P = 1.01\times10^5\ Nm^2$

$$I = \frac{p_0^2}{2\rho v} = \frac{(1.01\times10^5)^2}{2\times1.3\times332} = 1.18\times10^7 W/m^2$$

$$\therefore SL = 10\log\frac{I}{I_0} = 10\log\frac{10^7}{10^{-12}} = 190 dB$$

WE-29: ***The faintest sounds the human ear can detect at a frequency of 1000 Hz correspond to an intensity of about*** $1.00\times10^{-12} W/m^2$, ***which is called threshold of hearing. The loudest sounds the ear can tolerate at this frequency correspond to an intensity of about*** $1.00 W/m^2$, ***the threshold of pain. Determine the pressure amplitude and displacement amplitude associated with these two limits.***

Sol. $\Delta P_{max} = \sqrt{2\rho v I} \quad = \sqrt{2(1.20)(343)(1.00\times10^{-12})}$

$= 2.87\times10^{-5} N/m^2$

$$A = \frac{\Delta P_{max}}{\rho v \omega} = \frac{2.87\times10^{-5}}{(1.2)(343)(2\pi\times1000)}$$

$(\because \omega = 2\pi n)$; $= 1.11\times10^{-11} m$

WE-30: ***A firework charge is detonated many metres above the ground. At a distance of 400 m from the explosion, the acoustic pressure reaches a maximum of*** $10.0 N/m^2$***. Assume that the speed of sound is constant at 343 m/s throughout the atmosphere over the region considered, the ground absorbs all the sound falling on it, and the air absorbs sound energy at the rate of 7.00 dB/km. What is the sound level (in decibels) at 4.00 km from the explosion?***

Sol. $r = 400m, r^1 = 4000m,$

$\rho = 1.2 kg/m^3, v = 343 m/s$

$$I = \frac{\Delta P^2_{max}}{2\rho v} = \frac{10}{2(1.2)(343)} = 1.21\times10^{-2} W/m^2$$

as $I \propto \frac{1}{r^2} \Rightarrow \frac{I^1}{I} = \left(\frac{r}{r^1}\right)^2$

$$I^1 = \frac{I(400)}{4000} = 1.21\times10^{-3} W/m^2 \qquad \beta = 10\log\left(\frac{I}{I^1}\right) = \log\left(\frac{1.21\times10^{-3}}{1\times10^{-12}}\right) = 90.8\ dB$$

At a distance of 4 km from the explosion, absorption from the air will decrease the sound level by an additional amount,

$\Delta\beta = (7)(3.60) = 25.2\ dB$

At 4 km, the sound level will be

$\beta_f = \beta - \Delta\beta = 90.8 - 25.2 = 65.6\ dB$

Organ pipes

Organ pipe: An organ pipe is a cylindrical tube of uniform cross section in which a gas is trapped as a column.

Open pipe : If both ends of a pipe are open and a system of air is directed against an edge, standing longitudinal waves can be set up in the tube. The open end is a displacement antinode.

- Due to finite momentum, air molecules undergo certain displacement in the upward direction hence antinode takes place just above the open end but not exactly at the end of the pipe.
- Due to pressure variations, reflection of longitudinal wave takes place at open end and hence longitudinal stationary waves are formed in open tube.
- 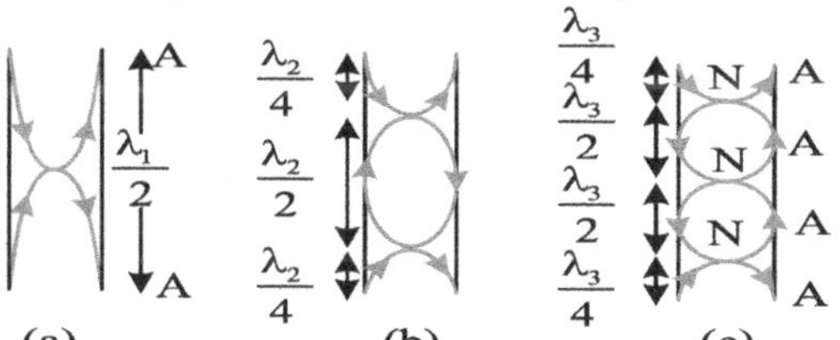

fig: a) For fundamental mode of vibrations or I harmonic

$L = \frac{\lambda_1}{2}$; $\therefore \lambda_1 = 2L$

$V = \lambda_1 n_1$; $\therefore V = 2Ln_1 \Rightarrow n_1 = \frac{V}{2L}$ ------ (1)

- **fig:b) For the second harmonic or first overtone,**

$L = \lambda_2$

$V = \lambda_2 n_2$ $\therefore V = Ln_2$ $\Rightarrow n_2 = \frac{2V}{2L}$ ------ (2)

- **fig:c) For the third harmonic or second overtone,**

  $L = 3\times\frac{\lambda_3}{2}\quad \therefore \lambda_3 = \frac{2}{3}L$

  $V = \lambda_3 n_3\quad \therefore V = \frac{2}{3}Ln_3\ \Rightarrow n_3 = \frac{3V}{2L}$ ------- (3)

- From (1), (2) and (3) we get,

  $n_1 : n_2 : n_3 ..... = 1:2:3:......$

  i.e. for a cylindrical tube, open at both ends, the harmonics excitable in the tube are all integral multiples of its fundamental.

- $\therefore$ In the general case, $\lambda = \frac{2L}{p}$, where $p = 1, 2, ...$

- p^{th} harmonic frequency $= \frac{V}{\lambda} = \frac{pV}{2l}$, where $p = 1, 2, ...$

  Closed pipe: If one end of a pipe is closed, then reflected wave is $180°$ out of phase with the wave. Thus the displacement of the small volume elements at the closed end must always be zero. Hence the closed end must be a displacement node.

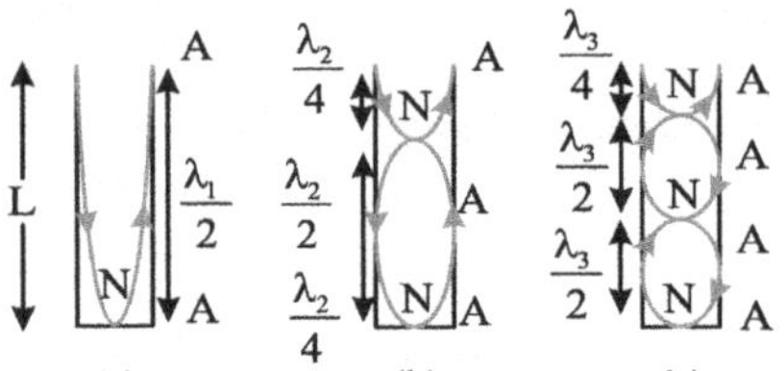

(a) (b) (c)

- **figure a) for the fundamental mode of vibration or I harmonic :**

  $L = \frac{\lambda_1}{4}\quad \therefore \lambda_1 = 4L$

  If n_1 is the fundamental frequency, then the velocity of sound waves is given as

  $V = \lambda_1 n_1\quad \therefore V = 4Ln_1\ \Rightarrow n_1 = \frac{V}{4L}$ ----- (1)

- **figure b) for third harmonic or first overtone.**

  $L = 3\times\frac{\lambda_2}{4},\ \therefore \lambda_2 = \frac{4}{3}L$

  $V = \lambda_2 n_2,\ \therefore V = \frac{4}{3}Ln_2 \Rightarrow n_2 = \frac{3V}{4L}$ ---- (2)

- **figure c) for fifth harmonic or second overtone.**

  $L = 5\times\frac{\lambda_3}{4},\ \therefore \lambda_3 = \frac{4}{5}L$

  $V = \lambda_3 n_3,\ V = \frac{4}{5}Ln_3\ \Rightarrow n_3 = \frac{5V}{4L}$ ---- (3)

  From (1), (2) and (3) we get,

  $n_1 : n_2 : n_3 ..... = 1:3:5:......$

- In the general case, $\lambda = \frac{4L}{(2p+1)}$, where $p = 0, 1, 2, ...$

- p^{th} harmonic frequency $= \frac{(2p-1)V}{4L}$,

  where p = 1, 2....

End Correction

Due to finite momentum of air molecules in organ pipe reflection takes place not exactly at open end but some what above it. Hence antinode is not formed exactly at the open end rather it is formed at a little distance away from open end outside it.

The distance of antinode form the open end is known as end correct (e).

It is given by e = 0.6 r , where r = radius of pipe.

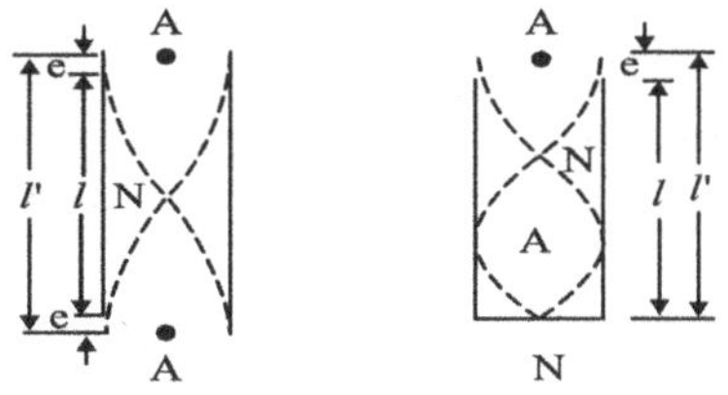

(A) Open pipe (B) Close pipe

Effective length in open organ pipe $l' = (l + 2e)$

Effective length in closed organ pipe $l' = (l + e)$

Note: When the end correction is considered, then i)the fundamental frequency of open pipe

$$n = \frac{V}{2(l+2e)} \Rightarrow n = \frac{V}{2(l+1.2r)}$$

ii) The fundamental frequency of closed pipe $n = \frac{V}{4(l+e)} \Rightarrow n = \frac{V}{4(l+0.6r)}$

Velocity of sound (Resonance column apparatus) :

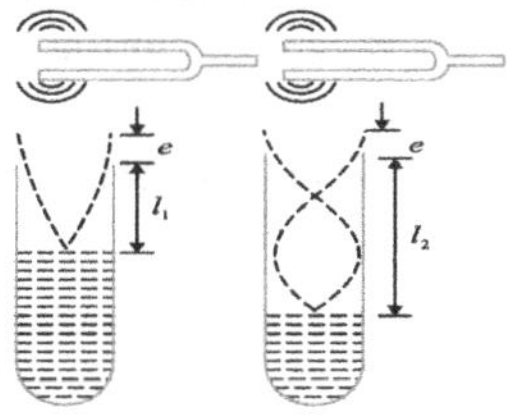

- If l_1, l_2 and l_3 are the first, second and third resonating lengths then $l_1 + e = \frac{\lambda}{4}$ ....(1)

$$l_2 + e = \frac{3\lambda}{4} \quad ...(2)$$

$$l_3 + e = \frac{5\lambda}{4} \quad --- (3)$$

- From equations (1) and (2)

1) $\lambda = 2(l_2 - l_1)$ 2) $V = n\lambda = 2n(l_2 - l_1)$

3) $\frac{(1)}{(2)} \Rightarrow e = \frac{l_2 - 3l_1}{2}$ 4) $l_3 - l_2 = l_2 - l_1 \Rightarrow l_3 = 2l_2 - l_1$

W.E-31: ***A tube of certain diameter and length 48 cm is open at both ends. Its fundamental frequency of resonance is found to be 320 Hz. The velocity of sound in air is 320 m/s. Estimate the diameter of the tube. One end of the tube is now closed. Calculate the frequency of resonance for the tube.***

Sol. $n_0 = \frac{v}{2[L+2e]} = \frac{v}{2[L+2\times 0.6r]} [as\, e = 0.6r]$

So substituting the given data,

$$320 = \frac{320\times 100}{2[48+1.2r]} (or)\, r = \frac{10}{6} cm$$

So, D = 2r = 2 × (10/6) = 3.33cm.
Now when one end is closed,

$$n_c = \frac{v}{4(L+0.6r)}$$

$$= \frac{320 \times 100}{4[48+0.6\times(10/6)]} = 163.3Hz$$

W.E-32: ***A tuning fork of frequency 340 Hz is vibrated just above a cylindrical tube of length 120 cm. Water is slowly poured in the tube. If the speed of sound in air is 340 m/s. Find the minimum height of water required for resonance. (v = 340m/s)***

Sol: $n = p\frac{v}{4L}$ with p = 1, 3, 5,............

So length of air column in the pipe

$L = \frac{pv}{4n} = 25p\,cm$ with p = 1,3,5,.....

i.e., L = 25cm, 75cm, 125cm
Now as the tube is 120 cm , so length of air column must be lesser than 120 cm, i.e., it can be only 25 cm or 75 cm. Further if h is the height of water filled in the tube,
L + h = 120 cm or h = 120 – L
So h will be minimum when L_{max} = 75cm
∴ $(h)_{min}$ = 120 – 75 = 45cm.

BEATS

- It is the phenomenon of periodic change in the intensity of sound when two waves of slightly different frequencies travelling in same direction superpose with each other.
- Maximum Intensity of sound(Waxing) is produced in the beats when constructive Interference takes place.
- Minimum Intensity of sound(Waning) is produced in the beats when destructive Interference takes place.

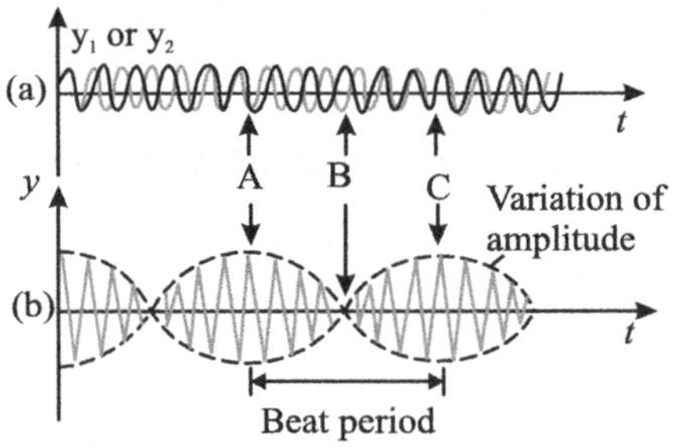

Analytical treatment of Beats:

- Equations of waves producing beats are given as $y_1 = a\sin\omega_1 t$ and $y_2 = a\sin\omega_2 t$ let $\omega_1 > \omega_2$
- Resultant wave equation is

$$y = y_1 + y_2 = 2a\cos\left(\frac{\omega_1-\omega_2}{2}\right)t\sin\left(\frac{\omega_1+\omega_2}{2}\right)t$$

$$y = A(t)\cos\left(\frac{\omega_1-\omega_2}{2}\right)t$$

Here $A(t) = 2a\sin\frac{(\omega_1+\omega_2)}{2}t$

- Amplitude is function of time. Frequency of variation of amplitude $= \frac{n_1-n_2}{2}$

- Frequency of resultant wave $=\frac{n_1+n_2}{2}$
- The variation in the intensity of sound between successive maxima or minima is called one beat.
- The number of beats per second is called beat frequency. If n_1 and n_2 are the frequencies of the two sound waves that interfere to produce beats then
  Beat frequency $= n_1 \sim n_2$
- The time period of one beat (or) the time interval between two successive maxima or minima is $\frac{1}{n_1 \sim n_2}$
- The time interval between a minima and the immediate maxima is $\frac{1}{2(n_1 \sim n_2)}$
- As the persistence of human hearing is about 0.1 sec, beats will be detected by the ear only if beat period is $\Delta t \geq 0.1 \sec$ or beat frequency
  $\Delta n = n_1 \sim n_2 \leq 10 Hz$
- Maximum number of beats that can be heard by a human being is 10 per second.
- If more than 10 beats are produced then no. of beats produced are same but no. of beats heard are zero
- If a_1, a_2 are amplitudes of two sound waves that interfere to produce beats then the ratio of maximum and minimum intensity of sound is, $\frac{I_{max}}{I_{min}} = \left(\frac{a_1+a_2}{a_1-a_2}\right)^2$

Uses of Beats:

i) To determine unknown frequency of a tuning fork with the help of a standard tuning fork.
ii) To tune the stretched string of a musical instrument to a particular frequency.
iii) To detect the presence of dangerous gases in mines.

Note:

1. When wax is added to the arms of one of the tuning forks then its frequency decreases.
   i.e. $n^1 < n$
2. When arms of one of the tuning forks are filed then its frequency increases. i.e., $n^1 > n$
3. The following table gives the relation for beats produced when sounded together under different conditions.

Fork	Frequency	Relation Δn when	
		$\Delta n^1 > \Delta n$	$\Delta n^1 < \Delta n$
Wax is added to 1st fork	$n_1^1 < n_1$	$\Delta n = n_2 - n_1$	$\Delta n = n_1 - n_2$
Wax is added to 2nd fork	$n_2^2 < n_2$	$\Delta n = n_1 - n_2$	$\Delta n = n_2 - n_1$
1st fork is filled	$n_1^1 < n_1$	$\Delta n = n_1 - n_2$	$\Delta n = n_2 - n_1$
2nd fork is filled	$n_2^1 < n_1$	$\Delta n = n_2 - n_1$	$\Delta n = n_1 - n_2$

W.E-33: ***The frequency of tunning fork 'A' is 250 Hz. It produces 6 beats/sec, when sounded together with another tunning fork B. If its arms are loaded with wax then it produces 4 beats/sec. Find the frequency of tuning fork B.***

Sol. $\Delta n = n_A \sim n_B = 6$ beats / sec

If wax is added to the tunning fork A then its frequency decreases. i.e., $n_A^1 < n_A$ and given $\Delta n^1 = 4$ beats / sec $< \Delta n$

This is possible when $n_A - n_B = \Delta n$

$\Rightarrow 250 - n_B = 6 \Rightarrow n_B = 244Hz$

W.E-34: ***A tunning fork of frequency of 512 Hz when sounded with unknown tunning fork produces 5 beats/sec. If arms of the unknown fork are filed then it produces only 3 beats / sec. Find the frequency of unknown tunning fork.***

Sol. $\Delta n = n_1 \sim n_2, n_1 = 512Hz$, $\Delta n = 5$ beats/sec

If arms of unknown fork are filed then its frequency increases.

i.e., $n_2^1 > n_2$ and given $\Delta n^1 = 3$ beats / sec.

This is possible when $\Delta n = n_2 - n_1$

$\Rightarrow 5 = n_2 - 512 \Rightarrow n_2 = 517Hz$

W.E-35: ***The lengths of two open organ pipes are l and $l + \Delta l\left(\Delta l << l\right)$. If v is the speed of sound, find the frequency of beats between them.***

Sol. Beat frequency $= n_1 - n_2 = \dfrac{v}{2l} - \dfrac{v}{2(l+\Delta l)}$

$$= \frac{v}{2l}\left[1 - \left(1 + \frac{\Delta l}{l}\right)^{-1}\right] \approx \frac{v}{2l}\left[1 - 1 + \frac{\Delta l}{l}\right] = \frac{v\Delta l}{2l^2}$$

W.E-36: ***If two sound waves,*** y_1=0.3sin596π[t-x/330] ***and*** y_2=0.5sin604π[t -x/330] ***are superposed, what will be the (a) frequency of resultant wave (b) frequency at which the amplitude of resultant waves varies (c) Frequency at which beats are produced. Find also the ratio of maximum and minimum intensities of beats.***

Sol. Comparing the given wave equation with

$y = A\sin\omega\left[t - (x/v)\right]\left[as\ k/\omega = 1/v\right]$ we find that here,

A_1=0.3, $\omega_1 = 2\pi n_1 = 596\pi \Rightarrow n_1$=298Hz

and A_2= 0.5, $\omega_2 = 2\pi n_2 = 604\pi \Rightarrow n_2$ = 302 Hz

a) The frequency of the resultant

$$n_{avg} = \frac{n_1 + n_2}{2} = \frac{(298 + 302)}{2} = 300Hz$$

b) The frequency at which amplitude of resultant wave varies:

$$n_A = \frac{n_1 - n_2}{2} = \frac{(298 - 302)}{2} = 2Hz$$

c) The frequency at which beats are produced $n_b = 2n_A = n_1 - n_2 = 4$Hz

d) The ratio of maximum to minimum intensities of beat

$$\frac{I_{max}}{I_{min}} = \frac{(A_1 + A_2)^2}{(A_1 - A_2)^2} = \frac{(0.3 + 0.5)^2}{(0.3 - 0.5)^2} = \frac{64}{4} = 16$$

W.E-37: ***The frequency of a tuning fork 'x' is 5% greater than that of a standard fork of frequency 'K'. The frequency of another fork 'y' is 3% less than that of 'K'. When 'x' and 'y' are vibrated together 4 beats are heard per second. Find the frequencies of x and y.***

Sol. Let the frequency of standard fork be K

$$n_x = K + \frac{5K}{100} = \frac{105K}{100}$$

$$n_y = K - \frac{3K}{100} = \frac{97K}{100}$$

$\Delta n = n_x - n_y \Rightarrow \quad 4 = \frac{105}{100}K - \frac{97}{100}K$

On solving, K = 50 Hz

The frequency of $x = \frac{105}{100} \times 50 = 52.5Hz$

Similarly frequency of $y = \frac{97}{100} \times 50 = 48.5Hz$

W.E-38: ***A string under a tension of 129.6 N produces 10 beats per sec when it is vibrated along with a tuning fork. When the tension in the string is increased to 160 N, it sounds in unison with the same tuning fork. Calculate the fundamental frequency of the tuning fork.***

Sol. Let 'n' be the frequency of fork.

The wire frequency would be $(n \pm 10)$

In case of a wire under tension $n \propto \sqrt{T}$

$\therefore \frac{n-10}{n} = \sqrt{\frac{129.6}{160}} \Rightarrow$ n = 100 Hz

W.E-39 ***Two open organ pipes 80 cm and 81 cm long found to give 26 beats in 10 sec, when each is sounding its fundamental note. Find the velocity of sound in air.***

Sol. Number of beats per second $\quad \Delta n = \frac{v}{2l_1} - \frac{v}{2l_2}$

$= \frac{26}{10} = \frac{v}{160} \sim \frac{v}{162} \Rightarrow 2.6 = \frac{2v}{160 \times 162}$

$\Rightarrow v = \frac{2.6 \times 160 \times 162}{2} = 33696 cms^{-1} \cong 337 ms^{-1}$.

Doppler's effect: Whenever there is a relative motion between a source of sound and the observer (listener), the frequency of sound heard by the observer is different from the actual frequency of sound emitted by the force.

The frequency observed by the observer is called the apparent frequency. It may be less than or greater than the actual frequency emitted by the sound source. The difference depends on the relative motion between the source and observer.

1. **When observer and source are stationary**

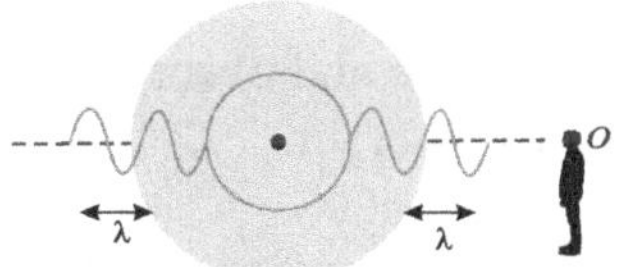

i) Sound waves propagate in the form of spherical wavefronts (shown as circles)
ii) The distance between two successive circles is equal to wavelength λ
iii) Number of waves crossing the observer = Number of waves emitted by the source

iv) Thus apparent frequency $(n') =$ actual frequency (n).

2. **When source is moving but observer is at rest**

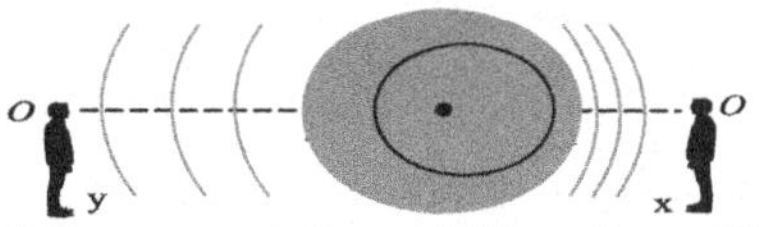

i) S_1, S_2, S_3 are the positions of the source at three different positions.
ii) Waves are represented by non-concentric circles, they appear compressed in the forward direction and spread out in backward direction.
iii) For observer (X)

Apparent wavelength $(\lambda') <$ Actual wavelength (λ)

$\Rightarrow$ Apparent frequency $(n') >$ Actual frequency (n)

For observer (Y) : $\lambda' > \lambda \Rightarrow n' < n$

3. **When source is stationary but observer is moving**

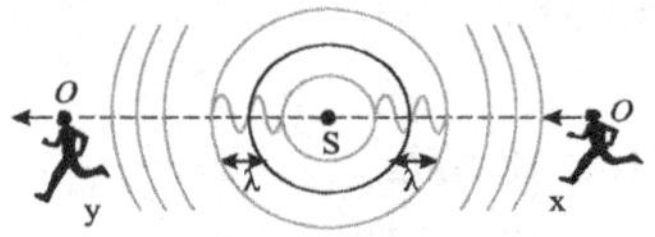

i) Waves are again represented by concentric circles.
ii) No change in wavelength received by either observer X or Y.
iii) Observer X (moving towards) receives wave fronts at shorter interval thus $n' > n$.
iv) Observer Y receives wavelengths at longer interval thus $n' < n$

4. **General expression for apparent frequency:** If v, v_o, v_s are the velocities of sound, observer, source respectively and velocity of medium is v_m then apparent frequency observed by observer when wind blows in the direction of v (from the source to observer)is given by

$n' = \left[\dfrac{(v+v_m) \pm v_o}{(v+v_m) \pm v_S}\right] n$ and in opposite direction of v

(from observer to source) $n' = \left[\dfrac{(v-v_m) \pm v_o}{(v-v_m) \pm v_S}\right] n$

If medium is stationary i.e., $v_m = 0$ then $n' = \left(\dfrac{v \pm v_0}{v \pm v_S}\right) n$

➢ **Sign convention for different situation**
i) The direction of v is always taken from source to observer.
ii) If the velocities v_o, v_s in the direction of v then positive +ve is taken.
iii) If the velocities v_o, v_s in the opposite direction of v then positive -ve is taken.

Note:- i) Doppler effect in sound is asymmetric. ii) Doppler effect in light is symmetric.
iii) Doppler's effect in vector form is written as

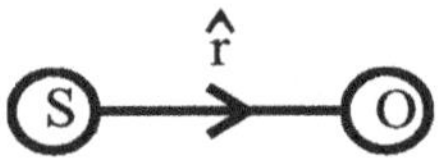

$n^1 = \left(\dfrac{v - \overline{v_0}.\hat{r}}{v - \overline{v_s}.\hat{r}}\right) n$

$\hat{r} =$ unit vector along the line joining source and observer $\overline{V} =$ Velocity of sound in the medium. Its direction is always taken from source to observer.

➢ **limitations of Doppler effect:**
i) Doppler effect is not observed if

a) $v_0 = v_s = 0$ (both are in rest)

b) $v_0 = v_s = 0$ and medium is alone in motion direction.

c) v_s is perpendicular to the line of sight

ii) Doppler effect is applicable only when,

$v_0 << v$ and $v_s << v$. (v is velocity of sound)

5. **Common Cases in Doppler's Effect**

➢ **Source is moving but observer at rest.**

1. Source is moving towards the observer

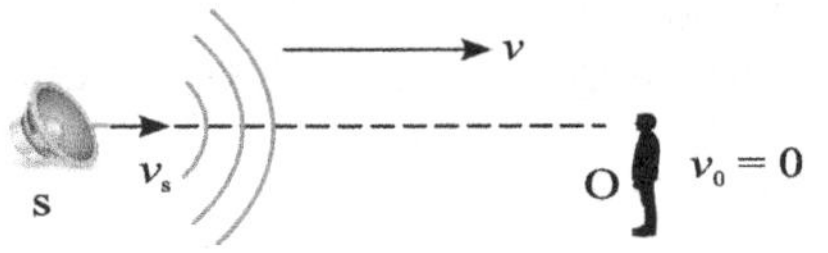

Apparent frequency $n' = n\left(\dfrac{v}{v - v_S}\right)$

Apparent wavelength $\lambda' = \lambda\left(\dfrac{v - v_S}{v}\right)$

2. Source is moving away from the observer.

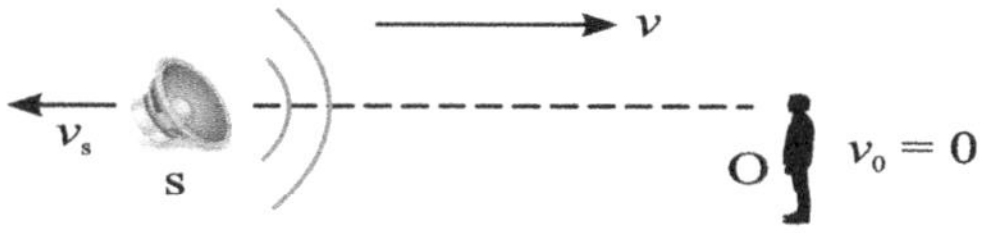

Apparent frequency $n' = n\left(\dfrac{v}{v + v_S}\right)$ Apparent wavelength $\lambda' = \lambda\left(\dfrac{v + v_S}{v}\right)$

➢ **Source is at rest but observer is moving.**

1. Observer is moving towards the source.

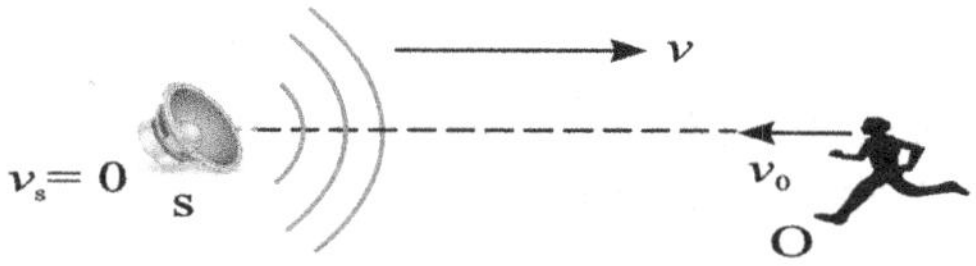

Apparent frequency $n' = n\left[\dfrac{v + v_0}{v}\right]$

Apparent wavelength $\lambda' = \dfrac{(v + v_0)}{n'} = \dfrac{(v + v_0)}{n\dfrac{(v + v_0)}{v}} = \dfrac{v}{n} = \lambda$

2. Observer is moving away from the source

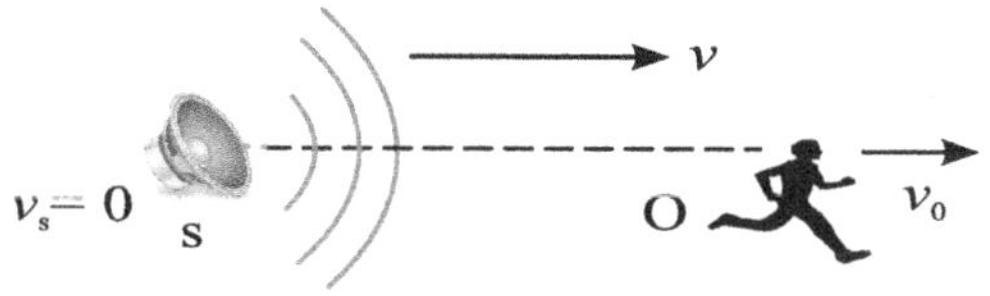

Apparent frequency $n' = n\left[\dfrac{v - v_O}{v}\right]$

Apparent wavelength $\lambda' = \lambda$

4. When source and observer both are moving

1. When both are moving towards each other

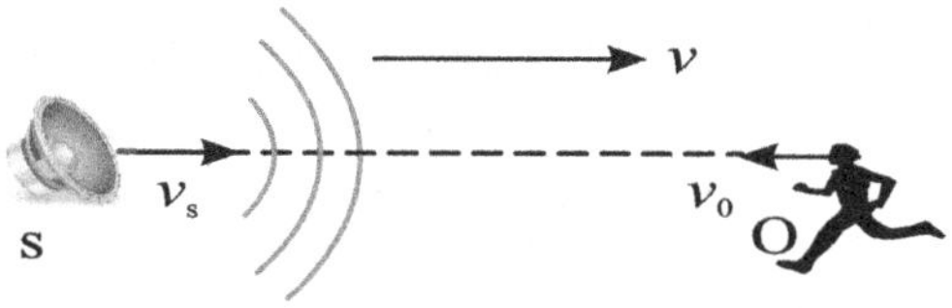

i) Apparent frequency $n' = n\left[\dfrac{v + v_S}{v - v_S}\right]$ ii) Apparent wavelength $\lambda' = \lambda\left(\dfrac{v - v_S}{v}\right)$

iii) Velocity of wave with respect to observer $= (v + v_0)$

2. When both are moving away from each other.

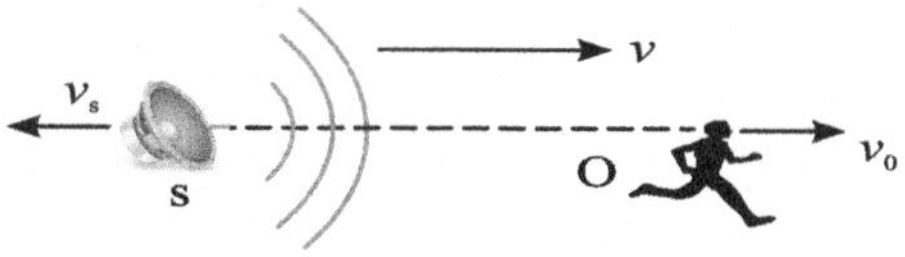

i) Apparent frequency $n' = n\left[\frac{v - v_O}{v + v_S}\right](n' < n)$ ii) Apparent wavelength $\lambda' = \lambda\left(\frac{v + v_S}{v}\right)(\lambda' > \lambda)$

iii) Velocity of waves with respect to observer $= (v - v_O)$

3. **When source is moving behind observer**

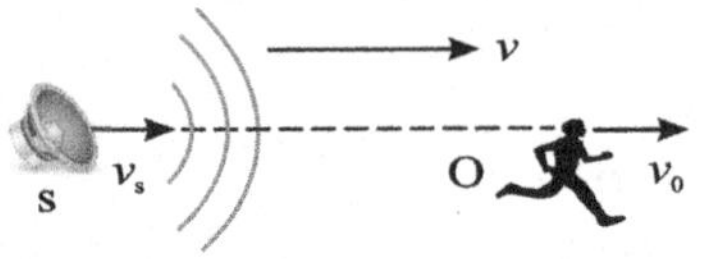

i) Apparent frequency $n' = n\left(\frac{v - v_O}{v - v_S}\right)$

a) If $v_O < v_S$, then $n' > n$ b) If $v_O > v_S$, then $n' < n$ c) If $v_O = v_S$ then $n' = n$

ii) Apparent wavelength $\lambda' = \lambda\left(\frac{v - v_S}{v}\right)$

iii) Velocity of waves with respect to observer $= (v - v_O)$

4. **When observer is moving behind the source**

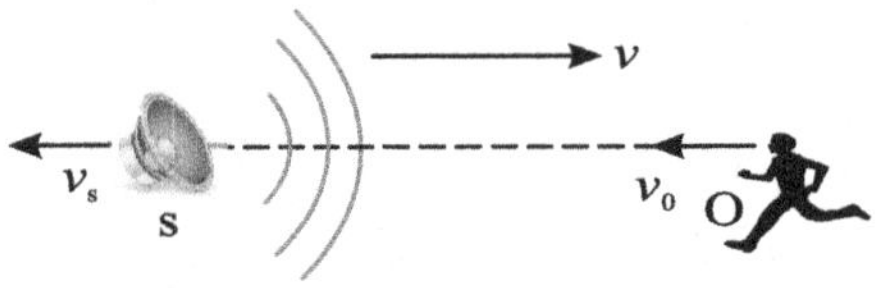

i) Apparent frequency $n' = n\left(\frac{v + v_0}{v + v_s}\right)$

a) If $v_0 > v_s$, then $n' > n$

b) If $v_0 < v_s$, then $n' < n$ c) If $v_0 = v_s$, then $n' = n$

ii) Apparent wavelength $\lambda' = \lambda\left(\frac{v + v_S}{v}\right)$

iii) The velocity of waves with respect to observer $= (v - v_O)$

Crossing

1. **Moving sound source crosses a stationary observer**

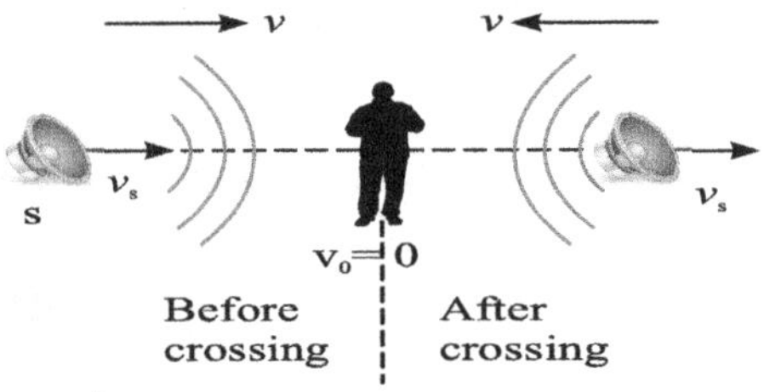

Apparent frequency before crossing

$$n'_{Before} = n\left[\frac{v}{v - v_S}\right]$$

Apparent frequency $n'_{After} = n\left[\frac{v}{v + v_S}\right]$

Ratio of two frequencies $\frac{n'_{Before}}{n'_{After}} = \left[\frac{v + v_S}{v - v_S}\right] > 1$

Change in apparent frequency $n'_{Before} - n'_{After} = n\left(\frac{v}{v-v_s} - \frac{v}{v+v_s}\right) = nv\left(\frac{2v_s}{v^2 - {v_s}^2}\right)$

If $v_s << v$ then $n'_{Before} - n'_{After} = \frac{2nv_S}{v}$

2. Moving observer crosses a stationary source

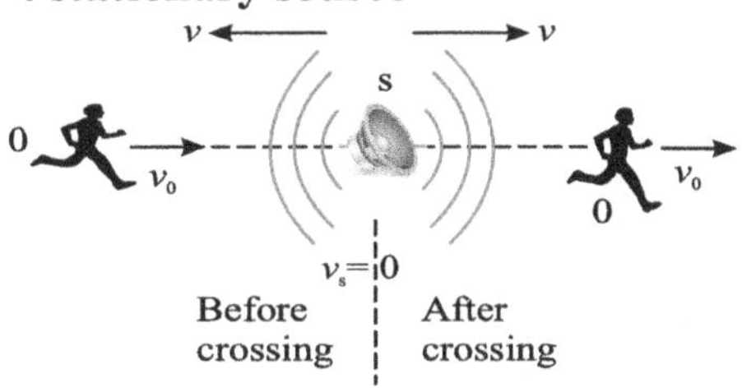

Apparent frequency before crossing $n'_{Before} = n\left(\frac{v+v_O}{v}\right)$

Apparent frequency after crossing $n'_{After} = n\left[\frac{v-v_O}{v}\right]$

Ratio of two frequencies $\frac{n'_{Before}}{n'_{After}} = \frac{v+v_O}{v-v_O}$

Change in apparent frequency $n'_{Before} - n'_{After} = \frac{2nv_O}{v}$

Some Typical Cases of Doppler' Effect

1. Moving car towards wall: When a car is moving towards a stationary wall as shown in figure. If the car sounds a horn, wave travels toward the wall and is reflected from the wall. When the reflected wave is heard by the driver, it appears to be of relatively high pitch, if we wish to measure the frequency of reflected sound.

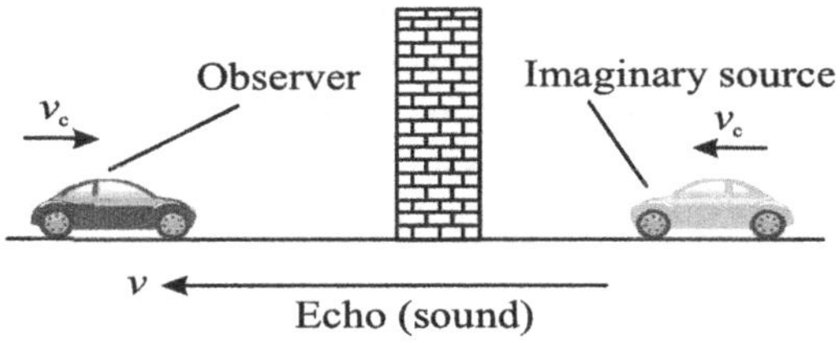

Here we assume that the sound which is reflected by the stationary wall is coming from the image of car which is at the back of it and coming towards it with velocity v_C. Now the frequency of sound heard by car driver be given as

$$n'_{direct} = n;\; n'_{reflected} = n\left(\frac{v+v_c}{v-v_c}\right)$$

No. of beats

$$\Delta n' = n'_{reflected} - n'_{direct} = \frac{2v_c n}{v-v_c}$$

Case (i): If the observer is at rest in between source and wall as shown

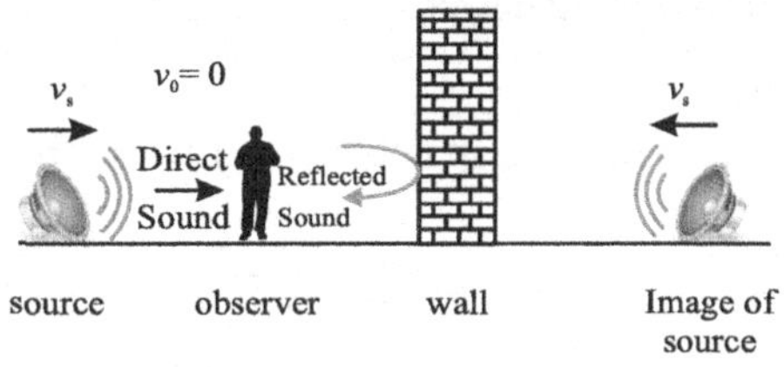

$$n^1_{\text{direct}} = \left(\frac{v}{v - v_S}\right)n \; ; \quad n^1_{\text{reflected}} = \left(\frac{v}{v - v_S}\right)n$$

No.of beats $\Delta n' = n'_{reflected} - n'_{direct} = 0$

Case (ii): If the source is in between observer and wall

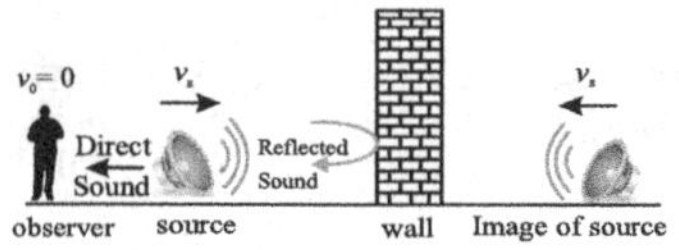

$$n^1_{direct} = \left(\frac{v}{v + v_S}\right)n \; ; \; n^1_{reflected} = \left(\frac{v}{v - v_S}\right)n$$

No. of beats $\Delta n' = n'_{reflected} - n'_{direct} = \left(\frac{v}{v - v_S}\right)n - \left(\frac{v}{v + v_S}\right)n = n\left(\frac{2v_s}{v^2 - v_s^{\,2}}\right)$

If $v_s << v$ then $\Delta n' = \dfrac{2nv_S}{v}$

Note: This method of images for solving problems of Doppler effect is very convenient but is used only for velocities of source and observer which are very small compared to the speed of sound and it should not be used frequently when the reflector of sound is moving.

2. **Moving target:** Let a sound source S and observer O are at rest (stationary). The frequency of sound emitted by the source is n and velocity of waves is v.

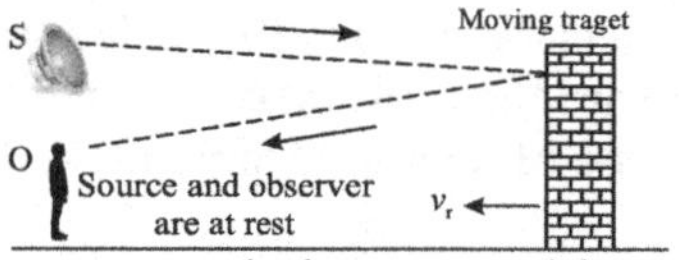

A target is moving towards the source and observer, with a velocity v_T. Our aim is to find out the frequency observed by the observer, for the waves reaching it after reflection from the moving target. The formula is derived by applying Doppler equations twice, first with the target as observer and then with the target as source.

The frequency n' of the waves reaching surface of the moving target (treating it as observer) will be

$$n' = \left(\frac{v + v_T}{v}\right)n$$

Now these waves are reflected by the moving target (which now acts as a source). Therefore the apparent frequency, for the real observer O will be $n'' = \dfrac{v}{v - v_T}n' \Rightarrow n'' = \dfrac{v + v_T}{v - v_T}n$

i) If the target is moving away from the observer, then $n' = \dfrac{v - v_T}{v + v_T}n$

ii) If target velocity is much less than the speed of sound, ($v_T << v$), then $n' = \left(1 + \dfrac{2v_T}{v}\right)n$, for approaching target and $n' = \left(1 - \dfrac{2v_T}{v}\right)n$ for receding target

3. **Transverse Doppler's effect**

i) If a source is moving in a direction making an angle θ w.r.t. the observer.

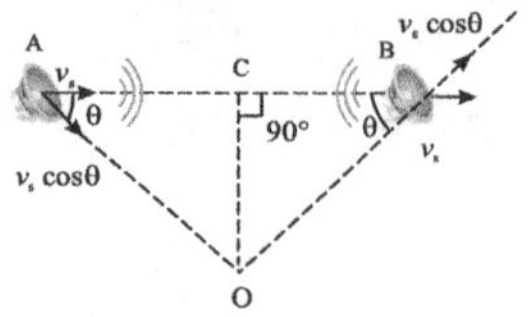

The apparent frequency herd by observer O at rest

At point A: $n' = \dfrac{nv}{v - v_S \cos\theta}$

As source moves along AB, value of θ increases, $\cos\theta$ decreases, n' goes on decreasing.

At point C:

$\theta = 90°$, $\cos\theta = \cos 90° = 0, n' = n$

At point B: The apparent frequency of sound becomes $n'' = \dfrac{nv}{v + v_s \cos\theta}$

ii) When two cars are moving on perpendicular roads: When car-1 sounds a horn of frequency n, the apparent frequency of sound heard by car-2 can be given as $n' = n\left[\dfrac{v + v_2\cos\theta_2}{v - v_1\cos\theta_1}\right]$

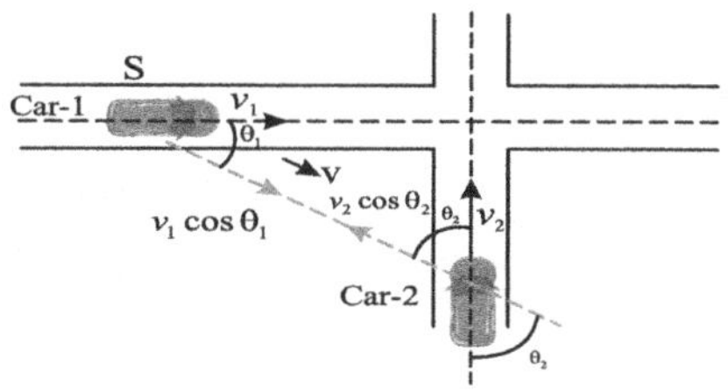

4. **Rotating source/observer:** Suppose that a source of sound/observer is rotating in a circle of radius r with angular velocity ω (Linear velocity $v_S = r\omega$)

i) When source / observer at rest at centre of circle and observer / source is rotating in a circle then the line of sight is perpendicular to the direction of motion of observer / source and hence no doppler effect.

$\therefore n^1 = n$

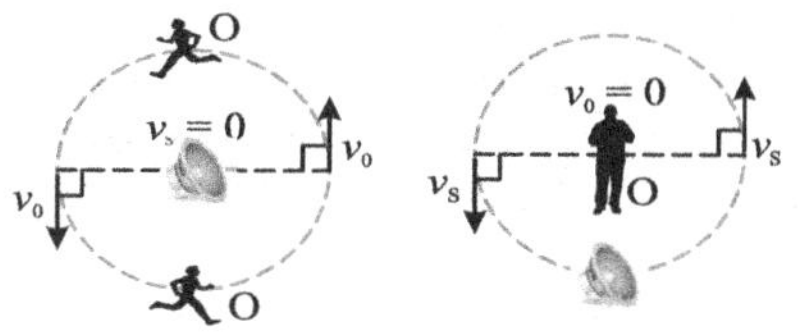

ii) **When source is rotating**

a) Towards the observer heard frequency will be maximum i.e., $n_{max} = \dfrac{nv}{v - v_S}$

b) Away from the observer heard frequency will be minimum and $n_{min} = \dfrac{nv}{v + v_S}$

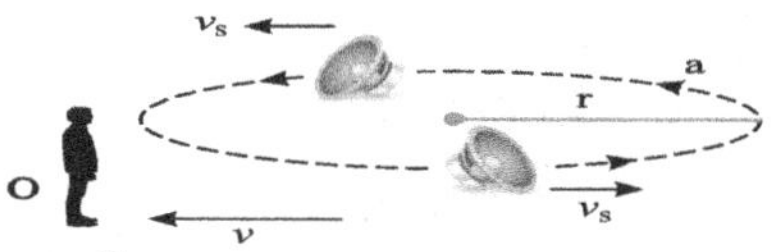

c) Ratio of maximum and minimum frequency

$$\frac{n_{max}}{n_{min}} = \frac{v + v_S}{v - v_S}$$

iii) **When observer is rotating**

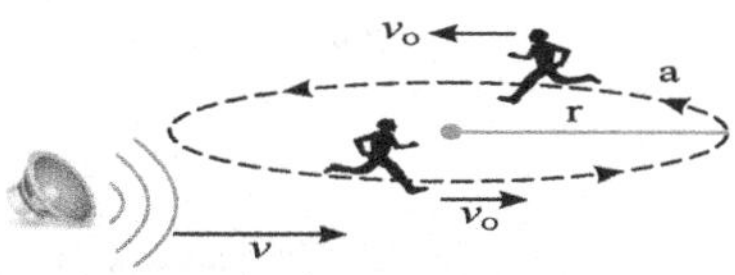

a) Towards the source heard frequency will be maximum

i.e., $n_{max} = n\left(\frac{v+v_0}{v}\right)$

b) Away from the source heard frequency will be minimum and $n_{min} = n\left(\frac{v-v_0}{v}\right)$

c) Ratio of maximum and minimum frequency $\frac{n_{max}}{n_{min}} = \frac{v+v_0}{v-v_0}$

5. **Doppler shift in RADAR:** A microwave beam is directed towards the aeroplane and is received back after reflection from it. If 'v' is the speed of the plane and 'n' is the actual frequency of the microwave beam then the frequency of the microwave beam then the frequency received by moving plane $n^1 = \left(\frac{c+v}{c}\right)n$

Now the plane act as a moving source, the frequency of the wave from it is $n^{11} = \left(\frac{c+v}{c-v}\right)n$

(c is velocity of microwave)

Change in frequency $\Delta n \approx \frac{2nv}{c}$

By measuring Δn, the speed 'v' can be obtained.

6. **Uses of Doppler effect:**
It is used in
a) SONAR
b) RADAR (Radio detection and ranging used to determine speed of objects in space)
c) To determine speeds of automobiles by traffic police. The technique is applied in the airports to guide the air crafts.
d) To determine speed of rotation of sun.
e) In Astrophysics, it is applied in the study of the saturn's rings and in the study of binary satrs. Here the doppler's shift in the frequency of light from the atronomical objects is measured.
f) Accurate navigation and accurate target bombing techniques.
g) Tracking earth's satellite.
h) In medicine, it is applied to study the velocity of blood flow in different parts of the body and the moment of the fetus in the woomb using ultra sound. The conditions of heart beat can be inferred by "echocardiogram" generated from this technique.

W.E-40 ***When a train is approaching the observer, the frequency of the whistle is 100 cps while when it has passed the observer, it is 50 cps. Calculate the frequency when the observer moves with the train.***

Sol. In case of approaching of source, $100 = \frac{nv}{v - v_S}$

while in case of recession of source, $50 = \frac{nv}{v + v_S}$

Which on simplification gives $n = \frac{200}{3} = 66.67\,Hz$

W.E-41: ***A car approaching a crossing at a speed of 20 m/s sounds a horn of frequency 500Hz when at 80m from the crossing. Speed of sound in air is 330 m/s. What frequency is heard by an observer 60 m from the crossing on the straight road which crosses car road at right angles ?***

Sol. The situation is as shown in figure

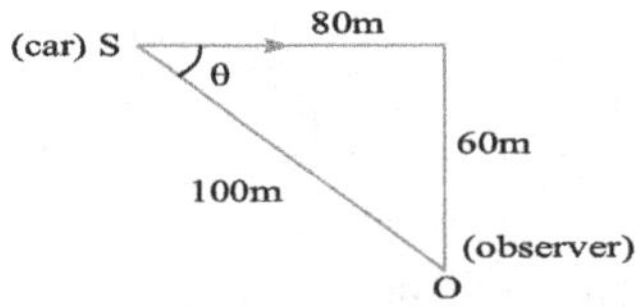

$\cos\theta = \frac{80}{100} = \frac{4}{5}$ ∴ Apparent frequency is

$$n_{app} = \frac{v}{v - v_S\cos\theta}n = \left(\frac{330}{330 - 20\times\frac{4}{5}}\right)(500) = 525.5 \text{ Hz}$$

W.E-42: ***A whistle of frequency 540 Hz rotates in a circle of radius 2 m at a linear speed of 30m/s. What is the lowest and highest frequency heard by an observer a long distance away at rest with respect to the centre of circle. Take speed of sound in air as 330 m/s. Can the apparent frequency be ever equal to actual ?***

Sol. Apparent frequency will be minimum when the source is at N and moving away from the observer.

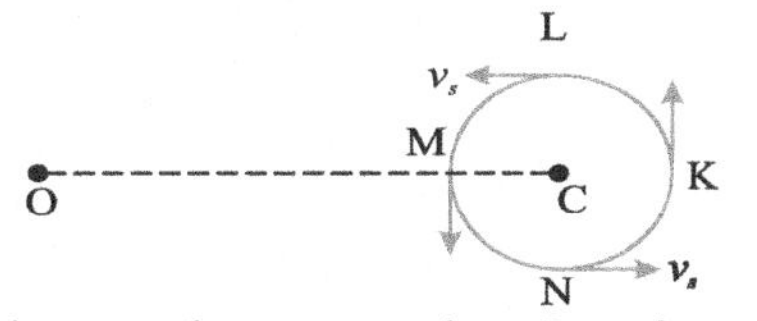

Apparent frequency will be maximum when source is at L and approaching the observer.

$$n'_{max} = \frac{v}{v - v_S}n = \left(\frac{330}{330 - 30}\right)(540) = 594Hz$$

Further when source is at M and K, angle between velocity of source and line joining source and observer is 90° (i.e., line of sight is perpendicular to v_s) or $v_S\cos\theta = v_S\cos 90° = 0$.

So, there will be no Doppler effect.

W.E-43: ***A source of sound is moving along a circular orbit of radius 3 m with an angular velocity of 10 rad/s. A sound detector located far away from the source is executing linear simple harmonic motion along the line BD with amplitude BC = CD = 6 m. The frequency of oscillation of the detector is $(5/\pi)$ rev /sec. The source is at the point A when the detector is at the point B. If the source emits a continuous sound wave of frequency 340 Hz, find the maximum and the minimum frequencies recorded by the detector [velocity of sound = 330 m/s].***

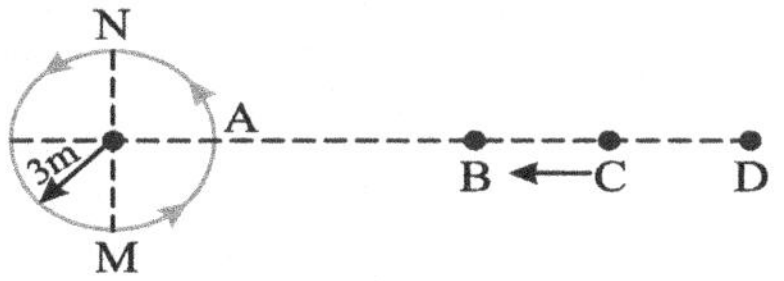

Sol. Time period of circular motion $T = (2\pi/\omega) = (2\pi/10)$ is same as that of SHM i.e., T=(1/n) $= (\pi/5)$, so both will complete one periodic motion in same time.

Further more source is moving on a circle, its speed $v_s = r\omega = 3 \times 10 = 30$m/s and as detector is executing SHM, $v_D = \omega\sqrt{A^2 - y^2} = 10\sqrt{6^2 - 0^2} = 60m/s$ i.e., detector is at C.

So n' will be maximum when both move towards each other. $n'_{max} = n\left(\frac{v + v_D}{v - v_S}\right)$ with v_D = max i.e.,

the source is at M and detector at C and moving towards B, so $n'_{max} = 340\left[\frac{330 + 60}{330 - 30}\right] = 442Hz$

Similarly n' will be minimum when both are moving away from each other i.e.,

$n'_{min} = n\left(\frac{v - v_D}{v + v_S}\right)$ with v_D = max i.e., the source is at N and detector at C but moving towards D, so

$$n_{min} = 340\left[\frac{330 - 60}{330 + 30}\right] = 255Hz$$

Chapter - 9

WAVE MOTION
(Part - B)

Echo (additional)

- The sound reflected by an obstacle which is heard by a listener is called an echo.
- Persistence of hearing is the minimum interval of time between two sound notes to distinguish them. Persistence of hearing is 0.1s
- A person is at a distance 'd' from a reflected surface (a wall, mountain etc). The person sounds a horn and hears its echo at the end of a time 't'. If V is the velocity of sound in air then.

$$d = \frac{Vt}{2}$$

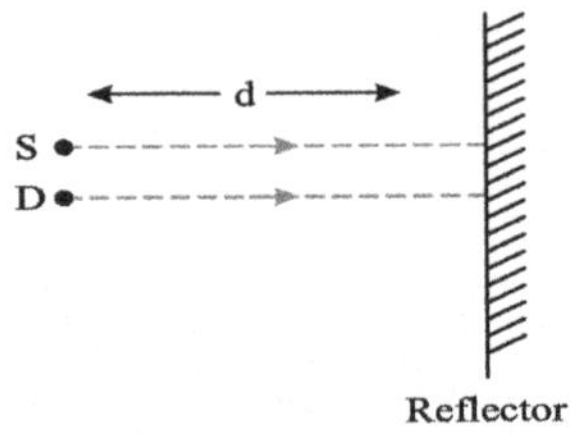

To hear a clear echo, the minimum distance of the obstacle,

$$d_{min} = \frac{V \times 0.1}{2} = \frac{V}{20}$$

If $V = 330\ ms^{-1}$ then $d_{min} = 16.5m$

If $V = 340\ ms^{-1}$ then $d_{min} = 17\ m$

CONCEPTUAL QUESTIONS

CHARACTERISTICS OF PROGRESSIVE WAVE

1. **When a wave is travelling in a medium, in that process, the following is/are transporting from one particle to other**

   1) energy 2) momentum 3) both 1 & 2 4) length

2. **A plane progressive wave cannot be represented by**

   1) $y = a \sin(\omega t \pm kx)$ 2) $y = a\sin 2\pi\left(\frac{t}{T} \mp \frac{x}{\lambda}\right)$

   3) $y = a \sin \frac{2\pi}{\lambda}(Vt \mp x)$ 4) $y = A\log x + B\log x$

3. **The speed of wave of time period T and propagation constant K is**

   1) $\frac{2\pi}{TK}$ 2) $\frac{TK}{2\pi}$ 3) $\frac{1}{TK}$ 4) $\frac{T}{K}$

4. **The phase change between incident and reflected sound wave from a fixed wall is**

   1) 0 2) π 3) 3π 4) 2π

5. **The phase change between incident and reflected sound wave from a free end is**

   1) 0 2) π 3) 3π 4) $\pi/2$

6. **Figure shows the shape of a string, the pairs of points which are in opposite phase is**

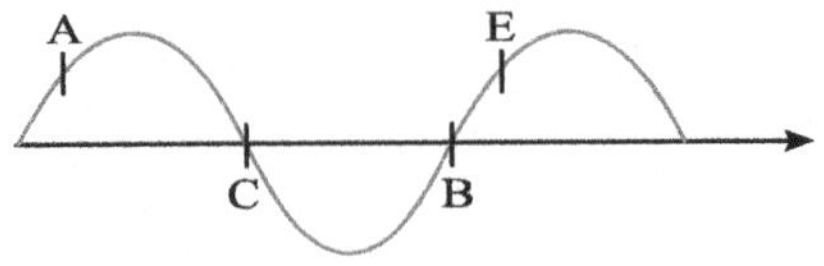

   1) A and B 2) B and C 3) C and E 4) A and E

7. **During propagation of longitudinal plane wave in a medium the two particles separated by a distance equivalent to one wavelength at an instant will be/have**

1) in phase, same displacement
2) in phase, different displacement
3) different phase, same displacement
4) different phase, different displacement

8. **The equation of a progressive wave is Y=a sin (ω t-kx), then the velocity of the wave is**

1) kω 2) k/ω 3) ω/k 4) aω

9. **When a progressive wave is propagating in a medium, at a given instant, two particles which are separated by three wave lengths will have.....**

1) Different displacement in same direction
2) Different displacement in opposite direction
3) Same displacement in opposite direction
4) Same displacement in same direction

10. **The speed of sound in a medium does not change with the change of**

1) frequency 2) wave length 3) pressure 4) density

11. **Phase difference between a particle at a compre-ssion and a particle at the next rarefaction is**

1) Zero 2) $\pi/2$ 3) π 4) $\pi/4$

12. **One similarity between sound and light waves is that.**

1) both can propagate in vacuum
2) both have same speed
3) both can show polarization
4) both can show interference

13. **When a body is undergoing undamped vibration, the physical quantity that remains constant is**

1) amplitude 2) velocity 3) acceleration 4) phase

SPEED OF A TRAVELLING WAVE

14. **At any instant a wave travelling along the string is shown in figure. Here, if point A is moving upward, the true statement is**

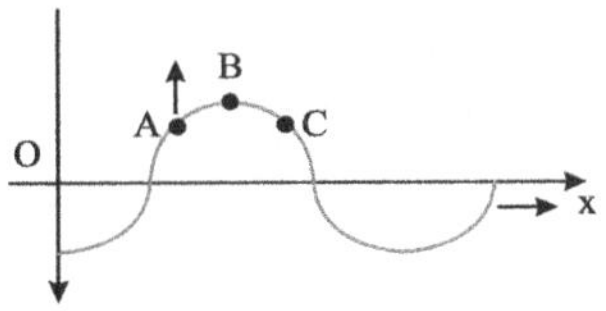

1) the wave is travelling to the right
2) the displacement amplitude of wave is equal to displacement of B at this instant
3) at this instant, C also directed upward
4) 1 and 3

15. **Transverse waves are produced in a long string by attaching its free end to a vibrating tuning fork. Figure shows the shape of a part of the string. The points in phase are**

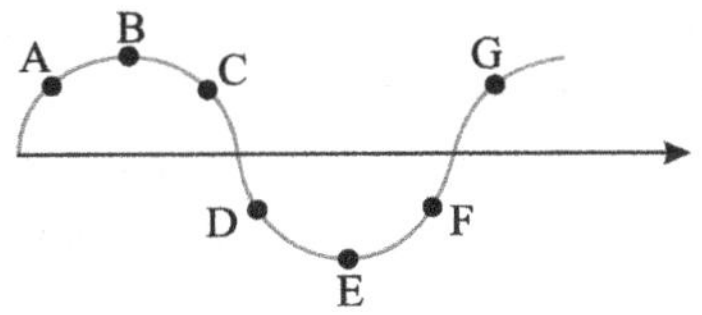

1) A and D 2) B and E 3) C and F 4) A and G

16. **A pulse in a rope approaches a solid wall and it gets reflected from it**

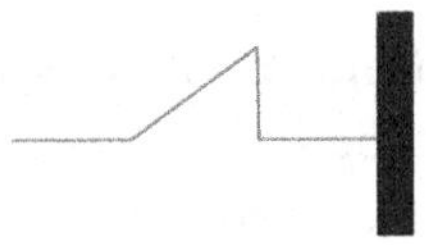

The wave pulse after reflection is best represented by _____

17. **A transverse wave is travelling along a string from left to right. The figure below represents the shape of the string at a given instant. At this instant the points have an upward velocity are (here X-wave displacement, Y-particle displacement)**

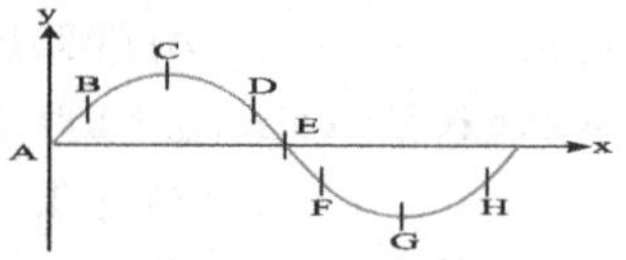

1) D,E,F 2) A,B,H 3) B,D,F 4) A,E,H

18. **A metal string is fixed between rigid supports. It is initially at negligible tension. It's Young's modulus is 'Y' density is ρ and coefficient of linear expansion is α . It is now cooled through a temperature 't', transverse waves will move along it with a speed.**

1) $\sqrt{\frac{Y\alpha t}{\rho}}$ 2) $Y\sqrt{\frac{\alpha t}{\rho}}$ 3) $\alpha\sqrt{\frac{Yt}{\rho}}$ 4) $t\sqrt{\frac{\rho}{Y\alpha}}$

PRINCIPLE OF SUPER POSITION, INTERFERENCE AND STATIONARY WAVES ON STRETCHED STRINGS.

19. **The interference phenomenon can take place**

1) in transverse wave 2) in longitudinal wave
3) in electromagnetic waves 4) in all waves

20. **For superposition of two waves, the following is correct**

1) they must have the same frequency and wavelength
2) they must have equal frequencies but may have unequal wavelengths
3) they must have the same wave-length, but may have different frequencies
4) they may have different wavelength and different frequencies

21. **At a certain instant a stationary transverse wave is found to have maximum kinetic energy. The appearance of the string at that instant is a**

1) sinusoidal shape with amplitude A/3 2) sinusoidal shape with amplitude A/2
3) sinusoidal shape with amplitude A 4) straight line

22. **When stationary waves are set up, pick out the correct statement from the following**

1) all the particles in the medium are in the same phase of vibration at all times and distances
2) the particles with an interval between two consecutive nodes are in phase, but the particles in two such consecutive intervals, are of opposite phase
3) the phase lag along the path of the wave increa-ses as the distance from the source increases
4) only antinodes are in same phase

23. **In a stationary wave along a string the strain is**

1) zero at the antinodes 2) maximum at the antinodes
3) zero at the nodes 4) maximum at the nodes

24. **In a stationary wave**

1) phase is same at all points in a loop 2) amplitude is same at all points
3) energy is constant at all points 4) temperature is same at all points

25. **A wave is represented by an equation; Y = A cos kx sin ωt , then**

1) it is a progressive wave with amplitude A
2) it is a progressive wave with amplitude A cos kx
3) it is a stationary wave with amplitude A 4) it is a stationary wave with amplitude A cos kx

26. **In a stationary wave**

1) pressure change is maximum at nodes 2) pressure change is maximum at antinodes
3) pressure change is minimum at nodes 4) amplitude is zero at all points

Wave Motion

27. A wire in sonometer experiment is vibrating in the third overtone. There are

1) two nodes, two antinodes
2) three nodes, three antinodes
3) four nodes, three antinodes
4) five nodes, four antinodes.

28. λ is maximum wavelength of a transverse wave that travels along a stretched wire whose two ends are fixed. The length of that wire is

1) 2λ
2) λ
3) $\lambda/2$
4) $3\lambda/2$

29. In the sonometer experment, a wire of density 'ρ' and radius 'a' is held between two bridges at a distance 'L' apart. Tension in the wire is 'T' the fundamental frequency of the wire will be

1) $\frac{1}{2L}\sqrt{\frac{\pi a^2}{T\rho}}$
2) $\frac{1}{2L}\sqrt{\frac{T\rho}{\pi a^2}}$
3) $\frac{1}{2L}\sqrt{\frac{T}{\pi a^2}}$
4) $\frac{1}{2L}\sqrt{\frac{T}{\pi a^2 \rho}}$

30. For a stretched string of given length, the tension 'T' is plotted on the X-axis and the frequency 'f' on the Y-axis. The graph is

1) rectangular hyperbola
2) straight line through the origin
3) parabola
4) straight line not through the origin

31. The equation of a stationary wave in a medium is given as $y = \sin \omega t \cos kx$. The length of a loop in fundamental mode is

1) $\frac{\pi}{2K}$
2) $\frac{\pi}{K}$
3) $\frac{2\pi}{K}$
4) $\frac{K}{\pi}$

32. A stretched string of length l, fixed at both ends, can sustain stationary waves of wavelength λ, correctly given by(P is number of loops)

1) $\lambda = \frac{l^2}{2p}$
2) $\lambda = \frac{p^2}{2l}$
3) $\lambda = 2lp$
4) $\lambda = \frac{2l}{p}$

33. A knife-edge divides a sonometer wire into two parts. The fundamental frequencies of the two parts are n_1 and n_2. The fundamental frequency of the sonometer wire when the knife-edge is removed will be

1) $n_1 + n_2$
2) $\frac{1}{2}(n_1 + n_2)$
3) $\sqrt{n_1 n_2}$
4) $\frac{n_1 n_2}{n_1 + n_2}$

SOUND AND VELOCITY OF SOUND

34. Pitch of sound primarily depend upon

1) intensity
2) frequency
3) quality
4) overtone

35. Quality of sound primarily depends upon

1) intensity
2) frequency
3) shape of the source
4) wave form

36. It is possible to recognize a person by hearing his voice even if he is hidden behind a solid wall. This is due to the fact that his voice

1) Has a definite pitch
2) Has a definite quality
3) Has a definite loudness
4) Can penetrate the wall

37. The relation between the intensity (I) of a wave and on the distance (r) from a line source is

1) $I \propto r^{-1}$
2) $I \propto r^{-2}$
3) $I \propto r$
4) $I \propto r^{-\frac{1}{2}}$

38. Loudness of sound primarily depend upon

1) intensity
2) frequency
3) shape of the source
4) overtone

39. According to Laplace correction, the propagation of sound in gas takes place under

1) isothermal condition
2) isobaric condition
3) isochoric condition
4) adiabatic condition

40. The velocity of sound is not affected by change in

1) temperature
2) medium
3) pressure
4) wavelength

41. **Velocity of sound in air at the given temperature**
1) decreases with increase in pressure
2) may increase on decrease with pressure
3) is independent of the variation in pressure
4) varies directly as the square root of pressure

42. **If the temperature of atmosphere is increased the following character of sound waves is effected**
1) amplitude 2) frequency 3) velocity 4) wavelength

43. **Velocity of sound is**
1) directly proportional to absolute temperature
2) directly proportional to square root of absolute temperature
3) inversely proportional to absolute temperature
4) inversely proportional to $\sqrt{T}$

ORGAN PIPES

44. **A closed pipe has certain frequency. Now its length is halved. Considering the end correction, its frequency will now become**
1) double 2) more than double 3) less than double 4) four times

45. **The fundamental frequency of a closed organ pipe is 'n'. Its length is doubled and radius is halved. Its frequency will become nearly.**
1) n/2 2) n/3 3) n 4) 2n

46. **When an air column of length 'l' open at both ends produces its fundamental note, the wavelength is given by**
1) $\lambda = l$ 2) $\lambda = 2l$ 3) $\lambda = 4l$ 4) $\lambda = 3l$

47. **In the case of closed end organ pipe**
1) the maximum possible wavelength is same as that of open end organ piple
2) the maximum possible wavelength is less than that of open end organ pipe
3) the maximum possible wavelength may be less than that of open end organ pipe
4) the maximum possible wavelength is greater than that of open end organ pipe

48. **In the case of standing waves in organ pipe, the value of $\frac{\partial y}{\partial x}$ at the open end is**
1) >0 2) <0 3) =0 4) =10

49. **The harmonics formed in air column in an organ pipe closed at one end are**
1) only odd
2) only even
3) both odd and even
4) neither odd nor even

50. **A tube with both ends closed has same set of natural frequency as**
1) one end closed organ pipe
2) both end open organ pipe
3) vibratory string fixed at both ends
4) vibratory string fixed at one end

51. **The frequency of the sound emitted by an organ pipe will increase if the air in it is replaced by**
a) hot air b) moist air c) hydrogen
1) a is true 2) a,b are true 3) b,c are true 4) a,b,c are true

52. **An empty vessel is filled partially with water natural frequency**
1) increase 2) decrease 3) remains unchanged 4) insufficient data

53. **End correction in a closed organ pipe of diameter 'd' is**
1) 0.6d 2) 1.2d 3) 0.3d 4) 2.4d

54. **If oil of density higher than that of water is used in place of water in a resonance tube, its frequency will**
1) increase 2) decrease 3) remain the same
4) depend upon the density of the material of tube

55. **If λ_1, λ_2 and λ_3 are wave lengths of the waves giving resonance with fundamental ,first and second over tones of a closed organ pipe .the ratio of wave lengths $\lambda_1 : \lambda_2 : \lambda_3$ is**

1) 1:2:3 2) $1:\frac{1}{3}:\frac{1}{5}$ 3) 1:3:5 4) 5:3:1

56. **In closed pipes, the positions of antinodes are obtained at ____**

1) $\frac{\lambda}{4}, \frac{3\lambda}{4}, \frac{5\lambda}{4}$ 2) $0, \frac{\lambda}{2}, \lambda$ 3) $\lambda, 2\lambda, 3\lambda$ 4) $2\lambda, 4\lambda, 6\lambda$

57. **An open pipe of length 'l' vibrates in fundamental mode. The pressure variation is maximum at**

1) 1/4 from ends 2) the middle of pipe
3) the ends of pipe 4) at 1/8 from ends of pipe

BEATS

58. **Beat phenomenon is physically meaningful only, if**

1) $|\omega_1 - \omega_2| >> |\omega_1 + \omega_2|$ 2) $|\omega_1 - \omega_2| << |\omega_1 + \omega_2|$ 3) $\frac{\omega_1}{\omega_2} < 17$ 4) $|\omega_1 + \omega_2| >> \frac{\omega_1}{\omega_2}$

59. **Beats are produced by the superposition of two waves of nearly equal frequencies. Of the following, the correct statement is**

1) all particles of the medium vibrate simple harmonically with frequency equal to the difference between the frequencies of component waves
2) the frequency of beats changes with the location of the observer
3) the frequency of beats changes with time
4) amplitude of vibration of particle at any point changes simple harmonically with frequency equal to one half of the difference between the component waves

60. **When beats are formed by two waves of frequencies n_1 and n_2, the amplitude varies with frequency equal to**

1) n_1-n_2 2) $2(n_1$-$n_2)$ 3) $(n_1$-$n_2)/2$ 4) $(n_1+n_2)/2$

61. **Two wires are producing fundamental notes of same frequency. The change in which of the following factors of one wire does not produce beats between them**

1) stretching force 2) diameter of the wire
3) material of the wire 4) amplitude of the vibrations

62. **Beats are the result of**

1) diffraction 2) destructive interference
3) constructive and destructive interference
4) superposition of two waves of nearly equal frequencies

63. **To hear beats, it is essential that the two sound waves in air should**

1) be travelling in opposite directions 2) be travelling in the same direction
3) have slightly different amplitude 4) have slightly different wavelengths

64. **When the two tuning forks of nearly same frequency are vibrated to produce beats, then the velocity of propagation of beats will be**

1) less than that of sound 2) depend upon the relative frequency
3) more than that of sound 4) equal to that of sound

65. **A certain number of beats are heard when two tuning forks of natural frequencies n_1 and n_2 are sounded together. The number of beats when one of the forks is loaded**

1) increases 2) decreases 3) remains same 4) may increase or decrease

66. **The frequency of sound reaching a stationary listener behind a moving source is**

1) lower than source frequency 2) higher than source frequency
3) zero 4) same as the frequency of the source

DOPPLER EFFECT

67. Doppler shift in frequency does not depend upon

1) the frequency of wave produced
2) the speed of the source
3) distance between source and observer
4) the speed of the observer

68. An observer is moving away from a source at rest. The pitch of the note heard by the observer is less because

1) the pitch of the source decreases
2) the velocity of sound in air increases
3) wave length of the wave becomes small
4) wavelength of the wave remains unchanged but observer receives less number of waves

69. Doppler effect is not applicable to

1) sound Waves
2) light Waves
3) radio Waves
4) matter Waves

70. In Doppler effect, when a source moves towards a stationary observer, the apparent increase in frequency is due to

1) increase in wavelength of sound received by observer
2) decrease in wavelength of sound received by obzerver
3) increase in number of waves received by observer in one second
4) decrease in number of waves received by observer in one second

71. When a source moves away from stationary observer with velocity v then apparent change in frequency is Δn_1. When an observer approaches the stationary source with same velocity v then change in frequency is Δn_2 then

1) $\Delta n_1 = \Delta n_2$
2) $\Delta n_1 > \Delta n_2$
3) $\Delta n_1 < \Delta n_2$
4) $\dfrac{\Delta n_1}{\Delta n_2} < 1$

72. The graph between distance between source and observer and apparent frequency in the case of Doppler's effect will be

1) 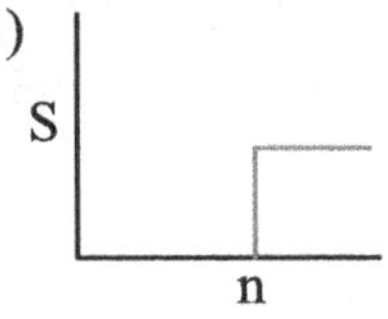

2) 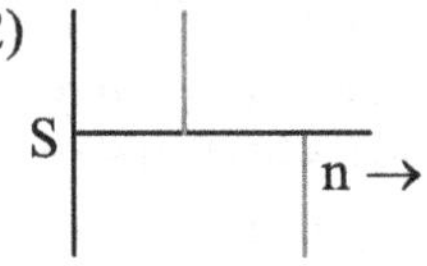

3) 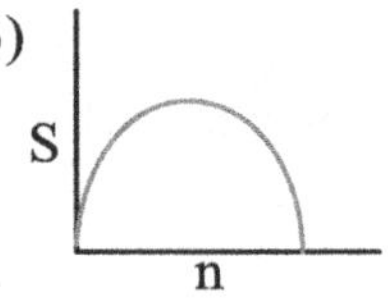

4) 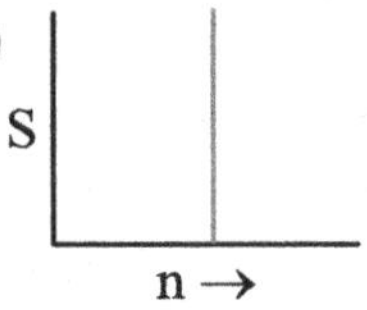

73. A source of sound moves towards a stationary listener. The apparent pitch of the sound is found to be higher than its actual value. This happens because

1) wavelength of sound waves decreases
2) wavelength of sound waves increases
3) the number of waves received by the listener increases
4) the number of waves received by the listener decreases

ASSERTION & REASON

In each of the following questions, a statement is given and a corresponding statement or reason is given just below it. In the statements, mark the correct answer as

1) Both Assertion and Reason are true and Reason is correct explanation of Assertion.

2) Both Assertion and Reason are true but Reason is not the correct explanation of Assertion.

3) Assertion is true but Reason is false.
4) Assertion is false but Reason is true.

74. Assertion (A): In transverse wave particle velocity is perpendicular to wave velocity.
Reason(R): In a longitudinal wave particle velocity is along the direction of propagation of wave.

75. Assertion (A): When a sound wave is propagating through the medium the acceleration of a particle is directly proportional to its displacement.

Reason(R): Wave velocity depends on the properties of the medium and is independent of time and position.

76. Assertion (A): When a transverse wave propagates along a string it transports energy in two forms : kinetic energy and potential energy.

Reason(R): A transverse pulse reflected from a fixed end gets inverted.

77. Assertion (A): When a wave is transmitted from one medium to another, the frequency does not change.

Reason(R): A wave transports energy and momentum along with matter.

78. Assertion (A): Mechanical waves require material medium for their propagation.

Reason(R): Transverse waves cannot be generated with in the liquids.

79. Assertion (A): The wave reflected from a denser medium has a phase change of π from the incident wave.

Reason(R): The rate of energy transfer by a wave is proportional to the frequency.

80. Assertion (A): A pulse traveling in one rope is reflected at the boundary with another rope. If the reflected pulse is not inverted, then the transmitted pulse is longer than the original pulse.

Reason(R): The speed of a transverse wave in an elastic stretched string is doubled if extension in the string is quadrapled.

81. Assertion (A): When a sound wave is propagating through the medium, the pressure fluctuations are $\pi/2$ out of phase with the displacements.

Reason(R): When a sound wave is propagating through the medium maximum pressure deviation occurs at the positions of zero displacement and no pressure change occur at the positions of maximum displacements.

82. Assertion (A): The speed of sound in moist air is more than that in dry air.

Reason(R): Speed of sound is independent of change in pressure at constant temperature.

83. Assertion (A): The Doppler formula is not symmetric with respect to the speed of source and the speed of observer.

Reason(R): Doppler effect is applicable for both mechanical as well as electromagnetic waves.

84. Assertion (A): In propagation of sound waves, pressure amplitude is proportional to the displacement amplitude.

Reason(R): Suppose the source and observer both are stationary and wind is blowing in a direction from source to observer, then the observer detects an apparent increase in frequency.

85. Assertion (A): In stationary wave, all the particles of the medium between two nodes vibrate in same phase with same frequency but with different amplitude.

Reason(R): In stationary wave, every particle of the medium vibrates with its own energy and it does not share or transmit it to any other particle.

86. Assertion (A): In the phenomena of superposition, each wave propagates as if the other wave were not present.

Reason(R): The superposition of waves is valid only when the amplitude of the wave is much less than the wave length and velocity of the wave is much longer than the particle velocity.

87. Assertion (A): A standing wave can be produced even if the component waves have different amplitude.

Reason(R): Only periodic waves can produce interference.

88. Assertion (A): Displacement node is a pressure antinode.

Reason(R): Standing waves can only be transverse.

89. **Two points P and Q have a phase difference of π when a traveling sine wave passes through the region.**

Assertion (A): P and Q move in opposite directions.

Reason(R): P oscillates at half the frequency of Q.

90. **Assertion (A): A transverse wave is traveling in a string. Equation of the wave is equal of shape of the string at an instant t.**

Reason(R): At a point beats frequency of n Hz is observed. It means, at that point, zero intensity is observed 2n times per second.

91. **Assertion (A): In case of longitudinal stationary wave, compressions and rarefractions are obtained in place of nodes & antinodes respectively.**

Reason(R): Two plane waves, one longitudinal and the other transverse having same frequency and amplitude are traveling in medium in opposite directions with same speed forms a stationary wave.

92. **Assertion (A): If all the particles of a string are oscillating in same phase, the string is resonating in its fundamental tone.**

Reason(R): A sound wave is propagating through the medium, when the particle moves in the opposite direction to the propagation of the wave, it is in a region of rarefaction.

93. **Assertion (A): In longitudinal progressive waves, when the particle is at the mean position, it is a region of maximum compression or rarefaction.**

Reason(R): In longitudinal progressive waves, compressions and rare fractions travel forward along the wave.

94. **Assertion (A): In progressive waves, the minimum distance between two particles vibrating in phase is equal to the wavelength.**

Reason(R): In progressive waves, all particles have the same amplitude, frequency and time period.

95. **Assertion (A): The intensity of sound increases with increase in the density of the medium.**

Reason(R): The sound heard is more intense in carbon dioxide in comparison to air.

96. **Assertion (A): The velocity of sound increases with increase of temperature.**

Reason(R): Sound wave travel through longer distances during night than during day.

97. **Assertion (A): The velocity of sound is generally greater in solids than in gases at NTP.**

Reason(R): The ratio of modulus of elasticity to density for solids is much greater than that for gases.

98. **Assertion (A): Beats can be observed by two light sources as in sound.**

Reason(R): To observe beats by two light sources, the phase difference between the two sources should change regularly.

99. **Assertion (A): When a vibrating tuning fork is placed on a sonometer, a large sound is heard.**
Reason(R): It is due to resonance.

100. **Assertion (A): When the star approaches the earth, the spectral lines are shifted towards the blue end of the spectrum.**

Reason(R): It is due to Doppler effect.

101. **Assertion (A): The change in frequency due to Doppler effect does not depend on separation between the source and the observer.**

Reason(R): When a listener and sound source are moving with the same velocity in the same direction, the apparent frequency as heard by listener increases.

102. Assertion (A): The Doppler effect can be observed in case of light and sound, but Doppler effect formula for light differs from that for sound.

Reason(R): Light does not require a material medium for its transmission whereas sound requires medium for its transmission.

103. Assertion (A): In a standing wave, the particles at nodes always remain in rest.

Reason(R): In a standing wave, all the particles cross their mean positions together.

104. Assertion (A): In a travelling wave, energy is transmitted from one region of space to other but in a standing wave the energy of one region is always confined that region.

Reason(R): In a travelling wave, the phases of nearly particles are always different, whereas in a standing wave all particles between two successive nodes move in phase.

105. Assertion (A):Solids can support both longitudinal and transverse waves, but only longitudinal waves can propagate in gases.

Reason(R): Solid possess elasticity of length as well as shape, but gases possess only Bulk modulus.

106. Assertion (A): A driver under water is unable to hear the sound produced in air.

Reason(R): The sound produced in air undergoes reflection at the water surface.

107. Assertion (A): We need two ears to locate the direction from which the sound is coming.

Reason(R): The phase difference between the sound reaching the two ears depends on the direction and distance of source.

108. Assertion (A): The sound waves do not exhibit the property of polarization.

Reason(R): The sound waves do not have the transverse nature.

109. Assertion (A): Two pulses are traveling along a uniform string in opposite directions. One of them is inverted pulse when compared with the other in shape. There is no point clamped between the pulses. Then they will have different speeds.

Reason(R): In a uniform string under a given tension, wave speed of any pulse is same.

110. Two waves in a string (all in SI units)are $y_1 = 0.6\sin(10t - 20x)$ **and** $y_2 = 0.4\sin(10t + 20x)$. **Mass per unit length of the string is** $10^{-2}\frac{kg}{m}$.

Assertion (A): Stationary wave can be formed by their superposition but net energy transfer through any section will be non-zero.

Reason(R): Their amplitudes are unequal.

111. Assertion (A): When a longitudinal pressure wave is reflected at the open end of an organ pipe, the compression pressure wave pulse becomes rarefraction pressure wave pulse during the reflection.

Reason(R): The phase of the wave changes by π **when reflected at the open end.**

112. Assertion (A): An open organ pipe of certain length have the same fundamental frequency as closed organ pipe of half the length

Reason(R): In the case of open organ pipe, at both the ends antinodes are formed, while in the closed organ pipe at one end antinode and at the other end node is formed

113. Assertion (A) When a traveling wave is sent along a particular string by oscillating one end, the speed of the wave will increase if we increase the frequency of oscillations.

Reason(R): If a traveling wave sent along a particular string by oscillating its one end, it is the wave length of the wave that decreases.

MATCH THE FOLLOWING

114. Match the following :

List – I	List – II
In a stretched string %	Change in frequency
a) Length increases by 2%	e) Decreases by 4%
b) Radius increases by 4%	f) Increases by 4%
c) Tension increases by 2%	g) Decreases by 4%
d) Density decreases by 2%	h) Increases by 8%
	i) Changes by 8%

The correct match is

1) a-g, b-e, c-h, d-f 2) a-h, b-i, c-g, d-h 3) a-e, b-g, c-f, d-h 4) a-f, b-i, c-e, d-g

115. Match the following :

List – I	List – II
a) Resonance	e) Law of conservation of energy
b) Reflection	f) Doppler effect is due to change in wave length
c) Source is in to number of waves	g) Doppler effect is due motion reaching the observer
d) Observer is forced	h) Special case of in motion vibrations
	i) Reverberation

The correct match is

1) a-e, b-h, c-g, d-i 2) a-f, b-g, c-e, d-h 3) a-g, b-h, c-e, d-f 4) a-h, b-i, c-f, d-g

116. I represent intensity of sound wave, A the amplitude and r the distance from the source. Then the match the following two comlumns.

Column - I	Column - II
a) Intensity due to a point source.	p) Proportional to $r^{-1/2}$
b) Amplitude due to a point source.	q) Proportional to r^{-1}
c) Intensity due to a line source	r) Proportional to r^{-2}
d) Amplitude due to a line source	s) Proportional to r^{-4}

1) a-r,b-q,c-q,d-p 2) a-p,b-q,c-r,d-s 3) a-q,b-q,c-s,d-r 4) a-s,b-p,c-q,d-r

117. Transverse waves are produced in a stretched wire. Both ends of the string are fixed. Let us compare between second overtone mode(in numerator) and fifth harmonic mode(in denominator). Match the following two columns.

Column - I	Column - II
(a) Frequency ratio	(p) 2/3
(b) Number of nodes ratio	(q) 4/5
(c) Number of antinodes ratio	(r) 3/5
(d) Wavelength ratio	(s) 5/3

1) a-r,b-p,c-r,d-s 2) a-q,b-p,c-s,d-s 3) a-r,b-s,c-q,d-p 4) a-s,b-p,c-r,d-p

118. A wave travels from a denser medium to rarer medium, then match the following two columns.

Column - I	Column - II
(a) speed of wave	(p) will increase
(b) wavelength of wave	(q) will decrease
(c) amplitude of wave	(r) will remain constant
(d) frequency of wave	(s) may increase or decrease

1) a – r; b – p; c – r; d – s 2) a – p; b – p; c – p; d – r

3) a – r; b – s; c – q; d – p 4) a – s; b – p; c – r; d – p

119. y-x graph of a transverse wave at a given instant is shown in figure. match the following two columns.

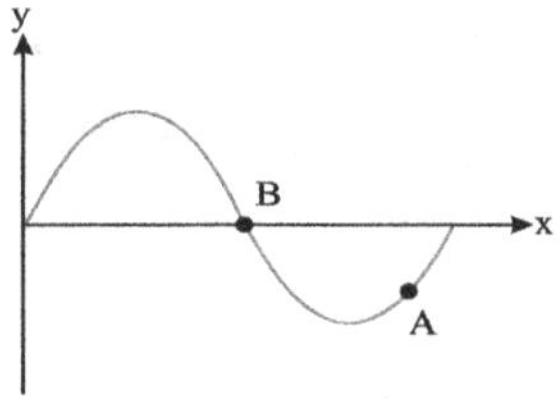

Column - I	**Column - II**	
(a) velocity of particle A	**(q) negative**	**(p) positive**
(b) acceleration of particle A		
(c) velocity of particle B	**(s) cen't tell**	**(r) zero**
(d) acceleration of particleB		

1) a-r,b-p,c-r,d-s 2) a-p,b-p,c-p,d-r 3) a-r,b-s,c-q,d-p 4) a-s,b-p,c-s,d-r

120. **Match the following**

Column-I	**Column-II**
A) Laplace equation	**P) humidity**
B) Newton equation	**Q)** $\sqrt{\frac{\gamma p}{d}}$
C) Speed of longitudinal wave depends on	**R) Temperature**
	S) isothermal process
	T) $\sqrt{\frac{P}{d}}$

	A	B	C		A	B	C
1)	P	Q	S,T	2)	Q	S,T	P,R
3)	P,Q	R,S	T	4)	Q,R	P,Q,R	R,S

121. A tuning fork 'P' of frequency 280 Hz produces 6 beats/s with unknown tuning fork 'Q'

Column-I	**Column-II**
A) P is waxed and number of beats decreases	**P) Frequency of 'Q' is 286 Hz**
B) Q is filed and numbered beats decreases	**Q) Frequency of 'Q' is 274 Hz**
C) P is filed and number of beats remains same	**R) Frequency of'Q' is 272 Hz**
D) 'Q' is filed and number of beats	**S) Frequency 'Q' is 288 Hz**

	A	B	C	D		A	B	C	D
1)	Q	R	S	R	2)	P	P	Q	Q
3)	P,Q	Q,R	R,S	P	4)	R,S	Q,R	S	R

122. Match the following

Column-I	**Column-II**
A) Beats	**P) Ratio of harmonics is 1:2:3**
B) open organ pipe	**Q) Transverse stationary waves**
C) string stretched at both ends	**R) Superposition of sound waves of nearly equal frequencies**
D) closed organ pipe	**S) longitudinal stationary waves**
	T) Interference in time

	A	B	C	D		A	B	C	D
1)	P,T	P,S	P,Q	S	2)	Q,R	S,T	R,S,T	Q
3)	S,T	Q,R,T	P	Q	4)	Q	P,Q	R,S	T

123. A string of length 1m stretched at both ends vibrating with frequency 300 Hz which is 3 times the fundamental frequency

Column-I	Column-II
A) Number of loops	P) $\frac{1}{3}m$
B) Number of antinodes	Q) 200 Hz
C) Distance between two successive	R) 1st overtone antinodes
D) 2nd harmonic	S) 3

	A	B	C	D		A	B	C	D
1)	Q	R,S	P,R	Q,S	2)	T	Q,R	P	S,T
3)	P,Q	R,S	P,R,T	Q	4)	S	S	P	Q,R

MORE THAN ONE OPTION IS CORRECT

124. A string of length l is stretched along the x-axis and is rigidly clamped at x=0 and x=l. Transverse vibrations are produced in the string. For n^{th} harmonic which of the following relations may represent the shape of the string at any time

a) $y=2A\cos\omega t\ \cos\left(\frac{n\pi x}{l}\right)$
b) $y=2A\sin\omega t\ \cos\left(\frac{n\pi x}{l}\right)$

c) $y=2A\cos\omega t\ \sin\left(\frac{n\pi x}{l}\right)$
d) $y=2A\sin\omega t\ \sin\left(\frac{n\pi x}{l}\right)$

1) c only 2) c and d only 3) a only 4) a, b, c and d

125. In case of superposition of waves (at x=0), $y_1=4\sin(1026\pi t)$ and $y_2=2\sin(1014\pi t)$

a) the frequency of resulting wave is 510 Hz

b) the amplitude of resulting wave varies at frequency 3 Hz

c)the frequency of beats is 6 per second

d) the ratio of maximum to minimum intensity is 9

The correct statements are

1) a, d 2) b, d 3) a, c, d 4) all

126. In case of stationary sound waves in air the correct statement(s) is a/are

A) each air particle vibrates with the same amplitude

B) amplitude is maximum for some particles and minimum for some other particles

C) the particles do not execute periodic motion D) the particles do not vibrate at all

1) A 2) B 3) C 4) B & C

127. The tension in a stretched string fixed at both ends is changed by 2%, the fundamental frequency is found to get changed by 15 Hz.

a) wavelength of the string of fundamental frequency does not change

b) velocity of propagation of wave changes by 2%

c) velocity of propagation of wave changes by 1% d) original frequency is 1500 Hz

1) c only correct 2) c and d are correct

3) a, c and d are correct 4) b and d are correct

128. A wave y = Acos(ωt – kx) passes through a medium. If v is the particle velocity and a is the particle acceleration then,

a) y, v and a all are in the same phase **b) y lags behind v by a phase angle of $\pi/2$**

c) a leads y by a phase angle of π **d) v leads a by a phase angle of $\frac{3\pi}{2}$**

1) a only correct 2) b and c are correct 3) a, b, c are correct 4) b and d are correct

129. The equation of the standing wave in a string clamped at both ends, vibrating in its third harmonic is given by $y = 0.4\sin(0.314x)\cos(600\pi t)$ where, x and y are in cm and t in sec.

a) the frequency of vibration is 300 Hz b) the length of the string is 30 cm

c) the nodes are located at $x = 0$, 10 cm, 30 cm

1) Only a is true 2) a, b are true 3) b, c are true 4) a, b, c are true

130. The equation of a stationary wave in a string is $y = (4mm)\sin\left[\left(3.14m^{-1}\right)x\right]\cos\omega t$.

Select the correct alternative(s) :

a) the amplitude of component waves is 2 mm b) the amplitude of component waves is 4mm

c) the smallest possible length of string is 0.5 m d) the smallest possible length of string is 1.0 m

1) a, c are correct 2) b, c are correct 3) a, d are correct 4) all are correct

CONCEPTUAL KEY

1) 3	2) 4	3) 1	4) 2	5) 1	6) 1	7) 1	8) 3	9) 4	10) 1
11) 3	12) 4	13) 1	14) 2	15) 4	16) 3	17) 1	18) 1	19) 4	20) 4
21) 4	22) 2	23) 4	24) 1	25) 4	26) 1	27) 4	28) 3	29) 4	30) 3
31) 2	32) 4	33) 4	34) 2	35) 4	36) 2	37) 2	38) 1	39) 4	40) 3
41) 3	42) 3	43) 2	44) 3	45) 1	46) 2	47) 4	48) 3	49) 4	50) 2
51) 4	52) 1	53) 3	54) 3	55) 2	56) 1	57) 2	58) 2	59) 4	60) 3
61) 4	62) 4	63) 4	64) 4	65) 4	66) 1	67) 3	68) 4	69) 4	70) 2
71) 3	72) 4	73) 1	74) 2	75) 2	76)2	77)3	78)2	79)3	80)4
81)1	82)2	83)2	84)3	85)3	86)2	87)3	88)3	89)3	90)2
91)3	92)2	93)3	94)2	95)3	96)3	97)1	98)1	99)1	100)1
101)3	102)1	103)3	104)2	105)1	106)1	107)1	108)1	109)4	110)1
111)1	112)2	113)4	114)1	115)4	116)1	117)1	118)2	119)4	120)2
121)2	122)1	123)4	124)2	125)1	126)2	127)3	128) 4	129)4	130)3

LEVEL - I (CW)

CHARACTERISTICS OF PROGRESSIVE WAVE

1. The equation y = 4 Sin π [200t-(x/25)] represents a transverse wave that travels in a stretched wire, where x, y are in cm and t in second. Its wavelength and velocity are

1) 0.5m, 25ms^{-1} 2) 0.5m, 50ms^{-1} 3) 1m, 50ms^{-1} 4) 1m, 25ms^{-1}

2. The equation $y = A\cos^2\left(2\pi nt - 2\pi\frac{x}{\lambda}\right)$ represents a wave with

1) amplitude $\frac{A}{2}$, frequency 2n and wavelength $\frac{\lambda}{2}$

2) amplitude $\frac{A}{2}$, frequency 2n and wavelength λ

3) amplitude A, frequency 2n, and wavelength 2λ

4) amplitude A, frequency n, and wavelength λ

3. A transverse wave along a string is given by $y = 2\sin\left(2\pi(3t - x) + \frac{\pi}{4}\right)$, where x and y are in cm and 't' is in second. The acceleration of a particle located at $x = 4cm$ at t =1s is

1) $36\sqrt{2}\pi^2\ cm/s^2$ 2) $36\pi^2\ cm/s^2$ 3) $-36\sqrt{2}\pi^2\ cm/s^2$ 4) $-36\pi^2\ cm/s^2$

4. A transverse wave is described by the equation $Y=Y_0 \text{ Sin } 2\pi(ft-x/\lambda)$. The maximum particle velocity is equal to four times the wave velocity if

1) $\lambda = \pi Y_0/4$ 2) $\lambda = \pi Y_0/2$ 3) $\lambda = \pi Y_0$ 4) $\lambda = 2\pi Y_0$

5. The frequency of a fork is 500 Hz. Velocity of sound in air is $350ms^{-1}$. The distance through which sound travel by the time the fork makes 125 vibrations is

1) 87.5 m 2) 700 m 3) 1400m 4) 1.75 m

6. A wave has a frequency of 120Hz. Two points at a distance of 9m apart have a phase difference of 1080°. The velocity of the wave is

1) 340m/s 2) 300m/s 3) 330m/s 4) 360m/s

7. Two simple harmonic motions are represented by the equations $y_1 = 0.1\sin\left(100\pi t + \frac{\pi}{3}\right)$ and $y_2 = 0.1\cos \pi t$. The phase difference of the velocity of particle 1 with respect to velocity of particle 2 at t = 0 is

1) $-\frac{\pi}{6}$ 2) $\frac{\pi}{3}$ 3) $-\frac{\pi}{3}$ 4) $\frac{\pi}{6}$

SPEED OF A TRAVELLING WAVE

8. The amplitude of a wave disturbance propagating in the positive x direction is given by $y = \frac{1}{1+x^2}$ at $t = 0$ in metres and $y = \frac{1}{1+(x-1)^2}$ at $t = 2s$ where x, y are in metres. If the shape of the wave does not changing with time, the velocity of the wave is

1) $2ms^{-1}$ 2) $0.5ms^{-1}$ 3) $3ms^{-1}$ 4) $1ms^{-1}$

9. If Young's modulus of the material of a rod is Y and density is ρ then time taken by sound wave to travel l length from bottom is

1) $l\sqrt{\frac{\rho}{Y}}$ 2) $l\sqrt{\frac{Y}{\rho}}$ 3) $\frac{1}{l}\sqrt{\frac{Y}{\rho}}$ 4) $\frac{1}{l}\sqrt{\frac{\rho}{Y}}$

10. The velocity of a transverse wave in a stretched wire is $100ms^{-1}$. If the length of wire is doubled and tension in the string is also doubled, the final velocity of the transverse wave in the wire is

1) 100 ms^{-1} 2) 141.4 ms^{-1} 3) 200 ms^{-1} 4) 282.8 ms^{-1}

11. A transverse wave propagating on a stretched string of linear density $3\times10^{-4}kgm^{-1}$ is represented by the equation $y = 0.2\,Sin(1.5x + 60t)$ where x is in metre and t is in seconds. The tension in the string in N.

1) 0.24 2) 0.48 3) 1.20 4) 1.80

12. The extension in a string, obeying Hooke's law is x. The speed of sound in the stretched string is V. If the extension in the string is increased to 1.5x the speed of the sound will be

1) 1.22V 2) 0.61V 3) 1.50V 4) 0.75V

PRINCIPLE OF SUPER POSITION, INTERFERENCE AND STATIONARY WAVES ON STRETCHED STRINGS.

13. Standing waves are produced in 10m long stret-ched wire. If the wire vibrates in 5 segments and wave velocity is 20m/s , then the freqency is

1) 2 Hz 2) 4 Hz 3) 5 Hz 4) 10Hz

14. The equation $y = 5\sin\left(\frac{\pi x}{25}\right)\cos(450t)$ represents the stationary wave setup in a vibrating sonometer wire, where x,y are in cm and t in second. The distances of second and third nodes from one end are(in cm).
1) 50, 75 2) 25, 50 3) 15, 50 4) 20, 50

15. A sonometer consists of two wires of lengths 1.5 m and 1m made up of different materials whose densities are 5 g/cc, 8g/cc and their respective radii are in the ratio 4:3. The ratio of tensions in those two wires if their fundamental frequencies are equal is
1) 5:3 2) 5:2 3) 2:5 4) 3:5

16. If the length of a stretched string is shortened by 40% and the tension is increased by 44%, then the ratio of the final and initial fundamental frequencies is:
1) 2:1 2) 3:2 3) 3:4 4) 1:3

17. The fundamental frequency of a stretched string with a weight of 9kg is 289 Hz. The weight required to produce its octave is
1) 9 kg wt 2) 16 kg wt 3) 25 kg wt 4) 36 kg wt

18. In an experiment it was found that string vibrates in n loops when a mass M is placed on the pan. The mass that should be placed on the pan to make it vibrate in 2n loops with same frequency (neglect the mass of pan) is
1) 2M 2) M/4 3) 4M 4) M/2

19. Transverse waves are generated in two uniform wires A and B of the same material by attaching their free ends to a vibrating source of frequency 200Hz. The area of cross section of A is half that of B while tension on A is twice than on B. The ratio of wavelengths of transverse waves in A and B is.
1) $1:\sqrt{2}$ 2) $\sqrt{2}:1$ 3) 1:2 4) 2:1

20. A string is stretched between fixed points separated by 75.0 cm. It is observed to have resonant frequencies of 420 Hz and 315 Hz. There are no other resonant frequencies between these two. Then, the lowest resonant frequency for this string is
1) 105 Hz 2) 1.05 Hz 3) 1005 Hz 4) 10.5 Hz

SOUND AND VELOCITY OF SOUND

21. The pressure of air increases by 100 mm of Hg and the temperature decreases by 1°C. The change in the speed of sound in air at STP is
1)61 ms^{-1} 2)61 mms^{-1} 3)61 cms^{-1} 4)0.61 cms^{-1}

22. The temperature at which the speed of sound will be same in oxygen as the speed in nitrogen at 15°C is (Densities are in the ratio 16:14)
1) 561°C 2) 56.1°C 3) 5.61°C 4) 5.061°C

23. The ratio of speed of sound in Nitrogen gas to that in Helium gas at 300 K is (assume both gases to be ideal)
1) $\sqrt{2}:\sqrt{7}$ 2) $1:\sqrt{7}$ 3) $\sqrt{3}:5$ 4) $\sqrt{6}:5$

24. A window whose area is 2 m^2 open on a street where the street noise results at the window an intensity level of 80 dB. How much acoustic power enters the window through sound waves?
1) $100\mu W$ 2) $200\mu W$ 3) $300\mu W$ 4) $400\mu W$

ORGAN PIPES

25. A cylindrical resonance tube open at both ends has a fundamental frequency 'f' in air. If half of the length is dipped vertically in water, the fundamental frequency of the air column will be (JEE-2012)
1) f/2 2) f 3) 3f/2 4) 2f

26. A closed organ pipe is vibrating in first overtone and is in resonance with another open organ pipe vibrating in third harmonic. The ratio of lengths of the pipes respectively is
1) 1 : 2 2) 4 : 1 3) 8 : 3 4) 3 : 8

27. A glass tube 1.0 m length is filled with water. The water can be drained out slowly at the bottom of the tube. If a vibrating tuning fork of frequency 500 c/s, is brought at the upper end of the tube, and the velocity of sound is 330 m/s, then the total number of resonances obtained will be

1) 4 2) 3 3) 2 4) 1

28. An open pipe and a closed pipe have same length. The ratio of pth overtones of two pipes is

1) $\frac{1}{p}$ 2) p 3) $\frac{2(p+1)}{2p+1}$ 4) $\frac{2p+1}{2(p+1)}$

BEATS

29. The natural frequency of a tuning fork P is 432 Hz. 3 beats/s are produced when tuning fork P and another tuning fork Q are sounded together. If P is loaded with wax, the number of beats increases to 5 beats/ s. The frequency of Q is

1) 429 Hz 2) 435 Hz 3) 437 Hz 4) 427 Hz

30. Two organ (open) pipes of lengths 50 cm and 51 cm produce 6 beats/s. Then the speed of sound is nearly

1) 300 ms^{-1} 2)306 ms^{-1} 3)303 ms^{-1} 4)350 ms^{-1}

31. A source x of unknown frequency produces 8 beats with a source of 250 Hz and 12 beats with a source of 270 Hz. The frequency of source x is

1) 258 Hz 2) 242 Hz 3) 262 Hz 4) 282 Hz

32. In an experiment it was found that when a sonometer in its fundamental mode of vibration and a tuning fork gave 5 beats when length of wire is 1.05 metre or 1 metre. The velocity of transverse waves in sonometer wire when its length is 1m

1) 400 m/s 2) 210 m/s 3) 420 m/s 4) 450 m/s

33. A sonometer has 25 forks. Each produces 4 beats with the next one. If the maximum frequency is 288 Hz, which is the frequency of last fork. The lowest frequency is

1) 72 Hz 2) 96 Hz 3) 128 Hz 4) 192 Hz

34. A tuning fork produces 6 beats/sec with sonometer wire when its tensions are either 169N or 196N. The freqency of that fork is

1) 162 Hz 2) 190Hz 3) 200Hz 4) 80Hz

35. In an open pipe when air column is 20 cm it is in resonance with tuning fork A. When length is increased by 2cm then the air column is in resonance with fork B. When A and B are sounded together 4 beats/sec are heard. Freqencies of A and B are respectively (in Hz)

1) 40, 44 2) 88,80 3) 80,88 4) 44,40

DOPPLER EFFECT

36. The speed at which a source of sound should move so that a stationary observer finds the apparent frequency equal to 11/12 of the original frequency

1) V/2 2) 2V 3) V/4 4) V/11

37. A whistling engine is approaching a stationary observer with a velocity of 110m/s. The velocity of sound is 330m/s. The ratio of frequencies as heard by the observer as the engine approaches and receedes is

1) 4:3 2) 4:1 3) 3:6 4) 2:1

38. Two aeroplanes 'A' and 'B' are moving away from one another with a speed of 720 kmph. The frequency of the whistle emitted by 'A' is 1100 Hz. The apparent frequency of the whistle as heard by the passenger of the aeroplane 'B' is.(velocity of sound in air is 350 ms^{-1}).

1) 300Hz 2) 400Hz 3) 500Hz 4) 600H

39. An engine is moving on a circular path of radius 100 metre with a speed of 20 metre per second. The frequency observed by an observer standing stationary at the centre of circular path when the engine blows a whistle of frequency 500 Hz is

1) more than 500 Hz 2) less than 500 Hz 3) 500 Hz 4) no sound is heard

40. The frequency of a radar is 780 MHz. The frequency of reflected wave from an aeroplane is increased by 2.6KHz. The velocity of the aeroplane is

1)2 km/s 2) 1 km/s 3) 0.5 km/s 4) 0.25 km/s

41. **An observer moves towards a stationary source of sound, with a velocity one-fifth of the velocity of sound. The percentage increase in the apparent frequency is**
1) 5% 2) 20% 3) zero 4) 0.5%

42. **A train is moving at $30ms^{-1}$ in still air. The frequency of the locomotive whistle is 500 Hz and the speed of sound is 345 ms^{-1}. The apparent wavelengths of sound infront of and behind the locomotive are respectively**
1) 0.63 m, 0.80 m 2) 0.63 m, 0.75 m 3) 0.60 m, 0.85 m 4) 0.60 m, 0.75 m

43. **A vehicle, with a horn of frequency n is moving with a velocity of 30 m/s in a direction perpendicular to the straight line joining the observer and the vehicle. The observer perceives the sound to have a frequency $(n+n_1)$. If the velocity of sound in air is 300 m/s, then**
1) $n_1 = 10n$ 2) $n_1 = 0$ 3) $n_1 = 0.1n$ 4) $n_1 = -0.1n$

44. **a source of sound is travelling towards stationary observer. The freqency of sound heard by the observer is 25% more than that of the actual freqency if speed of sound is V, that of the source is**
1) V/5 2) V/4 3) V/3 4) V/2

45. **A truck blowing horn of frequency 500 Hz travels towards a vertical mountain and driver hears echo of frequency 600Hz. If velocity of sound in air is 340m/s then speed of truck is**
1) 31 m/s 2) 41m/s 3) 51m/s 4) 21m/s

LEVEL -I (C. W) - KEY

1) 2	2) 1	3) 3	4) 2	5) 1	6) 4	7) 1	8) 2	9) 1	10) 2
11) 2	12) 1	13) 3	14) 2	15) 2	16) 1	17) 4	18) 2	19) 4	20) 1
21) 3	22) 2	23) 3	24) 2	25) 2	26) 1	27) 2	28) 3	29) 2	30) 2
31) 1	32) 3	33) 4	34) 1	35) 4	36) 4	37) 4	38) 1	39) 3	40) 3
41) 2	42) 2	43) 2	44) 1	45) 1					

LEVEL - I (HW)

CHARACTERISTICS OF PROGRESSIVE WAVE

1. **A longitudinal progressive wave is given by the equation $y = 5x10^{-2} \sin\pi$ (400 t + x). The amplitude and wave length of the wave are (y, x are in m)**
1) A = $5x10^{-2}$m, $\lambda = 2m$ 2) A = $5x10^{-2}$m, $\lambda = 3m$
3) A = $5x10^{-2}$m, $\lambda = 4m$ 4) A = $5x10^{-2}$m, $\lambda = 5m$

2. **The equation of a wave is $y = 4\sin\left\{\frac{\pi}{2}\left(2t + \frac{x}{8}\right)\right\}$ where y, x are in cm and time in seconds. The acceleration of particle located at $x = 8cm$ and $t = 1\sec$ is**
1) $4\pi^2\ cm/s^2$ 2) $-4\pi^2\ cm/s^2$ 3) $16\pi^2\ cm/s^2$ 4) $-16\pi^2\ cm/s^2$

3. **The equation of a transverse wave is y = a sin 2π [t-(x/5)], then the ratio of maximum particle velocity and wave velocity is**
1) $\frac{2\pi a}{\sqrt{5}}$ 2) $\frac{2\pi a}{5}$ 3) $\frac{a}{5}$ 4) $2\pi a\sqrt{5}$

4. **The frequency of a tuning fork is 256Hz. The velocity of sound in air is $344ms^{-1}$. The distance travelled (in metres) by the sound during the time in which the tuning fork completes 32 vibrations is**
1) 21 2) 43 3) 86 4) 129

5. **A progressive wave moves with a velocity of 36 m/s in a medium with a freqency of 200 Hz. The phase difference between two particles separeted by a distance of 1 cm is**
1) 40^0 2) $20rad$ 3) $\frac{\pi}{9}rad$ 4) $\left(\frac{\pi}{9}\right)^0$

6. A standing wave set up in a medium is $y = 4\cos\left(\frac{\pi x}{3}\right)\sin 40\pi t$ where x, y are in cm and t in sec The velocity of medium particle at x =6cm at t=1/8 sec is

1) 40π cm/s 2) 80π cm/s 3) 120π cm/s 4) -160π cm/s

SPEED OF A TRAVELLING WAVE

7. The equation of a wave pulse is given as $y = \frac{0.8}{(4x+5t)+4}$ the amplitude and velocity of the pulse are

1) 0.2 units, 1.25 units along -ve x-axis
2) 0.2 units, 1.25 units along +ve x-axis
3) 0.4 units,1.25 units along -ve x-axis
4) 0.4 units, 1.25 units along +ve x-axis

8. If Young's modulus of the material of a rod is $2\times10^{11} Nm^{-2}$ and density is 8000kg m^{-3}, the time taken by a sound wave to traverse 1m of the rod is

1) 1.11×10^{-4}s 2) 3×10^{-4}s 3) 2×10^{-4} s 4) 1×10^{-4}s

9. A string of length 10.0 m and mass 1.25 kg is stretched with a tension of 50 N. If a transverse pulse is created at one end of the string, the time taken by it to reach the other end is

1) 0.5 s 2) 1.0 s 3) 1.5 s 4) 2.0 s

10. The linear density of a vibrating string is 1.3×10^{-4} kg/m. The transverse wave propagating along the string is described by y = 0.021 sin (x+30t) where x is in meter and t is in second. The tension in the string is

1)0.12 N 2) 0.48 N 3)1.2N 4) 4.8N

11. The extension in a string, obeying Hooke's law is x. The speed of sound in the stretched string is V. If the extension in the string is increased to 2x then speed of sound will be

1) 1.5V 2) 4.14 V 3) 1.414 V 4) 2V

PRINCIPLE OF SUPER POSITION, INTERFERENCE AND STATIONARY WAVES ON STRETCHED STRINGS.

12. The speed of transverse waves in a stretched string is 700cm/s. If the string is 2m long, the frequency with which it resonates in fundamental mode is

1) (7/12)Hz 2) (7/4)Hz 3) 14Hz 4) (2/7)Hz

13. The length of a sonometer wire is 90 cm and the stationary wave setup in the wire is represented by an equation y = 6sin(π x/30) cos (250 π t) where x, y are in cm and t is in second. The distances of successive antinodes from one end of the wire are

1) 22.5 cm, 67.5 cm
2) 15 cm, 30 cm, 60cm
3) 15 cm, 45 cm,75cm
4) 30 cm, 45 cm, 60cm

14. A sonometer consists of two wires of same length, same material whose radii are in the ratio 2:3. The ratio of tension in two wire if their fundamental freqencies are equal is

1) 1:4 2) 2:3 3) 9:4 4) 4:9

15. The bridge of a sonometer is slightly displaced so that the length of wire is decreased by 0.5% and tension in the wire is increased by 1%. The fundamental frequency of wire

1) increases by 1%
2) decreases by 1%
3) increases by 1.5%
4) decreases by 1.5%

16. A segment of wire vibrates with a fundamental frequency of 450Hz under a tension of 9kg weight. Then tension at which the fundamental frequency of the same wire becomes 900Hz is

1) 36 kg wt 2) 27 kg wt 3)18 kg wt 4)72 kg wt

17. In an experiment, the string vibrates in 4 loops when 50 gm wt is placed in pan of weight 15 gm. To make the string vibrate in 6 loops the weight that has to be removed from the pan is approximately

1) 72 gm 2) 36 gm 3) 21 gm 4) 29 gm

18. **Two vibrating strings of the same material but lengths L and 2L have radii 2r and r respectively. They are stretched under the same tension. Both the strings vibrate in their fundamental modes, the one of length L with frequency v_1 and the other with frequency v_2. The ratio $\frac{v_1}{v_2}$ is given by**

1) 2 2) 4 3) 8 4) 1

19. **A stretched string of length 2m is found to vibrate in resonance with a tuning fork of frequency 420 Hz. The next higher frequency for which resonance occurs is 490 Hz. The velocity of the transverse wave along this string is**

1) 140 m/s 2) 360 m/s 3) 340 m/s 4) 280 m/s

20. **Two uniform stretched strings A and B, made of steel are vibrating under the same tension. If the first overtone of A is equal to the second overtone of B and if the radius of A is twice that of B, the ratio of the lengths of the strings is**

1) 2 : 3 2) 1 : 2 3) 1 : 3 4) 1 : 4

SOUND AND VELOCITY OF SOUND

21. **If the speed of sound is changed by 1 percent, the temperature of air near 0°C be changed is**

1) 5°C 2) 6°C 3) 5.5°C 4) 6.5°C

22. **The ratio of speed of sound wave in Neon to that in H_2O vapours at any temperature is**

1) 9/8 2) $3/2\sqrt{2}$ 3) 3/2 4) 3/4

23. **In a class of 100 students each shouting at 100 dB. Find noise level of class?**

1) 10dB 2) 100dB 3) 12dB 4) 120dB

ORGAN PIPES

24. **The air column in a pipe which is closed at one end will be in resonance with a vibrating tuning fork at a frequency 260Hz, if the length of the air column is (speed of sound in air $=330ms^{-1}$)**

1) 31.73cm 2) 62.5cm 3) 35.75cm 4) 12.5cm

25. **A cylinderical tube open both ends has a fundamental frequency 'n' in air. The tube is dipped vertically in water so that one - fourth of it is immersed in water. The fundamental frequency of air column is**

1) 3n 2) 2n/3 3) n/3 4) n

26. **An open pipe and a closed pipe are in resonance with each other with their first overtones. The ratio of their lengths are**

1) 4:3 2) 3:4 3) 1:3 4) 3:1

27. **A pipe of length 85cm closed from one end. Find then number of possible natural oscillations of air column in the pipe whose frequencies lie below 1250Hz. The velocity of sound in air is 340 m/s.**

1) 12 2) 8 3) 6 4) 4

28. **A tuning fork of frequency 340 Hz is vibrated just above a cylindrical tube of length 120 cm. Water is slowly poured in the tube. If the speed of sound in air is 340 ms^{-1}, then the minimum height of water required for resonance is**

1) 25 cm 2) 45 cm 3) 75 cm 4) 95 cm

29. **An organ pipe P_1, closed at one end and containing a gas of density ρ_1 is vibrating in its first harmonic. Another organ pipe P_2, open at both ends and containing a gas of density ρ_2 is vibrating in its third harmonic. Both the pipes are in resonance with a given tuning fork. If the compressibility of gases is equal in both pipes, the ratio of the lengths of P_1 and P_2 is (assume the given gases to be monoatomic)**

1) $\frac{1}{3}$ 2) 3 3) $\frac{1}{6}\sqrt{\frac{\rho_1}{\rho_2}}$ 4) $\frac{1}{6}\sqrt{\frac{\rho_2}{\rho_1}}$

BEATS

30. **When tuning forks A and B are sounded together 5 beats per second are heard. Frequency of A is 250 Hz. On loading A with wax 2 beats per second are produced with B. The frequency of B is**
    1) 255 Hz 2) 320 Hz 3) 245 Hz 4) 420 Hz
31. **Two open pipes of length 20cm and 20.1cm produces 10 beats/s. The velocity of sound in the gas is**
    1) 804 ms^{-1} 2) 402 ms^{-1} 3) 420 ms^{-1} 4) 330 ms^{-1}
32. **Two tuning forks have frequencies 200 Hz and x. When they are sounded together 4 beats / sec are heard. The value of x is**
    1) 200 Hz or 198 Hz 2) 196 Hz or 204 Hz 3) 205 Hz or 201 Hz 4) 200 Hz only
33. **A tuning fork of frequency 480 Hz produces 10 beats per second when sounded with a vibrating sonometer string. What must have been the frequency of the string if a slight increase in tension produces fewer beats per second than before?**
    1) 460 Hz 2) 480 Hz 3) 490 Hz 4) 470 Hz
34. **Five beats per second are produced on 21:20, then their frequencies will be**
    1) 105 Hz and 100 Hz 2) 105 Hz and 110 Hz
    3) 100 Hz and 105 Hz 4) 110 Hz and 105 Hz
35. **An accurate and reliable audio oscillator is used to standardise a tuning fork. When the oscillator reading is 514, two beats are hear per second. When the oscillator reading is 510, the beat frequency is 6Hz. The frequency of the tuning fork is**
    1) 506 2) 510 3) 516 4) 158
36. **25 tuning forks are arranged in decreasing order of frequency. Any two successive forks produce 3 beats/sec. If the freqency of the first tuning fork is the octave of last, then freqency of 21st fork is**
    1) 72Hz 2) 288Hz 3) 84Hz 4) 87Hz
37. **A tuning fork produces 4 beats/s with a sonometer wire when its lengths are 50 cm, 51cm. The frequency of that tuning fork is**
    1) 400 Hz 2) 404 Hz 3) 408 Hz 4) 412 Hz
38. **In a closed tube when air column is 20 cm it is in resonance with tuning fork A. When the length is increased by 2 cm then the air column is in resonance with tuning fork B. When A and B are sounded together they produce 8 beats per second. The frequencies of the tuning forks A and B are (in Hz)**
    1) 40, 44 2) 88, 80 3) 80, 88 4) 44, 40

DOPPLER EFFECT

39. **A train is approaching a station with a uniform velocity of 72 kmph and the frequency of the whistle of that train is 480 Hz. The apparent increase in the frequency of that whistle heard by a stationary observer on the platform is (Velocity of sound in air is340m/s)**
    1) 60 Hz 2) 45 Hz 3) 30 Hz 4) 15 Hz
40. **A train is travelling at 120 kmph and blows a whistle of frequency 1000Hz. The frequency of the note heard by a stationary observer if the train is approaching him and moving away from him are (Velocity of sound in air = 330 ms^{-1}).**
    1) 1112Hz,908Hz 2) 908Hz,1112Hz 3) 1080Hz,820Hz 4) 820Hz,1080Hz
41. **A source and an observer move away from each other with speed of 10m/s with respect to ground. Apparent frequency of the source is 1950Hz. The natural frequency of the source is (velocity of sound is 340m/s)**
    1) 2068Hz 2) 1832Hz 3) 1950Hz 4) 1650Hz
42. An observer is moving on a circular path of radius r with speed V_0 **around source kept at centre. The apparant freqency observed by observer is(n is actual freqency**
    1) greater than n 2) less than n 3) n 4) no sound is heard

43. A source of sound moves towards a listener with a velocity equal to that of sound. If the source emits n waves per second, then the listener moving away from the source with the same velocity receives

1) n waves per sec 2) 2n waves per sec 3) zero waves per sec 4) $n/2$

44. A source of sound and an observer are approaching each other with the same speed which is equal to $\frac{1}{10}$ times the speed of sound. The apparent change in the frequency of the source is

1) 22.2% increase 2) 22.2% decrease 3) 18.2% decrease 4) 18.2% decrease

45. A source of sound produces waves of wave legnth 48 cm. This source is moving towards north with speed 1/4 th that of sound .the apparent wave length of the waves to an observer standing south of the moving source will be

1) 60 cm 2) 72 cm 3) 48 cm 4) 96 cm

46. A whistle producing sound waves of frequencies 9500 Hz and above is approaching a stationary person with speed Vms^{-1}. The velocity of sound in air is 300 ms^{-1}. If the person can bear frequencies upto a maximum of 10,000 Hz, the maximum value of V upto which he can hear the whistle is

1) $15\sqrt{2}ms^{-1}$ 2) $\frac{15}{\sqrt{2}}ms^{-1}$ 3) $15ms^{-1}$ 4) $30ms^{-1}$

47. A whistle of frequency 540 Hz rotates in a horizontal circle of radius 2m at an angular speed of 15 rad/s. The highest frequency heard by a listener at rest with respect to the centre of circle (velocity of sound in air = $330ms^{-1}$)

1) 590Hz 2) 594Hz 3) 598Hz 4) 602Hz

48. If a source emitting waves of frequency f moves towards an observer with a velocity v/3 and the observer moves away from the source with a velocity v/4, the apparent frequency as heard by the observer will be (v = velocity of sound)

1) 9f/8 2) 8f/9 3) 3f/4 4) 4f/3

LEVEL - I (H. W) - KEY

1) 1	2) 1	3) 2	4) 2	5) 3	6)4	7)1	8) 3	9) 2	10) 1
11) 3	12) 2	13) 3	14) 4	15) 1	16) 1	17) 2	18) 4	19) 4	20) 3
21) 3	22) 2	23) 4	24) 1	25) 2	26) 1	27) 3	28) 2	29)4	30) 3
31) 1	32) 2	33) 4	34) 3	35) 3	36) 3	37) 2	38) 2	39) 3	40) 1
41) 1	42) 3	43) 3	44) 1	45) 1	46) 3	47) 2	48) 1		

LEVEL - II (CW)

CHARACTERISTICS OF PROGRESSIVE WAVE

1. The equation of progressive wave is $y = 0.01\sin(100t - x)$ where x,y are in meter and t in second, then

a) Velocity of wave is 50 m/s

b) Maximum velocity of particle is 1m/s

c) Wave length of wave is 2π meter.

1) only a,c are true 2) only a,b are true

3) only b,c are ture 4) a,b,c are true

2. A wave pulse on a string has the dimension shown in figure. The wave speed is $v = 1\ cm/s$. If point O is a free end. The shape of wave at time $t = 3s$ is

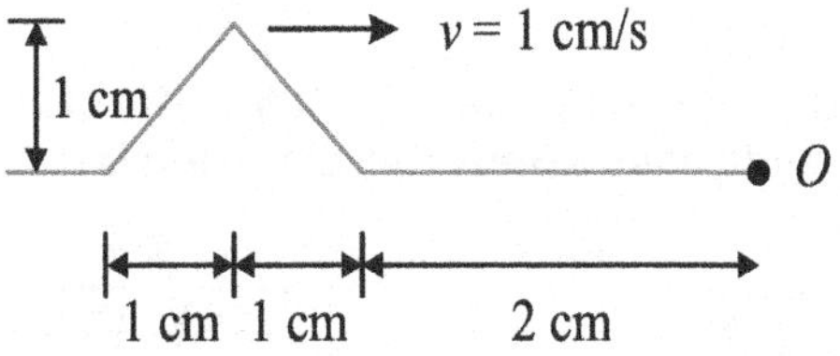

1) 

2) 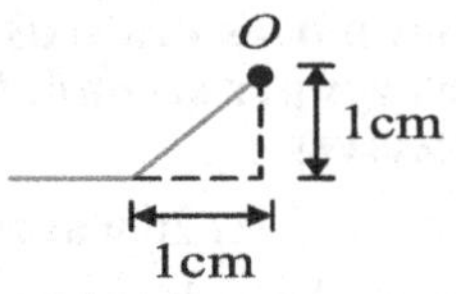

3) 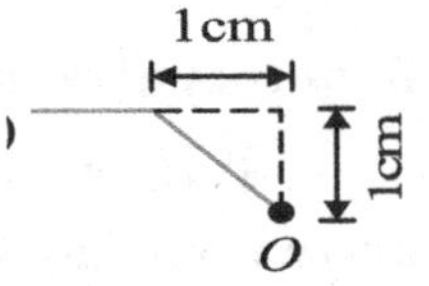

4) 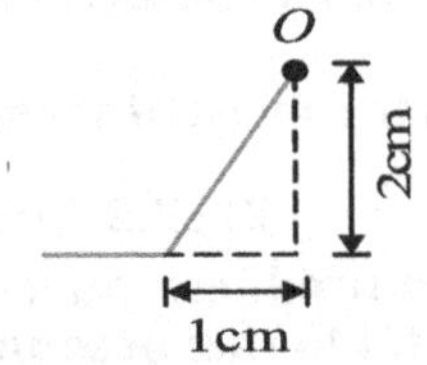

3. **Two sound waves are represented by $y_1 = \sin\omega t + \cos\omega t$ and $y_2 = \frac{\sqrt{3}}{2}\sin\omega t + \frac{1}{2}\cos\omega t$. The ratio of their amplitudes is**

1) 1:1 2) $\sqrt{3}:2$ 3) $2:\sqrt{3}$ 4) $\sqrt{2}:1$

4. **A wave of angular frequency ω propagates so that a certain phase of oscillation moves along x-axis, y-axis and z-axis with speeds c_1, c_2 and c_3 respectively. The propagation constant k is**

1) $\frac{\omega}{\sqrt{c_1^2+c_2^2+c_3^2}}\left(\hat{i}+\hat{j}+\hat{k}\right)$ 2) $\frac{\omega}{c_1}\hat{i}+\frac{\omega}{c_2}\hat{j}+\frac{\omega}{c_3}\hat{k}$

3) $\left(\omega\hat{i}+\omega\hat{j}+\omega\hat{k}\right)\frac{1}{c}$ 4) $\frac{\omega}{\left(c_1+c_2+c_3\right)}\left(\hat{i}+\hat{j}+\hat{k}\right)$

SPEED OF A TRAVELLING WAVE

5. **A uniform rope of length 12m and mass 6 kg hangs vertically from a rigid support. A block of mass 2kg is attached at the free end of the rope. A transverse pulse of wavelength 0.06m is produced at the lower end of the rope. The wavelength of the pulse when it reaches the top of the rope is**

1) 0.06 m 2) 0.12 m 3) 0.24 m 4) 0.03 m

6. **A string of length l hangs freely from a rigid support. The time required by a transverse pulse to travel from bottom to half length of the string is**

1) $\sqrt{lg}$ 2) $\sqrt{\frac{l}{g}}$ 3) $\sqrt{\frac{2l}{g}}$ 4) $2\sqrt{\frac{l}{g}}$

7. **A transverse wave is passing through a light string shown in the figure. The equation of wave is $y = A\sin(\omega t - kx)$. The area of cross-section of string is A and density is ρ. The hanging mass is**

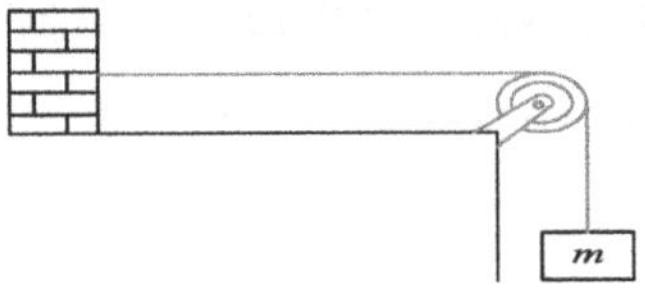

1) $A\omega$ 2) $\frac{\omega}{kg}$ 3) $\frac{\rho A\omega^2}{k^2 g}$ 4) $\frac{k^2 g}{\omega}$

8. **The equatin of a wave on a stirng of linear mass density 0.04 kg m^{-1} is given by**

$y = 0.02\sin\left[2\pi\left(\frac{t}{0.04(s)} - \frac{x}{0.50(m)}\right)\right]$, **The tension in the string is**

1) 6.25N 2) 4.0N 3) 12.5 4) 0.5N

9. **A string of length l is fixed at both ends and is μ. The ratio of magnitude of maximum velocity of particle and the magnitude of maximum acceleration is**

1) $\frac{1}{2\pi}\sqrt{\left(\frac{\mu l^2}{T}\right)}$ 2) $2\pi\sqrt{\left(\frac{\mu l^2}{T}\right)}$ 3) $\frac{1}{2\pi}\sqrt{\left(\frac{T}{\mu l^2}\right)}$ 4) $\frac{1}{4\pi}\sqrt{\left(\frac{\mu l^2}{T}\right)}$

PRINCIPLE OF SUPER POSITION, INTERFERENCE AND STATIONARY WAVES ON STRETCHED STRINGS.

10. **A Sound wave with an amplitude of 3 cm starts towards right from origin and gets reflected at a rigid wall after a second. If the velocity of the wave is 340 ms^{-1} and it has a wavelength of 2 m, the equations of incident and reflected waves respectively are:**
1) $y = 3 \times 10^{-2} \sin \pi (340t - x)$, $y = -3 \times 10^{-2} \sin \pi (340t + x)$ towards left
2) $y = 3 \times 10^{-2} \sin \pi (340t + x)$, $y = -3 \times 10^{-2} \sin \pi (340t + x)$ towards left
3) $y = 3 \times 10^{-2} \sin \pi (340t - x)$, $y = -3 \times 10^{-2} \sin \pi (340t - x)$ towards left
4) $y = 3 \times 10^{-2} \sin \pi (340t - x)$, $y = 3 \times 10^{-2} \sin \pi (340t + x)$ towards left

11. **Sound signal is sent through a composite tube as shown in figure. The radius of the semicircle is r. Speed of sound in air is V. The source of sound is capable to generate frequencies in the range f_1 to f_2 $(f_2 > f_1)$. If n is an integer then frequency for maximum intensity is given by**

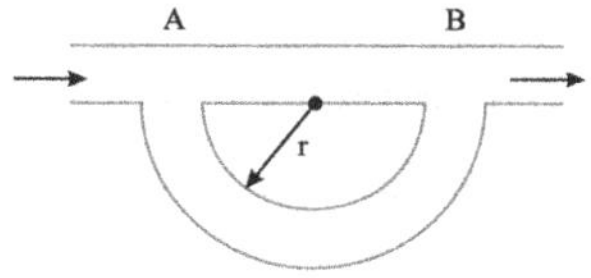

1) $\frac{nV}{r}$ 2) $\frac{nV}{r(\pi-2)}$ 3) $\frac{nV}{\pi r}$ 4) $\frac{nV}{(r-2)\pi}$

12. **Four simple harmonic vibrations.**

$y_1 = 8\cos\omega t$, $y_2 = 4\cos\left(\omega t + \frac{\pi}{2}\right)$

$y_3 = 2\cos(\omega t + \pi)$, $y_4 = \cos\left(\omega t + \frac{3\pi}{2}\right)$

are superimposed on one another. The resulting amplitude and phase are respectively.

1) $\sqrt{45}$ and $\tan^{-1}\left(\frac{1}{2}\right)$ 2) $\sqrt{45}$ and $\tan^{-1}\left(\frac{1}{3}\right)$

3) $\sqrt{75}$ and $\tan^{-1}(2)$ 4) $\sqrt{75}$ and $\tan^{-1}\left(\frac{1}{3}\right)$

13. **The length of a sonometer wire is 90 cm and the stationary wave setup in the wire is represented by an equation $y = 6\sin\left(\frac{\pi x}{30}\right)\cos(250t)$ where x, y are in cm and t is in second. The number of loops is**
1) 1 2) 2 3) 4 4) 3

14. **A sonometer is set on the floor of a lift. When the lift is at rest, the sonometer wire vibrates with fundamental frequency 256 Hz. When the lift goes up with acceleration $a = \frac{9g}{16}$, the frequency of vibration of the same wire changes to**
1) 512 Hz 2) 320 Hz 3) 256 Hz 4) 204 Hz

15. **Standing wave produced in a metal rod of length 1m is represented by the equation $y = 10^{-6}\sin\frac{\pi x}{2}\sin 200\pi t$ where x is in metre and t is in seconds. The maximum tensile stress at the mid-point of the rod is (Young's modulus of material of rod $= 10^{12}\ N/m^2$).**

1) $\frac{\pi}{2}\times 10^6\ N/m^2$ 2) $2\pi\times 10^6\ N/m^2$

3) $\frac{\pi}{2\sqrt{2}}\times 10^6\ N/m^2$ 4) $\frac{2\pi}{\sqrt{3}}\times 10^6\ N/m^2$

16. **An additional bridge is kept below a sonometer wire so that it is divided into two segments of lengths in the ratio 2 : 3 and n_1, n_2 are their respective fundamental frequencies. If the additional bridge is removed then the fundamental frequency of that sonometer wire is n, the ratio of n, n_1, n_2 is**
1) 2 : 3 : 5 2) 2 : 5 : 3 3) 4 : 9 : 25 4) 6 : 15 : 10
17. **A piano wire 0.5m long and mass 5gm is stretched by a tension of 400N. The number of highest overtone that can be heared by a person is**
1) 160 2) 99 3) 140 4) 120
18. **An iron load of 2 kg is suspended in air from the free end of a sonometer wire of length 1m. A tuning fork of frequency 256 Hz, is in resonance with $1/\sqrt{7}$ times the length of the sonometer wire. If the load is immersed in water, the length of the wire in metre that will be in resonance with the same tuning fork is (specific gravity of iron is 8)**
1) $\sqrt{8}$ 2) $\sqrt{6}$ 3) $1/\sqrt{6}$ 4) $1/\sqrt{8}$

SOUND AND VELOCITY OF SOUND

19. **A pressure of 100 kPa causes a decrease in volume of water by 5×10^{-3} percent. The speed of sound in water is**
1) 1414 ms^{-1} 2) 1000 ms^{-1} 3) 2000 ms^{-1} 4) 3000 ms^{-1}
20. **The speed of sound in hydrogen at STP is V. The speed of sound in a mixture containing 3 parts of hydrogen and 2 parts of oxygen at STP will be**
1) V/2 2) $V/\sqrt{5}$ 3) $\sqrt{7}V$ 4) $V/\sqrt{7}$
21. **A mirror of diatomic gases is obtained by mixin m_1 and m_2 masses of two gases, with velocities of sound in them c_1 and c_2 respectively. Determine the velocity of sound in the mixture of gases.**
1) $c=\sqrt{\dfrac{m_1c_1^2+m_2c_2^2}{m_1+m_2}}$ 2) $c=\sqrt{\dfrac{m_2c_1^2+m_1c_2^2}{m_1+m_2}}$
3) $c=\sqrt{\dfrac{m_2c_2+m_1c_2}{m_1+m_2}}$ 4) $c=m_2\sqrt{\dfrac{c_2^2+c_1^2}{m_1+m_2}}$

ORGAN PIPES

22. **A tube of certain diameter and of length 48cm is open at both ends. Its fundamental frequency of resonance is found to be 320Hz. If velocity of sound in air is 320ms^{-1} the diameter of the tube is**
1) 1.33cm 2) 2.33cm 3) 3.33cm 4) 4.33cm
23. **A closed organ pipe has length l. The air in it is vibrating in 3rd overtone with a maximum amplitude of A. Find the amplitude at a distance of $l/14$ from closed end of the pipe**
1) A 2) zero 3) $A/\sqrt{2}$ 4) $\sqrt{3}/2A$
24. **The freqency of a stretched uniform wire of certain length is in resonance with the fundamental frequency of closed tube. If length of wire is decreased by 0.5m, it is in resonance with first overtone of closed pipe. The initial length of wire is**
1) 0.5m 2) 0.75m 3) 1m 4) 1.5m
25. **An open pipe resonates to a frequency f_1 and a closed pipe resonates to a frequency f_2. If they are joined together to form a longer tube, then it will resonate to a frequency of (neglect end corrections)**
1) $\dfrac{f_1f_2}{2f_2+f_1}$ 2) $\dfrac{f_1f_2}{f_2+2f_1}$ 3) $\dfrac{2f_1f_2}{f_2+f_1}$ 4) $\dfrac{f_1+2f_2}{f_1f_2}$
26. **In a resonace air column experiment, first and second resonances are obtained at lengths of air columns l_1 and l_2, the third resonance will be obtained at a length of**
1) $2l_2-l_1$ 2) l_2-2l_1 3) l_2-l_1 4) $3l_2-l_1$

27. A pop-gun consits of a cylindrical barrel 3cm^2 in cross section closed at one end by a cork and having a well fitting piston at the other. If the piston is pushed slowly in, the cork is finally ejected, giving a pop, the frequency of which is found to be 512 Hz. Assuming that the initial distance between the cork and the piston was 25 cm and that there is no leaking of air, calculate the force required to eject the cork. Atmospheric pressure = 1kg.cm^2, v = 340 m/s (in kg.wt)

1) 1.5 2) 3 3) 6 4) 8

BEATS

28. A closed organ pipe and an open pipe of the same length produce 4 beats when they are set into vibrations simultaneously. If he length of each of them were twice their initial lengths, the number of beats produced will be

1) 2 2) 4 3) 1 4) 8

29. An air column in a tube 32 cm long, closed at one end, is in resonance with a tuning fork. The air column in another tube, open at both ends, of length 66 cm is in resonance with another tuning fork. When these two tuning forks are sounded together, they produce 8 beats per second. Then the frequencies of the two tuning forks are, (Consider fundamental frequencies only).

1) 250 Hz, 258 Hz 2) 240 Hz, 248 Hz 3) 264 Hz, 256 Hz 4) 280 Hz, 272 Hz

30. The string of a sonometer is divided into two parts using wedge. Total length of string is 1m and two parts differ by 2mm. When sounded together they produce 2 beats/sec. The freqencies of two parts are

1) 501Hz, 503Hz 2) 501Hz, 499Hz 3) 499Hz, 497Hz 4) 497Hz, 495Hz

31. The fundamental frequency of a sonometer wire of length ℓ is f_0. A bridge is now introduced at a distance of $\Delta\ell$ from the centre of the wire ($\Delta\ell << \ell$). The number of beats heard per second if both sides of the bridge are set to vibrate in their fundamental mode is

1) $\frac{8f_0\Delta\ell}{\ell}$ 2) $\frac{f_0\Delta\ell}{\ell}$ 3) $\frac{2f_0\Delta\ell}{\ell}$ 4) $\frac{4f_0\Delta\ell}{\ell}$

32. On vibrating an air column at 627^0 C and a tuning fork simultaneously, 6beats/sec are heard. The freqency of fork is less than that of air column. No beats are heard at -48^0 C. The freqency of fork is

1) 3Hz 2) 6Hz 3) 10Hz 4) 15Hz

33. A string 25cm long, having a mass 2.5gm is under tension. A pipe closed at one end is 40cm long. When the string is set vibrating in is in its fundamental frequency, 8 beats are heard per second. It is observed that decreasing the tension in the string decreases the beat frequency. If the speed of sound in air is 320ms^{-1}, the tension in the string is nearly

1) 27N 2) 54N 3) 13.5N 4) 108N

34 .Two identical piano wires have a fundamental frequency of 600 cycle per second when kept under the same tension. What fractional increase in the tension of one wires will lead to the occurrence of 6 beats per second when both wires vibrate simultaneously?

1) 0.01 2) 0.02 3) 0.03 4) 0.04

DOPPLER EFFECT

35. One train is approaching an observer at rest and another is receding him with same velocity 4 m/s. Both the trains blow whistles of same frequency of 243 Hz. The beat frequency in Hz as heard by the observer is: (Speed of sound in air =320 m/s)

1) 10 2) 6 3) 4 4) 1

36. A tuning fork of frequency 328 Hz is moved towards a wall at a speed of 2 ms^{-1}. An observer standing on the same side as the fork hears two sounds, one directly from the fork and the other reflected from the wall. Number of beats per second is (Velocity of sound in air 330 ms^{-1}).

1) 4 2) 5 3) 6 4) 7

37. The frequency of the sound of a car horn as recorded by an observer towards whom the car is moving differs from the frequency of the horn by 10%. Assuming the velocity of sound in air to be 330ms^{-1}, the velocity of the car is

1) 36.7ms^{-1} 2) 40 ms^{-1} 3) 30 ms^{-1} 4)33 ms^{-1}

38. **Two trains are approaching each other on parallel tracks with same velocity. The whistle sound produced by one train is heard by a passenger in another train. If actual frequency of whistle is 620Hz and apparent increase in its frequency is 100Hz, the velocity of one of the two trains is (Velocity of sound in air = 335 ms^{-1})**
1) 90kmph 2) 72 kmph 3) 54kmph 4) 36 kmph

39. **A girl swings in a cradle with period $\pi/4$ second and amplitude 2m. A boy standing infront of it blows a whistle of natural frequency 1000 Hz. The minimum frequency as heard by the girl is (Velocity of sound in air is 320 ms^{-1})**
1) 850 Hz 2) 1000 Hz 3) 750 Hz 4) 950 Hz

40. **The difference between apparent freqencies of a source of sound as percieved by a stationary observer during its approach and recession is 2% of the actual freqency. The speed of source is (V = 340m/sec.)**
1) 12m/s 2) 6.2m/s 3) 3.4m/s 4) 1.5m/s

41. **Two different sound sources $S_1 and\ S_2$ have freqencies in the ratio 1:2. Source S_1 is approaching towards observer and S_2 receding from same observer. Speeds of both $S_1 and\ S_2$ are V each and speed of sound air is 330 m/s. If no beats are heard by the observer then the value of V is**
1) 50 m/s 2) 75 m/s 3) 110 m/s 4) 125m/s

42. **A stationary source emitting sound of frequency 680Hz is at the origin. An observer is moving with the velocity $\sqrt{2}\left(\bar{i}+\bar{j}\right)$m/s at a certain instant. If the speed of sound in air is 340 m/s, then the apparent frequency received by him at that instant is**
1) 680Hz 2) 676Hz 3) 684Hz 4) either 676Hz or 684Hz

43. **A source S emitting sound of frequency 300Hz is fixed on block A which is attached to the free end of a spring S_A as shown in figure. The dectector D fixed on block B attached to free end of spring S_B detects this sound. The blocks A and B are simultaneously displaced towards each other through a distance of 1.0m and then left to vibrate. The maximum and minimum frequencies of sound detected by D, if the vibrational frequency of each block is 2Hz are (Velocity of sound v = 340m/s)**

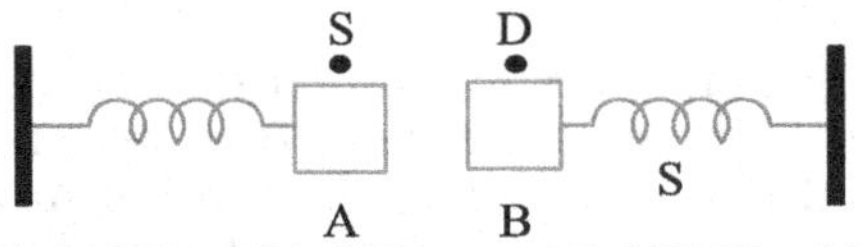

1) 378.6Hz, 223 Hz 2) 323Hz, 278.6 Hz 3) 178 Hz, 276 Hz 4) 420Hz, 220 Hz

44. **A locomotive approaching a crossing at a speed of 80 mile/hr. sounds a whistle of frequency 400 Hz when 1 mile from the crossing. There is no wind, and the speed of sound in air is 0.200mile/s. What frequency is heard by an observer 0.60 miles from the crossing on the straight road which crosses the railroad at right angles?**
1) 440Hz 2) 442Hz 3) 444Hz 4) 446Hz

45. **An observer is standing 500 m away from a verticall hill. Starting between the observer and the hill, a police van sounding a siren of frequency 1000 Hz moves towards the hill with a uniform speed. If the frequency of the sound heard directly from the siren is 970 Hz, the frequency of the sound heard after reflection from a hill (in Hz) is about (velocity of sound = 300 ms^{-1})**
1) 1042 2) 1032 3) 1022 4) 1012

LEVEL - II (C. W) - KEY

1) 3	2) 4	3) 4	4) 2	5) 2	6) 3	7) 3	8) 1	9) 1	10) 1
11) 2	12) 1	13) 4	14) 2	15) 3	16)4	17)2	18)4	19) 1	20) 4
21) 1	22) 3	23) 3	24) 2	25) 1	26) 1	27) 1	28) 1	29)3	30) 2
31) 1	32) 2	33) 4	34) 2	35) 2	36) 1	37) 3	38) 1	39) 4	40) 3
41) 3	42) 4	43) 2	44) 3	45) 2					

LEVEL - II (C. W) - HINTS

1. $V = \frac{\omega}{K}, (V_\rho)_{max} = A\omega, \quad \lambda = \frac{2\pi}{K} = 2\pi m$

2. In 3 sec, pulse reaches free end '0'
At free end crest reflects as crest
$\therefore$ net amplitude $= 1 + 1 = 2cm$

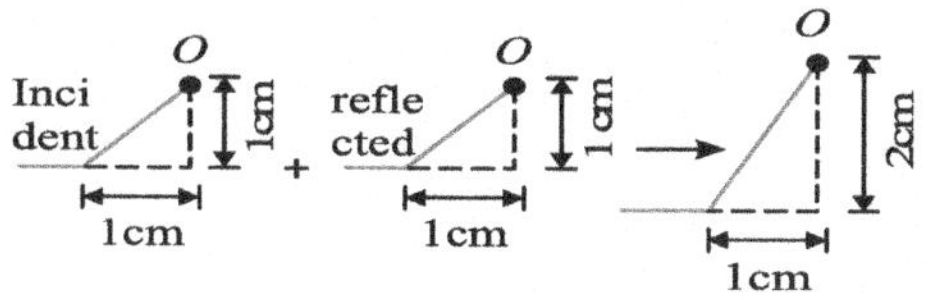

3. $y_1 = \sin \omega t + \cos \omega t = \sqrt{2} \sin\left(\omega t + \frac{\pi}{4}\right)$

$y_2 = \frac{\sqrt{3}}{2} \sin \omega t + \frac{1}{2} \cos \omega t = \sin\left(\omega t + \frac{\pi}{6}\right)$

$\frac{A_1}{A_2} = \frac{\sqrt{2}}{1}$

4. $\bar{k} = k_x \hat{i} + k_y \hat{j} + k_z \hat{k} = \frac{\omega}{c_1} \hat{i} + \frac{\omega}{c_2} \hat{j} + \frac{\omega}{c_3} \hat{k}$

5. Tension at the bottom =2kg.wt=T_b

Tension at the top =(2+6)=8 kgwt=T_t

$\frac{V_t}{V_b} = \frac{n\lambda_t}{n\lambda_b} = \sqrt{\frac{T_t}{T_b}}$

6. $v = 2\sqrt{\frac{x}{g}}$

7. $V^2 = \frac{Mg}{A\rho}; \frac{\omega^2}{K^2} = \frac{Mg}{A\rho}$

8. $T = V^2 \mu = \left(\frac{\omega}{k}\right)^2 \mu$

9. $\frac{(V_\rho)_{max}}{(a_\rho)_{max}} = \frac{A\omega}{A\omega^2} = \frac{1}{\omega} = \frac{1}{kv} = \frac{\lambda}{2\pi}\sqrt{\frac{\mu}{T}}$

here $\lambda = l$ for 2nd harmonic

10. $y_i = -A \sin(\omega t - kx)$

$K = \frac{2\pi}{\lambda} = \frac{2\pi}{2} = \pi m^{-1}, \ \omega = kv$

$y_i = 0.03 \sin(340\pi t - \pi x)$

As it is reflected $y_r = A \sin(wt + kx + \pi)$

11. Path differ $\Delta x = \pi r - 2r = (\pi - 2) r$

For nth maximum $\Delta x = n\lambda$ and $\lambda = v / f$

12. Resultant displacement along x-axis is

$x = y_1 - y_3 = 8 - 2 = 6$

Resultant displacement along y-axis is

$y = y_2 - y_4 = 4 - 1 = 3$

$A = \sqrt{x^2 + y^2}$ and $\tan\theta = \frac{y}{x}$

13. By comparing with $y = 2A\sin(kx)\cos(250t)$

$K = \frac{\pi}{30} \Rightarrow \lambda = \frac{2\pi}{k}, l = 90\text{cm}, l = \frac{p\lambda}{2}$

14. $n = \frac{1}{2\lambda}\sqrt{\frac{Mg}{\mu}} \Rightarrow n \propto \sqrt{g} \Rightarrow \frac{n'}{n} = \sqrt{\frac{g+a}{g}}$

15. Tensile stress $= Y \times \textit{tensile strain} = Y \times \left(\frac{dy}{dx}\right)$

Max tensile stress $= Y \times \left(\frac{dy}{dx}\right)_{\max}$

16. $n \alpha 1/l$

17. $20000 = x(n)$; calculate $(x-1)$

18. We know, $l \propto \sqrt{T}$

$\therefore \frac{l_{air}}{l_{water}} = \sqrt{\frac{T_{air}}{T_{water}}}$

19. $B = \frac{P}{\left(\frac{\Delta V}{V}\right)}, V = \sqrt{\frac{B}{\rho}}$

20. $\rho_{mix} = \frac{\rho_1 V_1 + \rho_2 V_2}{V_1 + V_2}, \frac{\rho_1}{\rho_2} = \frac{M_1}{M_2} = \frac{2}{32}$

$V \propto \frac{1}{\sqrt{\rho}} \Rightarrow \frac{Vmix}{V} = \sqrt{\frac{\rho}{\rho_{mix}}} = \sqrt{\frac{\rho_1}{7\rho_1}}$

21. Total no.of moles in mixture is $n = n_1 + n_2$

$\frac{m_1 + m_2}{M_{mix}} = \frac{m_1}{M_1} + \frac{m_2}{M_2}$

Here velocity of sound $C = \sqrt{\frac{\gamma P}{\rho}} = \sqrt{\frac{\gamma RT}{M}}$

$\Rightarrow M \propto \frac{1}{c^2} \Rightarrow M = \frac{K}{c^2}$

22. $\frac{\lambda}{2} = 48 + 2(0.3d)$ where d is diameter of the tube, but $\lambda = \frac{v}{n}$

23. $A' = A\cos kx$

$l = (2p+1)\frac{\lambda}{4} = \frac{7\lambda}{4} \Rightarrow \lambda = \frac{4l}{7}, K = \frac{2\pi}{\lambda}$

Calculate A' at $x = \frac{l}{14}$

24. $l_2 = l_1 - 0.5m$

$n_1 = \frac{V}{4L_1}, n_2 = \frac{3V}{4L} \Rightarrow \frac{n_1}{n_2} = \frac{1}{3}$........(1)

As $n \propto \frac{1}{l} \Rightarrow \frac{n_1}{n_2} = \frac{l_2}{l_1}$

25. 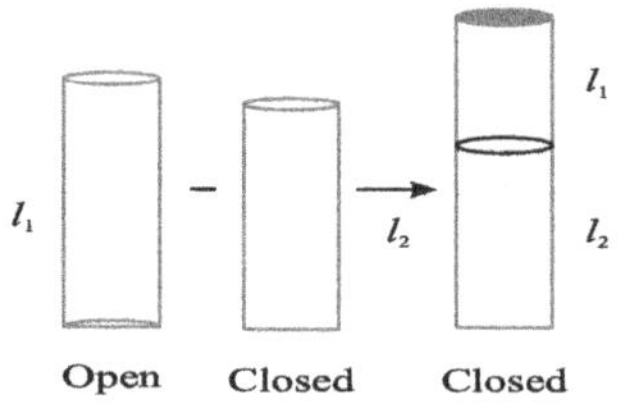

$\therefore l = l_1 + l_2 = \frac{V}{4f}$

26. $l_1 = \frac{\lambda}{4} + e,\ l_2 = \frac{3\lambda}{4} + e,\ l_3 = \frac{5\lambda}{4} + e$

$\Rightarrow l_3 - l_2 = l_2 - l_1$

27. $n = \frac{V}{4l'}$ 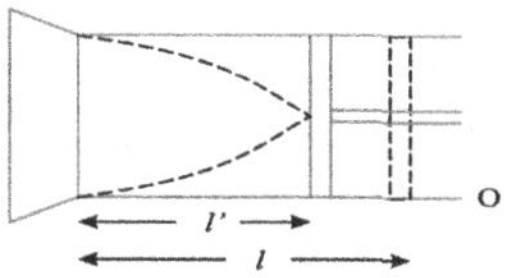

$l' = \frac{V}{4n}$ and $P'l = P'l' \Rightarrow P' = \frac{Pl}{l'}\ \therefore F = P'A$

28. $\Delta n = n_0 - n_C = \frac{V}{2l} - \frac{V}{4l} = \frac{V}{4l}$

$\Delta n \propto \frac{1}{l} \Rightarrow \frac{\Delta n'}{\Delta n} = \frac{l}{l'}$

29) $n_1 = \frac{v}{4l_1}$ and $n_2 = \frac{v}{2l_2}$ given $l_1 = 32$ cm,

$l_2 = 66$cm and $n_1 - n_2 = 8$ heat/s

30. $l_1 + l_2 = 100$cm and $l_2 - l_1 = 0.2$cm Adding $2l_2$=100.2 $\Rightarrow l_2 = 50.1cm$ and $l_1 = 49.9$cm

$n \propto \frac{1}{l} \Rightarrow \frac{n_1}{n_2} = \frac{l_2}{l_1}, \Delta n = n_1 - n_2$

31. 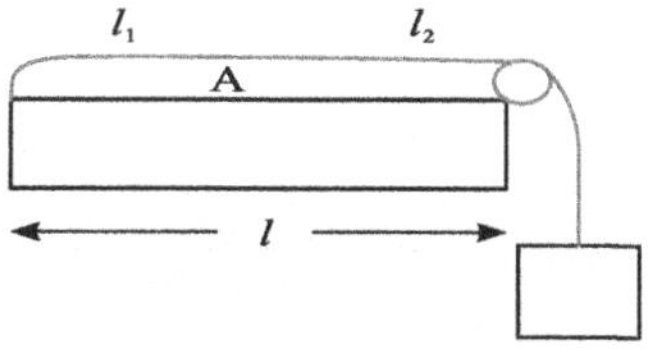

$n = \frac{1}{2l}\sqrt{\frac{T}{\mu}} = f_0$.....(1)

$l_1 = \frac{l}{2} - \Delta l, \qquad l_2 = \frac{l}{2} + \Delta l$

$\Delta n = n_1 - n_2$ and Δl^2 can be neglected

32. $n_1 - n = 6$ beats at $T_1 = 627 + 273 = 900K$
$n_2 = n$ at $T_2 = -48 + 273 = 225$ K

$$n \propto V \propto \sqrt{T} \Rightarrow \frac{n_1}{n_2} = \sqrt{\frac{T_1}{T_2}}$$

33. $n_S = \frac{1}{2l_S}\sqrt{\frac{T}{\mu}},\ n_C = \frac{V}{4L_C} = \frac{320}{4\times 40\times 10^{-2}}$

$\Rightarrow n_C = 200Hz$, $\Delta n = n_S \sim n_C$, $\Delta n' < \Delta n$ as $n \propto \sqrt{T} \Rightarrow n' < n$ as $T' < T$ this is possible when $n_S - n_C = \Delta n$, $\quad n_S = 200 + 8 = 208Hz$

34. $n \propto \sqrt{T}$ and $n' - n = 6$

35. $V_{S_1} = V_{S_2} = 4m/s,\ n = 243Hz,\ \Delta n \approx \frac{2nV_S}{V}$

36 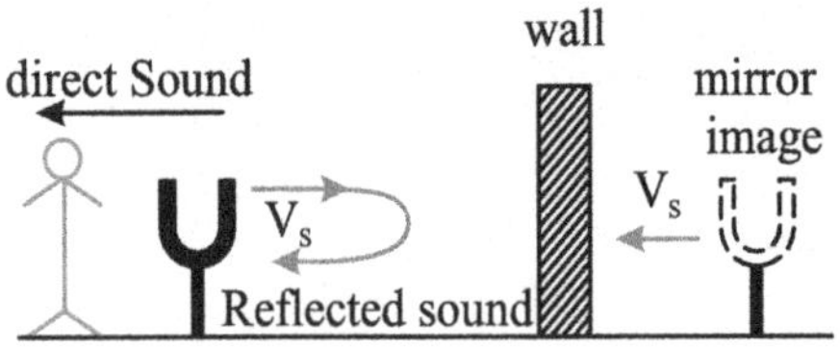

$$n'_{direct} = \left(\frac{V}{V+V_S}\right)n$$

The reflected sound appears to be coming from the mirror image of fork which moves towards observer

$$n'_{reflected} = \left(\frac{V}{V-V_S}\right)n$$

$$\Delta n = n'_{reflected} - n_{direct} \quad \Delta n = \frac{2nV_S}{V-V_S}$$

37. Source approaching stationary observer

$$n' = \frac{110}{100}n = n\left(\frac{V}{V-V_S}\right)$$

38. $n' = n\left(\frac{V+V_0}{V-V_S}\right)$

39. $V_0 = V_g = \omega A = \frac{2\pi A}{T} = \frac{2\pi\times 2}{\pi/4} = 16m/s$

$V_s = V_b = 0 \qquad n_{\min} = \left(\frac{V-V_0}{V}\right)n$

40. $\Delta n \approx \frac{2nV_S}{V} \Rightarrow \frac{\Delta n}{n} \approx \frac{2V_S}{V}$

41. 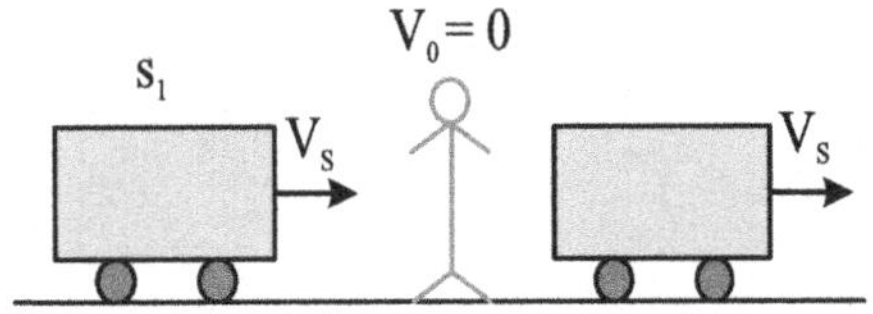

$$n_1^1 = n_1\left[\frac{V}{V-V_S}\right], \qquad n_2^1 = n_2\left[\frac{V}{V+V_S}\right]$$

$\Delta n = 0 \Rightarrow n_1^1 = n_2^1$

42. 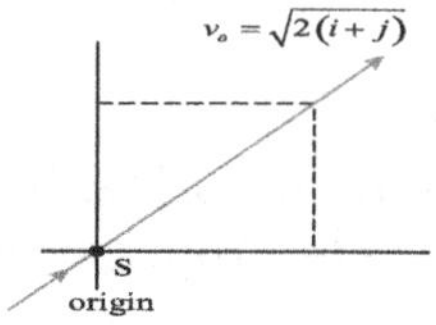

$V_0 = \sqrt{2}\sqrt{1^2+1^2} = 2m/s$ moving away from source or towards the source.

$\therefore n' = n\left(\dfrac{V \pm V_0}{V}\right)$

43. n = 300Hz ; $V_S = V_D = A\omega$

$n'_{max} = n\left(\dfrac{V+V_0}{V-V_S}\right)$, $n'_{min} = n\left(\dfrac{V-V_0}{V+V_S}\right)$

44. 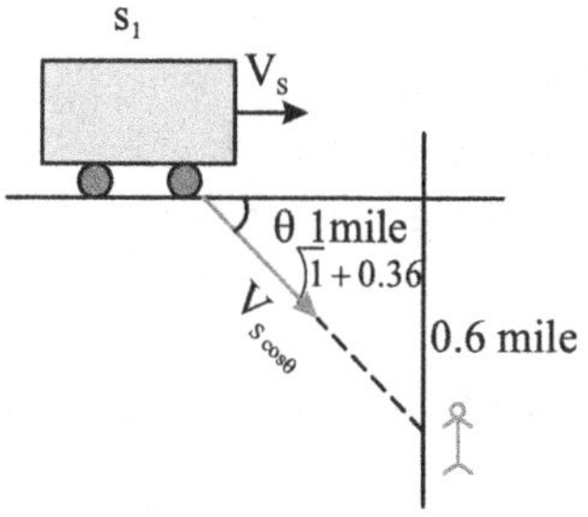

$V_S = \dfrac{80}{60\times160} mile/\sec$, $\cos\theta = \dfrac{1}{\sqrt{1.36}} = \dfrac{1}{1.166}$, $n' = n\left[\dfrac{V}{V - V_S\cos\theta}\right]$

45. $f_1 = f_0\left(\dfrac{v}{v+v_s}\right)$ and $f_2 = f_0\left(\dfrac{v}{v-v_s}\right)$

LEVEL - II (HW)

CHARACTERISTICS OF PROGRESSIVE WAVE

1. **The equation of a progressive wave is** $y = 0.05\sin\left(200t - \dfrac{x}{2}\right)$ **where x,y are in metres and t in seconds, then**

a) velocity of wave is 100 ms^{-1}

b) maximum velocity of particle in the wave is 10 ms^{-1} c) wavelength of wave is 4π m

1) only a and c are true 2) only b and c are true

3) only a and b are true 4) a,b,c are true

2. **A wave pulse on a string has the dimension at time** $t = 3s$ **is**

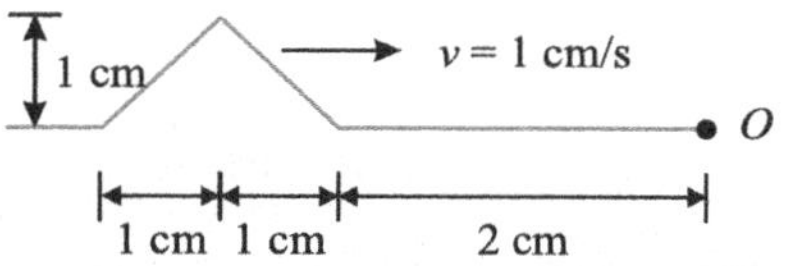

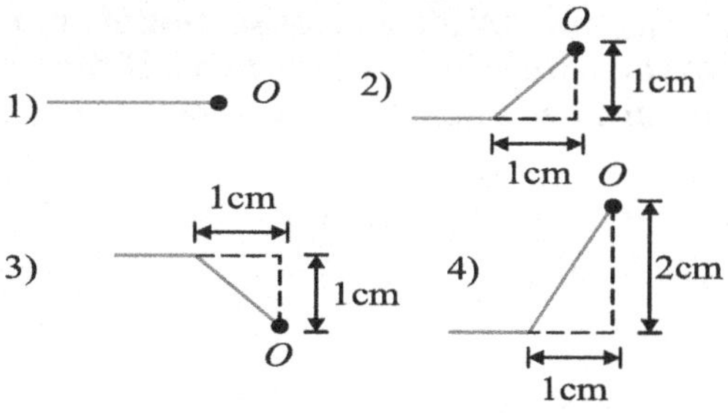

3. **Two sound waves are represented by $y_1 = \frac{1}{2}\sin\omega t + \frac{\sqrt{3}}{2}\cos\omega t$ and $y_2 = \frac{\sqrt{3}}{2}\sin\omega t + \frac{1}{2}\cos\omega t$. The initial phase difference between the two waves is**
1) 30° 2) 60° 3) 45° 4) 0°

4. **A wave of angular frequency 30 rad/sec propagates so that a certain phase of oscillation moves along x-axis,y-axis,z-axis with speeds 1m/s,2m/sand 2m/s respectively. The propagation constant K is**
1) $30\hat{i}+15\hat{j}+15\hat{k}$ 2) $10\hat{i}+10\hat{j}+10\hat{k}$
3) $30\hat{i}+30\hat{j}+30\hat{k}$ 4) $6\hat{i}+6\hat{j}+6\hat{k}$

SPEED OF A TRAVELLING WAVE

5. **A uniform rope of length 20m and mass 5kg is hanging vertically from a rigid support. A block of mass 4 kg is attached to the free end. The wave length of the transverse wave pulse at the lower end of the rope is 0.04 m. The wavelength of the same pulse as it reaches the top is**
1) 0.06 m 2) 0.12 m 3) 1.5 m 4) 2.2 m

6. **A uniform rope of mass 0.1 kg and length 2.45 m hangs from a ceiling**
i) The speed of transverse wave in the rope at a point 0.05 m distant from the lower end
ii) The time taken by a transverse wave to travel the full length of the rope are($g = 9.8\ m/s^2$)
1) 0.7 m/s, 1 s 2) 0.7 m/s, 2 s 3) 0.7 m/s, 4 s 4) 0.7 m/s, 6 s

7. **A travelling wave pulse is given by $y = \frac{10}{5+(x+2t)^2}$. The amplitude and velocity of the pulse propagating are**
1) 2 units, -2 units 2) 2 units, 2 units 3) 10 units, 5 units 4) 10 units, 10 units

8. **In the given arrangement, if hanging mass will be changed by 4%, then percentage change in the wave speed in string will be**

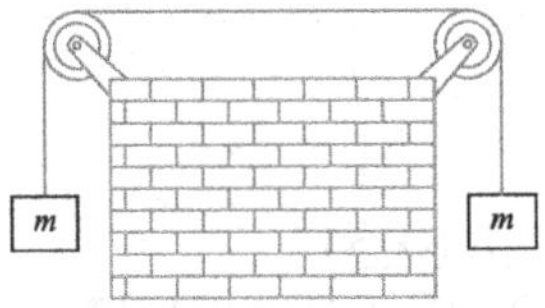

1) 2% 2) 8% 3) 3% 4) 4%

PRINCIPLE OF SUPER POSITION, INTERFERENCE AND STATIONARY WAVES ON STRETCHED STRINGS.

9. A string of length l **along x-axis is fixed at both ends and is vibrating in second harmonic. If at t = 0, y = 2.5mm for incident wave, the equation of standing wave is (T is tension and μ is linear density)**
1) $(2.5mm)\sin\left(\frac{2\pi}{l}x\right)\cos\left(2\pi\sqrt{\left(\frac{T}{\mu l^2}\right)}t\right)$ 2) $(5mm)\sin\left(\frac{\pi}{l}x\right)cos2\pi t$
3) $(5mm)\sin\left(\frac{2\pi}{l}x\right)\cos\left(2\pi\sqrt{\left(\frac{T}{\mu l^2}\right)}t\right)$ 4) $(5mm)\cos\left(\frac{2\pi}{l}x\right)\cos\left(2\pi\sqrt{\left(\frac{T}{\mu l^2}\right)}t\right)$

10. **A steel wire of length 1m, mass 0.1kg and uniform cross sectional area 10^{-6} m² is rigidly fixed at both ends. The temperature of the wire is lowered by 20°C. If the transverse waves are set up by plucking the string in the middle. Then the frequency of the fundamental mode of vibration is ($Y = 2\times10^{11}$ N/m², $\alpha = 1.21\times10^{-5}\ /°C$)**
1) 21Hz 2) 42Hz 3) 11Hz 4) 22Hz

11. **Two stereo speakers are separated by a distance of 2.4 m. A person stands at a distance of 3.2 m as shown directly in front of one of the speakers. The frequencies in audible range for which the listener will hear a minimum sound intensity are (Speed of the sound in air is 320 ms^{-1} and n is order of minimum)**

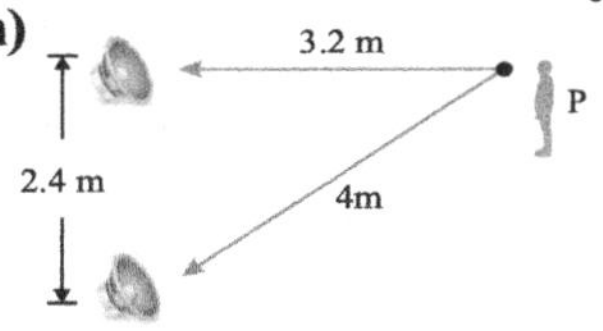

1) $160(2n+1)$ 2) $320(2n+1)$ 3) $200(2n+1)$ 4) $100(2n+1)$

12. **A sound wave of wavelength 32cm enters the tube as shown in the figure. Then the smallest radius 'r' so that a maximum of sound is heard at detector is**

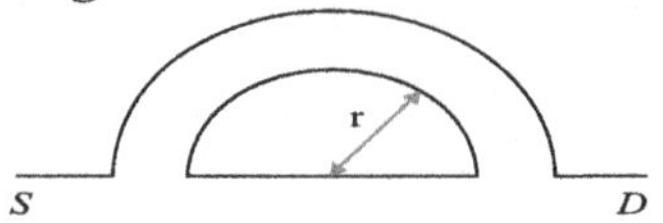

1) 7 cm 2) 14 cm
3) 21 cm 4) 28 cm

13. **A sonometer wire, with a suspended mass of M=1 kg, is in resonance with a given tuning fork. The apparatus is taken to the moon where the acceleration due to gravity is 1/6 that on earth. To obtain resonance on the moon, the value of M should be**

1) 1 kg 2) $\sqrt{6}$ kg 3) 6 kg 4) 36 kg

14. **A sonometer wire of length L is plucked at a distance L/8 from one end then it vibrates with a minimum frequency n. If the same wire plucked at a distance L/6 from another end the minimum frequency with which it vibrates is**

1) $\frac{\sqrt{3}}{2}n$ 2) $\frac{3}{2}n$ 3) $\frac{3n}{4}$ 4) $\frac{4n}{3}$

15. **A metal wire of linear mass density of 9.8 g/m is stretched with a tension of $10\ kgwt$ between two rigid supports 1m apart. The wire passes at its middle point between the poles of a permanent magnet and it vibrates in resonance when carrying an alternating current of frequency n. The frequency n of the alternating source is**

1) 50 Hz 2) 100 Hz 3) 200 Hz 4) 25 Hz

16. **A stretched wire of length 114 cm is divided into three segments whose frequencies are in the ratio 1 : 3 : 4, the lengths of the segments must be in the ratio**

1) 18 : 24 : 72 2) 24 : 72 : ;18 3) 24 : 18 : 72 4) 72 : 24 : 18

17. **If n_1, n_2 and n_3 are the fundamental frequencies of three segments into which a string is divided, then the original fundamental frequency n of the string with the same tension is given by**

1) $\frac{1}{n} = \frac{1}{n_1} + \frac{1}{n_2} + \frac{1}{n_3}$ 2) $\frac{1}{\sqrt{n}} = \frac{1}{\sqrt{n_1}} + \frac{1}{\sqrt{n_2}} + \frac{1}{\sqrt{n_3}}$

3) $\sqrt{n} = \sqrt{n_1} + \sqrt{n_2} + \sqrt{n_3}$ 4) $n = n_1 + n_2 + n_3$

SOUND AND VELOCITY OF SOUND

18: **The speed of sound in oxygen (O_2) at a certain temperature is 460 ms^{-1}. The speed of sound in helium (He) at the same temperature will be (assume both gases to be ideal)**

1)460 ms^{-1} 2)500 ms^{-1} 3)650 ms^{-1} 4)1420 ms^{-1}

19. **A pressure of P causes a decrease in volume of water by a fraction 'Q'. The speed of sound in water is (in CGS)**

1) $\sqrt{\frac{P}{Q}}$ 2) $\sqrt{PQ}$ 3) $\frac{P}{Q}$ 4) PQ

20. **Velocity of hydrogen at NTP is V. The velocity of sound in a mixture of hydrogen and oxygen in the ratio of 4:1 at NTP is**

1) $\frac{1}{5}V$ 2) $\frac{1}{4}V$ 3) $\frac{1}{3}V$ 4) $\frac{1}{2}V$

21. **A mixture of two diatomic gases exists in a closed cylinder. The volumes and velocities in the two gases are V_1, V_2 and c_1, c_2 respectively. Determine the velocities of the gases in the mixture.**

1) $C_1C_2\sqrt{\dfrac{V_1+V_2}{V_1c_2^2+V_2c_1^2}}$ 2) $C_1C_2\sqrt{\dfrac{V_1+V_2}{V_1c_1^2+V_2c_2^2}}$

3) $C_1C_2\sqrt{\dfrac{V_1+V_2}{V_1^2c_1+V_2c_2^2}}$ 4) $C_1C_2\sqrt{\dfrac{V_1+V_2}{V_1^2c_1+V_2^2c_2}}$

ORGAN PIPES

22. **The vibrating of four air columns are represented in the figure. The ratio of frequencies $n_p : n_q : n_r : n_s$ is**

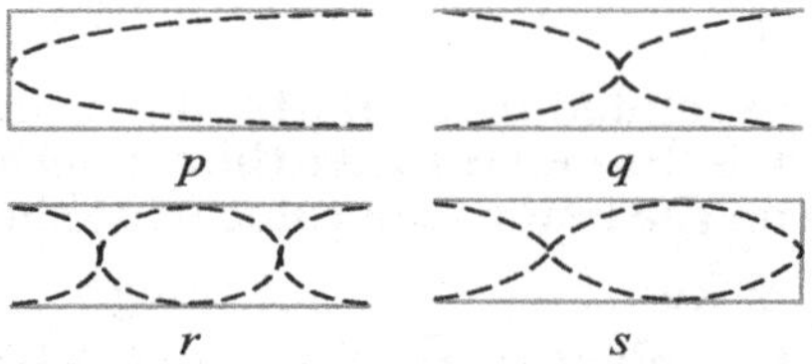

1) 12 : 6 : 3 : 5 2) 1 : 2 : 4 : 3 3) 4 : 2 : 3 : 1 4) 6 : 2 : 3 : 4

23. **An open pipe of length 24 cm is in resonance with a frequency 660 Hz in fundamental mode. The radius of pipe is ($V = 330\ ms^{-1}$)**

1) 3 cm 2) 0.83 cm 3) 3.5 cm 4) 2cm

24. An open organ pipe has length l. **The air in it vibrating in 3rd overtone with maximum amplitude A. The amplitude at a distance of $l/16$ from any open end is.**

1) A 2) Zero 3) $\dfrac{A}{\sqrt{2}}$ 4) $\dfrac{\sqrt{3}A}{2}$

25. **The frequency of a stretched uniform wire under tension is in resonance with the fundamental frequency of a closed tube. If the tension in the wire is increased by 8N, it is in resonance with first overtone of the closed tube. The initial tension in the wire is**

1) 16 N 2) 8 N 3) 4 N 4) 1 N

26. **An open pipe resonates with frequency 100Hz and a closed pipe resonates with frequency 50Hz. If they are joined to form a longer tube then it will resonate with freqency of (neglect end corrections)**

1) 25Hz 2) 50Hz 3) 75Hz 4) 100Hz

27. **In a resonance column, first and second resonance are obtained at depths 22.7 cm and 70.2 cm, the third resonance will be obtained at a depth of**

1)117.7 cm 2)92.9 cm 3)115.5 cm 4)113.5 cm

28. **A 'pop' gun consists a tube 25cm long closed at one end by a cork and the other end by a tightly fitted piston as shown. The piston is pushed slowly in. When the pressure rises to one and half times the atmospheric pressure, the cork is violently blow out. The frequency of the'pop'caused by its ejection is ($V = 340 m/s$)**

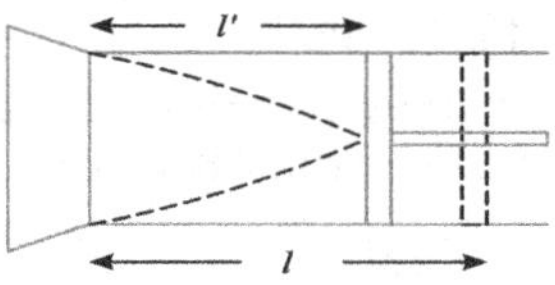

1) 510Hz 2) 1020Hz

3) 205Hz 4) 740Hz

BEATS

29. **A closed pipe and an open pipe of same length produce 2 beates, when they are set into vibration simultaneously in thier fundamental mode. If the length of the open pipe is halved, and that of closed pipe is doubled, and if they are vibrating in the fundamental mode, then the number of beats produced is**

1) 8 2) 4 3) 7 4) 2

30. **A closed pipe is suddenly opened and changed to an open pipe of same length. The fundamental frequency of the resulting open pipe is less than that of 3rd hoarmonic of the earlier closed pipe by 55Hz. Then, the value of fundamental frequency of the closed pipe is**

1) 165Hz 2) 100Hz 3) 55Hz 4) 220Hz

31. **A fork gives 5 beats/s with a 40 cm long sonometer wire. If the length of the wire is shortened by 1 cm, the number of beats/s is still the same. The frequency of the fork is**

1) 385 Hz 2) 320 Hz 3) 395 Hz 4) 400 Hz

32. **Two tuning forks A and B are sounded together and 7 beats/s are heard. A is in resonance with a 32 cm air column closed at one end and B is in resonance when length of air column is increased by 1 cm. The frequencies of forks A and B are**

1) 264 Hz, 256 Hz 2) 272 Hz, 264 Hz 3) 231 Hz, 224 Hz 4) 220 Hz, 512 Hz

33. **An organ pipe opened from both ends produces 5 beats per second when vibrated with a source of frequency 200Hz. The second harmonic of the same pipe produces 10 beats per second with a source of frequency 420Hz. The fundamental frequency of organ pipe is**

1) 195Hz 2) 205Hz 3) 190Hz 4) 210Hz

34. **When a vibrating tuning fork is placed on a sound box of a sonometer, 8 beats per second are heard when the length of the sonometer wire is kept at 101 cm or 100 cm. Then the frequency of the tuning fork is (consider that the tension in the wire is kept constant)**

1) 1616 Hz 2) 1608 Hz 3) 1632 Hz 4) 1600 Hz

35. **Two parts of a sonometer wire divided by a movable bridge differ in length by 0.2 cm and produce one beat per second, when sounded together. The total length of wire is 1m, then the frequencies are**

1) 250.5 and 249.5 Hz 2) 230.5 and 229.5 Hz
3) 220.5 and 219.5 Hz 4) 210.5 and 209.5 Hz

36. **On vibrating an air column at 27°C and a tuning fork simultaneously, 5 beats per second are produced. The frequency of the fork is less than that of air column. No beats are heard at -3°C. The frequency of the fork is**

1) 70 Hz 2) 147 Hz 3) 104 Hz 4) 92 Hz

37. **The wavelength of two sound notes in air are 40/195m and 40/193m. Each note produces 9 beats per second, separately with a third note of fixed frequency. The velocity of sound in air in m/s is**

1) 360 2) 320 3) 300 4) 340

DOPPLER EFFECT

38. **The velocity of a listener who is moving away from a stationary source of sound such that the listener notices 5% apparent decrease in frequency of sound is(Velocity of sound in air = 340 m/s)**

1) 12.5 ms^{-1} 2) 17 ms^{-1} 3) 25 ms^{-1} 4) 34 ms^{-1}

39. **Two trains are moving towards each other on parallel tracks at speeds of 144 kmph and 54 kmph. The first train sounds a whistle of freqency 600Hz. Freqency of the whistle as heard by a passenger in the second train is (V=340m/s)**

1) 510Hz 2) 610Hz 3) 710Hz 4) 810Hz

40. **A boy sitting on a swing which is moving to an angle of 30^0 from the vertical is blowing a whistle which is of frequency 1000 Hz. The whistle is 2 m from the point of support of the swing. If a girl stands infront of the swing, the maximum and minimum frequencies she will hear are (velocity of sound = 330 m/s, g = 9.8 m/s^2)**

1) 1000, 990 Hz 2) 1007, 1000 Hz 3) 1007,993 Hz 4) 1100, 900 Hz

41. **A source of sound produces waves of wave length 48cm. This source is moving towards north with speed V/4 where V is speed of sound. The apparant wavelength of the waves to an obsever standing south of the moving source will be**

1) 48 cm 2) 60 cm 3) 72 cm 4) 96 cm

42. **A siren of frequency n approaches a stationary observer and then receedes from the observer. If the velocity of source (V) << the velocity of sound (C), the apparent change in frequency is**

1) 2 n V/C 2) 2 n C/V 3) n/V 4) 2 VC/n

43. s_1 and s_2 are two sound sources of frequencies 338 Hz and 342 Hz respectively placed at a large distance apart. The velocity with which an observer should move from s_2 to s_1 so that he may hear no beats will be..(velocity of sound in air = 340 m/s)

1) 1 m/s   2) 2 m/s   3) 3 m/s   4) 4 m/s

44. A vehicle moving on a straight road sounds a whistle of frequency 256Hz while nearing a hill with a velocity $10ms^{-1}$. The number of beats per second observed by a person travelling in the vehicle is $(V = 330ms^{-1})$.

1) zero   2) 10   3) 14   4) 16

45. If a vibrating tuning fork of frequency 255Hz is approaching with a velocity 4m/s perpendicular to a wall. The number of beats produced per sec is (speed of sound in air = 340m/s)

1) 3   2) 4   3) 5   4) 6

46. Two sources A and B are sending notes of frequency 680Hz. A listener moves from A and B with a constant velocity u. If the speed of sound in air is 340 ms^{-1}, what must be the value of u so that he hears 10 beats per second?

1) 2.0 ms^{-1} 2) 2.5 ms^{-1} 3) 3.0 ms^{-1} 4) 3.5 ms^{-1}

47. A source of sound is travelling at $\frac{100}{3}ms^{-1}$ along a road, towards a point A. When the source is 3m away from A, a person standing at a point O on a road perpendicular to the track hears a sound of frequency 'v'. The distance of O from A at that time is 4m. If the original frequency is 640 Hz, then the value of v is (given velocity of sound $= 340ms^{-1}$)

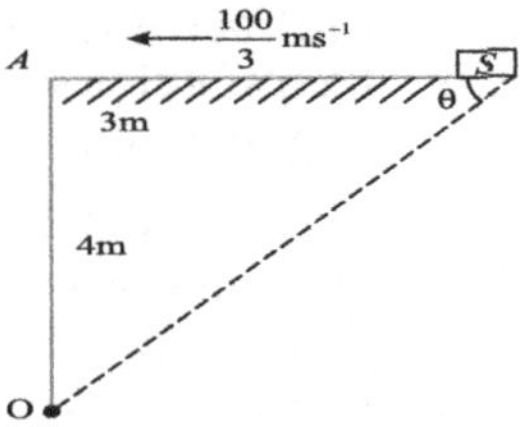

1) 620 Hz   2) 680 Hz

3) 720 Hz   4) 840 Hz

LEVEL - II (H. W) - KEY

1) 2	2) 1	3) 1	4) 1	5) 1	6) 1	7) 1	8)1	9) 3	10) 3
11) 3	12) 4	13) 3	14) 3	15)1	16) 4	17) 1	18) 4	19) 1	20) 4
21) 1	22) 2	23) 2	24) 3	25) 4	26) 1	27) 1	28) 1	29) 3	30) 3
31) 3	32) 3	33) 2	34) 2	35) 1	36) 4	37) 1	38) 2	39) 3	40) 3
41) 2	42) 1	43) 2	44) 1	45) 4	46) 2	47) 2			

LEVEL - II (H. W) -HINTS

1. By comparing with $y = A\sin(\omega t - kx)$ we get $\omega = 200rad/\sec, K = \frac{1}{2}m^{-1}, A = 0.05m$

a) $V = \frac{\omega}{K}$ b) $(V_\rho)_{max} = A\omega$ c) $\lambda = \frac{2\pi}{K}$

2. The wave reflected at fixed end undergo a phase different of πrad at the time t = 3sec, x = Vt = 3cm, A = 1–1 = 0

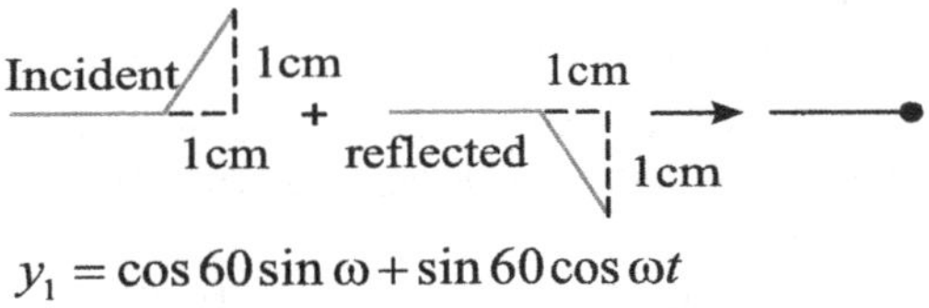

3. $y_1 = \cos 60 \sin \omega + \sin 60 \cos \omega t$

$= \sin(\omega t + 60°)$

$y_2 = \cos 30 \sin \omega t + \sin 30 \cos \omega t \quad = \sin(\omega t + 30°)$

$\therefore$ Phase different $\Delta\phi = \phi_1 - \phi_2 = \omega t + 60 - \omega t - 30 = 30°$

4. $K_x = \dfrac{\omega}{V_x}, K_y = \dfrac{\omega}{V_y}, K_z = \dfrac{\omega}{V_z}$

$\therefore \vec{K} = K_x\hat{i} + K_y\hat{j} + K_z\hat{k} = 30\hat{i} + 15\hat{j} + 15\hat{k}$

5. 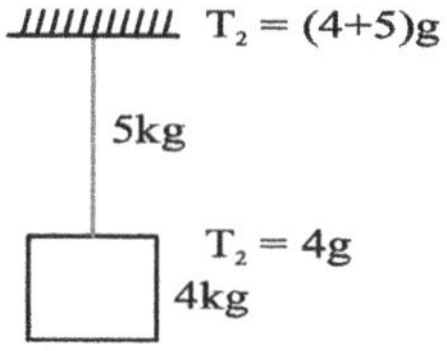

$V_1 \propto \sqrt{T} \propto \lambda \Rightarrow \dfrac{\lambda_2}{\lambda_1} = \sqrt{\dfrac{T_2}{T_1}}$

6. i) $V = \sqrt{\dfrac{T}{\mu}} = \sqrt{\dfrac{(\mu x)g}{\mu}} = \sqrt{gx}$ ii) $t = 2\sqrt{\dfrac{L}{g}} = 2\sqrt{\dfrac{2.45}{9.8}} = 1\,\text{sec}$

7. $A = \dfrac{10}{5} = 2units$

For same particle displacement $y, x + 2t$ is constant. $\therefore$

$\dfrac{dx}{dt} + 2 = 0 \Rightarrow v = \dfrac{dx}{dt} = -2$

8. When m is changed by 4% the tension is also changed by 4% and $v \alpha T^{1/2}$

9. Equation of standing wave fixed at both the ends is $y = 2a\sin(kx)\cos(\omega t)$

Put a = 2.5 mm

$K = \dfrac{2\pi}{\lambda}$ and $\lambda = l$ (in 2nd harmonic)

$\omega = KV = \dfrac{2\pi}{\lambda}\sqrt{\dfrac{T}{\mu}} = 2\pi\sqrt{\dfrac{T}{\mu l^2}}$

10. $n = \dfrac{1}{2L}\sqrt{T/\mu} = \dfrac{1}{2L}\sqrt{\dfrac{YA\alpha\Delta\theta}{\mu}}$

11. Path difference $\Delta x = 4 - 3.2 = 0.8m$

For minima $\Delta x = (2n+1)\dfrac{\lambda}{2}$

12. Path difference $\Delta x = n\lambda$ for maximum

n = 1 for r = going to be least $\therefore \pi r - 2r = \lambda$

13. $n = \dfrac{1}{2l}\sqrt{\dfrac{T}{\mu}}$ To get resonance with same fork

$\Rightarrow T = Mg = \text{const} \Rightarrow M_2 g_2 = M_1 g_1$

14. $n \propto p \Rightarrow \dfrac{n_2}{n_1} = \dfrac{p_2}{p_1}$

If the string is pluncked at $\dfrac{L}{2p}$ then it vibrates with p loops

$$\therefore p_1 = 4, p_2 = 3 \Rightarrow \frac{n_2}{n_1} = \frac{3}{4} \Rightarrow n_2 = \frac{3}{4}n$$

15. $n = \dfrac{1}{2l}\sqrt{\dfrac{T}{\mu}}$

16. $l \propto \dfrac{1}{n} \Rightarrow l_1 : l_2 : l_3 = \dfrac{1}{1} : \dfrac{1}{3} : \dfrac{1}{4}$

17. $l \propto \dfrac{1}{n} \Rightarrow l = l_1 + l_2 + l_3$

18. $v = \sqrt{\dfrac{\gamma RT}{M}}$

19. $V = \sqrt{\dfrac{B}{\rho}}$ $\qquad \dfrac{\Delta V}{V} = -Q \Rightarrow B = \dfrac{-P}{\left(\dfrac{\Delta V}{V}\right)} = \dfrac{P}{Q}$ and $\rho = 1gm/cc \quad \therefore V = \sqrt{\dfrac{P}{Q}}$

20. $\dfrac{\rho_1}{\rho_2} = \dfrac{M_1}{M_2} = \dfrac{2}{32} = \dfrac{1}{16} \Rightarrow \rho_2 = 16\rho_1$

$$\rho_{mix} = \frac{\rho_1 V_1 + \rho_2 V_2}{V_1 + V_2} = \frac{\rho_1(4) + 16\rho_1(1)}{4+1} \quad \Rightarrow \rho_{mix} = \frac{20\rho_1}{5} = 4\rho_1$$

$$V \propto \frac{1}{\sqrt{\rho}} \Rightarrow \frac{V_{mix}}{V_1} = \sqrt{\frac{\rho_1}{\rho_{mix}}} = \sqrt{\frac{\rho_1}{4\rho_1}}$$

21. $\rho_{mix} = \dfrac{\rho_1 V_1 + \rho_2 V_2}{V_1 + V_2}$

Velocity of sand $C \propto \dfrac{1}{\sqrt{\rho}} \Rightarrow \rho \propto \dfrac{1}{C^2} \Rightarrow \dfrac{\rho_1}{\rho_2} \dfrac{C_2^2}{C_1^2}$

$$\Rightarrow \rho_1 = \frac{C_2^2}{C_1^2} . \rho_2 \Rightarrow \rho_{mix} = \frac{\frac{C_2^2}{C_1^2}\rho_2 V_1 + \rho_2 V_2}{V_1 + V_2}$$

$$\rho_{mix} = \left(\frac{C_2^2 V_1 + C_1^2 V_2}{V_1 + V_2}\right)\frac{\rho_2}{C_1^2} \Rightarrow \frac{C_{mix}}{C_2} = \sqrt{\frac{\rho_2}{\rho_{mix}}}$$

22. $n_p = \dfrac{v}{\lambda_p} = \dfrac{v}{4l}$ $\qquad n_q = \dfrac{v}{\lambda_q} = \dfrac{v}{2l}$ $\qquad n_r = \dfrac{v}{\lambda_r} = \dfrac{v}{l}$ $\qquad n_s = \dfrac{v}{\lambda_s} = \dfrac{3v}{4l}$

23. $n = \dfrac{V}{2(l + 1.2r)}$

24. A' = A Cos Kx
3rd overtone in open pipe is 4th
Harmonic $\therefore l = \dfrac{4\lambda}{2} \Rightarrow \lambda = \dfrac{l}{2}$

$x = \dfrac{l}{16}$ $\qquad \therefore A' = A\cos\left(\dfrac{2\pi}{\lambda}.x\right)$

25. $T_2 = T_1 + 8$

$$n_1 = \frac{V}{4l}, n_2 = \frac{3V}{4l}$$

$$n \propto \sqrt{T} \Rightarrow \frac{n_1}{n_2} = \sqrt{\frac{T_1}{T_2}} \Rightarrow \frac{1}{3} = \sqrt{\frac{T_1}{T_1 + 8}}$$

26. 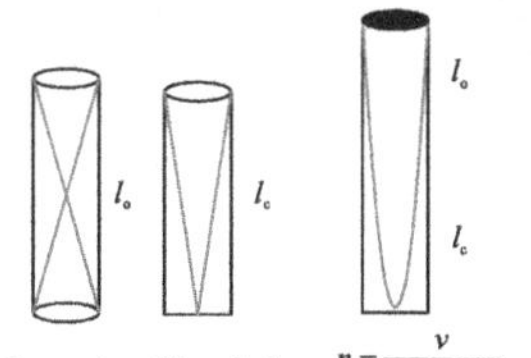

$$n_1 = \frac{V}{2l_0} \qquad n_2 = \frac{V}{4l_C}$$

$$\Rightarrow l_0 = \frac{V}{2n_1} \qquad l_C = \frac{V}{4n_2}$$

$$\therefore l_0 + l_C = \frac{V}{4n}$$

$$\frac{V}{2n_1} + \frac{V}{4n_2} = \frac{V}{4n} \qquad \Rightarrow n = \frac{n_1 n_2}{n_1 + 2n_2}$$

27. $\Rightarrow l_3 = 2l_2 - l_1$

28. 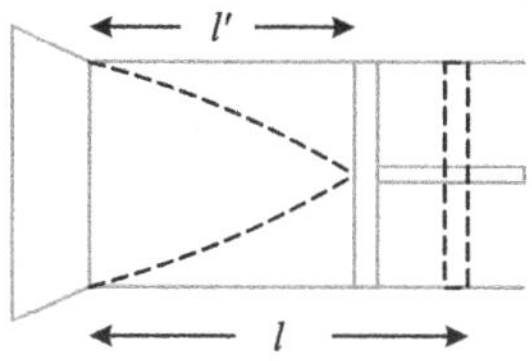

$$n = \frac{V}{4l'} \text{ and } P'l' = Pl$$

$$1.5Pl' = P \times 25 \times 10^{-2} \qquad \Rightarrow l' = \frac{25 \times 10^{-2}}{1.5}$$

29. $\frac{n_0}{n_c} = 2 \Rightarrow n_o = 2n_c$ and $n_0 - n_c = 2$

$\Rightarrow n_c = 2$ and $n_0 = 4$

$n \propto \frac{1}{\ell} \Rightarrow n'_o = 8$ and $n'_c = 1$

30: $\frac{n_{C1}}{n_{03}} = \frac{3}{2} \Rightarrow n_{C1} = \frac{3}{2} n_{O3}$

$n_{C1} = n_{O3} - 55$

31. $\frac{l_2}{l_1} = \frac{39}{40} \Rightarrow \frac{n_1}{n_2} = \frac{l_2}{l_1} = \frac{39}{40}$

$\Rightarrow n_1 < n < n_2$

$n - n_1 = 5$ ....(1)

$n_2 - n = 5$ ....(2)

$n_2 - n_1 = 10$

32. $\frac{n_1}{n_2} = \frac{l_2}{l_1} = \frac{33}{32}$ and $n_1 - n_2 = 7$

33. $n_1 = \frac{V}{2l}, n_2 = \frac{2V}{2l}$ $\Rightarrow \frac{n_1}{n_2} = \frac{1}{2} \Rightarrow n_2 = 2n_1$

Given that, $n_{S_2} \Rightarrow 2n_{S_1} \Rightarrow n_2 n_{S_2}$ & $n_{S_1} < n_1$

$\therefore n_1 - n_{S_1} = 5$

34. $n_1 l_1 = n_2 l_2$

$(n + 8)\ 100 = (n - 8)\ 101$

35. $l_1 + l_2 = 100cm$ $l_2 - l_1 = 0.2cm$

$l_2 = 50.1cm, l_1 = 49.9cm$ $n_1 - n_2 = 1$

36. $n_1 - n_2 = 5$ at $T_1 = 27 + 273 = 300K$

$n_2 = n$ at $T_2 = -3 + 273 = 270$ K

$n \propto V \propto \sqrt{T} \Rightarrow \frac{n_1}{n_2} = \sqrt{\frac{T_1}{T_2}} = \sqrt{\frac{300}{270}} = \sqrt{\frac{10}{9}}$

37. $\frac{\lambda_1}{\lambda_2} = \frac{n_2}{n_1} = \frac{40/193}{40/195} = \frac{195}{193}$

$\Rightarrow n_2 > n_3 > n_1 \quad \Rightarrow n_2 - n_3 = 9..........(1)$

$n_3 - n_1 = 9..........(2)$

$(1) + (2) \Rightarrow n_2 - n_1 = 18$

38. $n' = n\left(\frac{V - V_0}{V}\right)$ Here $n' = \frac{95}{100}n$

39. $n' = n\left(\frac{V + V_0}{V - V_S}\right)$

40. 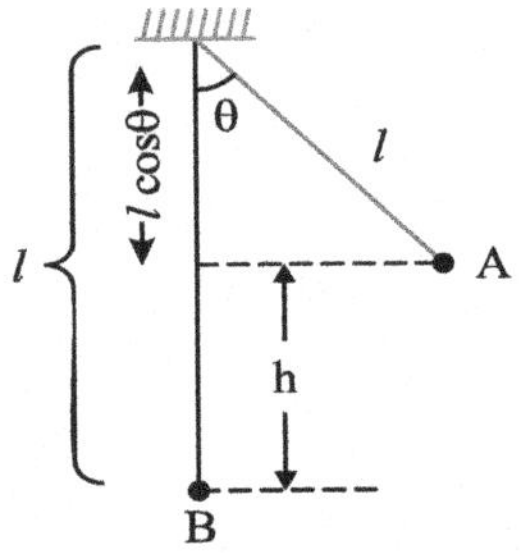

$h = l - l\cos\theta = [1 - \cos\theta]$

$p.E_A = kE_B \Rightarrow mgh = \frac{1}{2}mv^2$

$\Rightarrow V_S = \sqrt{2gh} = \sqrt{2gl(1 - \cos\theta)}$

$n_{max} = n\left[\frac{V}{V - V_S}\right]$ $n_{min} = n\left[\frac{V}{V + V_S}\right]$

41. $\lambda' = \left(\frac{V+V_S}{V}\right)\lambda$

42. $n' = n\left(\frac{V}{V-V_S}\right) = n\left(\frac{C}{C-V}\right)$

$n'' = n\left(\frac{C}{C+V}\right)$

$\Delta n'' = n' - n'' = nC\left[\frac{1}{C-V} - \frac{1}{C+V}\right]$

$= nC\left[\frac{2V}{C^2 - V^2}\right]$

V^2 can be neglected as V<<C $\therefore \Delta n' = \frac{2nV}{C}$

43. 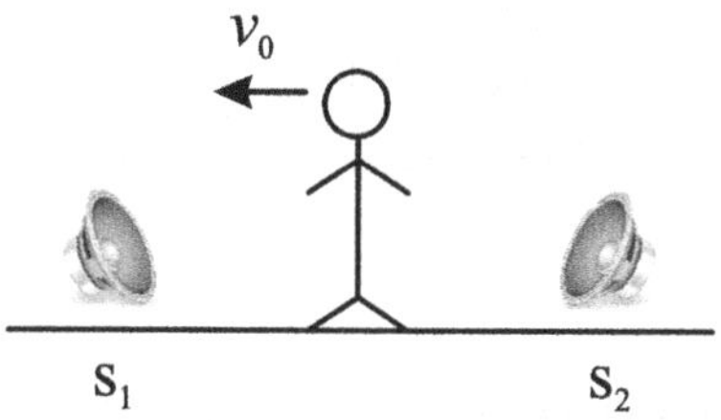

$n_1^1 = n_1\left[\frac{V+V_0}{V}\right], n_2^1 = n_2\left[\frac{V-V_0}{V}\right]$

no beat are heard $\Rightarrow n_1^1 = n_2^1$

$n_1\left(\frac{V+V_0}{V}\right) = n_2\left(\frac{V-V_0}{V}\right)$

44. $\Delta n = \frac{2nV_S}{V-V_S} = \frac{2\times 256\times 10}{320} = 16$

The no. of beats produced is 16 per second but no. of beats neard (observed) by observer 'O' as maximum no. of beats heard is only 10 per second.

45. $\Delta n = \frac{2nV_S}{V-V_S} = \frac{2\times 255\times 4}{336} = 6 beats/\sec$

46. $n' = n\left(\frac{v-v_0}{v+v_s}\right)$, $n'' = n\left(\frac{v+v_0}{v-v_s}\right)$ and $n'' - n' = 10$

47. $n_1 = \frac{V}{V - V_s\cos\theta}\times n$

LEVEL - III

1. **The transverse displacement $y(x,t)$ of a wave on a string is given by $y(x,t) = e^{-(ax^2+bt^2+2\sqrt{ab}xt)}$. This represents a**

1) Wave moving in -ve x direction with speed $\sqrt{b/a}$

2) Wave moving in +ve x direction with speed $\sqrt{b/a}$

3) Standing wave of frequency $\sqrt{b}$

4) Standing wave of frequency $\frac{1}{\sqrt{b}}$

2. **A plane progressive wave is shown in the adjoining phase diagram. The wave equation of this wave, if its position is shown at t=0, is**

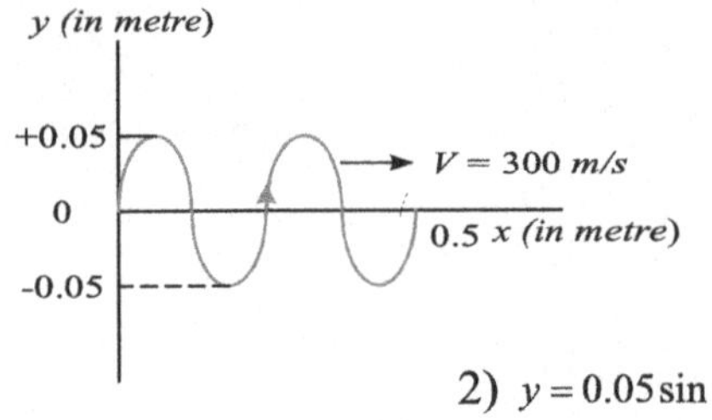

1) $y = 0.05\sin 2\pi(300t - x)$ 2) $y = 0.05\sin 2\pi(300t + x)$

3) $y = 0.05\sin 8\pi(300t + x)$ 4) $y = 0.05\sin 8\pi(300t - x)$

3. **A plane progressive wave of frequency 25 Hz, amplitude $2.5\times10^{-5}m$ and initial phase zero moves along the negative x-direction with a velocity of 300 ms^{-1}. A and B are two points 6m apart on the line of propagation of the wave. At any instant the phase difference between A and B is θ. The maximum difference the displacement of the particles at A and B is Δ, then**

1) $\theta = \pi$ 2) $\theta = 0$ 3) $\Delta = 0$ 4) $\Delta = 5\times10^{-5}\ m$

4. **A wave pulse starts propagating in positive x- direction along a non-uniform wire of length 10 m with mass per unit length given $m = m_0 + \alpha x$ and under a tension of 100 N. The time taken by the pulse to travel from the lighter end (x=0) to the heavier end is (given $m_0 = 10^{-2}\,kgm^{-1}$ and $\alpha = 9\times10^{-3}\,kgm^{-2}$) (in sec)**

1) 4.66 2) 0.226 3)7.133 4) 5.24

5. **A long rubber tube having mass 0.9kg is fastened to a fixed support and the free end of the tube is attached to a chord which passes over a pulley and supports an object, with a mass of 5kg as shown in figure. If the tube is struck by a transverse blow at one end, the time required for the pulse to reach the other end is**

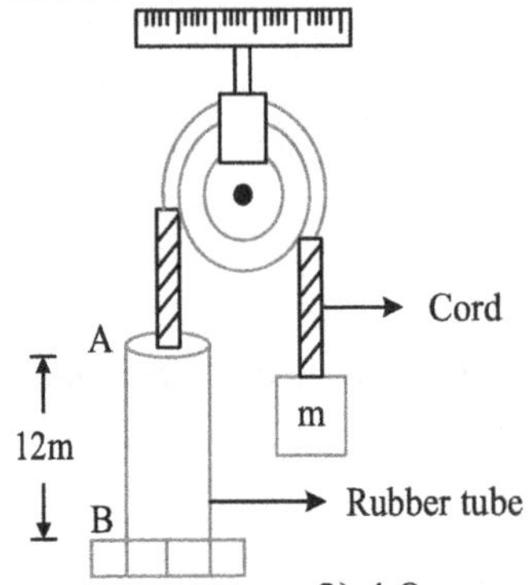

1) 5s 2) 0.48s 3) 4.8s 4) 3.2s

6. **A wave represented by $y = 100\sin(ax + bt)$ is reflected from a dense plane at the origin. If 36% of energy is lost and rest of the energy is reflected then the equation of the reflected wave will be**

1) $y = -8.1\sin(ax - bt)$ 2) $y = 8.1\sin(ax + bt)$ 3) $y = -80\sin(ax - bt)$ 4) $y = -10\sin(ax - bt)$

7. **In a sonometer wire, the tension is maintained by suspending a 20kg mass from the free end of the wire. The fundamental frequency of vibration is 300 Hz.**

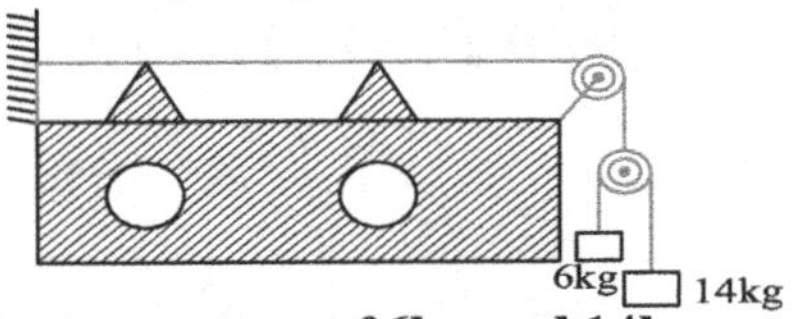

If the tension is provided by two masses of 6kg and 14kg suspended from a pulley as show in the figure the fundamental frequency will

1) stll remain 300 Hz 2) become larger 3) become smaller

4) decrease in the present situation and increase if the suspended masses of 6kg and 14kg are interchanged.

8. **The length of the wire shown in figure between the pulley is 1.5m and its mass is 12 gm. Find the frequency of vibration with which the wire vibrates in two loops leaving the middle point of the wire between the pulleys at rest.**

1) 35 Hz 2) 40 Hz

3) 70 Hz 4) 80 Hz

9. **A rod PQ of length 'L' is hung from two identical wires A and B. A block of mass 'm' is hung at point R of the rod as shown in figure. The value of 'x' so that the fundamental mode in wire A is in resonance with first overtone of B is**

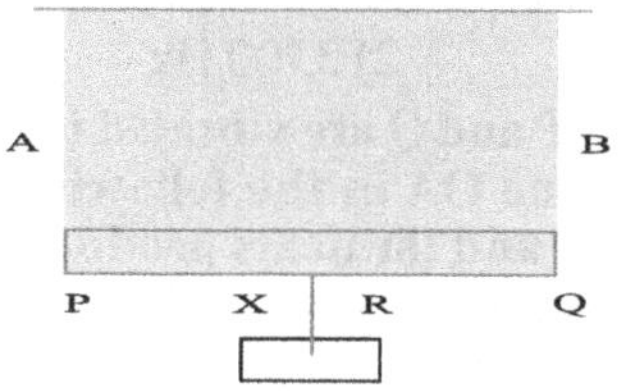

1) $\frac{4L}{5}$ 2) $\frac{L}{4}$

3) $\frac{L}{5}$ 4) $\frac{2L}{3}$

10. **Two wires are fixed on a sonometer. Their tensions are in the ratio 8:1, their lengths are in the ratio 36:35, the diameters are in the ratio 4:1 and densities are in the ratio 1:2. Find the frequencies of the beats produced if the note of the higher pitch has a frequnecy of 360 /s.**

1) 20 Hz 2) 10 Hz 3) 30 Hz 4) 40 Hz

11. **The vibrations of a string fixed at both ends are represented by**

$y = 16Sin\left(\frac{\pi x}{15}\right)Cos(96\pi t)$. **Where 'x' and 'y' are in cm and 't' in seconds. Then the phase difference between the points at x = 13 cm and x = 16 in radian is**

1) $\pi/5$ 2) π 3) 0 4) $2\pi/5$

12. **Two metallic strings A and B of different materials are connected in series forming a joint. The strings have similar cross-sectional area. The length of A is $l_A = 0.3m$ and that of B is $l_B = 0.75m$. One end of the combined string is tied with a support rigidly and the other end is loaded with a block of mass m passing over a frictionless pulley. Transverse waves are setup in the combined string using an external source of variable frequency. The total number of** antinodes at this frequency with joint as node is (the densities of A and B are 6.3×10^3 **kg m^{-3} and 2.8×10^3 kgm^{-3} respectively)**

1) 5 2) 8 3) 9 4) 6

13. **A source of sound emits waves isotropically in three dimensions. If the intensity at a distance r_0 from source is I_0, at what distance from the source is the intensity $0.100\ I_0$?**

1) 1.16 r_0 2) 2.16 r_0 3) 3.16 r_0 4) 4.16 r_0

14. **How long will it take sound waves to travel the distance l between the points A and B if the air temperature between them varies linearly from T_1 to T_2? The velocity of sound propagation in air is equal to $v = \alpha\sqrt{T}$, where α is constant.**

1) $t = \frac{2l}{\alpha\left[\sqrt{T_2} + \sqrt{T_1}\right]}$ 2) $t = \frac{4l}{\alpha\left[\sqrt{T_1} + \sqrt{T_2}\right]}$ 3) $t = \frac{4l}{\alpha\left[\sqrt{T_1}\sqrt{T_2}\right]}$ 4) $t = \frac{2l}{\alpha\left[\sqrt{T_1 + T_2}\right]}$

15. **Air column of 20 cm length in a resonance tube resonates with a certain tuning fork when sounded at its upper open end. The lower end of the tube is closed and adjustable by changing the quantity of mercury filled inside the tube. The temperature of the air is 27°C. The change in length of the air column required, if the temperature falls to 7°C and the same tuning fork is again sounded at the upper open end is nearly**

1) 1 mm 2) 7 mm 3) 5 mm 4) 13 mm

16. **AB is a cylinder of length 1m fitted with a thin flexible diaphragm C at middle and two other thin flexible diaphragms A and B at the ends as shown. The portions AC and BC contain hydrogen and oxygen gases respectively. The diaphragms A and B are set into vibrations of the same frequency. The minimum frequency of these vibrations for which diaphragm C is a node is (Under the conditions of the experiment the velocity of sound in hydrogen is 1100 m/s and oxygen 300 m/s)**

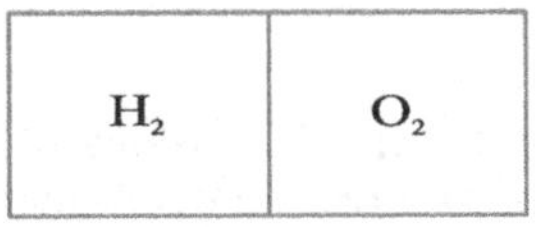

1) 1100 Hz 2) 3300 Hz 3) 1650 Hz 4) 1500 Hz

17. **Two tuning forks P and Q are vibrated together. The number of beats produced are represented by the straight line OA in the following graph. After loading Q with wax again these are vibrated together and the beats produced are represented by the line OB. If the frequency of P is 341 Hz, the frequency of Q will be ___**

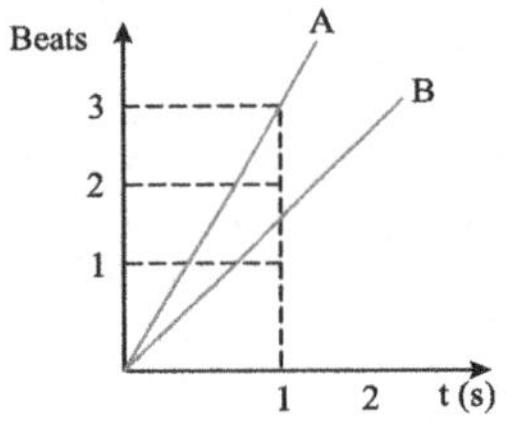

1) 341 Hz 2) 338 Hz 3) 344 Hz 4) 330 Hz

18. **A driver in a stationary car blows a horn which produces monochromatic sound waves of frequency 1000 Hz normally towards a reflecting wall. The wall approaches the car with a** speed of 3.3 ms^{-1}. **(velocity of sound v = 336m/s]**

1) the frequency of sound reflected from wall and heard by the driver is 1000 Hz
2) the frequency of sound reflected from wall and heard by the driver is 980 Hz.
3) the percentage increase in frequency of sound after reflection from wall is 2%.
4) the percentage decrease in frequency of sound after reflection from wall is 2%.

19. **A sources of sonic oscillations with frequenccy n=1700Hz and a receiver are located on the same normal to a wall. Both the source and receiver are stationary, and the wall recedes from the source with velocity u=6.0 cm/s. Find the beat frequency registred by th receiver. The velocity of sound is equal to v=340 m/s.**

1) 0.2 Hz 2) 0.3 Hz 3) 0.4 Hz 4) 0.6 Hz

20. **A source of oscillations S is fixed to the riverbed of a river with stream velocity v. Two receivers R_1 and R_2 are fixed also to the riverbed. If the source generates frequency f_s, what frequencies are received by receivers R_1 and R_2.**

1) f_s 2) 1.2 f_s 3) 1.4 f_s 4) 1.6 f_s

21. **A train A crosses a station with a speed of 40 m/s and whistles a short pulse of natural frequency $n_0 = 596 Hz$. Another train B is approaching towards the same station with the same speed along a parallel track. Two tracks are d = 99m apart. When train A whistles, train B is 152m away from the station as shown in fig. If velocity of sound in air $v = 330 m/s$, calculate frequency of the pulse heard by driver of train B.**

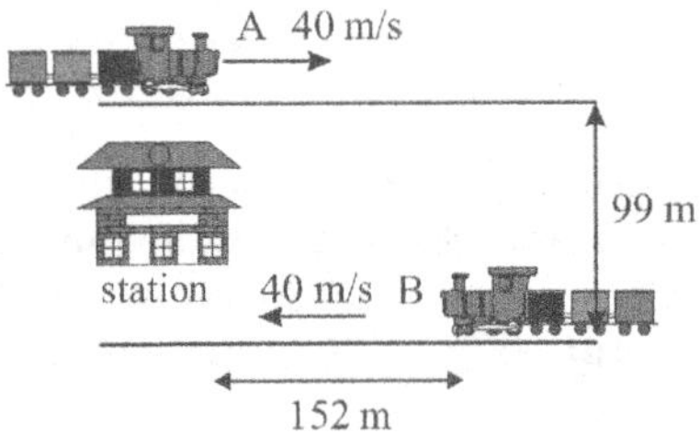

1) 724 Hz 2) 660 Hz 3) 742 Hz 4) 427 Hz

22. **A wave travelling along the x-axis is described by the equation y(x, t) = 0.005 cos$(\alpha x - \beta t)$. If the wavelength and the time period of the wave are 0.08m and 2.0 s respectively, then α and β in appropriate units are**

1) $\alpha = 25.00\pi, \beta = \pi$ 2) $\alpha = \frac{0.08}{\pi}, \beta = \frac{2.0}{\pi}$ 3) $\alpha = \frac{0.04}{\pi}, \beta = \frac{1.0}{\pi}$ 4) $\alpha = 12.50\pi, \beta = \frac{\pi}{2.0}$

23. **A travelling wave represented by $y = A\sin(\omega t - kx)$ is superimposed on another wave represented by $y = A\sin(\omega t + kx)$. The resultant is**

1) a standing wave having nodes at $x = \left(n + \frac{1}{2}\right)\frac{\lambda}{2}$, where n = 0,1,2

2) a wave travelling along +x direction 3) a wave travelling along –x direction

4) a standing wave having nodes at $x = \frac{n\lambda}{2}$ where n = 0,1,2

24. **While measuring the speed of sound by performing a resonance column experiment, a student gets the first resonance condition at a column length of 18 cm during winter. Repeating the same experiment during summer, she measures the column length to be x cm for the second resonance. Then**

1) 18 > x 2) x > 54 3) 54 > x > 36 4) 36 > x > 18

25. **A motor cycle starts from rest and accelerating along a straight path at 2ms^{-2}. At the starting point of the motor cycle, there is a stationary electric siren. How far has the motor cycle gone when the driver hears the frequency of the siren at 94% of its value when the motor cycle was at rest? (Speed of sound = 330 ms^{-1})**

1) 49 m 2) 98 m 3) 147 m 4) 196 m

LEVEL - III - KEY

1) 1	2) 4	3) 1, 4	4) 2	5)2	6)3	7)3	8)3	9)3	10) 2
11) 2	12)2	13)3	14)1	5)2	16) 3	17) 3	18)3	19) 4	20) 1
21) 1	22) 1	23) 1	24) 2	25) 2					

LEVEL - III - HINTS

1. $y(x,t) = e^{-(\sqrt{ax}+\sqrt{bt})^2}$

2. $y = a\sin(\omega t - Kx) = a\sin\left[2\pi\left(nt - \frac{x}{\lambda}\right)\right]$

The terms can be get from the phase diagram

3. $\lambda = \frac{V}{n}, \ \phi = \frac{2\pi}{\lambda}x$

4. $V = \frac{dx}{dt} = \sqrt{\frac{T}{m}} = \sqrt{\frac{T}{m_0 + \alpha x}}$ $\therefore \left(\sqrt{m_0 + \alpha x}\right)dx = \sqrt{T}.dt$

$\int_0^{10}\left(\sqrt{m_0 + \alpha x}\right)dx = \sqrt{T}\int_0^t dt$

5. $v = \sqrt{\frac{T}{\mu}} = \sqrt{\frac{Mg}{\mu}}$ and $t = \frac{l}{V}$

6. $I_r = I_I - \frac{36}{100}I_i = \frac{64}{100}I_i$

As $A \propto \sqrt{I} \Rightarrow \frac{A_r}{A_i} = \sqrt{\frac{I_r}{I_i}} = \sqrt{\frac{64}{100}} = \frac{8}{10}$ $\Rightarrow A_r = \frac{8}{10}A_i = \frac{8}{10}\times 100 = 80 units$

Due to reflection at rigid end, an additional phase II is created. $\therefore y_r = A_r \sin(ax - bt) + \pi$

7. Let a be the accelertion in the cord.

Then $T - 6g = 6a$

$14g - T = 14a$

Adding both, $8g = 20a \Rightarrow a = \frac{8g}{20}$

$\therefore T = \frac{6\times 8g}{20} + 6g = 6g\left(\frac{8}{20}+1\right)$

The total effective tension in the sonometer wire T'=2T

When in the first case the tension was 20g. Hence, the effective tension is reduced.

Since, $f = \frac{1}{2L}\sqrt{\frac{T'}{m}}$

8. $f_p = \frac{P}{2l}\sqrt{T/\mu} \Rightarrow f_2 = \frac{2}{2l}\sqrt{\frac{Mg}{m/l}}$

9. $f_A = 2f_B$; $\frac{1}{2l}\sqrt{\frac{T_A}{\mu}} = \frac{2}{2l}\sqrt{\frac{T_B}{\mu}} \Rightarrow T_A = 4T_B$

$\vec{\tau} = \vec{r}\times\vec{f} = \text{constant}$; $\frac{T_A}{T_B} = 4 = \frac{l-x}{x}$

10. $\frac{n_1}{n_2} = \frac{\ell_2}{\ell_1}\sqrt{\frac{T_1}{T_2}\cdot\frac{\mu_2}{\mu_1}}$

11. $K = \frac{2\pi}{\lambda} = \frac{\pi}{15} \Rightarrow \lambda = 30\,cm$

$X_1 = 13\,cm$ and $X_2 = 16\,cm$ are in adjacent loops and $\Delta\phi = \pi$

12. 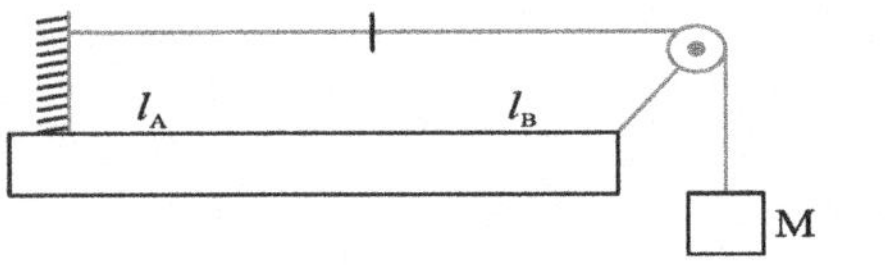

$n = \frac{P}{2l}\sqrt{\frac{T}{m}} = \text{constant}$

where, P = number of loops (or antinodes)

$n = \frac{P}{2lr}\sqrt{\frac{T}{\rho\pi}} = \text{constant}$ $\quad \frac{P}{l\sqrt{\rho}} = \text{constant} \Rightarrow P \propto l\sqrt{\rho}$

$\frac{P_1}{P_2} = \frac{l_1}{l_2}\sqrt{\frac{\rho_1}{\rho_2}} = \frac{0.3}{0.7}\sqrt{\frac{6.3\times10^{+3}}{2.8\times10^{+3}}} \Rightarrow P_1 + P_2 = 8$

13. $I = \frac{E}{At} = \frac{E}{4\pi r^2}t \Rightarrow I \propto \frac{1}{r^2} \Rightarrow \frac{I_2}{I_1} = \left(\frac{r_1}{r_2}\right)^2$

$\frac{0.1I_0}{I_0} = \left(\frac{r_0}{r_2}\right)^2 \Rightarrow r_2 = \sqrt{10}\,r_0 = 3.16r_0$

14. $V_{avg} = \frac{V_1+V_2}{2} = \frac{1}{2}\propto\sqrt{T_1} + \propto\sqrt{T_2}$

$$t=\frac{l}{V_{avg}}=\frac{l}{\frac{1}{2}\left[\alpha\sqrt{T_1}+\alpha\sqrt{T_2}\right]}=\frac{2l}{\alpha\left[\sqrt{T_1}+\sqrt{T_2}\right]}$$

15. $l_1=20cm, l_2=3, \Delta t=?$

$$n=\frac{V}{4l}=cent \Rightarrow V\propto l..........(1)$$

Velocity of sound in air $V=\sqrt{\frac{\gamma RT}{M}}$

$$\Rightarrow V\propto\sqrt{T}...............(2) \therefore l\propto\sqrt{T} \Rightarrow \frac{\Delta l}{l}=\frac{1}{2}\frac{\Delta T}{T}$$

16. As diaphragm C is a node, A and B will be antinodes. Thus each part behave as closed organ pipe. Their fundamental frequency are

$$n_H=\frac{V_H}{4l_H}=\frac{1100}{4\times0.5}=550Hz \quad n_0=\frac{V_0}{4l_0}=\frac{330}{4\times0.5}=150Hz$$

As $n_H\neq n_0$, the common frequency of system is written as

$$n=pn_H=qn_0\Rightarrow\frac{p}{q}=\frac{3}{11}$$

17. $n_Q=341\pm3$

18. 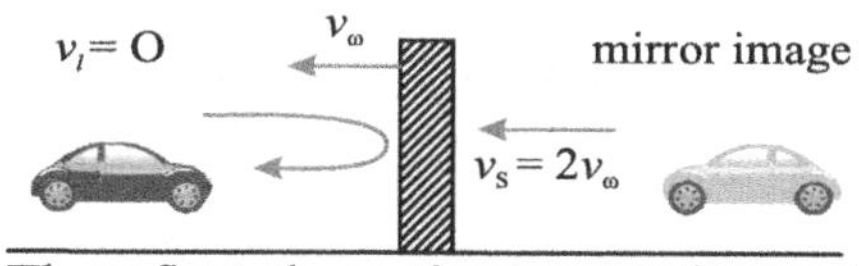

The reflected sound appears to be coming from mirror image which appears to move towards driver with velocity double the velocity of wall.

$$\therefore V_C=V_\omega=6.6m/s$$

$$n'_{reflected}=n\left[\frac{V}{V-V_S}\right] \quad n'_{reflected}>n \quad \therefore\frac{n'_r}{n}=\frac{V}{V-V_S}$$

$$\Rightarrow\frac{n'_r-n}{n}\times100\%=\frac{V_S}{V-V_S}\times100\%$$

19. $n_1=n_0\left[\frac{V}{v+u}\right], n_2=n_0\left[\frac{v-u}{v}\right], \Delta n=n_1-n_2$

20. $f_R=f_S$
since no doppler effect due to motion of medium

21. 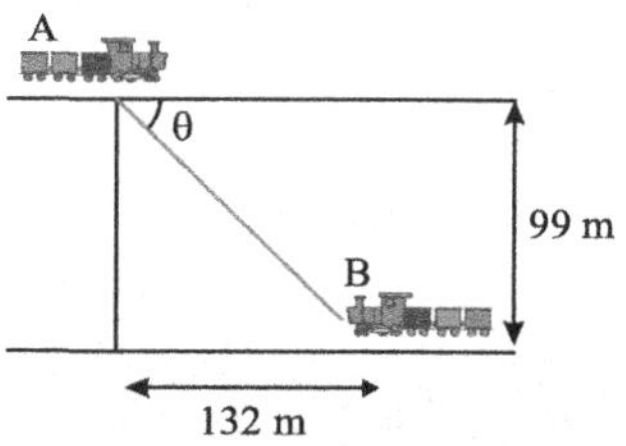

The distance of train B from station when its driver hear the pulse is $152-40t$.

The distance travelled by the pulse is $\sqrt{(152-40t)^2+(99)^2}=vt=330t\Rightarrow t=0.5s$

The distance of train B from station when its driver hear the pulse is $152-40t=132$

$$\tan\theta=\frac{99}{132}\Rightarrow\theta=37^0;\; n=\left(\frac{v+v_0\cos\theta}{v-v_s\cos\theta}\right)n_0$$

22. As general equation is $y(x,t) = a\cos(kx - \omega t)$

where, $k = \frac{2\pi}{\lambda}$ and $\omega = \frac{2\pi}{T}$

23. By superposition principle,

$y = y_1 + y_2 = A\sin(\omega t - kx) + A\sin(\omega t + kx)$

$y = 2A\sin\omega t\cos kx$

Clearly, it is equation of standing wave for position of nodes y = 0.

i.e., $x = (2n+1)\frac{\lambda}{4} \Rightarrow \left(n + \frac{1}{2}\right)\frac{\lambda}{2}, n = 0,1,2,3$

24. $l_1 = 18$ cm . $f = \frac{v_1}{4l_1}$ [for first resonances] and $f = \frac{3v_2}{4l_2}$ [for second resonance]

where, l_2 = x according to given situation and

$\Rightarrow \frac{3v_2}{4l_2} = \frac{v_1}{4l_1}$ (or) $l_2 = 3l_1 \times \frac{v_2}{v_1}$ $\Rightarrow$ x = 54 × (A quantity greater than 1)

also $v_1 < v_2$ as during summer temperature would be higher. So, x > 54

25. For motor cycle, u = 0, a = $2ms^{-2}$

Observer is in motion and source is at rest, then

apparent frequency, $n' = n\frac{V - V_0}{V + V_S}$ and $S = \frac{V^2 - u^2}{2a}$

LEVEL - IV
MATCHING TYPE QUESTIONS

1. A train T horns a sound of frequency f. It is moving towards a wall with speed $\frac{1}{4}$th the speed of sound. There are three observers O_1, O_2 and O_3 as shown. Match the following two columns.

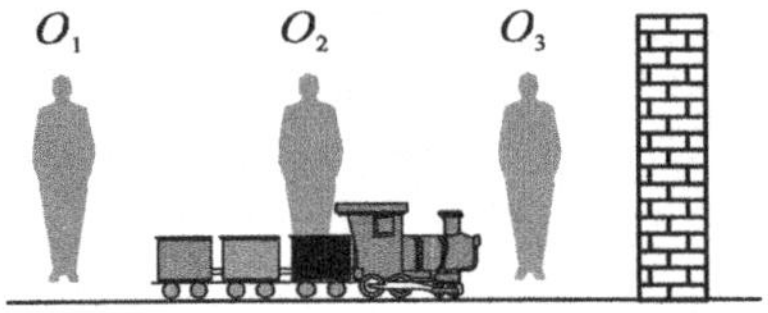

Column - I Column - II

Column - I	Column - II
a) Beat frequency Observed to O_1	p) $\frac{2}{3}$f
b) Beat frequency observed to O_2	q) $\frac{8}{15}$f
c) Beat frequency observed to O_3	r) f
d) If train moves in oppoisite direction with the same speed then beat frequency observed to O_3	s) zero

1) a – q; b – s; c – r; d – p
2) a – q; b – p; c – s; d – s
3) a – r; b – s; c – q; d – p
4) a – s; b – p; c – r; d – p

2. **In each of the four situations of Column-I a stretched string or an organ pipe is given along with required data. In case of strings the tension in string is $T = 102.4N$ and the mass per unit length of string is 1 g/m. Speed of sound in air is 320 m/s. Neglect and corrections. The frequencies of resonance are given in column II. Match each situation in column-I with the possible resonance frequencies given in column-II**

Column-I	**Column-II**
A) String fixed at both end	P) 320 Hz
B) String fixed at one end and free at other end	Q) 480 Hz
C) Open organ pipe	R) 640 Hz
D) Closed organ pipe	S) 800 Hz

A) String fixed at both end

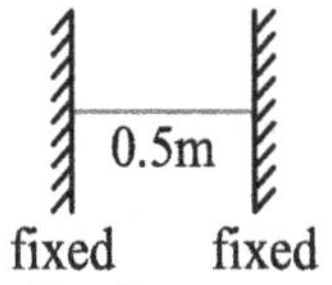

B) String fixed at one end and free at other end

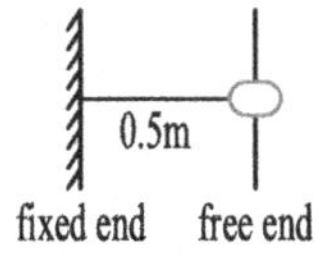

C) Open organ pipe

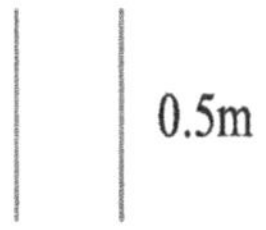

D) Closed organ pipe

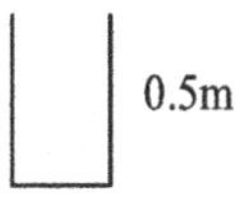

	A	B	C	D		A	B	C	D
1)	P,R	Q,S	P,R	Q,S	2)	P,S	Q,R	P,Q	Q,S
3)	Q,R	P,Q	Q,S	R,S	4)	P,R	Q,R	S,R	R,P

COMPREHENSION QUESTIONS

Paragraph - I

A heavy but uniform rope of length L is suspended from a ceiling

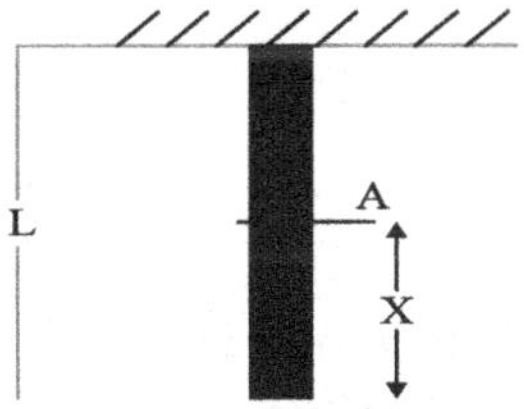

3. **Find the velocity of transverse wave travelling on the string as a funcition of the distance(x) from the lower end**

1) $L\sqrt{\frac{g}{x}}$ 2) $x\sqrt{\frac{g}{L}}$ 3) $\sqrt{gx}$ 4) $\sqrt{gx^2}$

4. **If the rope is given a sudden sideways jerk at the bottom, how long will it take for the pulse to reach the ceiling ?**

1) $\sqrt{\frac{2L}{g}}$ 2) $\sqrt{\frac{L}{2g}}$ 3) $\frac{1}{2}\sqrt{\frac{L}{g}}$ 4) $2\sqrt{\frac{L}{g}}$

5. **A particle is dropped from the ceiling at the same instant the bottom end is given the jerk. where will the particle meet the pulse measured from bottom ?**

1) $\frac{L}{2}$ 2) $\frac{L}{3}$ 3) $\frac{L}{6}$ 4) $\frac{L}{4}$

Paragraph - II

The vibrations of a string of length 600cm fixed at both ends are represented by the equation.

$y = 4\sin\left[\frac{\pi x}{15}\right].\cos\left[90\pi.t\right]$ **where 'x' and 'y' are in cm. and 't' in seconds**

6. **What is the maximum displacement of a point x = 5 cm ?**

1) $\sqrt{3}$cm 2) $2\sqrt{3}$ cm 3) $\frac{\sqrt{3}}{2}$cm 4) $\frac{3\sqrt{3}}{2}$cm

7. **Where are the nodes located along the string?**

1) 15cm, 25cm 2) 20cm, 40cm 3) 15cm, 45cm 4) 10cm, 30cm

8. **What is the velocity of the particle at** $x = 7.5\,\text{cm}$ **at t = 0.25 s**

1) zero 2) 1/2 cm/s 3) 0.25 cm / s 4) 1 cm / s

Paragraph - III

Two sources $s_1 \& s_2$ **seperated by 2m, vibrate according to equation** $y_1 = 0.03\sin\pi t$ **and** $y_2 = 0.02\sin\pi t$ **where** y_1, y_2 **and t are in M.K.S units. They send out waves of velocity 1.5m/s. Calculate the amplitude of the resultant motion of the particle co-linear with** $s_1 \& s_2$ **and located at a point**

9. **To the right of** S_2

1) 0.0265m 2) 0.0365m 3) 0.0165m 4) 0.0465m

10. **To the left of** S_2

1) 0.0265m 2) 0.0365m 3) 0.0165m 4) 0.0465m

11. **In the middile of** S_1 **and** S_2

1) 0.25m 2) 0.05m 3) 1m 4) 2m

Paragraph - IV

A Source emits sound waves of frequency 1000 HZ .The source moves to the right with a speed of 32 m\s relative to ground , on the right a reflecting surface moves towards left with a speed of 64 m/s relative to the ground. The speed of sound in air is 332 m/s

12. **Find the wave length of sound in ahead of the source.**

1) 0.1 m 2) 0.2 m 3) 0.3 m 4) 0.4m

13. **Find the number of waves arriving per second which meets the refelcting surface**

1) 1320 2) 1220 3) 1120 4) 1020

14. **Find the wavelength of reflected waves**

1) 0.1m 2) 0.2m 3) 0.3m 4) 0.4m

PassageV :

When a compositewireismadeby joining twowiresasshown in figureand possiblefrequencies of thiswireisasked (both endsfixed) then the lowest frequency is that at which individual lowest frequencies of the two wires are equal.

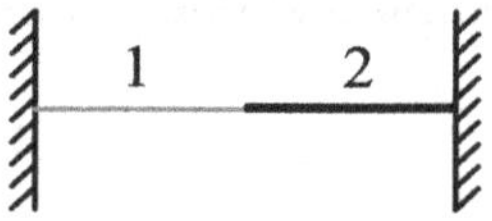

In the figure given : $l_1 = l_2 = l,\ \mu_1 = \frac{\mu_2}{9} = \mu.$

15. The lowest frequency such that the junction is a node is

1) $\frac{1}{2l}\sqrt{\frac{T}{\mu}}$ 2) $\frac{1}{l}\sqrt{\frac{T}{\mu}}$ 3) $\frac{4}{l}\sqrt{\frac{T}{\mu}}$ 4) $\frac{2}{l}\sqrt{\frac{T}{\mu}}$

16. The lowest frequency such that the junction is an antinode is

1) $\frac{1}{4l}\sqrt{\frac{T}{\mu}}$ 2) $\frac{3}{4l}\sqrt{\frac{T}{\mu}}$ 3) $\frac{5}{4l}\sqrt{\frac{T}{\mu}}$ 4) $\frac{7}{4l}\sqrt{\frac{T}{\mu}}$

Passage VI:

Two speakers S_1 & S_2 driven by the same amplifiers are placed at y=1m and y=-1m. The speakers vibrate in phase at 600 Hz. A man stands at a point on x-axis at a very large distance form the origin and starts moving parallel to y-axis. The speed of sound in air is 330 m/s.

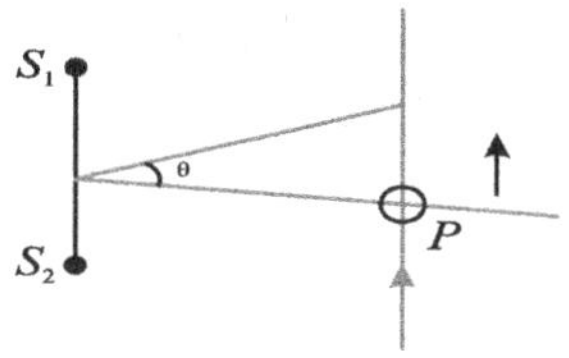

17. The angle θ at which intensity of sound drop to a minimum for the first time

1) 4° 2) 20.5° 3) 7.9° 4) 35°

18. The angle θ at which he will hear maximum intensity for first time?

1) 2° 2) 10° 3) 16° 4) 25°

19. If he continous to walk along the same line how many more maxima can he hear

1) one 2) two 3) 5 4) 10

Passage VII:

A detector is moving in a circular path of radius r in anticlock wise direction with a constant angular velocity ω as shown in the figure. At time t=0, it starts from the location shown at A, assuming source at rest,

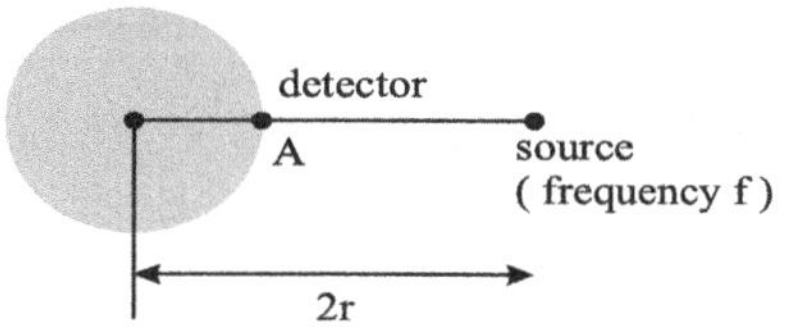

20. The frequency as received by the detector when it rotates by an angle $\frac{\pi}{2}$

1) $\left(\frac{V - r\omega}{V}\right)f$ 2) $\left(\frac{V - \frac{2}{\sqrt{5}}r\omega}{V}\right)f$ 3) $\left(\frac{V + r\omega}{V}\right)f$ 4) $\left(\frac{V + \frac{2}{\sqrt{5}}r\omega}{V}\right)f$

21. The time at which the detector will hear the maximum frequency for the 1st time

1) $\frac{\pi}{2\omega}$ 2) $\frac{3\pi}{2\omega}$ 3) $\frac{5\pi}{3\omega}$ 4) $\frac{2\pi}{3\omega}$

22. The time interval between minimum and maximum frequency as received by the detector

1) $\frac{\pi}{\omega}$ 2) $\frac{3\pi}{2\omega}$ 3) $\frac{4\pi}{3\omega}$ 4) $\frac{5\pi}{3\omega}$

LEVEL - IV - KEY

1)2	2)1	3)3	4)4	5)2	6)2	7)3	8)1	9)1	10)1
11)2	12)3	13)1	14)2	15)1	16)1	17)3	18)3	19)2	20)2
21)3	22) 3								

LEVEL - IV - HINTS

3. $T_A = \frac{M}{L}.x.g \Rightarrow v = \sqrt{\frac{T}{\mu}} = \sqrt{\frac{\frac{m}{L}.xg}{\frac{M}{L}}} = \sqrt{xg}$

4. $-\int_0^{\ell} dx = \int_0^{t} v.dt$

5. $L - x = \frac{1}{2}g\left[2\sqrt{\frac{x}{g}}\right]^2$

6. $y = 2a\sin kx, \cos wt$

7. nodes are at $(0,15,30,45,60)$ cm

8. $\frac{dy}{dt} = 0$ at $x = 7.5\,cm \;\&\; t = 0.25\,sec$

9. $\Delta\phi = \frac{4\pi}{3}$, $A = \sqrt{A_1^2 + A_2^2 + 2A_1A_2\cos\phi}$

10. $\Delta\phi = \frac{4\pi}{3}$, $A = \sqrt{A_1^2 + A_2^2 + 2A_1A_2\cos\phi}$

11. $\Delta\phi = 0$, $A = 0.05m$

12. $\lambda^1 = \frac{v - v_s}{f}$

13. $f^1 = \frac{f}{\lambda^1}$

14. $\lambda^{11} = \frac{v - v_o}{f^1}$

15. $f_1 = \frac{1}{2l}\sqrt{\frac{T}{\mu}}, \frac{2}{2l}\sqrt{\frac{T}{\mu}}, \frac{3}{2l}\sqrt{\frac{T}{\mu}}$ etc. (just like open pipe)

$f_2 = \frac{1}{2l}\sqrt{\frac{T}{9\mu}}, \frac{2}{2l}\sqrt{\frac{T}{9\mu}}, \frac{3}{2l}\sqrt{\frac{T}{9\mu}}$ etc.

or $f_2 = \frac{1}{6l}\sqrt{\frac{T}{\mu}}, \frac{1}{3l}\sqrt{\frac{T}{\mu}}, \frac{1}{2l}\sqrt{\frac{T}{\mu}}$ etc.

We see that $\frac{1}{2l}\sqrt{\frac{T}{\mu}}$ is the lowest frequency at which f_1 and f_2 both are equal. This corresponds to fundamental frequency (or first harmonic) of wire-1 and second overtone (or third harmonic of wire-2).

16. $f_1 = \frac{1}{4l}\sqrt{\frac{T}{\mu}}, \frac{3}{4l}\sqrt{\frac{T}{\mu}}, \frac{5}{4l}\sqrt{\frac{T}{\mu}}$ etc. (just like closed pipe)

$f_2 = \frac{1}{3(4l)}\sqrt{\frac{T}{\mu}}, \frac{3}{3(4l)}\sqrt{\frac{T}{\mu}}, \frac{5}{3(4l)}\sqrt{\frac{T}{\mu}}$ etc or $f_2 = \frac{1}{12l}\sqrt{\frac{T}{\mu}}, \frac{1}{4l}\sqrt{\frac{T}{\mu}}, \frac{5}{12l}\sqrt{\frac{T}{\mu}}$ etc.

Therefore, the lowest frequency is $\frac{1}{4l}\sqrt{\frac{T}{\mu}}$.

17. $\frac{xd}{D} = \frac{(2n-1)\lambda}{2}$ for minimum (n=1) $\frac{xd}{D} = \frac{\lambda}{2}$

$\theta = \frac{x}{D} = \frac{\lambda}{2d}\,radian$ or $\frac{\lambda}{2d}\left(\frac{180^o}{\pi}\right) = 7.9^o$

18. $\frac{xd}{D} = n\lambda \Rightarrow$ $\frac{x}{D} = \frac{\lambda}{d} = 16^o$

19. $Sin\theta = \frac{x}{D} \le 1$ and $\frac{n\lambda}{d} \le 1 \Rightarrow n \le \frac{d}{\lambda} \Rightarrow \frac{xd}{D} = n\lambda$ $\Rightarrow \frac{x}{D} = \frac{n\lambda}{d}$

20. 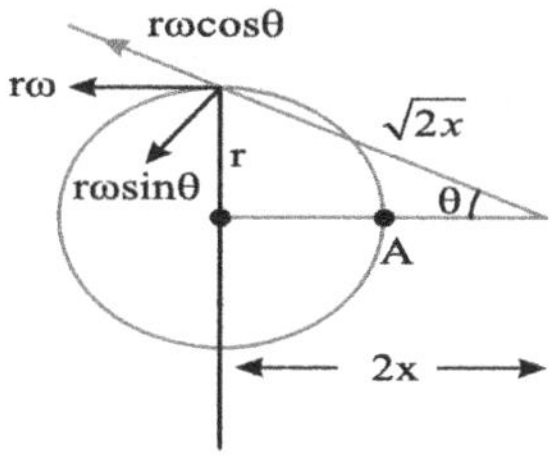

$F_{app} = \left(\frac{V - r\omega\cos\theta}{V}\right)f$

21. 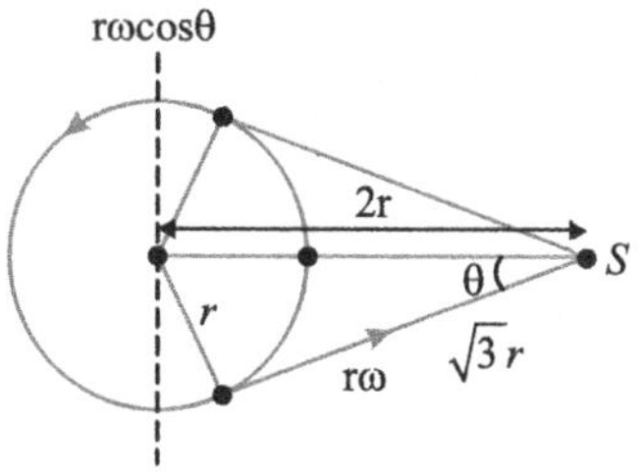

$\cos\theta = \frac{\sqrt{3}}{2} \Rightarrow \theta = \frac{\pi}{6}$

Total angle the detector moves is $\frac{3\pi}{2} + \frac{\pi}{6} = \frac{5\pi}{3}$

$\therefore time = \frac{5\pi}{3\omega}$

22. Time interval between minimum & maximum $= \left(\frac{\pi}{6} + \pi + \frac{\pi}{6}\right) = \frac{4\pi}{3\omega}$

Chapter - 10

RAY OPTICS

RAY OPTICS

SYNOPSIS

1. INTRODUCTION

i) Light is a form of energy which causes sensation of vision.

ii) Light not only makes things visible but also rises temperature and exerts pressure on them.

iii) Light travels with a speed of 3 x $10^8 ms^{-1}$ in vacuum. The speed of light is a scalar.

iv) Nature and propagation of light is explained by following theories.

2. THEORIES OF LIGHT:

i) Newton's Corpuscular Theory :

a) Light consists of tiny particles called corpuscles.

b) Corpuscles travel in straight lines.

c) Corpuscles are elastic in nature

d) Colour of light is due to different sizes of corpuscles.

e) When the corpuscles approach close to the surface of a material medium, then the path of the corpuscles is modified.

f) Reflection of light is due to the force of repulsion applied by the particles of the denser medium on the corpuscles.

g) Refraction of light is due to the force of attraction of the particles of the denser medium on the corpuscles.

h) This theory could explain reflection, refraction and dispersion.

i) This theory could not explain the phenomena of interference, diffraction and polarization.

Failures of theory:

j) According to this theory, the velocity of corpuscles increases as the temperature of source increases. But there is no such dependence of velocity light on temperature of source

k) Explanation of simultaneous reflection and refraction is not convincing

l) According to this theory, the velocity of light in denser medium must be greater than the velocityof light in rarer medium, but Foucault's rotating mirror experment showed that the velocity of light in denser medium is less than the velocity of light in rarer medium.

m) When a souce of light emits corpuscles, in due course of time the mass of the source of light should decrease. Which is not true experimentally.

ii) Huygen's wave theory :

a) Light is a form of wave (mechanical) which travels in a hypothetical medium called ether.

b) Ether was assumed to be a continuous medium which is homogeneous and isotropic with large elasticity and extremely low density.

c) The speed of light is less in denser medium than that in a rarer medium

d) A wave front is the locus of all particles vibrating in the same phase.

e) Every point on a wave front can be considered as secondary source of light from which spherical wavelets are spread out in forward direction with speed of light.

f) Huygen's principle helps to locate a wavefront at any future time if its present position is known.

g) The new wave front is the envelope of all the wavelets.

h) This theory could explain the phenomena like reflection, refraction of light, interference and diffraction.

Limitations of Huygen's wave theory of light:

i) It cannot explain the phenomenon of polarization.

j) It fails to explain the black body radiation, photoelectric effect and compton effect.

k) This theory fails to explain the absence of longitudinal component of light.

l) Ether was never discovered and Michelson -Morley experiment ruled out the existance of ether.

III) Electromagnetic wave theory :

a) Maxwell proposed that the light is propagated in the form of electromagnetic waves.

b) Electromagnetic waves are generated by an accelerating charges.

c) An accelerated charge produces sinusoidal time varying electric field which is associated with sinusoidal time varying magnetic field. The two fields are mutually perpenducular to each other constituting an electromagnetic wave.

d) If the electromagnetic wave is propagating along x - axis, then the electric and magnetic fields are represented by

$$E = E_0 \sin\omega\left(t - \frac{x}{c}\right), B = B_0 \sin\omega\left(t - \frac{x}{c}\right)$$

where E_0, B_0 are maximum values of electric and magnetic fields.

e) Sinusoidal time varying elelctric and magnetic fields are perpendicular to each other and also perpendicular to the direction of propagation of the wave i.e electromagnetic waves are transverse in nature.

f) If varying electric and magnetic fields are $\vec{E}$ and $\vec{B}$, then the direction of propagation of veclocity vector $\vec{V}$ is in the direction of $\vec{E} \times \vec{B}$.

g) Electromagnetic waves do not require material medium for their propagation.

h) The relation between the maximum values of the electric and magnetic fields is $E_0 = CB_0$ where 'C' is velocity of light in vacuum.

i) The average electric energy density is equal to the average magnetic energy density.

j) Intensity of an electromagnetic wave is equal to the product of average electric or magnetic energy density and velocity of light.

k) Accroding to Maxwell, speed of light in a medium is $V = \frac{1}{\sqrt{\mu\varepsilon}}$ and in vacuum $C = \frac{1}{\sqrt{\mu_0\varepsilon_0}}$

l) According to Weiner, photographic action of light is due to electric field only and not due to magnetic field. Hence electric field vector represents light vector.

m) Wave length range of visible light is 400 nm to 800 nm and frequency range is 4×10^{14}Hz to 8×10^{14}Hz.

n) This theory could explain the absence of longitudinal components in light waves.

o) This theory could explain the phenomena of light except photo–electric effect, compton effect etc.,

IV) Quantum theory :

a) Radiation emitted by luminous bodies consists of definite indivisible packets of energy called quanta.

b) Radiation of certain frequency ν emitted by a luminous body is made up of quanta of definite energy (E) given by $\boxed{E = h\upsilon}$

Where h is the plank's constant

h= 6.63×10^{-34} J – s

c) Phenomenon arising from interaction of light with matter (photoelectric effect, compton effect etc) are explained by quantum theory but not by wave theory.

d) Light has dual nature. It exhibits wave nature when interacts with another wave or during propagation, but exhibits particle nature when interacts with matter.

3. REFLECTION OF LIGHT :

i) Incident ray, reflected ray and normal drawn to the surface lie in the same plane.

ii) Angle of incidance (i)=Angle of reflection (r).

iii) Angle of deviation (d) = 180 – 2i

iv) For normal incidence, the deviation is 180^0

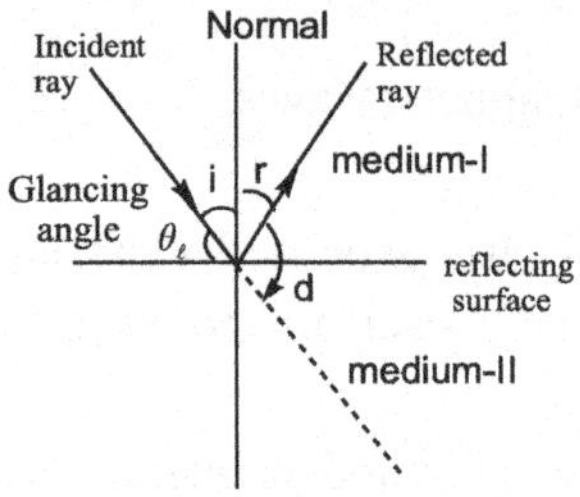

4. ABSOLUTE REFRACTIVE INDEX OF A MEDIUM:

i) According to Newton, the absolute refractive index of a medium is the ratio of speed of light in vacuum to that in medium.

$$\mu = \frac{\text{speed of light in vacuum (c)}}{\text{speed of light in a medium (v)}}$$

$$\therefore \; {}_{vacuum}\mu_{medium} = \frac{c}{v} = \frac{\lambda_{vacuum}}{\lambda_{medium}}$$

ii) Absolute refractive index of medium varies with

a) Wavelength of light or colour of light.

$\left(\mu_{violet} > \mu_{red} \text{ as } \lambda_{violet} < \lambda_{red}\right)$

b) Temperature of medium. In general, with increase of temperature, refractive index of the medium decreases.

iii) A medium in which the speed of light is more is known as optically rarer medium and a medium in which the speed of light is less, is known as optically denser medium.

5. REFRACTION OF LIGHT :

i) When a ray of light travels from one medium into another medium, it bends from its original path at the plane of separation of the two media due to change in the speed of light. This is called refraction of light.

ii) The angles made by the incident and refracted rays with the normal are called angle of incidence (i) and angle of refraction (r).

iii) a) When a ray of light goes from a rarer medium to a denser medium, it bends towards the normal and r <i.

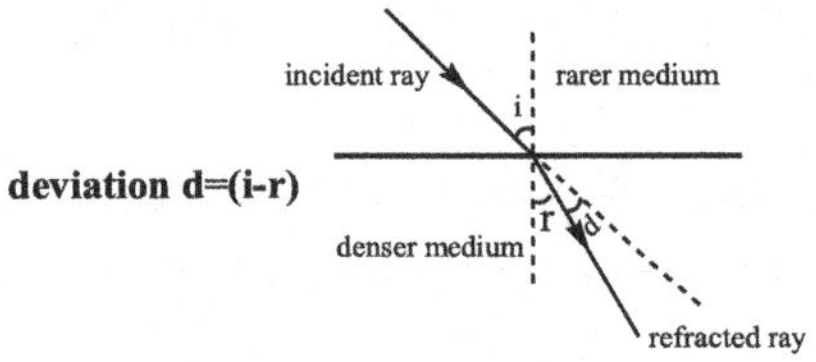

b) When a ray of light goes from a denser medium to a rarer medium, it bends away from the normal and r > i.

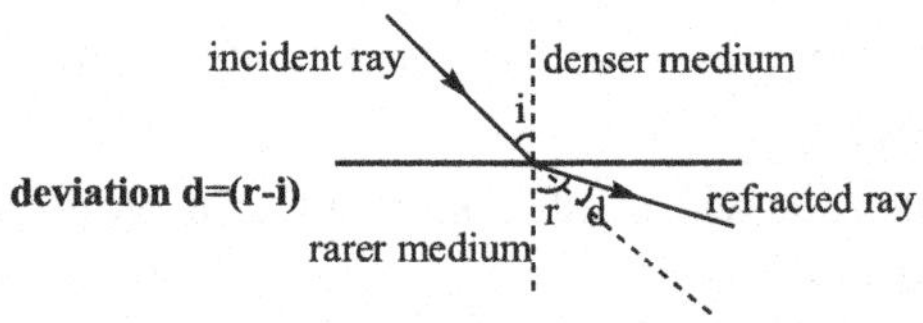

6. **SNELL'S LAW :**

i) When a ray of light travels from one medium to another medium, then the product of refractive index(μ) of a medium and the sine of angle(θ) between normal and the ray in that medium is constant. i.e, $\boxed{\mu \sin\theta = \text{constant}}$

ii) If θ_1 and θ_2 are the angles made by a ray of light with the normal in the two media of refractive indices μ_1 and μ_2 respectively, then

$\boxed{\mu_1 \sin\theta_1 = \mu_2 \sin\theta_2}$

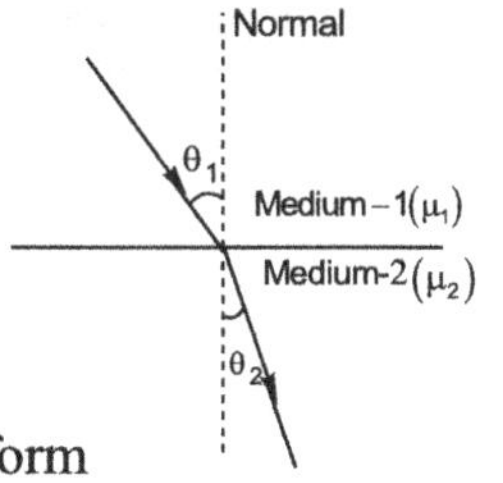

iii) Snell's law in vector form

$\boxed{\mu_1\left(\hat{i}\times\hat{n}\right) = \mu_2\left(\hat{r}\times\hat{n}\right)}$ where $\hat{i}, \hat{r}$ & $\hat{n}$ are the unit vectors along the direction of incident ray, refracted ray and the normal into the incident medium respectively.

iv) If a ray of light passes from vacuum to a medium of refractive index μ then, $\boxed{\mu = \dfrac{\sin i}{\sin r}}$

7. **RELATIVE REFRACTIVE INDEX :**

i) When a ray of light travels from medium (1) to medium (2), then the refractive index of medium (2) with respect to medium (1) is

i.e. $\boxed{{}_1\mu_2 = \dfrac{\mu_2}{\mu_1} = \dfrac{V_1}{V_2} = \dfrac{\lambda_1}{\lambda_2}}$

Where v_1 and v_2 are the speeds of light in medium (1) and medium (2) respectively.

ii) According to this principle if a light ray is reversed, it always retraces its path.

i.e, The refractive index of the second medium with respect to the first medium is equal to the reciprocal of the refractive index of first medium with respect to second medium.

$\boxed{{}_1\mu_2 = \dfrac{1}{{}_2\mu_1} = \dfrac{\mu_2}{\mu_1}}$

iii) If a ray of light travels a distance 'x' in time 't' in a medium of refractive index μ then

$\boxed{\mu = \dfrac{ct}{x}}$

iv) If x_1 and x_2 are the distances traveled by light in two media of absolute refractive indicies μ_1 and μ_2 respectively ***in a given time***, then $\boxed{\mu_1 x_1 = \mu_2 x_2}$

v) If a ray of light takes t_1 and t_2 times in two media of absolute refractive indices μ_1 and μ_2 respectively to travel *the* ***same distance*** then $\boxed{\mu_1 t_2 = \mu_2 t_1}$

vi) When a ray of light passes from one medium to another, its frequency remains same . Its velocity changes and hence wavelength changes.

vii) Refraction does not take place in the following situations.

i) For normal incidence $i = r = 0$.

ii) When refractive indices of two media are equal $\left(\mu_1 = \mu_2 \Rightarrow \angle i = \angle r\right)$.

viii) **OPTICAL PATH :** It is defined as the distance travelled by light in vacuum in the same time in which it travels a given path length in a medium.
If 'x' is the geometrical path length of a ray in a medium of refractive index μ then its optical path length in vacuum is μx.

ix) **LATERAL SHIFT :** Let a ray of light be incident on a parallel glass plate of thickness 't' at an angle 'i'. The emergent ray does not undergo deviation and dispersion but shifts parallel to the direction of incident ray. The perpendicular distance between incident and emergent rays is called **Lateral shift (x).**

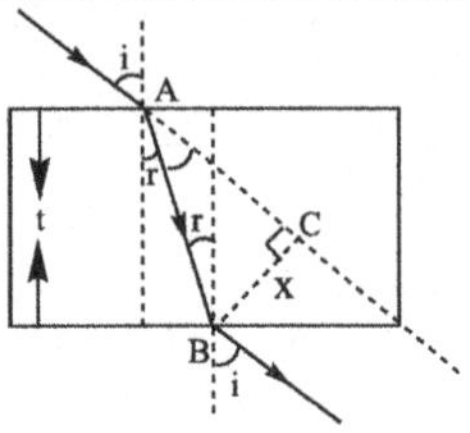

a) the distance traveled by light ray inside the glass plate is $\boxed{AB = \frac{t}{\cos r}}$,

r = angle of refraction in the glass plate.

b) the lateral shift $\boxed{x = \frac{t \sin(i-r)}{\cos r}}$

c) the time taken to cover the distance AB inside the slab is = $\frac{path\ length}{speed} = \frac{AB}{C_g}$

($C_g \rightarrow$ speed of light in the glass slab)

x) When a glass slab is placed in the path of a beam of divergent light, then the point of divergence shifts towards the glass slab by an amount $t\left[1-\frac{1}{\mu}\right]$

xii) When a glass slab is placed in the path of a beam of convergent light, then the point of convergence of light shifts away by an amount $t\left[1-\frac{1}{\mu}\right]$

Where t is the thickness of slab, μ is Refractive index of material of the slab.

xiii) As the light ray enters from one medium to another medium, if reflected and refracted rays are perpendicular to each other, then
(i = angle of incidence)

a) $\boxed{\mu = \tan i}$, if light enters from rarer to denser medium

b) $\boxed{\mu = \cot i}$, if light enters from denser to rarer medium

where μ is the refractive index of denser medium with respect to rarer medium.

xiv) **Real and apparent depths–Apparent shift :** If the object and an observer are situated in different media, then for the observer the object appears to be displaced from it's real position due to refractions.

Case (1) : Object in denser medium and observer in rarer medium

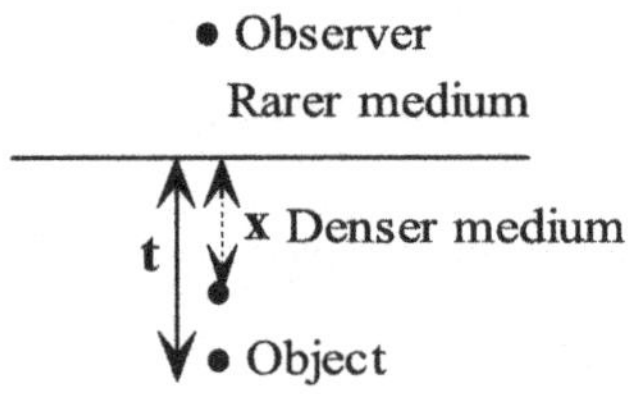

a) Apparent depth < Real depth

b) $_r\mu_d = \dfrac{\text{Real depth}(t)}{\text{apparent depth}(x)}$

c) Apparent shift $= s = t - x \Rightarrow \boxed{s = t\left(1 - \dfrac{1}{_r\mu_d}\right)}$

d) Three liquids of refractive indices μ_1, μ_2, μ_3 are filled in a vessel. Their thicknesses are d_1, d_2 & d_3 respectively. The apparent depth of an object placed at the bottom when seen from the top of the vessel will be

$$x = \frac{d_1}{\mu_1} + \frac{d_2}{\mu_2} + \frac{d_3}{\mu_3}$$

Total apparent shift 'S' is given by

$$S = d_1\left[1 - \frac{1}{\mu_1}\right] + d_2\left[1 - \frac{1}{\mu_2}\right] + d_3\left[1 - \frac{1}{\mu_3}\right]$$

Case (2) : Object in rarer medium and observer in denser medium

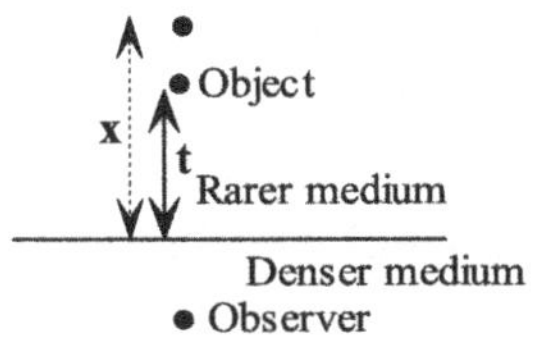

a) Apparent depth > Real depth

b) $_r\mu_d = \dfrac{\text{Apparent depth}(x)}{\text{Real depth}(t)}$

c) Apparent shift $= s = x - t \Rightarrow \boxed{s = t\left({}_r\mu_d - 1\right)}$

8. CRITICAL ANGLE AND TIR:

i) When a ray of light passes from optically denser medium to optically rarer medium, the angle of incidence in denser medium for which angle of refraction in rarer medium becomes 90^0 is called critical angle (C).

Applying Snell's law

$$\mu_d \sin C = \mu_r \sin 90^0 \Rightarrow \boxed{_r\mu_d = \frac{1}{\sin C}}$$

$$\boxed{\sin C = \frac{\mu_{rarer}}{\mu_{denser}} = \frac{V_{denser}}{V_{rarer}} = \frac{\lambda_{denser}}{\lambda_{rarer}}}$$

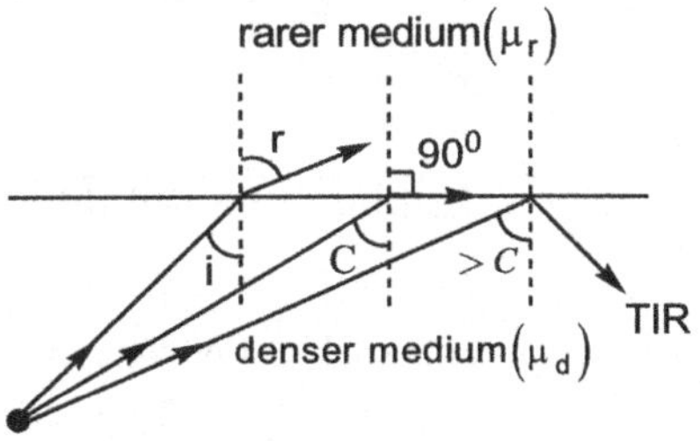

ii) The critical angle for a given pair of media varies with the colour of light and temperature .
$\Rightarrow (C)_{Red} > (C)_{Violet}$

iii) Generally, the critical angle of a medium is defined for the mean colour, yellow light.

iv) ***TIR :-*** When a ray of light passes from denser to rarer medium such that the angle of incidence in denser medium is greater than the critical angle, then the light is totally reflected back into the same denser medium according to laws of reflection. This phenomeon is called total internal reflection.

Conditions for TIR :-

a) The ray must travel from denser medium to rarer medium.

b) The angle of incidence in denser medium must be greater than the critical angle.

v) In total internal reflection of light, there is no loss of energy due to either absorption or by refraction.

vi) Examples for total internal reflection :

a) A well cut diamond glitters.
b) Mirages at hot places or deserts.
c) Looming at cold places
d) Shining of air bubble in water.
e) Working of optical fibre

vii) ***Field of vision of fish :*** - A fish at a depth 'h' from the surface of water of refractive index μ can see the outer world through an inverted cone with

a) Vertex angle = 2C

b) Radius of the circular base of the cone formed on surface of water is given by

$$\boxed{r = h \tan C = \frac{h}{\sqrt{\mu^2 - 1}}}.$$

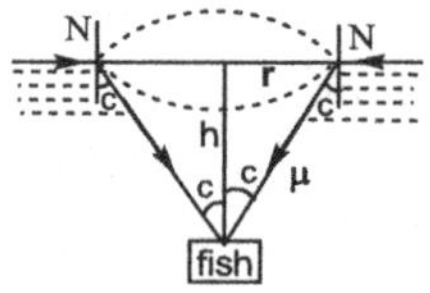

c) For water $2C = 98^0$

and $r = \dfrac{3h}{\sqrt{7}}$

9. OPTICAL FIBRE :

i) It works on the principle of total internal reflection of light.

ii) The inner cylindrical material made of glass or plastic which is the light guiding region is called core. Its radius is about few microns.

iii) The signal confines itself to the core during its propagation. It is also called as 'light pipe'.

iv) The core is enveloped by a concentric cylinder called cladding.

v) Core is optically denser than cladding.

$\mu_{core} > \mu_{cladding}$, generally $\mu_{core} = 1.7,\ \mu_{cladding} = 1.6$

vi) If μ_1 and μ_2 are the refractive indices of core and cladding, the launching angle of incidence for which light will be confined in the core is $\boxed{i_L \le \sin^{-1}\left(\sqrt{\mu_1^2 - \mu_2^2}\right)}$.

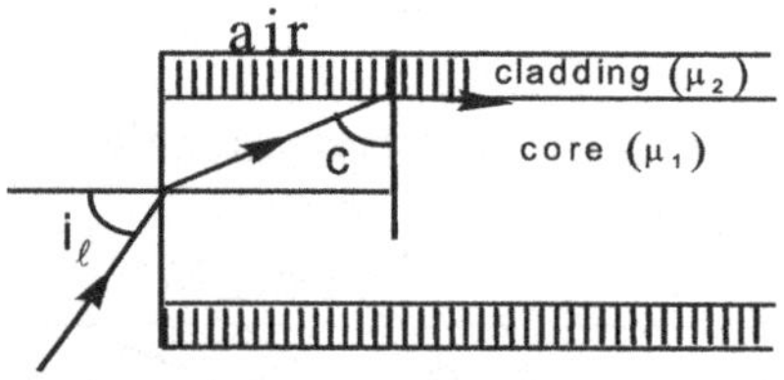

vii) Optical fibres are used : -

i) In communicatons, in medical equipments such as Laproscope and endoscope to observe unreachable parts in human body.

ii) As sensors in industry to measure temperature , pressure etc.

viii) Optical fibres are flexible, light weight and non – corrosive.

ix) Optical fibres are immune to interference from lightening and external electro magnetic radiations.

10. PRISM :

i) Prism is a transparent piece having two refracting surfaces with non-zero angle between them.

ii) The angle between two refracting faces at the refracting edge of the prism is called its apex angle or refracting angle or angle of the prism(A).

iii) The angle between the direction of emergent ray and the direction of incident ray is called the angle of deviation(d).

iv) AB , AC are refracting faces,

BC is the base.

i_1 = angle of incidence

i_2 = angle of emergence

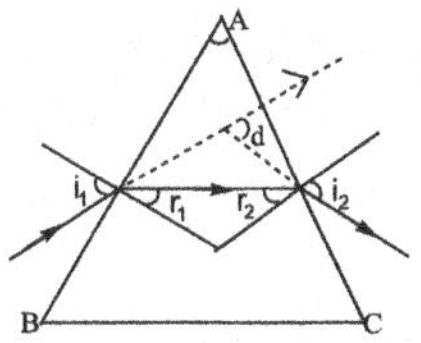

r_1, r_2 are angles of refraction inside the prism.

a) Angle of prism $\boxed{A = r_1 + r_2}$

b) Angle of deviation $\boxed{d = i_1 + i_2 - A}$

c) Refractive index of material of prism in air is $\boxed{\mu = \dfrac{\sin i_1}{\sin r_1} = \dfrac{\sin i_2}{\sin r_2}}$

v) The angle of deviation for a given prism and for a given colour of light depends upon the angle of incidence.

vi) As angle of incidence is gradually increased, initially the angle of deviation decreases reaches, a minimum value and then increases.

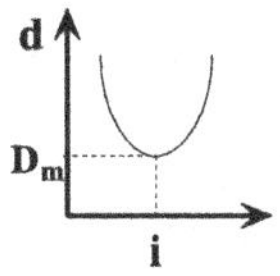

vii) The i - d graph is parabola.

viii) Under minimum deviation condition,

a) $i_1 = i_2 = i$, $r_1 = r_2 = r$, $d = D_m$

b) ray passes symmetrically w.r.to the faces of the prism or parallel to the base of the prism.

c) $D_m = 2i - A$ or $i = \dfrac{A + D_m}{2}$, and $r = A/2$

d) $\boxed{\mu = \dfrac{\sin\left(\dfrac{A + D_m}{2}\right)}{\sin(A/2)}}$, $\mu \rightarrow$ Refractive index of prism w.r.to the surroundings

ix) For prism made of denser medium and placed in rarer medium, the ray inside the prism bends towards the base of the prism.

x) For prism made of rarer medium and placed in denser medium, the ray inside the prism bends away from the base of the prism.

xii) When a glass prism is immersed in water, angle of deviation decreases.

xii) ***Normal incidence – grazing emergence :***

If the incident ray falls normally on the prism and grazes from the second surface, then

a) $i_1 = r_1 = 0$, $i_2 = 90^0$ and $r_2 = C = A$

b) $A = C = \sin^{-1}\left(\frac{1}{\mu}\right)$

c) Deviation $d = 90 - C$

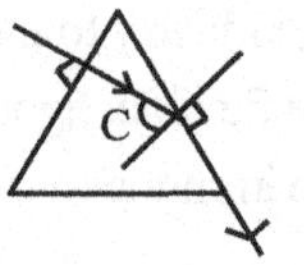

xiii) ***Grazing incidence – grazing emergence :***

If the incident ray falls on the prism with grazing incidence and grazes from the second surface, then

(i) $i_1 = i_2 = 90$

(ii) $r_1 = r_2 = C$

(iii) Angle of prism $A = 2C$

(iv) Deviation $d = 180 - 2C$

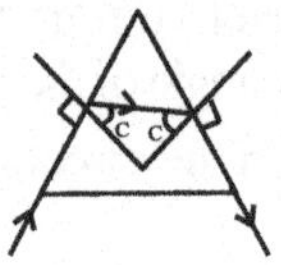

xiv) In order to have an emergent ray, the maximum angle of the prism is 2C, Where C is the critical angle of the prism w.r.t. the surrounding medium. ***2C is called the limiting angle of the prism.***

xv) If A>2C or $\mu > \operatorname{cosec}(A/2)$ then for any angle of incidence, light enters into the prism and undergoes total internal reflection at second surface of the prism. (no emergent ray)

xvi) The maximum angle of incidence at first face of the prism of angle A and refractive index μ for which the ray just undergoes total internal reflection at second face of the prism is

$$\boxed{i_{max} = \sin^{-1}\left[\left(\sqrt{\mu^2 - 1}\right)\sin A - \cos A\right]}$$

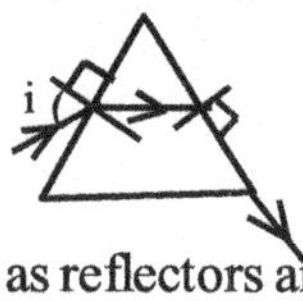

xvii) Prisms with A>2C are used as reflectors and called total reflecting prisms.

a) Rays are deviated through 90^0

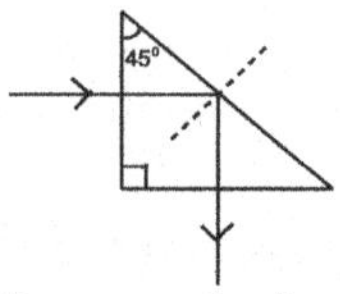

b) Rays are deviated through 180^0

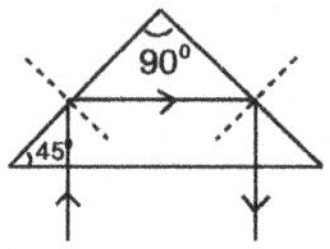

c) TIR prisms are used as ***erecting prisms***. (bright image is formed with 100% intensity due to reflection)

xviii) For a small angled prism ($A < 10^0$) placed in air

a) $i_1 = \mu r_1$, $i_2 = \mu r_2$

b) Angle of deviation = $\boxed{d = (\mu - 1)A}$

xix) For a small angled prism of refractive index μ_g placed in a medium of refractive index μ_m, then the angle of deviation is given by

$$\boxed{d_m = \left(\frac{\mu_g}{\mu_m} - 1\right)A}$$

11. DISPERSION :

i) The seperation of composite beam of light into its constituent colours is called dispersion and the medium which causes dispersion is known as dispersive medium.

ii) Refractive index of a medium is different for different colours of light $\Rightarrow$ different deviations for different colours $\Rightarrow$ dispersion.

iii) When a beam of white light is passed through a prism, then the angle between incident beam and emergent beam is called angle of deviation (δ)

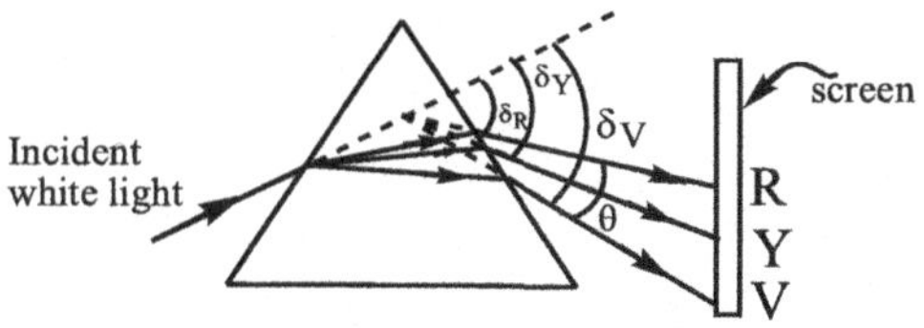

iv) For a given angle of incidence of composite beam of white light
  i) The angle of refraction is most for red and least for violet ($r_V < r_R$).
  ii) The angle of deviation is most for violet and least for red ($d_V > d_R$).
  iii) The refractive index is most for violet and least for red $(\mu_v > \mu_R)$.

v) The difference in the angles of deviations of any pair of colours is called angular dispersion for those two colours.

eg : – For violet and red

angular dispersion = $\boxed{\theta = (\delta_V - \delta_R)}$

vi) The ratio between angular dispersion and mean angle of deviation of a pair of colours is called the dispersive power of the medium (ω) for that pair of colours.

$$\boxed{\omega = \frac{\theta}{\delta_y} = \frac{\delta_V - \delta_R}{\left(\frac{\delta_V + \delta_R}{2}\right)}}$$

vii) The deviation of yellow ray is taken as mean deviation of violet and red. $\boxed{\therefore \omega = \frac{\delta_v - \delta_R}{\delta_y}}$

viii) For small angled prisms $\delta = (\mu - 1)A$
  a) Angular dispersion $= \theta = \delta_V - \delta_R = (\mu_V - \mu_R)A$
  b) Dispersive power (ω) of medium for violet and red is

$$\boxed{\omega = \frac{\theta}{\delta_y} = \frac{\delta_V - \delta_R}{\delta_y} = \frac{\mu_V - \mu_R}{\mu_y - 1}}, \mu_y = \frac{\mu_v + \mu_r}{2}$$

ix) For small angled prisms, the ***dispersive power*** of pair of colours depends only on nature of the material of the prism.

x) ***Angular dispersion*** of pair of colours depends on the nature of material and angle of prism.

xi) A single prism cannot produce dispersion without deviation (or) deviation without dispersion. This can be achieved by using two prisms with mutual inversion.

xii) ***Deviation without dispersion :*** Two prisms of different materials can be combined in such a way that no dispersion takes place with finite deviation (achromatic combination)

i,e $\theta + \theta^{|} = 0$ (or) $\omega\delta + \omega^{|}\delta^{|} = 0$ (or)

$(\mu_V - \mu_R)A + (\mu^{|}_V - \mu^{|}_R)A^{|} = 0$

and net deviation $= \delta + \delta^{|} = (\mu - 1)A + (\mu^{|} - 1)A^{|}$

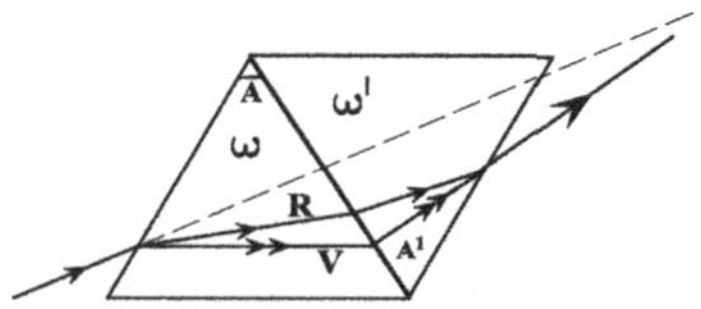

xiii) ***Dispersion with out deviation:*** Two prisms of different materials can be combined in such a way that no deviation takes place with finite dispersion.

$\delta + \delta^{l} = 0$ (or) $(\mu - 1)A + (\mu^{l} - 1)A^{l} = 0$

(or) $\frac{\theta}{\omega} + \frac{\theta^{l}}{\omega^{l}} = 0$

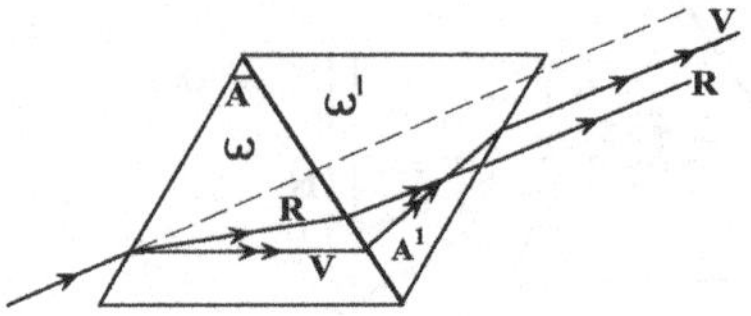

and net dispersion $= \theta + \theta^{l} = \omega\delta + \omega^{l}\delta^{l}$

$$= \theta\left(1 - \frac{\omega^{l}}{\omega}\right)\left(\because \delta^{l} = -\delta\right)$$

12. REFRACTION AT SPHERICAL SURFACE :

i) ***Law of refraction :*** Figure shows the geometry of formation of image (I) of an object (o) on the principal axis of the spherical surface with centre of curvature C, and radius of curvature R. When light ray enters from a medium of refractive index μ_1 to another medium of refractive index μ_2, then we have

$$\boxed{\frac{\mu_2}{v} - \frac{\mu_1}{u} = \frac{\mu_2 - \mu_1}{R}}$$

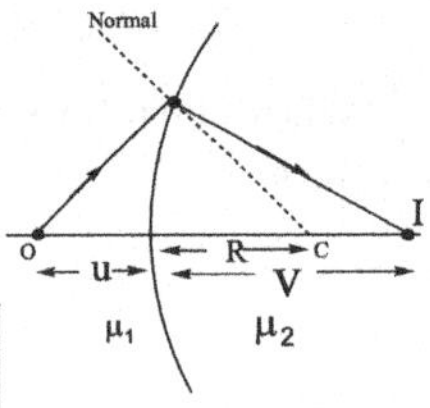

ii) Lens formula $\boxed{\frac{1}{f} = \frac{1}{v} - \frac{1}{u}}$

where 'f' is focal length, 'u' is object distance and 'v' is image distance.

iii) Lens maker's formula for thin lens.

$$\boxed{\frac{1}{f} = \left(\frac{\mu_2}{\mu_1} - 1\right)\left(\frac{1}{R_1} - \frac{1}{R_2}\right)}$$

μ_2 = refractive index of material of lens.

μ_1 = refractive index of surrounding medium.

iv) Sign convention :

a) All the distances are measured from optic centre.

b) The distances measured along the direction of incident ray are taken as poisitive (+Ve)

c) The distances measured in the opposite direction of incident ray are taken as negative (–Ve)

v) For diverging meniscus

$$\frac{1}{f} = -(\mu - 1)\left(\frac{1}{R_1} - \frac{1}{R_2}\right), \text{ (if } R_1 < R_2)$$

vi) **For converging meniscus**

$$\frac{1}{f} = (\mu - 1)\left(\frac{1}{R_1} - \frac{1}{R_2}\right), \text{(if } R_1 < R_2)$$

vii) **Convex lens (Converging lens)**

$$\frac{1}{f} = (\mu - 1)\left(\frac{1}{R_1} + \frac{1}{R_2}\right)$$

$$\frac{1}{f} = (\mu - 1)\frac{2}{R} \text{ (for equi convex } R_1 = R_2 = R)$$

$$\frac{1}{f} = \frac{(\mu - 1)}{R} \text{ (for plano convex)}$$

viii) **Concave lens (diverging lens)**

$$\frac{1}{f}=-(\mu-1)\left(\frac{1}{R_1}+\frac{1}{R_2}\right)\Rightarrow\frac{1}{f}=\frac{-(\mu-1)2}{R}$$

(for equi concave lens $R_1=R_2=R$)

$$\frac{1}{f}=\frac{-(\mu-1)}{R}$$ (for plano concave lens)

ix) Focal length of a lens depends upon
a) colour of light
b) nature of the medium of the lens
c) nature of the medium in which the lens is submerged
d) temperature of the medium
e) 'f' is maximum for red and minimum for violet colour of light $\left(\frac{1}{f}\propto\left({}_1\mu_2-1\right)\right)$

x) If a lens made of material of refractive index μ_g is placed in air, its focal length is f_a and if immersed in a liquid of refractive index μ_ℓ then its focal length becomes f_ℓ then $\frac{f_\ell}{f_a}=\frac{\left(\frac{\mu_g}{\mu_a}-1\right)}{\left(\frac{\mu_g}{\mu_\ell}-1\right)}$

μ_g = refractive index of the lens material
μ_ℓ = refractive index of the liquid
a) If $\mu_g>\mu_\ell$ then $f_\ell>f_a$. i.e, the nature of lens remains unchanged.
b) If $\mu_g<\mu_\ell$ then f_ℓ and f_a have opposite signs and the nature of lens changes.
c) If $\mu_g=\mu_\ell$ then $f_\ell=\infty$ it means the lens behaves as a plane glass plate and it cannot be distinguished from the liquid.

xi) Water drop in air behaves as a convergent lens.
xii) Air bubble in water or glass behaves as a divergent lens.
xiii) Water drop in glass behaves as divergent lens.
xiv) The focal power (p) of a lens is numerically equal to the reciprocal of its focal length in meters.
a) when f is measured ***in meters,***

then $\boxed{p=\frac{1}{f}}$ dioptre.

b) when f is measured ***in centimeters,***

$\boxed{p=\frac{100}{f}}$ dioptre.

xv) S.I. unit of focal power is ***'Dioptre' (D).***
xvi) Power of a convex lens is positive and for a concave lens is negative.

xvii) If two lenses are separated by a distance 'd' then the effective focal length of the combination is given by

$\boxed{\frac{1}{f}=\frac{1}{f_1}+\frac{1}{f_2}-\frac{d}{f_1f_2}}$. (proper sign must be given to f_1 and f_2)

xviii) If two lenses of focal powers P_1, P_2 are separated by a distance 'd' (in meters) then effective focal power of the combination is given by (proper sign must be given to P_1 and P_2)

$$\boxed{P = P_1 + P_2 - dP_1 P_2}$$

xix) If two convex lenses of focal lengths f_1, f_2 are separated by a distance 'd' such that the combination will act as parallel sided glass slab, then $\boxed{d = f_1 + f_2}$

xx) An equi-convex lens of focal length 'f' is cut into two equal plano-convex lenses. When they are arranged as shown; their equivalent focal length is unchanged.

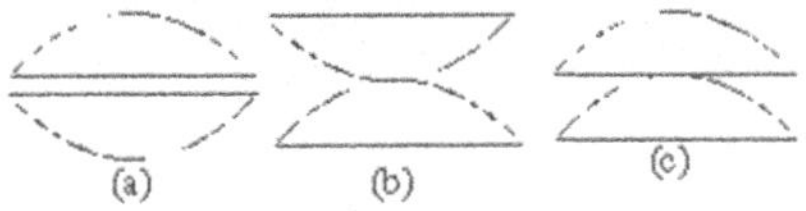

(Treat lenses in contact in these figures)

Focal length of each small lens becomes 2f.

$$\therefore \frac{1}{F} = \frac{1}{2f} + \frac{1}{2f} = \frac{1}{f} \therefore F = f.$$

xxi) If a lens of focal length 'f' is cut into two equal parts and arranged as shown, then their effective focal length is given by

a) f

b) f/2

c) Infinite

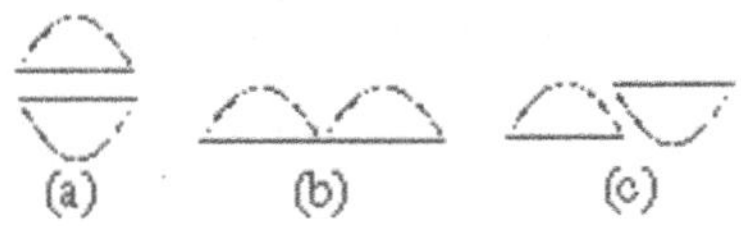

xxii) When a lens is made up of "n" different materials as shown in the figure. No. of images formed by the lens is "n".

13. ABERRATIONS :

a) The images formed by lenses have some defects. These defects may be in shape, size, position and colour. Such defects are called Aberrations.

b) The lens makers formula is applicable for thin lenses and for paraxial rays only.

c) According to lens makers formula, point objects illuminated by monochromatic light should form point images. But the images formed by lenses are not as expected, because all the rays are not paraxial. Such defects are called monochromatic aberrations.

d) Monochromatic aberrations are of different types. (1) Spherical aberration. (2) Astigmatism. (3) Coma, (4) Curvature. (5) Distortion.

In addition to the monochromatic aberrations, we also observe that when an object is illuminated by white light the image formed is coloured which is called chromatic aberration.

i) **spherical aberration :**

a) When a ray of light parallel to principal axis is incident on a lens at a height h from the principal axis it is deviated through an angle δ as shown in Fig. It meets the principal axist at I. It can be seen that in ΔOAI.

$$\tan \delta = \frac{h}{OI}$$

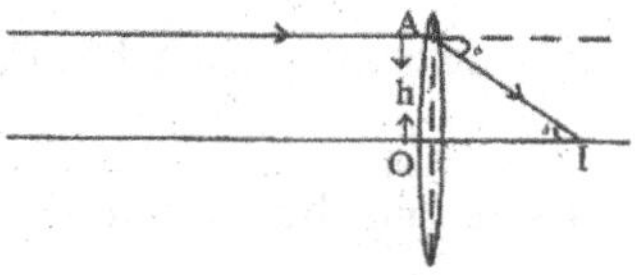

b) Paraxial rays are deviated less and marginal rays are deviated more. Hence, the focal length of a lens is greater for paraxial rays than for marginal rays.

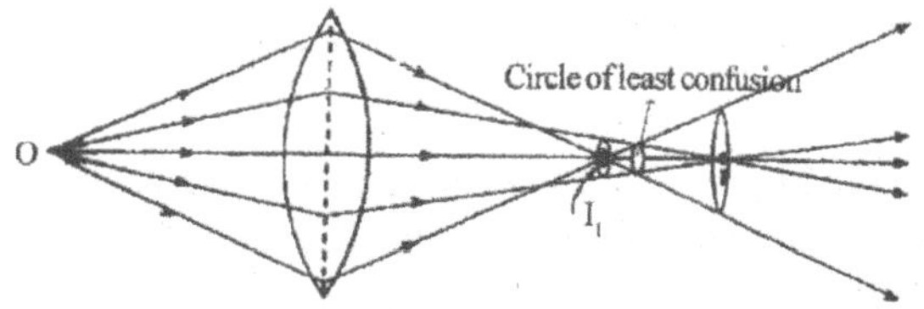

c) If a point object 'O' is palced on the axis of a convex lens of large aperture its image due to paraxial rays forms at I. The image due to marginal rays is formed at I_1. Thus, for a point object 'O' on the axis the image is not a point image but it is a line I I_1 along tha axis as shown in fig.

d) This formation of line image along the principal axis is called spherical aberration. I I_1 is a measure of longitudinal aberration.

e) If a screen is palced perpendicualr to the principal axis at I, the image is in the form of a circle with a bright centre and the intensity gradually decreases towards the edge. When the screen is at I_1, the centre is faint and the outer edge is bright. In between, there will be a circualr patch of light of almost uniform intensity. This circle is called the circle of least confusion. The radius of the circle of least confusion is a measure of lateral or transverse spherical aberration.

f) The spherical aberration produced by a convex lens is taken to be positive and by a concave lens is negative. So, the longitudinal spherical aberrations are opposite for convex and concave lenses.

ii) **Reducing the Spherical aberration :**

a) Spherical aberration occurs only when the aperture of the lens is large. This can be minimised by using "stops". The function of a stop is to prevent the light rays to pass through and to decrease the effictive aperture of lens. If the central pencil of rays is stopped, then the marginal rays give sharp image.

b) To decrease the defect, tht total deviation for the marginal rays should be brought about gradually and by equal amounts at a large number of refracting surfaces.
The radii of curvature of the surfaces of a lens are so chosen as to bring about equal deviations at the two surfaces. Such a lens is called a crossed lens. For crown glass of refractive index 1.50, the crossed lens for parallel rays is a convex lens of radii of curvature in the ratio 1 : –6 with the more curved surface facing the incident light.

c) A system of two convergent lenses of focal lengths f_1 and f_2 separated by a distance d equal to the numerical difference between the two focal lengths can be used to minimise spherical aberration. The reason is that the system gives equal increments of deviation to light rays that are initially parallel to the principal axis.

d) Spherical aberration can also be reduced by using a suitable combination of convex and concave lenses becuase a convex lens has positive defect and a concave lens has negative defect.

iii) **chromatic aberration :**

a) The refractive index, μ depends upon the colour of the light which is large for violet colour. Hence, the focal length for violet colour is less than all other colours.

b) When white light sent out by a point object on the axis of a lens is brought to focus the image formed is coloured and spread along the axis from I_1 to I_2 as shown in Fig.

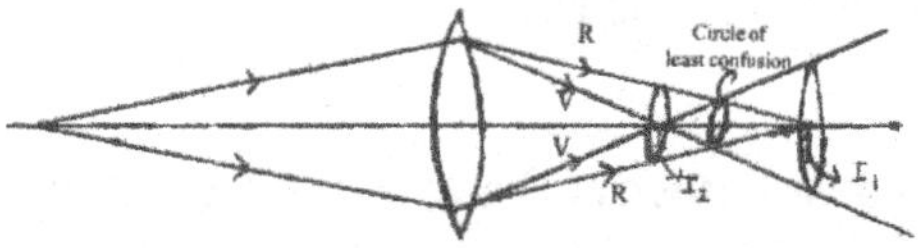

c) If a screen is palced perpendicular to the principal axis at I_1, the image is in the form of a circle with red center, periphery violet and the remaining colours between them.

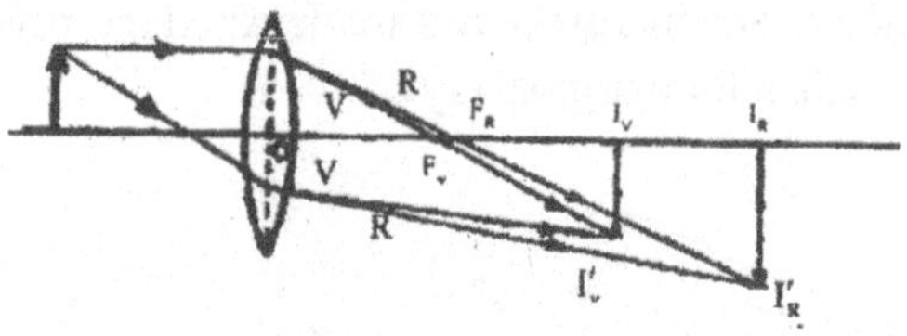

d) For an extended object the images formed by light rays of different colours are of different sizes as shown in Fig. The separation between the images formed by extreme colours of visible light i.e., $(I_R - I_V)$ is called the ***axial or longitudinal chromatic aberration.***

e) $(I_R I_R^1 - I_V I_V^1)$ is called ***transverse or lateral chromatic aberration***.

vi) **Reducing the chromatic aberration :**

a) A converging lens of smaller focal length. f_1 and small dispersive power, ω_1(crown glass) and diverging lens of larger focal length, f_2 and larger dispersive power, ω_2(flint glass) are kept in contact to reduce the chromatic aberration. This combination is called achromatic doublet. It satisfies the condition.

$$\frac{\omega_1}{f_1} + \frac{\omega_2}{f_2} = 0$$

b) A combination of two converging lenses of the same material separated by a distance of half the sum of the focal lengths of the two lenses minimizes chromatic aberration.

14. Optical Instruments :

i) The nearest point at which an object is seen clearly by the eye is called the "near point" of the eye.

ii) The least distance upto which on object can be clearly seen by a naked eye is called the least distance of distant vision. That is $D = 25$ cm for normal eye.

iii) The farthest point from an eye at which an object is distinctly seen is called far point, for a normal eye it is theoretically at infinity.

iv) <u>**Visual angle :**</u> It is the angle subtended by an object at the eye.

It is maximum when the object is at the least distance of distinct vision.

It 'h' is the height of the object placed at the near point, then the visual angle is

$$\theta_0 = \frac{h}{D} (\because \text{the angle is small})$$

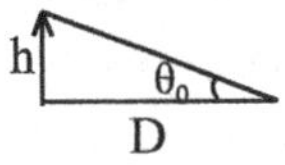

15. MICROSCOPES :

It is an optical instrument used to see very small objects. It's magnifying power is given by

$$m = \frac{\text{Visual angle with instrument}}{\text{Visual angle when object is placed at least distance of distinct vision}}$$

$$m = \frac{\tan\beta}{\tan\alpha} \Rightarrow m \approx \frac{\beta}{\alpha}$$

16. SIMPLE MICROSCOPE :

i) It is a single convex lens of lesser focal length.

ii) It is also called magnifying glass or reading lens.

iii) Magnification $m = D/u$.

iv) Magnification when final image is formed at D and ∞ (i.e. m_D and m_∞)

$$m_D = \left(1 + \frac{D}{f}\right)_{max} \text{ and } m_\infty = \left(\frac{D}{f}\right)_{min}$$

v) If the lens is kept at a distance a from the eye then $m_D = 1 + \frac{D-a}{f}$ and $m_\infty = \frac{D-a}{f}$

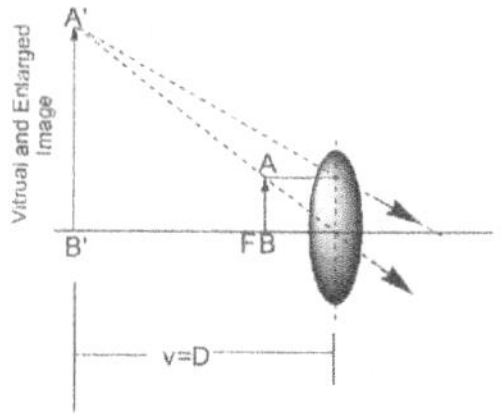

17. COMPOUND MICROSCOPE :

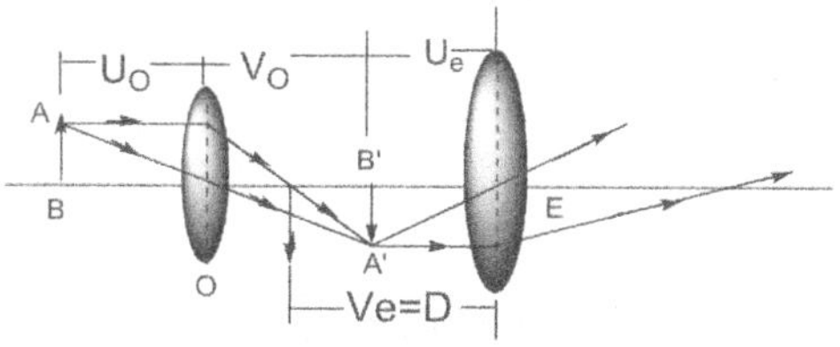

i) It consists of two converging lenses called objective and eye lens.

ii) $f_{\text{eye lens}} > f_{\text{objective}}$ and $(\text{diameter})_{\text{eye lens}} > (\text{diameter})_{\text{objective}}$

iii) Intermediate image is real and enlarged.

iv) Final image is magnified, virtual and inverted.

v) u_0 = Distance of object from objective (o),

υ_0 = Distance of image (A/B/) formed by objective from objective,

u_e = Distance of A/B/ from eye lens,

υ_e = Distance of final image from eye lens,

f_0 = Focal length of objective,

f_e = Focal length of eye lens.

vi) Magnification $m = \frac{\upsilon_0}{u_0}\left(\frac{D}{u_e}\right)$ and the length of the tube is $L = \upsilon_0 + u_e$.

vii) ***Final image formed at D :*** Magnification $m_D = -\frac{\upsilon_0}{u_0}\left(1+\frac{D}{f_e}\right)$ and length of the microscope tube (distance between two lenses) is $L_D = \upsilon_0 + u_e$.

Generally object is placed very near to the principal focus of the objective hence $u_0 \cong f_0$. The eye piece is also of small focal length and the image formed by the objective is also very near to the eye piece. So $\upsilon_0 \cong L_D$, the length of the tube.

Hence, we can write $m_D = \frac{-L}{f_0}\left(1+\frac{D}{f_e}\right)$

viii) ***Final image formed at*** ∞

$m_\infty = -\frac{\upsilon_0}{u_0}.\frac{D}{f_e} \approx \frac{LD}{f_0 f_e}$ and length of tube $L_\infty = \upsilon_0 + f_e$

ix) For large magnification of the compound microscope, both f_0 and f_e should be small.

x) The magnifying power of the compound microscope may be expressed as $m = m_0 \times m_e$; where m_0 is the magnification of the objective and m_e is magnification of eye piece.

18. ASTRONOMICAL TELESCOPE (REFRACTING TYPE)

i) By it is used to see heavenly bodies are seen.

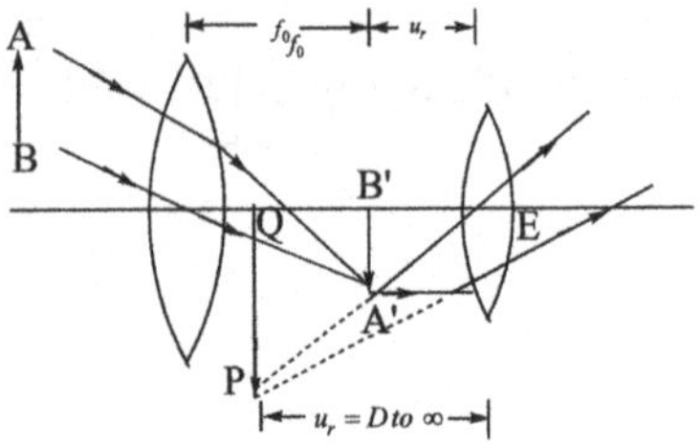

ii) $f_{objective} > f_{eyelens}$ and $d_{objective} > d_{eyelens}$
(d = diameter)

iii) Intermediate image is real, inverted and small (diminshed).

iv) Final image is virtual, inverted and diminshed.

v) Magnification : $m = -\frac{f_0}{u_e}$ and length. $L = f_0 + u_e$

vi) Magnification : $m_D = -\frac{f_0}{f_e}\left(1+\frac{f_e}{D}\right)$ and $m_\infty = -\frac{f_0}{f_e}$ (for normal adjustment)

vii) Length : $L_D = f_0 + u_e$ and $L_\infty = f_0 + f_e$

19. TERRESTRIAL TELESCOPE :

i) It is used to see distant objects on the earth.

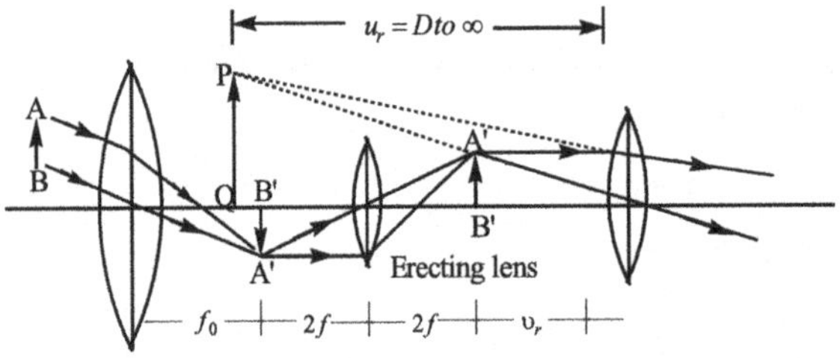

ii) It consists of three converging lens : objective, eye lens and erecting lens.

iii) Final image is virtual, erect and diminshed.

iv) ***Magnification :*** $m = \frac{f_0}{u_e}$

$m_D = \frac{f_0}{f_e}\left(1+\frac{f_e}{D}\right)$ and $m_\infty = \frac{f_0}{f_e}$

(for normal adjustment)

v) ***Length :*** $L = f_0 + 4f + f_e$, $L_D = f_0 + 4f + u_e$ and $L_\infty = f_0 + 4f + f_e$ (f = focal length of erecting lens)

20. EYE PIECES

a) When a single lens is used as an eye-piece, all the rays of light after refraction through the objective may not pass through the eye piece. Due to this, the field of view decreases. The image also suffers from spherical and chromatic aberrations.

b) A combination of plano convex lenses is used as an eyepiece. Then the lens on the side of the objective is called field lens and the other as eye lens.

c) The field lens increases the field of view and eye lens acts as a magnifier.

i) RAMSDEN'S EYEPIECE

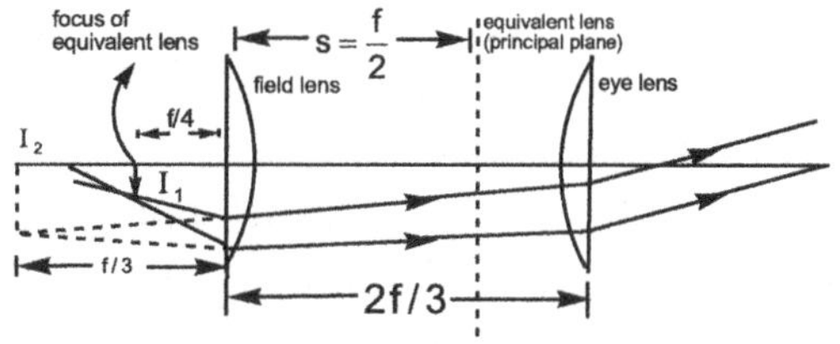

a) It consists of two plano convex lenses with their curved surfaces facing each other.
b) They have same focal length f, separated by a distance of 2f/3.
c) Effective focal length is 3f/4
d) In a combination of lenses, distances must be measured from the principal plane. Its distance (s) from the field lens is given by. $\boxed{S = \frac{F\ d}{f_e}}$ where F=effective focal length, d=2f/3 distance between the lenses, f_e=focal length of eyelens.
e) Position of equivalent lens or principal plane is at a distance of f/2 behind the field lens.
f) Focal plane is at a distance of 3f/4 from the principal plane and infront of the field lens.
g) The principal focus of equivalent lens is at a distance of f/4 before the field lens.
h) Objective produces the image I_1 at the focus of equivalent lens (f/4 before field lens)
i) I_1 behaves as real object to field lens. As it is within its focus, a virtual image I_2 is formed at a distance of f/3 before the field lens.
j) I_2 acts as an object to the eye lens. As I_2 is at its focus, final image is formed at infinity.
k) Cross - wires are placed where the image I_1 is formed due to the objective i.e, f/4 before the field lens or 11f/12 from the eye lens.
l) This can be used for measurements.
m) This can be used as a magnifier, because first principal plane lies outside the eye-piece.
n) This is called a positive - eyepiece since the image due to the objective forms in front of the field lens.
o) The spherical and chromatic aberrations are not eleminated.
p) The spherical aberration is reduced since the total deviation produced by the eye piece is shared by four surface. By taking each lens as an achromat, the chromatic aberration can also be reduced.
q) These eyepieces are used in measuring instruments like travelling microscope, spectrometer etc.

ii) HUYGENS' EYE PIECE

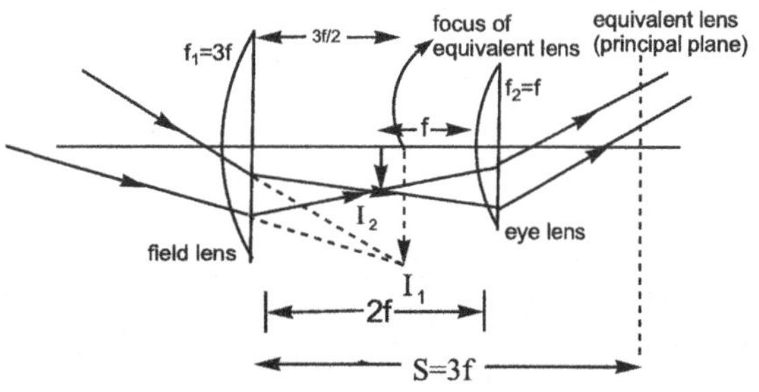

a) It consists of two plano-convex lenses of focal lengths 3f and f.
b) The lens with greater focal length is called field lens.
c) The two lenses are separated by a distance of 2f.
d) The convex surfaces face towards the object or the plane surfaces face towards the eye.
e) The effective focal length of the eyepiece is 3f/2

f) In a combination of lenses, the distances must be measured from the principal plane. Its distance (s) from the field lens is given by. Here s = 3f. $s = \frac{F\,d}{f_e}$ where F=effective focal length, d=2f distance between the lenses, f_e=focal length of eyelens.

g) The distance of equivalent lens is '3f' from field lens and 'f' from eye lens.

h) The principal focus of the combination is at f/2 from the eye lens.

i) The objective should form the image I_1 at the focus of equivalent lens. ie., at distance of 3f/2 behind the field lens.

j) I_1 behaves as object to field lens and it produces a real image I_2 at the focus of eyelens. ie., at the mid point of the two lenses.

k) I_2 behaves as object to eyelens, due to which the final image is formed at infinity.

l) The field lens must always produce the image at the focus of eyelens.

m) As the image due to the objective is behind the field lens, the eyepiece is called negative eyepiece.

n) As the separation $d = f_1 \sim f_2$, the spherical abberation is minimized and as $d = \frac{f_1 + f_2}{2}$, the chromatic abberration is also minimized.

o) The image is magnified by both the lenses where as scale is magnified by single lens.

p) Cross wires and the final image are magnified disproportionately and hence measurements are not relaible.

q) This is used in biological instruments where cross wires measurements are not required.

21. SPECTRA :

i). The coloured pattern obtained on the screen after dispersion of light is called spectrum.

ii). The spectrum without overlapping of colours is called pure spectrum.

iii). The spectrum with overlapping of colours is called impure spectrum (eg : Rainbow)

iv). Conditions for obtaining pure spectrum are

a) A source of light in the form of a narrow slit.

b) A convex lens is used for making the incident beam of parallel rays.

c) A prism must be set in the minimum deviation position at least for mean colour yellow.

d) Another convex lens is used to focus the dispersed light such that rays of a particular colour converge at a particular point only.

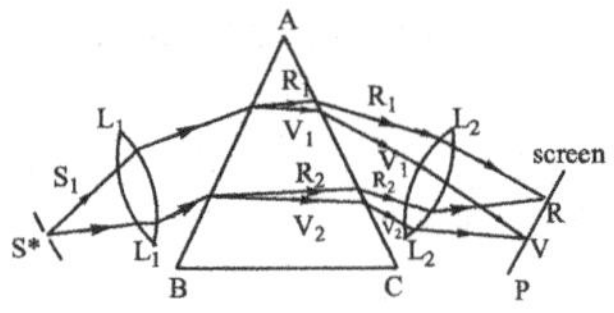

v) Spectra are of two types

a) Emission spectra

b) Absorption spectra

vi) Emission spectra is due to emission of radiation by substances when they are excited. In emission spectra, bright lines appear against dark back ground. It is further classified into two types.

a) Continuous spectra

b) Discontinuous spectra

vii) Continuous spectra contains all the wavelengths from violet to red without any dark space.

viii) Continuous spectra are emitted by

a) Incandescent solids or liquids

b) Dense luminous gas flame. (As the flame contains incandescent carbon particles)

ix) In continuous spectra, as the molecules are close together in solids, liquids and dense gas flame, the energy levels of atoms are closely spaced resulting in the overlap of different wavelengths one over the other.

x) Discontinuous spectra contains lines (or) bands of different colours with dark space between them. It is further classified into two types.

a) Line spectra

b) Band spectra

xi) Line spectra is due to transition of electron in atom from a higher energy level (E_2) to a lower energy level (E_1) and the difference in energy levels is released in the form of spectral line of frequency. $\boxed{\upsilon = \frac{E_2 - E_1}{h}}$

xii) Line spectra is due to individual atoms or ions which are in excited state, it is also called atomic spectra.

xiii) Line spectra are produced by

a) Atoms of luminous gases (or) vapour when electric discharge passes through them.

b) Burning the salts in the Bunsen burner flame.

c) Iron, copper and alloys like brass taken in the form of rods and an electric arc is struck between them.

xiv) Band spectra is due to molecules in excited state, it is also called molecular spectra.

xv) Band spectra consists of several bright bands inter spaced with dark spaces. One end of the band appears sharp called head and the other end gradually shades off.

xvi) The total energy of a molecule is regarded as the sum of translational, rotational, vibrational, electric and nuclear energies.

xvii) Translational energy of a molecule is not quantized, hence a molecule cannot produce line spectrum.

xviii)As rotational, Vibrational and electronic energies are quantized, fine lines are observed in the band.

xix) When light is transmitted through a material, some of the incident radiaiton is absorbed by the material and dark lines appear at the corresponding wavelengths producing absorption spectrum.

xx) A substance absorbs only that radiation which it can emit in emission spectrum.

xxi) In absorption spectrum, dark lines appear against bright background, where as in emission spectrum, bright lines appear against dark back ground.

xxii) When light from photosphere of sun passes through chromosphere gases and vapour it absorps light of certain wavelengths

(Which they emit at higher temperatures).

The corresponding missing wavelengths appear as dark lines called as fraunhoffer lines.

xxiii) Fraunhoffer lines give information regarding the nature of the elements present in the sun's atmosphere.

xxiv) Helium is discovered using the analysis of fraunhoffer lines, apart from nearly 60 elements present in the sun.

xxv) Spectral analysis is used.

a) To determine chemical composition even if very small sample is available.

b) To determine the abundance of an element in a sample (eg :– in mineral exploration)

c) In astronomy to study chemical composition of sun, stars, and galaxies.

(**Eg** :– Their temperature and motion)

d) In metallurgy, machine tool production, as it is a highly sensitive, simple and quick process.

EXERCISE - I

1. **The serious drawback of Huygen's wave theory of light is the**
   1) concept of wave front 2) concept of light rays
   3) ether postulate 4) concept of secondary wavelets

2. **Electromagnetic theory was failed to explain**
   1) reflection 2) refraction 3) total internal reflection 4) stability of atom

3. **Quantum theory of light was put forward by Planck to explain**
   1) Photo electric effect 2) Compton effect 3) Black body radiation 4) Raman effect

4. **The colour of light is determined by its**
   1) velocity 2) frequency 3) amplitude 4) intensity

5. **μ_a, μ_b, μ_c are refractive indices of the media A, B, C respectively, so that $\mu_a > \mu_b > \mu_c$. The critical angle arises when a ray of light travels from**
   1) C to A 2) C to B 3) B to A 4) A to C

6. **A, B and C are three optical media of respective critical angles C_1, C_2, and C_3. Total internal reflection of light can occur from A to B and also from B to C but not from C to A. Then the correct relation between critical angles is**
   1) $C_1 > C_2 > C_3$ 2) $C_1 = C_2 = C_3$ 3) $C_3 > C_1 > C_2$ 4) $C_1 < C_2 < C_3$

7. **Which of the following is used in optical fibres ?**
   1) refraction 2) scattering 3) diffraction 4) total internal reflection

8. **Consider the telecommunication through optical fibres. Which of the following statements is not true ?**
   1) Optical fibres can be of graded refractive index.
   2) Optical fibres are subjected to electromagnetic interference from outside.
   3) Optical fibres may have homogeneous core with a suitable cladding.
   4) Optical fibres have extremely low transmission loss.

9. **If ϵ_0 and μ_0 are respectively, the electric permitivity and the magnetic permeability of free space ϵ and μ the corresponding quantities in a medium, then the refractive index of the medium is**
   1) $\dfrac{\sqrt{\mu\epsilon}}{\mu_0\epsilon_0}$ 2) $\sqrt{\dfrac{\mu\epsilon}{\mu_0\epsilon_0}}$ 3) $\dfrac{\sqrt{\mu_0\epsilon_0}}{\mu\epsilon}$ 4) $\dfrac{\sqrt{\mu\mu_0}}{\epsilon\epsilon_0}$

10. **A plane glass slab is kept over various coloured letters ; the letter which appears least raised is :**
    1) blue 2) violet 3) green 4) red

11. **Relation between critical angles of water and glass is :**
    1) $C_w > C_g$ 2) $C_w < C_g$ 3) $C_w = C_g$ 4) $C_w = C_g = 0$

12. **The phenomenon used in optical fibers for transmission of light energy is**
    1) Total internal reflection 2) Scattering 3) Diffraction 4) Refraction

13. **A beam of light composed of red and green rays is incident obliquely at a point on the face of rectangular glass slab. When coming out on the opposite parallel face, the red and green rays emerge from :**
    1) Two points propagating in two different non parallel directions.
    2) Two points propagating in two different parallel directions.
    3) One point propagating in two different directions.
    4) One point propagating in the same direction.

14. **By placing the prism in minimum deviation position, images of the spectrum.**
1) Becomes inverted 2) Becomes broader 3) Becomes distinct 4) Becomes intensive

15. **In the position of minimum deviation when a ray of yellow light passes through the prism, then its angle of incidence is**
1) Less than the emergent angle 2) Greater than the emergent angle
3) Sum of angle of incidence and emergent angle is 90^0
4) Equal to the emergent angle

16. **When a white light passes through a hollow prism, then**
1) There is no dispersion and no deviation. 2) Dispersion but no deviation.
3) Deviation but no dispersion. 4) There is dispersion and deviation both

17. **The refractive index of a material of a prism of angles $45^0 - 45^0 - 90^0$ is 1.5. The path of the ray of light incident normally on the hypotenuse side is shown in**

1) 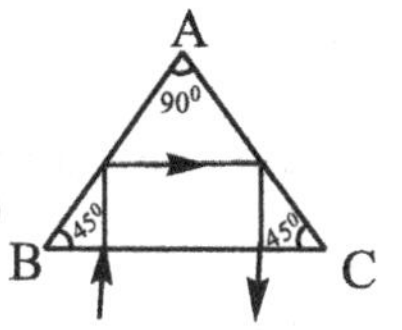

2) 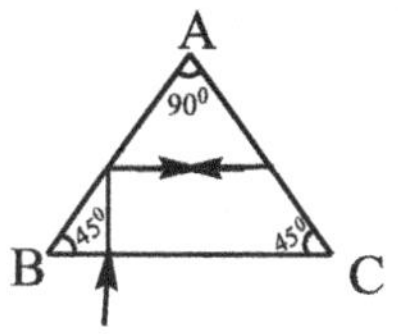

3) 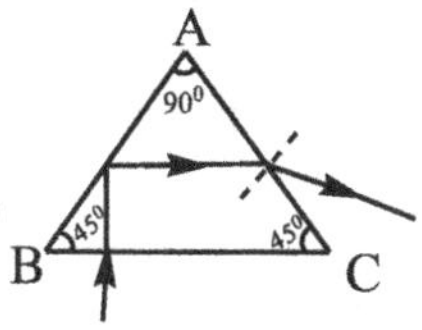

4) 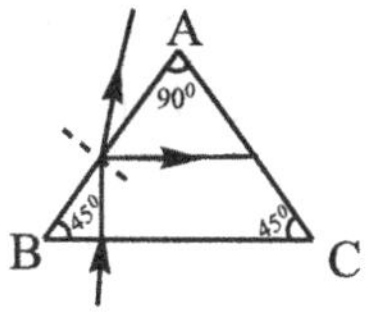

18. **When white light enters a prism, it gets split into its constituent colours. This is due to**
1) High density of prism material.
2) Refractive index, is different for different wavelengths.
3) Diffraction of light 4) Velocity changes for different frequencies.

19. **Dispersion can take place for**
1) Transverse waves only but not for longitudinal waves.
2) Longitudinal waves only but not for transverse waves.
3) Both transverse and longitudinal waves. 4) Neither transverse nor longitudinal waves.

20. **The refractive index of glass is 1.520 for red light and 1.525 for blue light. Let D_1 and D_2 are the angles of minimum deviation for red and blue respectively in a prism of the glass. Then**
1) $D_1 < D_2$ 2) $D_1 > D_2$ 3) $D_1 = D_2$
4) D_1 can be less than or greater than dependent upon the angle of prism.

21. **In dispersion without deviation.**
1) The emergent rays of all the colours are parallel to the incident ray.
2) Yellow coloured ray is parallel to the incident ray
3) Only red coloured ray is parallel to the incident ray
4) All the rays are parallel, but not parallel to the incident ray.

22. **When white light passes through the achro-matic combination of prisms, then what is observed.**
1) Only deviation 2) Only dispersion
3) Deviation and dispersion 4) None of the above

23. Dispersive power depends on the following:

1) material of the prism
2) shape of the prism
3) size of the prism
4) size, shape and material of the prism

24. In visible region, the dispersive powers and the mean angular deviations for crown and flint glass prisms are ' ω ',' ω^1 ' and 'd, d^1 ' respec-tively. The condition for getting deviation without dispersion when the two prisms are combined is

1) $\sqrt{\omega d}+\sqrt{\omega^1 d^1}=0$
2) $\omega^1 d+\omega d^1=0$
3) $\omega d+\omega^1 d^1=0$
4) $(\omega d)^2+(\omega^1 d^1)^2=0$

25. Lens maker's formula is applicable to

1) Thin lenses and paraxial rays which subtend very small angles with the principal axis
2) Thick lenses and paraxial rays which subtend very small angels with the principles axis
3) Thin lenses and for marginal rays
4) Thick lenses and for marginal rays

26. The focal length of a lens depends upon

1) nature of the material of lens
2) colour of light
3) the medium in which lens is placed
4) all of the above

27. If focal length of a double convex lens for red light is f_R, its focal length for the violet light is

1) f_R
2) greater than f_R
3) less than f_R
4) $2f_R$

28. The inability of a lens to form a point image of a point object on the axis is called

1) Spherical aberration
2) Coma
3) Chromatic aberration
4) Astigmatism

29. Spherical aberration is due to the

1) more deviation of par-axial rays
2) more deviation of marginal rays
3) dispersion
4) field lens

30. A plano convex lens is effectively used in decreasing spherical aberration

1) with its plane surface towards the incident rays
2) with its curved surface towards the incident rays
3) with its surface having less radius of curvature towards the ray from objective
4) both 2 and 3

31. In Huygen's eye piece

1) Chromatic aberration is not eliminated
2) Spherical aberration is completely eliminated
3) Focal length of field lens and eye lens are equal
4) Cross wires can be provided

32. In an eye piece to eliminate the spherical abberation, the required condition is

1) The distance between the field lens and eye lens is equal to the difference of their focal lengths
2) The distance between the field lens and eye lens is equal to the average of their focal lengths
3) The distance between the field lens and eye lens is equal to the sum of their focal lengths
4) The distance between the field lens and eye lens is equal to the product of their focal lengths

33. Spherical aberration in a thin lens can be reduced by

1) Using monochromatic light
2) Using achromatic combination
3) Increasing the size of the lens
4) Using a circular annular mask

34. Chromatic aberration is due to

1) Dependence of focal length of a lens on the wave length of light used
2) Dependence of focal length of a lens on the distance of the parallel rays from the axis
3) Both 1 and 2
4) Neither 1 nor 2

35. Chromatic aberration can be eliminated by

1) two lenses having same dispersive powers
2) two lenses having different dispersive powers
3) an achromatic doublet
4) using any one of the above arrangement

36. If f_r, f_y and f_v are the focal lengths of red, yellow and violet colours, then the longitudinal chromatic aberration for parallel rays is

1) $f_r + f_v$ 2) $f_r - f_v$ 3) $\frac{f_r - f_v}{f_y}$ 4) $\frac{f_r + f_v}{2}$

37. An achromatic combination of lenses produces

1) Images in black and white
2) Coloured images
3) Images unaffected by variation of refractive index with wavelength
4) Highly enlarged images

38. To remove the chromatic aberration, combination of lenses should be such that (F_r & F_v focal lengths for red and violet colours)

1) $F_R + F_V = 0$ 2) $F_R > F_V$ 3) $F_R < F_V$ 4) $F_R - F_V = 0$

39. A widely used achromatic doublet is designed with

1) Convex lens of low focal length made of crown glass and concave lens of greater focal length made of flint glass
2) Convex lens of greater focal length made of crown glass and concave lens of low focal length made of flint glass
3) Concave lens of low focal length made of crown glass and convex lens of greater focal length made of flint glass
4) Concave lens of greater focal length made of crown glass and convex lens of low focal length made of flint glass

40. A combination of two thin lenses of same material with focal lengths f_1 and f_2 arranged on a common axis minimizes chromatic aberration, if the distance between them is

1) $\frac{(f_1 + f_2)}{4}$ 2) $\frac{(f_1 + f_2)}{2}$ 3) $(f_1 - f_2)$ 4) $2(f_1 + f_2)$

41. For which of the following colour, the magnifying power of a microscope will be maximum

1) White colour 2) Red colour 3) Violet colour 4) Yellow colour

42. When the length of a microscope tube increases, its magnifying power.

1) Decreases 2) Increases 3) Does not change 4) May decrease or increase

43. In a compound microscope, if the objective produces an image I_0 and the eye piece produces an image I_e, then

1) I_0 is virtual but I_e is real
2) I_0 is real but I_e is virtual
3) I_0 and I_e are both real
4) I_0 and I_e are both virtual

44. The magnifying power of a simple microscope can be increased, if we use an eye piece of

1) Higher focal length 2) Smaller focal length 3) Higher diameter 4) Smaller diameter

45. Finger prints are observed by the use of

1) Telescope 2) Microscope 3) Gallilean telescope 4) Concave lens

46. In order to increase the magnifying power of a compound microscope

1) The focal lengths of the objective and the eye piece should be small.
2) Objective should have small focal length and the eye piece large.
3) Both should have large focal lengths.
4) The objective should have large focal length and eye piece should have small.

47. If the focal length of the objective lens is inreased, then

1) Magnifying power of microscope will increase, but that of telescope will decrease.
2) Magnifying power of microscope and telescope both will inrease.
3) Magnifying power of microscope and telescope both will decrease.
4) Magnifying power of microscope will decrease, but that of telescope will increase.

48. Resolving power of a microscope depends upon

1) The focal length and aperture of the eye lens
2) The focal lengths of the objective and the eye lens
3) The apertures of the objective and the eye lens
4) The wavelength of light illuminating the object

49. The objective of a compound microscope is essentially

1) A concave lens of small length and small aperture.
2) Convex lens of small focal length and large aperture.
3) Convex lens of large focal length and large aperture.
4) Convex lens of small focal length and small aperture.

50. In a compound microscope, the cross–wires are fixed at the point

1) Where the image is formed by the objective
2) Where the image is formed by the eye piece.
3) Where the focal point of the objective lies
4) Where the focal point of the eye piece lies

51. In a compound microscope, the intermediate image is

1) Virtual, erect and magnified
2) Real, erect and magnified
3) Real, inverted and magnified
4) Virtual, erect and reduced

52. If the red light is replaced by blue light illuminating the object in a microscope then the resolving power of the microscope.

1) Decreases 2) Increases 3) Gets halved 4) Remains unchanged

53. The angular magnification of a simple microscope can be increased by increasing

1) Focal length of lens 2) Size of object 3) Aperture of lens 4) Power of lens

54. The separation between two microscopic particles is measured P_A and P_B by two different lights of wavelengths $2000A^0$ and $3000A^0$ respectively, then

1) $P_A > P_B$ 2) $P_A < P_B$ 3) $P_A < 3/2P_B$ 4) $P_A = P_B$

55. The image formed by an objective of a compound microscope is

1) Virtual and enlarged
2) Virtual and diminished
3) Real and diminished
4) Real and enlarged

56. The magnifying power of a telescope can be increased by

1) Increasing focal length of the system.
2) Fitting eye piece of high power.
3) Fitting eye piece of low power.
4) Increasing the distance of objects.

57. A photograph of the moon was taken with a telescope. Later on, it was found that a housefly was sitting on the objective lens of the telescope. In photograph

1) The image of housefly will be reduced
2) There is a reduction in the intensity of the image.
3) There is an increase in the intensity of the image.
4) The image of the housefly will be enlarged.

58. For a telescope to have large resolving power, then the

1) Focal length of its objective should be large
2) Focal length of its eye piece should be large
3) Focal length of its eye piece should be small
4) Aperture of its objective should be large

59. To increase the magnifying power of telescope (f_0 = focal length of the objective and f_e = focal length of the eye lens)

1) f_0 should be large and f_e should be small
2) f_0 should be small and f_e should be large
3) f_0 and f_e both should be large
4) f_0 and f_e both should be small

60. If the telescope is reversed i.e. seen from the objective side

1) Object will appear very small
2) Object will appear very large
3) There will be no effect on the image formed by the telescope
4) Image will be slightly greater than the earlier one

61. In Gallilean telescope, the final image formed is

1) Real, erect and enlarged
2) Virtual, erect and enlarged
3) Real, inverted and enlarged
4) Virtual, inverted and enlarged

62. An astronomical telescope has a large aperture to

1) Reduce spherical aberration.
2) Have high resolution.
3) Increase span of observation
4) Have low dispersion.

63. The image due to the objective is formed infront of the field lens in

1) Huygen's eyepiece
2) Ramsden's eyepiece
3) Both 1 and 2
4) Neither 1 nor 2

64. In Ramsden's eyepiece the lenses are so arranged that their curved surfaces

1) Face each other
2) Face towards the eye position
3) Face towards the objective
4) One faces the objective the other faces the eye position.

65. Ramden's eye piece is

1) Positive eye piece
2) Negative eyepiece
3) Either positive or negative
4) Neither positive nor negative

66. In Ramsden eyepiece, the focal length of each lens is 'F'. The distance of the image formed by the objective lens from the eye lens is

1) $\frac{11F}{12}$
2) $\frac{13F}{14}$
3) $\frac{12F}{13}$
4) $\frac{14F}{15}$

67. The eyepiece that is used for measurement purpose is

1) Huygen's eyepiece
2) Ramsden's eyepiece
3) Both 1 and 2
4) Neither 1 nor 2

68. In Huygens eye piece :

1. The cross wires are outside the eyepiece
2. Condition for chromaticism is satisfied
3. Condition for minimum spherical aberration is not satisfied
4) the image formed by the objective is a virtual image

69. In Huygen's eyepiece the lenses are so arranged that their curved surfaces

1) Face each other
2) Face towards the eye position
3) Face towards the objective
4) One faces the objective the other faces the eye position

70. The eyepiece which satisfies the conditions of minimum spherical and chromatic abberation is

1) Huygen's eyepiece
2) Ramsden's eyepiece
3) Both
4) None

71. Which of the following is the negative eyepiece

1) Huygen's eyepiece
2) Ramsden's eyepiece
3) Both
4) None

72. **In Huygen's eyepiece, the image formed by the objective would have been**
1) Real and lie some where between the field lens and eyelens
2) Virtual and lies midway between the field lens and eyelens
3) Real and infront of the field lens 4) Real and behind the eyelens

73. **Light from a carbon arc passes through mercury vapour, and then enters a spectro-scope. The spectrum will be**
1) Continuous spectrum of carbon arc 2) Line spectrum of mercury
3) Absorption spectrum of mercury 4) Emission spectrum of carbon arc

74. **The nature of line spectra**
1) electronic 2) vibrational 3) rotational 4) None

75. **An incandescent filament emits a spectrum which is a**
1) line spectrum 2) band spectrum
3) continuous spectrum 4) characteristic spectrum

76. **Spectrum produced by vapour in atomic state is**
1) Continuous spectrum 2) Band spectrum 3) Absorption spectrum 4) Line spectrum

77. **Mercury vapour lamp is a source of**
1) Line spectrum 2) Band spectrum
3) Continuous spectrum 4) Monochromatic light

78. **Fraunhofer lines are produced due to the absorption of light in**
1) Chromosphere of sun 2) Photosphere of sun
3) Atmosphere of earth 4) Ionosphere of earth

79. **Fraunhofer lines are due to**
1) absorption by chromosphere of the light emitted by photosphere
2) absorption by photosphere of light emitted by chromosphere
3) light emitted by photosphere 4) light emitted by chromosphere

80. **The element that was discovered in Solar spectrum by studying Fraunhoffer's lines is.**
1) Hydrogen 2) Helium 3) CO_2 4) Oxygen

81. **Wave theory cannot explain the phenomena of**
A) Polarization **B) Diffraction**
C) Compton effect **D)Photoelectric effect**
Which of the following is correct ?
1) A and B 2) B and D 3) C and D 4) D and A

82. **The Spectram obtained by analysing sunlight as it possess through prism.**
1) Continuous spectra 2) Band spectra
3) Absorption line spectra 4) Emission line spectra

83. **Solar spectrum is an example of**
1) line emission spectrum 2) band absorption spectrum
3) line absorption spectrum 4) continuous emission spectrum

84. **The two lenses of an achromatic doublet should have**
1) equal powers 2) equal dispersive powers
3) equal ratio of their power and dispersive power
4) sum of the products of their power and dispersive power should be equal to zero

85. **Statement A:- According to Maxwell's electromagnetic wave the theory, changing electric field produces a magnetic field that varies in space and time and vice-versa**

Statement B:- According to EM wave the theory, energy propagates in a direction perpendicular to the direction variation of both the electric and magnetic fields

1) A & B are true　　2) A & B are false
3) A is true & B is false　　4) A is false & B is true

86. **A) Newton's theory assumes that the source of light loses mass as it emits corpuscles**

B) According to Newton's theory, the velocity of light in denser medium should be greater than that in air

C) Focault's experiment proved that the velocity of light is more in rarer medium than in denser medium

1) Only A & B are true　　2) A, B & C are true
3) Only A & C are true　　4) Only B & C are true

87. **A) Newton's theory proposes that the velocity of the corpuscles increases as the temperature of the source increases**

B) It is observed that the velocity of light is independent of temperature of the source

1) Both A and B are true　　2) Both A and B are false
3) A is true and B is false　　4) A is false and B is true

88. **A) According to Huygen the light waves travels in all directions in a hypothetical medium called ether**

B) The hypothetical medium ether was supposed to have very large elasticity and extremely low density

1) A is false and B is true　　2) Both A and B are false
3) A is true and B is false　　4) Both A and B are true

89. **A) The wavefornt is the locus of all particles the mediumvibrating in the same phase**

B) The forward envelop of the secondary wavelets at any instant gives the new wave front

1) Both A and B are true　　2) Both A and B are false
3) A is true and B is false　　4) A is false and B is true

90. **A) Michelson - morley experiment negated the effect of ether on the motion of the earth and the existance of ether totally**

B) Wave theory of light fails to explain why the light has no longitudinal component

1) A is true and B is false　　2) Both A and B are false
3) Both A and B are true　　4) A is false and B is true

91. **A) At every point in the wave at a given instant of time the electric and magnetic field strengths are proportional and are always in same phase**

B) According to Maxwell theory $V = \frac{1}{\sqrt{\mu\varepsilon}}$

C) According to Wiener experiment, the photographic action is due to the electric field $\vec{E}$ and not due to magnetic field $\vec{B}$

1) Only A and B are true　　2) A, B and C are true
3) Only B and C are true　　4) Only A and C are true

92. **Condition for no deviation when light travels from one medium to another is**

a) angle of incidence is zero

b) refractive index of two media must be same

1) a is only true　　2) b is only true
3) a & b are both false　　4) a & b are both true

93. **A converging lens is used to form an image on a screen. When the upper half of the lens is convered by an opaque, screen then**

a) Half the image will disappear b) Complete image will be formed
c) Intensity of the image will decrease d) Intensity of image will increase.
1) a & d are correct 2) b & d are correct 3) a & c are correct 4) b & c are correct

94. **A) Coma is a type of monochromatic aberration**

B) The measurements with Huygen's eyepiece are reliable

1) A is true and B is false 2) Both A and B are false
3) Both A and B are true 4) A is false and B is true

95. **a) Cross wires can be arranged in Ramsden's eye piece**

b) Crossed lens is used to reduce chromatic aberration

c) The condition for minimum spherical and minimum chromatic aberration is satisfied in Hygens eye piece

d) Chromatic abberation is one of mono-chromatic abberations

1) a and d are correct 2) b and c are correct 3) a and c are correct 4) b and d are correct

96. **In eyepieces**

A) Field lens increases the field of view B) Eye lens acts as a magnifier
1) Both A and B are true 2) Both A and B are false
3) A is true and B is false 4) A is false and B is true

97. **A) In Ramsden's eyepiece, the scale and the image are magnified proportionally, therefore measurements would be trust worthy**

B) Ramsden's eyepiece minimizes the spherical aberration, because the total devia-tion is shared by the four refracting surfaces

1) A is true and B is false 2) Both A and B are false
3) Both A and B are true 4) A is false and B is true

98. **Consider the following two statements A and B and identify the correct choice in the given answers**

A) In Huygens eye piece, the two plane surfaces of two Plano convex lenses face towards the eye

B) In Huygens eye piece, the focal length of the field lens is equal to the focal length of eye piece.

1) both A and B are true 2) A is true but B is false
3) A is false but B is true 4) both A and B are false

99. **Consider the following two statements A and B and identify the correct choice in the given answers.**

A) The curved surfaces of plano convex lenses in Ramsden's eye piece face each other

B) The focal length of eye lens in Ramsden's eye piece is more than that of field lens.

1) both A and B are true 2) both A and B are false
3) A is true and B is false 4) A is false and B is true

100. **A) Line spectra are obtained from atoms of luminous gases or vapors when electric discharge passes through them.**

B) Line spectra of solids such as salts are obtained by burning the salts in the Bunsen burner flame.

C) To obtain line spectra of iron, copper and alloys like brass can be taken in the form of rods and an arc can be struck between them.

1) A, B, C are true 2) Only A, B are true 3) Only A, C are true 4) Only B, C are true

101. **A) Spectro chemical analysis is used to determine the abundance of an element in a given sample**

B) The temperature and motion of sun, stars and galaxies in space can be determined from spectral analysis

1) A is false and B is true 2) Both A and B are false

3) A is true and B is false 4) Both A and B are true

102. **Consider the following two statements A and B and identify the correct choice in the given answers.**

A) Line spectra is due to atoms in gaseous state

B) Band spectra is due to molecules

1) both A and B are false 2) A is true but B is false

3) A is false and B is true 4) Both A and B are true

103. **Consider the following statements:**

A) line spectrum is characteristic of atomic spectra

B) band spectrum is characteristic of molecular spectra

1) both A and B are correct 2) A is correct but B is wrong

3) A is wrong but B is correct 4) both A and B are wrong

ASSERTION AND REASON

In each of the following questions, a statement of Assertion (A) is given followed by a corresponding statement of Reason (R) just below it. Of the statements, mark the correct answer.

1) If both assertion and reason are true and reason is the correct explanation of assertion.

2) If both assertion and reason are true but reason is not the correct explanation of assertion.

3) If assertion is true but rreason is false 4) If assertion is false but reason is true.

104. **(A) : Light not only makes the things visible, but also raises their tempe- rature and exerts pressure on them**

(R) : Light is a form of energy and travels through space

105. **(A) : Light waves are transverse waves**

(R) : The electric field vector $\vec{E}$ and magnetic field vector $\vec{B}$ oscillates in a direction perpendicular to the velocity vector $\vec{V}$ [1]

106. **(A) : The twinkling of star is due to reflection of light.**

(R) : The velocity of light changes while going from one medium to the other.

107. **(A) : The images formed by total internal reflections are much brighter than those formed by mirrors or lenses.**

(R) : There is no loss of intensity in total internal reflection.

108. **(A) : Propagation of light through an optical fiber is due to total internal reflection taking place at the core - clad interface.**

(R) : Refractive index of material of core of optical fiber is greater than that of air.

109. **(A) : The focal length of lens changes when red light is replaced by blue light.**

(R) : The focal length of lens does not depend on colour of light used.

110. **(A) : A single lens produces a coloured image of an object illuminated by white light.**

(R) : The refractive index of material of lens is different for different wavelengths of light.

111. **(A) : Spherical abberration occur in lenses of larger aperture.**

(R) : The two rays, paraxial and marginal rays focus at different points.

112. (A) : Optical instruments are assembled using lenses with very wide apertures
(R) : More collected light produces bright images.

113. (A) : There is no dispersion of light refracted through a rectangular glass slab.
(R) : Dispersion of light is the phenomenon of splitting of a beam of white light into its constituent colours.

114. (A) : A beam of white light gives a spectrum on passing through a hollow prism.
(R) : The speed of light inside the hollow prism is same as that of incident.

115. (A) : Dispersion of light occurs, because velocity of light in a material depends upon its colour.
(R) : The dispersive power depends only upon the material of the prism, but not upon the refracting angle of the prism.

116. (A) : When white light passes through prism, different colours under go different deviations
(R) : The deviations are different for different colours due to variation of refractive index with colour

117. (A) : Resolving power of both microscope and telescope depends on the wavelength of light used.
(R) : Resolving power of lens is the ability to resolve the two images so that they are distinctly identified

118. (A) : Resolving power of an optical instrument is resciprocal of its limit of resolution
(R) : Smaller the distance between two point objects, the instrument can resolve, higher is its resolving power

119. (A) : The resolving power of a telescope is more, if the diameter of the objective lens is more
(R) : Objective lens of large diameter collects more light

120. (A) : A molecule cannot produce line spectrum
(R) : The translational energy of a molecule is not quantized.

121. (A) : Fine lines are observed in the band spectrum.
(R) : The rotational, vibrational and electronic energies of molecule are quantized.

122. (A) : Using Huygen's eye piece, measurements can be taken but are not correct
(R) : The cross wires, scale and final image are not magnified propotionately, because the image of the object is magnified by two lenses, where as the cross wire scale is magnified by one lens only.

ASCENDING OR DESCENDING:

123. Arrange the following colours in increasing order of their critical angles with respect to glass
a) red b) Yellow c) Blue d) violet
1) a, b, c, d 2) a, c, b, d 3) d, c, b, a 4) a, d, b, c

124. Arrange the following values for a Ramsden's eyepiece in decreasing order
a) Separations between the lenses
b) Distance of image formed by objective with respect to field lens
c) Distance of image formed by field lens with respect to field lens
d) Distance between eye lens and image due to field lens
1) a, b, c, d 2) a, d, b, c 3) c, d, b, a 4) d, a, c, b

125. **A light ray is travellying between two media as given below. The angle of incidence on the boundary in all the cases is 30^0. Identify the correct sequence of increasing order of angles of refraction,**

a) Air to water b) Water to glass c) Glass to water

(Refractive indices of glass and water are respectively 3/2 and 4/3)

1) a, b, c 2) b, c, a 3) c, a, b 4) a, c, b

126. **In a spectrometer experiment, three points A, B, C with same angle of prism but of different materials of refractive indices $m_A = 1.33$, $m_B = 1.55$, $m_C = 1.44$ are used. The corresponding angles of minimum deviation D_A, D_B, D_C are measured will be such that**

1) $D_A > D_B > D_C$ 2) $D_A < D_B < D_C$ 3) $D_A < D_C < D_B$ 4) $D_A > D_C > D_B$

MATCHING

127. **Match the following in case of refraction in (i_1 = angle of incidence, i_2 = angle of emergence) prism**

List - I	List - II
a) For normal incidence	e) D == 90^0 - A
b) For normal emergence	f) D = $i_2 + 90^0$ - A
c) For grazing incidence	g) D = i_2 - A
d) For a ray at normal Incedence and emergent ray grazing the second face	h) D = i_1 - A

1) a→g, b→h, c→f, d→e 2) a→f, b→e , c→g, d→h
3) a→h, b→e, c→h, d→h 4) a→g, b→e, c→f, d→h

128. **Compare the final images formed by different optical instruments**

List - I	List - II
a) Simple Microscope	e) Magnefied, virtual and Inverted
b) Compound Microscope	f) Virtual, errect and smaller
c) Astronomical telescope	g) Virtual and enlarged
d) Terrestrial telescope	h) Virtual, inverted and small

1) a →g, b →e, c→h, d→f 2) a→h, b→g, c→e, d →f
3) a →g, b→h, c→g, d→e 4) a→ f, b→h, c→g, d→c

129.

LIST - 1	LIST - 2
a) Achromatic doublet	e) Reduction of aberrations
b) Ramden's eye piece	f) positive and negative lenses
c) Huygen's eye piece	g) biological instrument
d) eyepiece	h) travelling microscope

1) a→f; b→e; c→g; d→h 2) a→h; b→f; c→g; d→e
3) a→g; b→f; c→h; d→e 4) a→f; b→h; c→g; d→e

130.

LIST - 1	LIST - 2
a) Impure spectrum	e) molecules
b) line spectrum	f) high dense gases
c) band spectrum	g) atoms
d) continuous spectrum	h) rainbow

1) a→f; b→g; c→h; d→e 2) a→g; b→h; c→e; d→f
3) a→h; b→g; c→e; d→f 4) a→h; b→g; c→f; d→e

131. LIST-1 / LIST - 2

LIST-1	LIST - 2
a) Burning candle	e) Line spectrum
b) Sodium vapours	f) Continuous spectrum
c) Bunsen flame	g) Band spectrum
d) Dark lines in solar spectrum	h) Absorption spectrum

1) $a\to g; b\to e; c\to f; d\to h$ 2) $a\to f; b\to e; c\to g; d\to h$
3) $a\to h; b\to g; c\to f; d\to e$ 4) $a\to e; b\to f; c\to g; d\to h$

132. **Match the appropriate pairs from List - I and List - II :**

List - I	List - II
a) Nitrogen molecules	e) Continuous spectrum
b) Incandescent solids	f) Absorption spectrum
c) Fraunhofer lines	g) Band spectrum
d) Electric arc between iron rods	h) Emission spectrum

1) $a\to g; b\to e; c\to f; d\to h$ 2) $a\to f; b\to e; c\to h; d\to g$
3) $a\to h; b\to e; c\to f; d\to g$ 4) $a\to e, b\to g; c\to h; d\to f$

ANSWERS

1) 3	2) 4	3) 3	4) 2	5) 4	6) 4	7) 4	8) 2	9) 2	10) 4
11) 1	12) 1	13) 2	14) 3	15) 4	16) 1	17) 1	18) 2	19) 3	20) 1
21) 2	22) 1	23) 1	24) 3	25) 1	26) 4	27) 3	28) 1	29) 2	30) 4
31) 2	32) 1	33) 4	34) 1	35) 3	36) 2	37) 3	38) 4	39) 1	40) 2
41) 3	42) 2	43) 2	44) 2	45) 2	46) 1	47) 4	48) 4	49) 4	50) 1
51) 3	52) 2	53) 4	54) 2	55) 4	56) 2	57) 2	58) 4	59) 1	60) 1
61) 2	62) 2	63) 2	64) 1	65) 1	66) 1	67) 2	68) 2	69) 3	70) 1
71) 1	72) 1	73) 1	74) 1	75) 3	76) 4	77) 1	78) 1	79) 1	80) 2
81) 3	82) 3	83) 3	84) 4	85) 1	86) 2	87) 1	88) 4	89) 1	90) 3
91) 2	92) 4	93) 4	94) 1	95) 3	96) 1	97) 3	98) 2	99) 3	100) 1
101) 4	102) 4	103) 1	104) 2	105) 1	106) 4	107) 1	108) 2	109) 3	110) 1
111) 1	112) 1	113) 2	114) 4	115) 2	116) 1	117) 2	118) 1	119) 2	120) 1
121) 1	122) 1	123) 3	124) 4	125) 1	126) 3	127) 1	128) 1	129) 4	130) 3
131) 2	132) 1								

EXERCISE - II(A)

Refraction :

1. **If refractive indices of glass and water are respectively 3/2 and 4/3, refractive index of glass w.r.t. water is**
   1) 2 2) 8/9 3) 9/8 4) 1/2
2. **Absolute refractive index of a medium is 'X'. Refractive index of same medium w.r.t. to air is 'Y' and absolute refractive index of air is 'Z'. Then, the relation between them is**
   1) $X=Y/Z$ 2) $Y=X/Z$ 3) $Z=Y/X$ 4) $Y=1/XZ$
3. **If the refractive index of diamond is 2.4 find the velocity of light in diamond.($c=3\times10^8$ m/s)**
   1) 1.25×10^8 m/s 2) 2.25×10^8 m/s 3) 1.5×10^8 m/s 4) 4.25×10^8 m/s

4. **The refractive indices of glycerine and diamond with respect to air are 1.4 and 2.4 respectively. Calculate the speed of light in glycerine and diamond. From these results, calculate the refractive index of diamond with respect to glycerine. ($c = 3 \times 10^8$ m/s)**

1) 2.143×10^8 m/s; 1.250×10^8 m/s; 1.714
2) 1.143×10^8 m/s; 1.250×10^8 m/s; 1.714
3) 2.143×10^8 m/s; 2.250×10^8 m/s; 1.714
4) 2.143×10^8 m/s; 1.250×10^8 m/s; 1.514

5. **White light forms a continuous spectrum of minimum wavelength 3800 A^0 corresponding to violet light and maximum wavelength 7800A^0 corresponding to red light. Find the minimum and maximum frequencies of white light. ($c = 3 \times 10^8$ m/s).**

1) 3.846×10^{14} Hz; 7.895×10^{14} Hz
2) 2.546×10^{14} Hz; 6.985×10^{14} Hz
3) 4.846×10^{14} Hz; 8.895×10^{14} Hz
4) 1.546×10^{14} Hz; 5.985×10^{14} Hz

6. **Find the time taken by light to travel a distance of 3cm in glass. ($c = 3 \times 10^8$ m/s and refractive index of water = 4/3)**

1) 1.333×10^{-10}S 2) 2.333×10^{-10}S 3) 3.333×10^{-10}S 4) 4.333×10^{-10}S

7. **A ray of light passes normally through a slab $(\mu = 1.5)$ of thickness 't'. If the speed of light in vacuum be 'C', then time taken by the ray to go across the slab will be**

1) $\frac{t}{C}$ 2) $\frac{3t}{2C}$ 3) $\frac{2t}{3C}$ 4) $\frac{4t}{9C}$

8. **A monochromatic light passes through a glass slab $(\mu = 1.5)$ of thickness 9 cm in time t_1. If it takes a time t_2 to travel the same distance through water $(\mu = 4/3)$. The value of $(t_1 - t_2)$ is**

1) 5×10^{-11}sec 2) 5×10^{-8}sec 3) 2.5×10^{-10}sec 4) 5×10^{-10}sec

9. **For water $\mu = 4/3$ and the velocity of light in vacuum is 3×10^8 ms^{-1}, the time taken for light to travel a distance of 450 m in water will be (nearly)**

1) 1.5 μs 2) 1 μs 3) 2 μs 4) 3 μs

10. **A glass slab of thickness 4 cm contains the same number of waves as 5 cm of water when both are traversed by the same monochromatic light. If the refractive index of water is 4/3, then that of glass is**

1) 5/3 2) 5/4 3) 16/15 4) 3/2

11. **The wavelength of light in vacuum is 5000A^0. When it travels normally through diamond of thickness 1.0mm find the number of waves of light in 1.0mm of diamond. (Refractive index of diamond = 2.417)**

1) 4834 waves 2) 5834 waves 3) 4384 waves 4) 6834 waves

12. **The refractive index of denser medium with respect to rarer medium is 1.125. The difference between the velocities of light in the two media is 0.25×10^8 m/s. Find the velocities of light in the two media and their refractive indices. ($c = 3 \times 10^8$m/s).**

1) 2.0×10^8 m/s; 2.25×10^8 m/s; 1.500; 1.333
2) 2.5×10^8 m/s; 2.25×10^8 m/s; 1.500; 1.333
3) 2.0×10^8 m/s; 2.25×10^8 m/s; 1.333; 1.500
4) 2.25×10^8 m/s; 2.0×10^8 m/s; 1.500; 1.333

13. **The optical path of a monochromatic light is the same if it goes through 2.00cm of glass or x cm of ruby. If the refractive index of glass is 1.510 and that of ruby is 1.760 find the value of x.**

1) 1.716 cm 2) 1.525 cm 3) 2.716 cm 4) 2.525 cm

14. **A ray of light is incident on a glass slab making an angle of 60^0 with the surface. Calculate the angle of refraction in glass if the refractive index of glass is 1.5**

1) $19^0 28'$ 2) 30^0 3) 45^0 4) 50^0

15. **A ray of light incidents on a refracting surface at 30^0 with the surface. If the angle of refraction is 45^0, refractive index of the medium is**

1) $\sqrt{3}$ 2) $\frac{\sqrt{3}}{2}$ 3) $\sqrt{\frac{3}{2}}$ 4) $2\sqrt{3}$

16. **A beam of light is incident on a glass slab of refractive index 1.6 making an angle of 30^0 with the surface. Find the ratio of the widths of the refracted beam to the incident beam.**

1) 1.682:1 2) 1:1.682 3) 1.247:1 4) 1:1.247

17. **If $\hat{i}$ denotes a unit vector along an incident ray, $\hat{r}$ a unit vector along the refracted ray into a medium of refractive index 'μ' and $\hat{n}$ a unit vector normal to the boundary of the media directed towards the incident medium, then the law of refraction can be written as**

1) $\hat{i}.\hat{n} = \mu\left(\hat{r}.\hat{n}\right)$ 2) $\hat{i} \times \hat{n} = \mu\left(\hat{n} \times \hat{r}\right)$ 3) $\hat{i} \times \hat{n} = \mu\left(\hat{r} \times \hat{n}\right)$ 4) $\mu\left(\hat{i} \times \hat{n}\right) = \hat{r} \times \hat{n}$

18. **Calculate the refractive index of glass with respect to air if its critical angle is $40^0 12'$.**

1) 1.549 2) 1.459 3) 1.249 4) 1.149

19. **The velocities of light in two different media are 2×10^8 m/s and 2.5×10^8 m/s respectively. The critical angle for these media is**

1) $\sin^{-1}\left(\frac{1}{5}\right)$ 2) $\sin^{-1}\left(\frac{4}{5}\right)$ 3) $\sin^{-1}\left(\frac{1}{2}\right)$ 4) $\sin^{-1}\left(\frac{1}{4}\right)$

20. **The critical angle for refraction from glass to air is 42^0 and that from water to air is 48^0. Find the critical angle for refraction from glass to water.**

1) $61^0 3'$ 2) $51^0 3'$ 3) $61^0 2$ 4) $63^0 3'$

21. **Light takes time t_1 to travel a distance x_1 in vacuum and the same light takes time t_2 to travel a distance x_2 in a medium. The critical angle for that medium is**

1) $\sin^{-1}\left(\frac{x_2 t_2}{x_1 t_1}\right)$ 2) $\sin^{-1}\left(\frac{x_1 t_2}{x_2 t_1}\right)$ 3) $\sin^{-1}\left(\frac{x_1 t_1}{x_2 t_2}\right)$ 4) $\sin^{-1}\left(\frac{x_2 t_1}{x_1 t_2}\right)$

22. **Wave length of light in denser medium is $4000A^0$, it is grazing into a rarer medium. If critical angle for the pair of media is $\text{Sin}^{-1}\left(\frac{2}{3}\right)$ then the wave length of light in rarer medium is**

1) $4000A^0$ 2) $2666A^0$ 3) $8000A^0$ 4) $6000A^0$

23. **If the angle of incidence is twice the angle of refraction in a medium of refractive index 'μ' then the angle of incidence is**

1) $\cos^{-1}\left(\frac{\mu}{2}\right)$ 2) $\sin^{-1}\left(\frac{\mu}{2}\right)$ 3) $2\sin^{-1}\left(\frac{\mu}{2}\right)$ 4) $2\cos^{-1}\left(\frac{\mu}{2}\right)$

24. **A ray of light from a denser medium strikes a rarer medium. The reflected and refracted rays make an angle of 90^0 with each other. The angles of reflection and refraction are r and $r^{|}$. The critical angle would be**

1) $\sin^{-1}(\tan r)$ 2) $\tan^{-1}(\sin r)$ 3) $\sin^{-1}(\tan r^{|})$ 4) $\tan^{-1}(\sin r^{|})$

25. **The refractive index of the core of an optical fibre is μ_2 and that of the cladding is μ_1. The angle of incidence on the face of the core so that the light ray just under goes total internal reflection at the cladding is**

1) $\sin^{-1}\left(\frac{\mu_1}{\mu_2}\right)$ 2) $\sin^{-1}\sqrt{\mu_2^2 - \mu_1^2}$ 3) $\sin^{-1}\sqrt{\mu_2 - \mu_1}$ 4) $\sin^{-1}\sqrt{\mu_1^2 + \mu_2^2}$

26. **A light ray travels from a denser medium to a rarer medium. If the critical angle of the denser medium with respect to rarer medium is 'C', the maximum possible deviation of any ray will be**

1) $\pi - C$ 2) $\frac{\pi}{2} + C$ 3) 2C 4) $\pi - 2C$

27. **A point source of light is placed at the bottom of a water lake. If the area of the illuminated circle on the surface is equal to 3 times the square of the depth of the lake. The refractive index of water is**

1) $\sqrt{\pi + 1}$ 2) $\sqrt{\frac{\pi}{3} + 1}$ 3) $\frac{\pi}{3} + 1$ 4) $\frac{\pi}{4} + 1$

28. **A ray of light is incident at an angle of 60^0 on a $\sqrt{3}$ cm thick plate $(\mu = \sqrt{3})$. The shift in the path of the ray as it emerges out from the plate is (in cm)**

1) 1 2) 1.2 3) 0.5 4) 1.8

29. **A mark at the the bottom of a liquid appears to rise by 0.1 m. The depth of the liquid is 1 m. The refractive index of the liquid is**

1) 1.33 2) 9/10 3) 10/9 4) 1.5

30. **An under water swimmer is at a depth of 12 m below the surface of water. A bird is at a height of 18 m from the surface of water, directly above his eyes. For the swimmer the bird appears to be a distance from the surface of water equal to $(\mu_w = 4/3)$**

1) 24 m 2) 12 m 3) 36 m 4) 18 m

31. **A microscope is focussed on a coin lying at the bottom of a beaker. The microscope is now raised up by 1 cm. To what depth should the water be poured in to the beaker so that coin is again in focus $(\mu_w = 4/3)$**

1) 1 cm 2) 4/3 cm 3) 3 cm 4) 4 cm

Prism :

32. **The refractive index of a prism for amonochromatic wave is $\sqrt{2}$ and its refracting angle is 60^0. For minimum deviation the angle of incidence will be**

1) 30^0 2) 45^0 3) 60^0 4) 75^0

33. **The minimum deviation produced by a hollow prism filled with a certain liquid to be 30^0. The light ray is also found to be refracted at angle 30^0. The refractive index of the liquid is**

1) $\sqrt{2}$ 2) $\sqrt{3}$ 3) $\sqrt{\frac{3}{2}}$ 4) $\frac{3}{2}$

34. **A triangular prism of glass is shown in the figure. A ray incident normally to one face is totally reflected. If $\theta = 45^0$, the refractive index of glass is**

1) < 1.41
2) = 1.41
3) > 1.41
4) 1.732

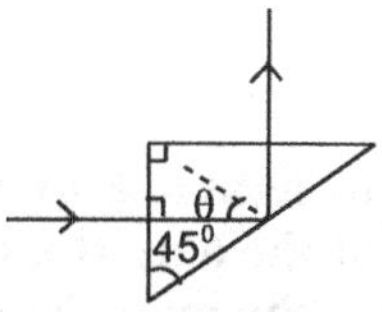

35. **For a prism of refractive index 1.732, the angle of minimum deviation is equal to the angle of the prism. The angle of prism is**

1) 30^0 2) 45^0 3) 60^0 4) 80^0

36. **Under minimum deviation condition in a prism, if a ray is incident at an angle 30^0, the angle between the emergent ray and the second refracting surface of the prism is**

1) 0^0 2) 30^0 3) 45^0 4) 60^0

37. A ray of light is incident at 60^0 on a prism of refracting angle 30^0. The emerging ray is at an angle 30^0 with the incident ray. The value of refractive index of the prism is : (2002 M)

1) $\frac{\sqrt{3}}{4}$ 2) $\frac{\sqrt{3}}{2}$ 3) $\sqrt{3}$ 4) $2\sqrt{3}$

38. A ray of light passes through an equilateral prism in such a manner that the angle of incidance is equal to the angle of emergence and each of these angles equal to 3/4 of angle of the prism. The angle of deviation is

1) 45^0 2) 39^0 3) 60^0 4) 30^0

39. If a light ray incidents normally on one of the faces of the prism of refractive index '2' and emergent ray just grazes the second face of the prism, then the angle of deviation is

1) 0^0 2) 30^0 3) 60^0 4) 90^0

40. Find the refractive index of the material of refracting angle $59^0 30'$ when the angle of minimum deviation is $36^0 44'$.

1) 1.5 2) 1.25 3) 1.33 4) 1.66

41. A ray of light is incident on one of the refracting surfaces of a prism of refracting angle 60^0. The emergent ray grazes the other refracting surface. Find the refractive index of the material of the prism.

1) 1.155 2) 1.75 3) 1.25 4) 1.00

42. A ray of light suffers minimum deviation while passing through an equilateral prism of refractive index 1.5. Calculate the angle of deviation, angle of incidence and angle of refraction.

1) $37^0 10'; 48^0 35'; 30^0$ 2) $33^0 10'; 48^0 35'; 30^0$
3) $37^0 10'; 48^0 45'; 30^0$ 4) $37^0 10'; 40^0 35'; 30^0$

43. A ray enters normally one face of an equilateral prism of crown glass. From which face does it emerge? Find the angle of deviation. (Refractive index of crown glass = 1.5.)

1) The ray undergoes total internal reflection and emerges perpendicular. Angle of deviation is 60^0.
2) The ray undergoes total internal reflection and emerges perpendicular. Angle of deviation is 120^0.
3) The ray undergoes total internal reflection and emerges perpendicular. Angle of deviation is 30^0.
4) The ray undergoes total internal reflection and emerges perpendicular. Angle of deviation is 90^0.

44. A thin prism of 4^0 angle gives a deviation of 2.4^0. What is the value of refractive index of the material of the prism ?

1) 1.6 2) 1.5 3) 1.3 4) 1.2

45. The refractive indices of violet and red light are 1.54 and 1.52 respectively. If the angle of prism is 10^0, then the angular dispersion is

1) 0.02 2) 0.2 3) 3.06 4) 30.6

46. A glass prism deviates the red and blue rays through 10^0 and 12^0 respectively. A second prism of equal angle deviates them through 8^0 and 10^0 respectively. Their dispersive powers are in the ratio

1) 11 : 9 2) 9 : 11 3) 3 : 2 4) 1 : 1

47. A parallel beam of white light falls on a convex lens. Images of blue, yellow and red light are formed on other side of the lens at a distance of 0.20 m, 0.205 m and 0.214 m respectively. The dispersive power of the material of the lens will be

1) $\frac{619}{1000}$ 2) $\frac{9}{200}$ 3) $\frac{14}{205}$ 4) $\frac{5}{214}$

48. The refractive indices of crown glass prism for C, D and F lines are 1.527, 1.530 and 1.535 respectively. Find the dispersive power of the crown glass prism.

1) 0.01509 2) 0.05109 3) 0.02108 4) 0.03402

Ray Optics

Lens maker's formula & Aberrations :

49. The refractive index of the material of a double convex lens is 1.5 and its focal length is 5cm. If the radii of curvature are equal, the value of the radius of curvature (in cms) is (2007 E)

1) 5.0 2) 6.5 3) 8.0 4) 9.5

50. A double convex lens of refractive index 1.5 has radii of 20 cm. Incident rays of light parallel to the axis will come to converge at a distance from the lens is

1) 20 cm 2) 10 cm 3) 40 cm 4) 30 cm

51. A diverging meniscus lens of radii of curvatures 25 cm and 50 cm has a refractive index 1.5. Its focal length is (in cm)

1) – 50 2) –100 3) 100 4) 50

52. Four lenses are made of glass and the radius of curvature of each face is given below. Which will have the greatest focal power.

1) 10 cm convex and 15 cm convex
2) 5 cm convex and 10 cm concave
3) 15 cm convex and plane
4) 20 cm convex 30 cm concave

53. Radius of curvature of first surface of double convex lens is three times that of the other. If focal length of the lens is 30 cm and refractive index of the lens is 3/2, then radius of curvature of the first surface is

1) 20 cm 2) 40 cm 3) 60 cm 4) 80 cm

54. A thin liquid convex lens is formed in glass. Refractive index of liquid is 4/3 and that of glass is 3/2. If 'f' is the focal length of the liquid lens in air, its focal length and nature in the glass is

1) f, convex 2) f, concave 3) 2f, concave 4) 3f, concave

55. The focal length of a convex lens of refractive index 1.5 is 'f' when it is placed in air. When it is immersed in a liquid it behaves as a converging lens and its focal length becomes xf $(x > 1)$. The refractive index of the liquid is

1) > 3/2 2) < 3/2 and > 1 3) < 3/2 4) = 3/2

56. A thin equiconvex lens is made of glass of refrctive index 1.5 and its focal length is 0.2 m. If it acts as a concave lens of 0.5 m focal length when dipped in a liquid, the refractive index of liquid is (2004 M & 2005 M)

1) 17/8 2) 15/8 3) 13/8 4) 9/8

57. The focal lengths of a lens are in the ratio 8 : 3 when it is immersed in two different liquids of refractive indeces 1.6 and 1.2 respectively. The refractive index of the material of the lens is

1) 1.25 2) 1.5 3) 1.8 4) 2

58. A thin equiconvex lens is made of glass of refractive index 1.5 and its focal length in air is 0.2 m. If it acts as a concave lens of 0.5 m focal length when dipped in a liquid, the velocity of light in the liquid is

1) $1.2 \times 10^8 ms^{-1}$ 2) $1.6 \times 10^8 ms^{-1}$ 3) $1.8 \times 10^8 ms^{-1}$ 4) $2.4 \times 10^8 ms^{-1}$

59. A thin double convex lens is cut into two equal pieces A and B by a plane containing principal axis. The piece 'B' is further cut into two more pieces 'C' and 'D' by another plane perpendicular to the principal axis. If the focal power of the original lens is 'P', then those of A and C are

1) $P, \frac{P}{4}$ 2) $P, \frac{P}{2}$ 3) $\frac{P}{2}, 2P$ 4) $\frac{P}{2}, \frac{P}{4}$

60. **A double convex lens of focal length 30 cm is made of glass. When it is immersed in a liquid of refractive index 1.4, the focal length is found to be 126 cm. The critical angle between glass and the liquid is**

1) $\sin^{-1}\left(\frac{3}{4}\right)$ 2) $\sin^{-1}\left(\frac{4}{5}\right)$ 3) $\sin^{-1}\left(\frac{7}{13}\right)$ 4) $\sin^{-1}\left(\frac{7}{8}\right)$

61. **Two thin convex lenses of focal lengths 20 cm and 5 cm respectively, are placed at a distance d. If a parallel beam incident on the first lens emerges as a parallel beam from the second lens, then the value of 'd' is**

1) 5 cm 2) 15 cm 3) 20 cm 4) 25 cm

62. **A convex lens of power + 6D is placed in contact with a concave lens of power – 4D. The nature and focal length of the combination is**

1) Concave, 25 cm 2) Convex, 50 cm 3) Concave, 20 cm 4) Convex, 100 cm

63. **Two thin lenses of focal length f_1 and f_2 are seperated by a distance 'd' between them. If the combination satisfies the condition for no chromatic and no spherical aberration then, $f_1 : f_2$ is**

1) 3 : 1 2) 1 : 2 3) 2 : 3 4) 4 : 1

64. **The power of an achromatic convergent lens of two lenses is + 2D. The power of convex lens is + 5D. The ratio of dispersive power of convex and concave lens will be**

1) 5 : 3 2) 3 : 5 3) 2 : 5 4) 5 : 2

65. **A convex lens of focal length 10 cm is combined with a concave lens of focal length 20 cm to form an achromatic combination. The convex lens is made of a material of dispersive power 0.016. Find the dispersive power of the material used for concave lens.**

1) 0.032 2) 0.016 3) 0.008 4) 0.004

Optical Instruments :

66. **A convergent lens of power 16D is used as a simple microscope. The magnification produced by the lens, when the final image is formed at least distance of distinct vision is**

1) 6 2) 4 3) 7 4) 5

67. **The maximum magnification that can be obtained with convex lens of focal length 2.5cm is (the least distance of distinct vision is 25 cm)**

1) 10 2) 0.1 3) 62.5 4) 11

68. **The objective lens of a compound microscope produces magnification of 10. In order to get an over all magnification of 100 when image is formed at 25 cm from the eye, the focal length of the eye lens should be (in cm)**

1) 4 2) 10 3) 25/9 4) 9

69. **Four lenses A, B, C and D of power +100D, -50D, 20D and 5D. Which lenses will you use to design a compound microscope for best magnification ?**

1) A and C 2) B and D 3) C and D 4) A and B

70. **A microscope consists of two convex lenses of focal lengths 2.0 cm and 6.25 cm placed 15.0 cm apart. Where must the object be placed so that the final virtual image is at a distance of 25 cm from the eye ?**

1) 2.5 cm 2) 2 cm 3) 1.5 cm 4) 3 cm

71. **The two lenses of a compound microscope are of focal lengths 2 cm and 5 cm. If an object is placed at a distance of 2.1 cm from the objective of focal length 2 cm the final image forms at the least distance of distinct vision of a normal eye. Find the distance between the objective and eyepiece**

1) 46.17 cm 2) 42 cm 3) 4.17 cm 4) 40 cm

72. The linear magnification is produced by a telescope having an objective of 50cm and eyepiece of 5cm focal length when used to read a scale 3m away is

1) 1.2 2) 2 3) 0.6 4) 4

73. A telescope, consisting of an objective of focal length 60 cm and a single eye lens of focal length 15 cm is focused on a distant object in such a way that parallel rays comes out from the eye lens. If the object subtends an angle 2^0 at the objective, the angular width of the image is

1) 10^0 2) 8^0 3) 5^0 4) $1/6^0$

74. An astronomical telescope has an angular magnification of magnitude 5 for distant objects. The seperation between the objective and the eye piece is 36 cm and the final image is formed at infinity. The focal length f_0 of the objective and the focal length f_e of the eye piece are

1) $f_0 = 45$ cm, $f_e = -9$ cm 2) $f_0 = 7.2$ cm, $f = 5$ cm
3) $f_0 = 50$ cm, $f_e = 10$ cm 4) $f_0 = 30$ cm, $f_e = 6$ cm

75. If tube length of astronomical telescope is 105 cm and magnifying power is 20 for normal setting, the focal length of the objective is

1) 100 cm 2) 10 cm 3) 20 cm 4) 25 cm

76. The focal lengths of the eyepiece and the objective of an astronomical telescope are 2 cm and 100 cm respectively. Find the magnifying power of the telescope for normal adjustment and the length of the telescope.

1) 50; 102 cm 2) 100 ; 204 cm 3) 25 ; 62 cm 4) 75 ; 125 cm

Eye Pieces :

77. The distance between the image formed by the objective and the image formed by the field lens in Ramsden's eye piece is (f = focal length of eye lens)

1) $\frac{11f}{12}$ 2) $\frac{f}{12}$ 3) $\frac{3f}{2}$ 4) $\frac{3f}{4}$

78. If focal length of eye lens of Ramsden's eye piece is 20 cm, distance of cross – wires from field lens is

1) 15 cm 2) $\frac{55}{3}$ cm 3) 5 cm 4) 7.5 cm

79. The distance between the two images one formed by the objective and the other formed by the field lens in Huygen's eye–piece is (f is focal length of eye lens)

1) $\frac{f}{2}$ 2) $\frac{2f}{3}$ 3) $\frac{f}{4}$ 4) $\frac{11f}{12}$

80. The effective focal length of a Huygen's eye piece is 'f', then seperation between the two lenses and focal length of field lens are

1) $\frac{2f}{3}, 3f$ 2) $2f, \frac{3f}{2}$ 3) $\frac{4f}{3}, 2f$ 4) $\frac{3f}{4}, 2f$

81. The ratio of the distance of the image formed due to the objective from field lens to the effective focal length in Huygen's eye piece is

1) 1 : 1 2) 2 : 1 3) 2 : 3 4) 3 : 1

82. The refractive index of a material of a plano concave lens is 5/3, the radius of curvature is 0.3 m. The focal length of the lens in air is

1) – 0.45 m 2) – 0.6 m 3) – 0.75 m 4) – 1.0 m

83. A glass slab of thickness 8 cms contains the same number of waves as 10 cms long path of water when both are traversed by the same monochromatic light. If the refractive index of water is 4/3, the refractive index of glass is

1) 5/3 2) 5/4 3) 16/15 4) 3/2

84. **In an experiment, with the position of the object fixed, a student varies the position of a convex lens and for each position, the screen is adjusted to get a clear image of the object. A graph between the object distance u and the image distance v, from the lens, is plotted using the same scale for the two axes. A straight line passing through the origin and making an an angle of 45^0 with the x-axis meets the experimental curve at P. The coordinates of P will be:**

1) $\left(\frac{f}{2},\frac{f}{2}\right)$ 2) (f,f) 3) $(4f,4f)$ 4) $(2f,2f)$

85. **A transparent solid cylindrical rod has a refractive index of $2/\sqrt{3}$. It is surrounded by air. A light ray is incident at the midpoint of one end of the rod as shown in the figure.**

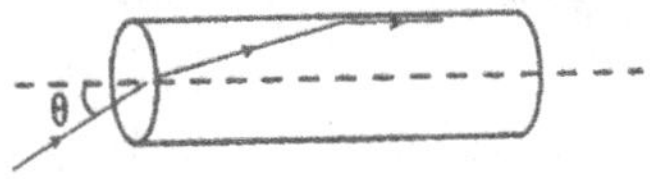

The incident angle θ for which the light ray grazes along the wall of the rod is

1) $\sin^{-1}\left(\frac{\sqrt{3}}{2}\right)$ 2) $\sin^{-1}\left(\frac{2}{\sqrt{3}}\right)$ 3) $\sin^{-1}\left(\frac{1}{\sqrt{3}}\right)$ 4) $\sin^{-1}\left(\frac{1}{2}\right)$

ANSWERS

1) 3	2) 2	3) 1	4) 1	5) 1	6) 1	7) 2	8) 1	9) 3	10) 1
11) 1	12) 1	13) 1	14) 1	15) 3	16) 1	17) 3	18) 1	19) 2	20) 1
21) 4	22) 4	23) 4	24) 1	25) 2	26) 4	27) 2	28) 1	29) 3	30) 1
31) 4	32) 2	33) 1	34) 3	35) 3	36) 4	37) 3	38) 4	39) 3	40) 1
41) 1	42) 1	43) 1	44) 1	45) 2	46) 2	47) 3	48) 1	49) 1	50) 1
51) 2	52) 1	53) 3	54) 4	55) 2	56) 2	57) 4	58) 2	59) 2	60) 4
61) 4	62) 2	63) 1	64) 2	65) 1	66) 4	67) 4	68) 3	69) 1	70) 1
71) 1	72) 1	73) 2	74) 4	75) 1	76) 1	77) 2	78) 3	79) 1	80) 3
81) 1	82) 1	83) 1	84) 4	85) 3					

EXERCISE - II(B)

Refraction :

1. **V_w and V_g are velocities of light in water and glass respectively. Refractive index of water w.r.t. glass is**

1) $\frac{V_w}{V_g}$ 2) $\sqrt{V_w V_g}$ 3) $\frac{V_g}{V_w}$ 4) $V_g . V_w$

2. **The velocity of light in glass whose refractive index with respect to air is 1.5 is $2 \times 10^8 ms^{-1}$. In a certain liquid the velocity of light is found to be 2.5×10^8 ms^{-1}. The refractive index of the liquid with respect to air is**

1) 0.64 2) 0.80 3) 1.20 4) 1.44

3. **The wavelength of blue light in air is 4500 A^0. What is its frequency? If the refractive index for blue light is 1.55 in glass what will be the wavelength of blue light glass? ($c = 3 \times 10^8$ m/s).**

1) 6.667×10^{14} Hz; 2903 $\overset{0}{A}$ 2) 5.667×10^{14} Hz; 2903 $\overset{0}{A}$

3) 6.667×10^{14} Hz; 3903 $\overset{0}{A}$ 4) 5.667×10^{14} Hz; 3903 $\overset{0}{A}$

4. **The refractive index of glass with respect to water is 9/8. If the velocity and wavelength of light in water are 2.25×10^8 m/s and 5400 A^0 respectively find the velocity and wavelength of light in glass.**

1) 2.0×10^8 m/s, 4800 A^0 2) 1.0×10^8 m/s, 6075 A^0

3) 2.0×10^8 m/s, 6075 A^0 4) 1.0×10^8 m/s, 4800 A^0

5. **When light of wavelength 5000 A^0 in vacuum passes through water what will be wavelength and frequency if the refractive index of water is 4/3 ?**

1) $3750\overset{0}{A}$ Hz ; 6×10^{14} Hz 2) $2750\overset{0}{A}$ Hz ; 5×10^{14} Hz

3) $3750\overset{0}{A}$ Hz ; 5×10^{14} Hz 4) $2750\overset{0}{A}$ Hz ; 6×10^{14} Hz

6. **Find the ratio of the time taken by light to travel a distance of 100m in air and water.** $\left(c = 3\times10^8 m/s;\ \mu_{air} = 1.0008;\ \mu_{water} = 1.33.\right)$

1) 0.7525 : 1 2) 1: 0 .7525 3) 1.7525 : 1 4) 1: 1.7525

7. **Velocity of light in air is $3\times10^8\,ms^{-1}$. If it takes 2.66×10^{-6} sec to travel in a medium of refractive index 1.33, the distance it travels in that medium is**

1) 400 m 2) 600 m 3) 500 m 4) 1000 m

8. **The same colour of light takes t_1 sec and t_2 sec to travel the same distance 'x' in two media 'A' and 'B' respectively. Refractive index of medium 'A' w.r.t to 'B' is**

1) $\frac{xt_1}{t_2}$ 2) $\frac{t_2}{xt_1}$ 3) $\frac{t_2}{t_1}$ 4) $\frac{t_1}{t_2}$

9. **The distances travelled by a ray of light in two media, in the same time are in the ratio of 2 : 3. The ratio of their refractive indices is**

1) 2 : 3 2) 3 : 2 3) 4 : 9 4) 9 : 4

10. **The refractive index of glass plate is 3/2. Then the correct thickness of glass plate that will permit the same number of wavelengths as that by an 18 cm long column of water is** $(\mu_w = 4/3)$

1) 12 cm 2) 16 cm 3) 18 cm 4) 24 cm

11. **When light of wavelength 4000 $\overset{0}{A}$ in vacuum travels through the same thickness in air and vacuum the difference in the number of waves is one. Find the thickness.** $(\mu_{air} = 1.0008)$.

1) 0.5 mm 2) 1 mm 3) 1.5 mm 4) 2 mm

12. **A electromagnetic radiation of frequency n, wavelength λ travelling with velocity 'V' in air, enters a glass slab of refractive index μ. The frequency, wavelength and velocity of radiation in the glass slab will be respectively**

1) $n,\frac{\lambda}{\mu},V$ 2) $\frac{n}{\mu},\lambda,\frac{V}{\mu}$ 3) $n,\frac{\lambda}{\mu},\frac{V}{\mu}$ 4) $\frac{n}{\mu},\frac{\lambda}{\mu},\frac{V}{\mu}$

13. **A ray of light travels 200 km in a given time in a medium of refractive index 1.6. Its optical path in air is (in km)**

1) $\frac{200}{1.6}$ 2) $\frac{1.6}{200}$ 3) 320 4) 640

14. **A ray of light is incident on a glass slab making an angle of 30^0 with the normal to the surface. Calculate the angle of refraction in glass and the velocity of light in glass if the refractive index of glass and the velocity of light in air are 1.5 and 3.0×10^8 m/s respectively.**

1) $19^0 28'; 2.0\times10^8 m/s$ 2) $30^0; 2.0\times10^8 m/s$

3) $50^0; 2.0\times10^8 m/s$ 4) $40^0; 2.0\times10^8 m/s$

15. A ray of light is travelling from medium 'A' into a rarer medium B. The angle of incidence is 45^0 and the angle of deviation is 15^0. The refractive index of medium A w.r.to B is

1) $\sqrt{\frac{3}{2}}$ 2) $\frac{\sqrt{3}}{2}$ 3) $\frac{1}{\sqrt{2}}$ 4) $\sqrt{\frac{2}{3}}$

16. A beam of light is incident on a glass slab of refractive index 1.5 at an angle of incidence of 45^0. Find the ratio of the width of refracted beam to the incident beam.

1) 1.682:1 2) 1:1.682 3) 1.247:1 4) 1:1.247

17. The X – Z plane separates two media A and B of refractive indices $\mu_A = 1.5$ and $\mu_B = 2$. A ray of light travels from A to B, its direction in two media are given by unit vectors. $\hat{u}_A = a\hat{i} + b\hat{j}$ and $\hat{u}_B = c\hat{i} + d\hat{j}$ respectively, then

1) $\frac{a}{c} = \frac{4}{3}$ 2) $\frac{a}{c} = \frac{3}{4}$ 3) $\frac{b}{d} = \frac{4}{3}$ 4) $\frac{b}{d} = \frac{3}{4}$

18. The refractive index of water relative to air is 1.33 and of glass relative to water is 1.166. What are the critical angles for water and glass relative to air ?

1) $48^0 45'; 40^0 10'$ 2) $40^0 10'; 48^0 45'$ 3) $46^0 45'; 40^0 10'$ 4) $48^0 45'; 42^0 10'$

19. Refractive indices of glass and water are 3/2 and 4/3 respectively. The critical angle between them is

1) $\sin^{-1}\left(\frac{9}{8}\right)$ 2) $\sin^{-1}\left(\frac{8}{9}\right)$ 3) $\sin^{-1}\left(\frac{2\sqrt{2}}{3}\right)$ 4) $\sin^{-1}\left(\frac{3}{\sqrt{8}}\right)$

20. The critical angle for refraction from medium -1 to air is θ_1 and that from medium - 2 to air is θ_2. If medium -2 is denser than medium -1. Find the critical angle for refraction from medium - 2 to medium.

1) $\sin^{-1}\left(\frac{\sin\theta_2}{\sin\theta_1}\right)$ 2) $\sin^{-1}\left(\frac{\sin\theta_1}{\sin\theta_2}\right)$ 3) $\sin^{-1}(\sin\theta_2)$ 4) $\sin^{-1}(\sin\theta_1)$

21. When a light ray is refracted from one medium into another, the wave length changes from 4500^0A to 3000^0A. The critical angle for a ray from second medium to first medium is

1) $\sin^{-1}\left(\frac{2}{13}\right)$ 2) $\cos^{-1}\left(\frac{2}{3}\right)$ 3) $\tan^{-1}\left(\frac{3}{2}\right)$ 4) $\tan^{-1}\left(\frac{2}{\sqrt{5}}\right)$

22. When a ray of light enters from one medium to another then its velocity in second medium becomes doubled. The maximum value of angle of incidence so that total internal reflection may not take place will be

1) 60^0 2) 90^0 3) 30^0 4) 180^0

23. A ray of light entering from air to glass $(\mu = 1.5)$ is partly reflected and partly refracted. If the incident and reflected rays are at right angles to each other, the angle of refraction is

1) $\sin^{-1}\left(\sqrt{\frac{2}{13}}\right)$ 2) $\sin^{-1}\left(\frac{\sqrt{2}}{3}\right)$ 3) $\sin^{-1}\left(\frac{2}{\sqrt{13}}\right)$ 4) $\sin^{-1}\left(\frac{2}{\sqrt{3}}\right)$

24. A ray of light incidents at an angle of incidence 60^0 on a refracting surface. If the reflected and refracted rays are perpendicular to each other, the refractive index of the medium is

1) $\sqrt{3}$ 2) $\frac{\sqrt{3}}{2}$ 3) $2\sqrt{3}$ 4) $\sqrt{2}$

25. **A light ray falls on a square glass slab on horizontal face with incident angle 45^0. What must be the refractive index of glass, if total internal reflection occurs at the vertical face**

1) $\frac{3}{2}$ 2) $\frac{\sqrt{3}}{2}$ 3) $\sqrt{\frac{3}{2}}$ 4) $\sqrt{2}$

26. **Refractive index of water with respect to air is $\sqrt{2}$. A light ray is incident on the surface at an angle 60^0 travelling through water. The deviation of light ray is**

1) 30^0 2) 120^0 3) 0^0 4) 60^0

27. **A fish looking up through water sees outside world contained in a circular horizon. If the refractive index of water is 4/3 and fish is 12 cm below the surface, the radius of circle in cm is**

1) $36\sqrt{5}$ 2) $4\sqrt{5}$ 3) $36/\sqrt{7}$ 4) $36\sqrt{7}$

28. **A ray of light is incident on a glass plate. The light ray travels distance of 5 cm inside the glass plate before emerging out of the glass plate. If the incident ray suffers a deviation of 30^0, the perpendicular distance between incident and the emergent ray is**

1) 5 cm 2) 2.5 cm 3) 7.5 cm 4) 10 cm

29. **Each quarter of a vessel of depth H is filled with liquids of the refractive indices μ_1, μ_2, μ_3 and μ_4 from the bottom respectively. The apparent depth of the vessel when looked normally is**

1) $\frac{H}{4}(\mu_1+\mu_2+\mu_3+\mu_4)$ 2) $\frac{H}{4}\left(\frac{1}{\mu_1}+\frac{1}{\mu_2}+\frac{1}{\mu_3}+\frac{1}{\mu_4}\right)$

3) $\frac{\mu_1+\mu_2+\mu_3+\mu_4}{4H}$ 4) $\frac{H}{2}\left(\frac{1}{\mu_1}+\frac{1}{\mu_2}+\frac{1}{\mu_3}+\frac{1}{\mu_4}\right)$

30. **A fish is at a depth of 12 cm in water is viewed by an observer on the bank of a lake. To what height the image of the fish is raised $(\mu_w = 4/3)$**

1) 9 cm 2) 12 cm 3) 16 cm 4) 3 cm

31. **A glass slab of thickness 3 cm and refractive index 3/2 is placed on ink mark on a piece of paper. For a person looking at the mark at a distance 2 cm above it, the distance of the mark will appear to be**

1) 3 cm 2) 4 cm 3) 4.5 cm 4) 5 cm

Prism :

32. **The angle of minimum deviation measured with a prism is 30^0 and the angle of prism is 60^0. The refractive index of prism material is**

1) $\sqrt{2}$ 2) 2 3) 3/2 4) 4/3

33. **When light of wavelenth 'λ' is incident on an equilateral prism kept in its minimum deviation position, it is found that the angle of deviation equals the angle of the prism itself. The refractive index of the material of the prism is**

1) $\sqrt{3}/2$ 2) $\sqrt{3}$ 3) 2 4) $\sqrt{2}$

34. **The critical angle between an equilateral prism and air is 45^0. If the incident ray is perpendicular to the refracting surface then**

1) After deviation it will emerge from the second refracting surface

2) It is totally reflected from the second surface and emerges out perpendicularly from the third face into air.

3) It is totally reflected from the second and third refracting surfaces and finally emerges out from the first surface

4) It is totally reflected from all the sides of prism and never emerges out

35. When light rays are incident on a prism at an angle of 45^0, the minimum deviation is obtained. If refractive index of the material of the prism is $\sqrt{2}$, then the angle of prism will be
1) 30^0 2) 45^0 3) 60^0 4) 90^0

36. A prism of refractive index 'μ' and angle 'A' is placed in the minimum deviation position. If the angle of minimum deviation is A, then the value of A in terms of 'μ' is
1) $\sin^{-1}\left(\frac{\mu}{2}\right)$ 2) $\sin^{-1}\sqrt{\frac{\mu-1}{2}}$ 3) $2\cos^{-1}\left(\frac{\mu}{2}\right)$ 4) $\cos^{-1}\left(\frac{\mu}{2}\right)$

37. A ray is incident at an angle of incidence 'i' on one surface of a prism of small angle 'A' and emerges normally from opposite surface. If refractive index of the material of the prism is 'μ' then the angle of incidence is nearly equal to
1) $\frac{A}{\mu}$ 2) $\frac{A}{2\mu}$ 3) μA 4) $\frac{\mu A}{2}$

38. A ray incident at 15^0 on one refracting surface of a prism of angle 60^0, suffers a deviation of 55^0. The angle of emergence is
1) 95^0 2) 45^0 3) 30^0 4) 100^0

39. If a light ray incident on one of the faces of the prism of refractive index 1.414 with grazing incidence and after refraction the emergent ray just grazes the second surface of the prism, then the angle of deviation is
1) 0^0 2) 90^0 3) 30^0 4) 45^0

40. Light falls at normal incidence on one face of a glass prism of refractive index 1.5. What will be the angle of emergence when the angle of the prism is 40^0 ?
1) $74^0 37'$ 2) 60^0 3) 50^0 4) 30^0

41. A glass prism of refractive index 1.5 is placed in water of refractive index 1.33. What will be the minimum value of the angle of the prism so that it will not be possible to have any emergent ray ?
1) 125^0 2) 90^0 3) 130^0 4) 100^0

42. Find the angle of emergence and deviation which light is incident at an angle of 45^0 on one of the refracting faces of an equilateral prism of refractive index 1.52.
1) $54^0 15'; 39^0 15'$ 2) $52^0 15'; 39^0 15'$ 3) $54^0 15'; 34^0 15'$ 4) $52^0 15'; 34^0 15'$

43 A ray of light incident on an equilateral prism shows minimum deviation of 30^0. Calculate the speed of light through the glass prism.
1) 2.121×10^8 m/s 2) 1.50×10^8 m/s 3) 1.25×10^8 m/s 4) 1.75×10^8 m/s

44. An equilateral prism of refractive index 1.53 is placed in water of refractive index 1.33. Calculate the angle of minimum deviation in water.
1) $10^0 12'$ 2) $8^0 12'$ 3) $11^0 12'$ 4) $7^0 12'$

45. If the angle of a thin prism is 4.5^0 and refractive index 1.52, the deviation produced by the prism is
1) 2^0 2) 3^0 3) 2.34^0 4) 0.76^0

46. Two small angled prisms A and B deviate the blue rays by 7^0 and 9^0 and the red rays by 5^0 and 7^0 respectively. Which prism has a greater dispersive power ?
1) Prism A 2) Prism B 3) Same for both Prism A & B 4) none of these

47. The refractive indices of flint glass prism for C, D and F lines are 1.790, 1.795 and 1.805 respectively. The dispersive power of the flint glass prism is
1) 0.01587 2) 0.01887 3) 0.01187 4) 0.01387

48. The refractive index of the material of the prism for violet colour is 1.69 and that for red is 1.65. if the refractive index for mean colour is 1.66, the dispersive power of the material of the prism is
1) 0.66 2) 0.06 3) 0.65 4) 0.69

Ray Optics

Lens maker's formula & Aberrations :

49. A plano convex lens of focal length 100 cm has a convex surface of radius of curvature 60 cm. The refractive index of the lens is

1) 1.5 2) 1.6 3) 1.7 4) 1.3

50. A double convex lens has radii of curvature of 1 m and 2 m. What should be the refractive index of the medium so that the focal length of the lens is 50 cm ?

1) $3\frac{1}{3}$ 2) $2\frac{1}{3}$ 3) $4\frac{1}{3}$ 4) 1.33

51. A diverging meniscus lens of 1.5 refractive index has surfaces of radii 3 and 4cm. The position of the image, if an object is placed 12cm in front of the lens, is

1) 7cm 2) –8cm 3) 9cm 4) 10cm

52. If R_1 and R_2 are the radii of curvature of double convex lens made of same material, the lens with more focal length is

1) $R_1 = 20$ cm, $R_2 = 10$ cm
2) $R_1 = R_2 = 20$ cm
3) $R_1 = R_2 = 10$ cm
4) $R_1 = R_2 = 5$ cm

53. The focal length of an equi – convex lens is greater than the radius of curvature of any of surfaces. Then the refractive index of material of lens is

1) Greater than 0 but less than 1.5
2) Greater than 1.5 but less than 2
3) Greater than 2 but less than 2.5
4) Greater than 2.5 but less than 3.0

54. A double convex lens of refractive index 3/2 is placed in a liquid of refractive index 4/3. If the focal length of the lens in air is 40 cm, in water it will be

1) 12.5 cm 2) 60 cm 3) – 100 cm 4) 1.6 m

55. A concave lens of glass, refractive index 1.5 has both surfaces of same radius of curvature R. On immersion in a medium of refractive index 1.75, it will behave as a

1) Convergent lens of focal length 3.5 R
2) Convergent lens of focal length 3.0 R
3) Divergent lens of focal length 3.5 R
4) Divergent lens of focal length 3.0 R

56. A thin converging lens of refractive index 1.5 has a focal power of 5D. When this lens is immersed in a liquid, it acts as a diverging lens of focal length 100 cm. The refractive index of the liquid is

1) 11/6 2) 9/5 3) 5/3 4) 2

57. A converging crown glass lens has a focal length 20 cm for the violet rays. Its focal length for red rays is (μ_V=1.56 and μ_R=1.50)

1) 20.82 cm 2) 22.40 cm 3) 22.85 cm 4) 24.85 cm

58. A convex lens of focal length 0.15 m is made of refractive index 3/2. When it is placed in a liquid, its focal length increases by 0.225m. The refractive index of the liquid is

1) 7/4 2) 5/4 3) 9/4 4) 3/2

59. An equiconvex lens of glass of focal length 0.1 m is cut along a plane perpendicular to principal axis in to two equal parts. The ratio of focal length of new lenses formed is

1) 1 : 1 2) 1 : 2 3) 2 : 1 4) 2 : 3

60. A diverging lens of focal length 10 cm having refractive index 1.5 is immersed in a liquid of refractive index 3. The focal length and nature of the lens in liquid is

1) 10 cm, convergent
2) 10 cm, divergent
3) 18 cm, convergent
4) 72 cm, divergent

61. A thin convex lens of focal length 10 cm is placed in contact with a concave lens of same material and same focal length. The focal length of the combination is

1) Zero 2) Infinity 3) 10 cm 4) 20 cm

62. **Three lenses in contact have a combined focal length of 12 cm. When the third lens is removed, the combined focal length is 60/ 7cm. The third lens is**

1) Converging lens of focal length 30 cm   2) Converging lens of focal length 60 cm
3) Diverging lens of focal length 30 cm   4) Converging lens of focal length 60 cm

63. **Two plano - convex lenses are separated by a distance of 2 cm to minimize spherical aberration. If the equivalent focal length of the combination is 6 cm find the focal lengths of the lenses.**

1) 12 cm, 10 cm   2) 10 cm, 12 cm   3) 24 cm, 20 cm   4) 6 cm, 5 cm

64. **The dispersive powers of crown and flint glasses are 0.02 and 0.04 respectively. In an achromatic combination of lenses, the focal length of flint glass is 40 cm. The focal length of crown glass lens will be**

1) –20 cm   2) +20 cm   3) –10 cm   4) +10 cm

65. **An achromatic doublet is a converging lens of equivalent focal length 100 cm. If the dispersive powers of the material of the lenses are in the ratio 1 : 2 find the focal lengths of the two lenses.**

1) +56 cm; -100 cm   2) - 56 cm; +100 cm
3) +46 cm; -100 cm   4) +56 cm; -120 cm

Optical Instruments :

66. **The focal length of a convex lens is 10 cm. Find its magnifying power when it is used as a magnifying glass to form the image at (i) near point and (ii) far point.**

1) 3.5 ; 2.5   2) 2.5 ; 3.5   3) 2.5 ; 1.5   4) 1.5 ; 2.5

67. **A magnifying glass is made of a combination of a convergent lens of power 20D and and divergent lens of power 4D. If the least distance of distinct vision is 25cm. The magnifying power is**

1) 7   2) 5   3) 3   4) 8

68. **The magnification produced by the objective lens of a compound microscope is 25. The focal length of eye piece is 5 cm and it forms find image at least distance of distinct vision. The magnifying power of the compound microscope is**

1) 19   2) 31   3) 150   4) $\sqrt{150}$

69. **A converging lens of 2.5cm focal length is used as a simple microscope producing virtual image at 25cm from the eye. The position of the object from the lens is (nearly)**

1) 1cm   2) 2cm   3) 2.2cm   4) 4cm

70. **A compound microscope is of magnifying power 100. The magnifying power of its eyepiece is 4. Find the magnification of its objective.**

1) 25   2) 20   3) 15   4) 30

71. **The two lenses of a compound microscope are of focal lengths 2 cm and 5 cm. If an object is placed at a distance of 2.1 cm from the objective of focal length 2 cm the final image forms at the final image forms at the least distance of distinct vision of a normal eye. Find the magnifying power of the microscope.**

1) 20   2) 6   3) 120   4) 60

72. **The minimum magnifying power of a telescope is M. If the focal length of its eyelens is halved, the magnifying power will become**

1) M   2) 2M   3) 3M   4) 4M

73. **The focal length of objective and eyelens of a astronomical telescope are respectively 20 cm and 5 cm. Final image is formed at least distance of distinct vision. The magnifying power will be**

1) – 4.8   2) – 4.0   3) 4.8   4) 4.0

74. The magnifying power of an astronomical telescope for relaxed vision is 16 and the distance between the objective and eyelens is 34 cm. Then the focal length of objective and eyelens will be respectively

1) 17 cm, 17 cm 2) 20 cm, 14 cm 3) 32 cm, 2 cm 4) 30 cm, 4 cm

75. An astronomical telescope of ten – fold angular magnification has a length of 44 cm. The focal length of the objective is

1) 4 cm 2) 40 cm 3) 44 cm 4) 440 cm

76. The magnifying power of an astronomical telescope for normal adjustment is 10 and the length of the telescope is 110cm. Find the magnifying power of the telescope when the image is formed at the least distance of distinct vision for normal eye.

1) 14 2) 48 3) 28 4) 52

Eye Pieces :

77. The distance between eyelens and crosswires in a Ramsden's eyepiece which has a field lens of focal length 1.2 cm is

1) 1.1 cm 2) 1.2 cm 3) 0.8 cm 4) 0.3 cm

78. In a Ramsden's eye–piece distance of cross–wires before eyelens is 'x'. Then distance of the crosswires from the field lens is

1) $\frac{11x}{4}$ 2) $\frac{4x}{11}$ 3) $\frac{3x}{11}$ 4) $\frac{11x}{3}$

79. The effective focal length of Huygen's eyepiece is 6 cm. The distance between the two images formed in the Huygen's eye piece is

1) 4 cm 2) 2 cm 3) 6 cm 4) 1 cm

80. The equivalent focal length of Huygen's eyepiece is 'f'. The focal lengths of the field lens and eyelens respectively

1) 2f and $\frac{2f}{3}$ 2) f and $\frac{f}{3}$ 3) 3f and f 4) 6f and 2f

81. The focal length of eyelens in Huygen's eye piece is 'f'. It is used such that the position of image formed by the objective lens is at a distance of

1) f behind field lens 2) f/4 behind field lens
3) 3f/2 behind field lens 4) 2f/3 before field lens

ANSWERS

1) 3	2) 3	3) 1	4) 1	5) 1	6) 1	7) 2	8) 4	9) 2	10) 2
11) 1	12) 3	13) 3	14) 1	15) 1	16) 3	17) 1	18) 1	19) 2	20) 1
21) 4	22) 3	23) 3	24) 1	25) 3	26) 4	27) 3	28) 2	29) 2	30) 4
31) 2	32) 1	33) 2	34) 2	35) 3	36) 3	37) 3	38) 4	39) 2	40) 1
41) 1	42) 1	43) 1	44) 1	45) 3	46) 1	47) 2	48) 2	49) 2	50) 2
51) 2	52) 2	53) 1	54) 4	55) 1	56) 3	57) 2	58) 2	59) 1	60) 1
61) 2	62) 3	63) 1	64) 1	65) 1	66) 1	67) 2	68) 3	69) 3	70) 1
71) 3	72) 2	73) 1	74) 3	75) 2	76) 1	77) 1	78) 3	79) 2	80) 1
81) 3									

EXERCISE – III

Refraction :

1. **The difference in the number of wavelengths is one, when yellow light propagates through air and vacuum columns of same thickness. The thickness of air column is (refractive index of air = 1.0003 and wavelength of light in vacuum = 6000^0A)**

   1) 1.8 mm 2) 2 mm 3) 2 cm 4) 2.2 cm

2. **Two beams of light are incident normally on water $(\mu = 4/3)$. If one of the beams passes through a glass $(\mu = 3/2)$ slab of height 'h' has shown in the figure, the time difference for both the beams for reaching the bottom is (C is velocity of light is vacuum)**

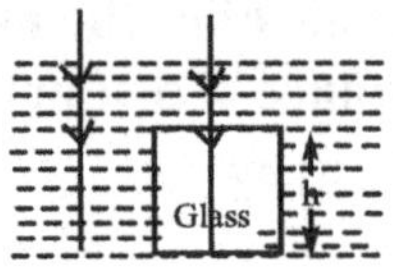

   1) Zero

   2) $\frac{h}{3C}$

   3) $\frac{6h}{C}$ 4) $\frac{h}{6C}$

3. **A ray of light entering from air to glass $(\mu = 1.5)$ is partly reflected and partly refracted. If the reflected and refracted rays are at right angles to each other, the angle of refraction is**

   1) $\sin^{-1}\left(\sqrt{\frac{2}{13}}\right)$ 2) $\sin^{-1}\left(\frac{\sqrt{2}}{13}\right)$ 3) $\sin^{-1}\left(\frac{2}{\sqrt{13}}\right)$ 4) $\sin^{-1}\left(\frac{3}{\sqrt{13}}\right)$

4. **A ray of light is incident on a rectangular plate at an angle of incidence 60^0. The light ray suffers a deviation which is 25% of the angle of incidence. The refractive index of the glass will be**

   1) $\sqrt{3}$ 2) $\sqrt{2}$ 3) $\sqrt{3/2}$ 4) 1.5

5. **A glass cube of edge 1 cm and $\mu = 1.5$ has a spot at the centre. The area of the cube face that must be covered to prevent the spot from being seen is (in cm^2)**

   1) $\sqrt{5}\,\pi$ 2) 5π 3) $\frac{\pi}{\sqrt{5}}$ 4) $\frac{\pi}{5}$

6. **There is a water film formed on a glass block. A light ray is incident on water film from air at an angle 60^0 with the normal. The angle of incidence on glass slab is $(\mu_g = 1.5, \mu_w = 4/3)$**

   1) $\sin^{-1}\left(\frac{3\sqrt{3}}{8}\right)$ 2) $\sin^{-1}\left(\frac{1}{\sqrt{3}}\right)$ 3) $\sin^{-1}\left(\frac{4\sqrt{3}}{9}\right)$ 4) $\sin^{-1}\left(\frac{9\sqrt{3}}{16}\right)$

7. **A ray of light travels in the way as shown in the figure. After passing through water, the ray grazes along the water air interface. The value of μ_g interms of 'i' is $\left(\mu_w = 4/3\right)$**

   1) $\frac{1}{\sin i}$ 2) $\frac{3}{4\sin i}$

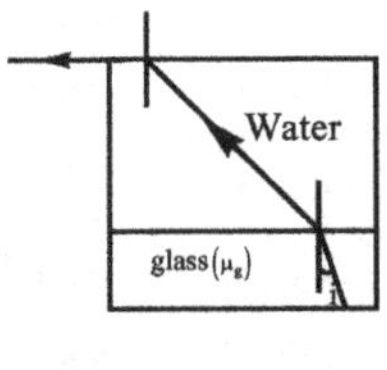

   3) $\frac{4}{3\sin i}$ 4) sin i

8. **A transparent cube of 15 cm edge contains a small air bubble. Its apparent depth when viewed through one face is 6 cm and when viewed through the opposite face is 4 cm. Then the refractive index of the material of the cube is,**

   1) 2 2) 2.5 3) 1.6 4) 1.5

9. **A layer of oil 3 cm thick is floating on a layer of coloured water 5 cm thick. The refractive index of the coloured water is 5/3 and the apparent depth of the two liquids is 36/7cm. Then the refractive index of the oil is,**

1) 7/4 2) 5/4 3) 7/5 4) 5/3

10. **The image of point 'P' when viewed from top of the slabs will be**

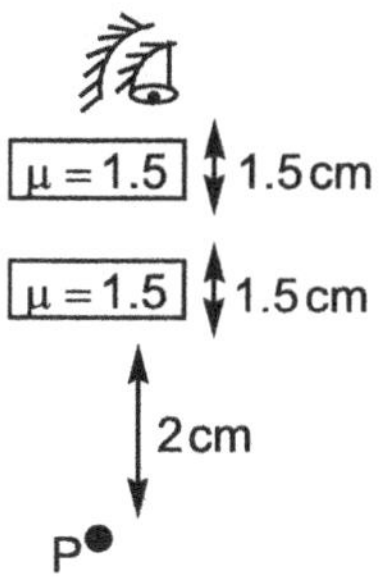

1) 2 cm above P

2) 1.5 cm above P

3) 2 cm below P

4) 1 cm above P

11. **'n' transparent slabs of refractive index 1.5 each having thickness 1cm, 2cm, 3cm... ..ncm are arranged one over another. A point object is seen through this combination from top with perpendicular light. If the shift of the object by combination is 5cm. Then the value of 'n' is**

1) 5 2) 4 3) 3 4) 6

12. **A fish rising vertically to the surface of water in a lake uniformly at the rate of 3 ms^{-1} observes a bird diving vertically towards the water at a rate of 9 ms^{-1} vertically above it. If the refractive index of water is 4/3, the actual velocity of the dive of the bird is (in ms^{-1})**

1) 6 2) 4.5 3) 1.5 4) 2

Prism :

13. **The angle of minimum deviation for a 75^0 prism of dense glass is found to be 45^0 when in air and 15^0 when immersed in certain liquid. The refractive index of the liqud is,**

1) $\frac{\sqrt{3}}{2}$ 2) $\frac{3}{2}$ 3) $\sqrt{\frac{3}{2}}$ 4) $\sqrt{3}$

14. **A certain prism of refracting angle 60^0 and of refractive index '2' is immersed in a liquid of refractive index $\sqrt{2}$, then the angle of minimum deviation will be**

1) 30^0 2) 45^0 3) 60^0 4) 75^0

15. **A ray of monochromatic light is incident on one refracting face of a prism of angle 75^0. It passes through the prism and is incident on the other face at the critical angle. If the refractive index of the material of the prism is $\sqrt{2}$, the angle of incidence on the first face of the prism is,**

1) 30^0 2) 45^0 3) 60^0 4) 0^0

16. **The light ray is incident at an angle of 60^0 on a prism of angle 45^0. When the light ray falls on the other surface at 90^0, the refractive index of the material of the prism μ and angle of deviation 'd' are given by**

1) $\mu = \sqrt{2}, d = 30^0$ 2) $\mu = 1.5, d = 15^0$ 3) $\mu = \frac{\sqrt{3}}{2}, d = 30^0$ 4) $\mu = \sqrt{\frac{3}{2}}, d = 15^0$

17. **A ray of light is incident normally on one of the face of a prism of angle 30^0 and refractive index $\sqrt{2}$. The angle of deviation will be**

1) 26^0 2) 0^0 3) 30^0 4) 15^0

18. **When a glass prism of refracting angle 60^0 is immersed in a liquid its angle of minimum deviation is 30^0. The critical angle of glass with respect to the liquid medium is**

1) 30^0 2) 45^0 3) 60^0 4) 50^0

19. Angle of prism is 'A' and its one surface is silvered. Light rays falling at an angle of incidence 2 A on first surface return back through the same path after suffering reflection at second silvered surface. Refractive index of the material of the prism is

1) 2 SinA 2) 2 CosA 3) 1/2 CosA 4) tanA

20. A ray of light is incident on the face AB of a glass prism ABC having the vertex angle A equal to 30^0. The face AC is silvered and a ray of light incident on the face AB retraces its path. If the refractive index of the material of prism is $\sqrt{3}$ find the angle of incidence on the face AB.

1) 60^0 2) 30^0 3) 45^0 4) 25^0

21. A ray of light incident normally on an isosceles right angled prism travels as shown in the figure. The refractive index of the prism must be greater than

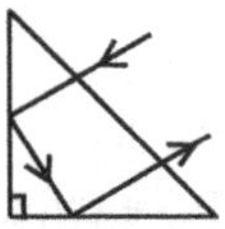

1) $\sqrt{2}$ 2) $\sqrt{3}$ 3) 1.5 4) 2

22. The refracting angle of a prism is A and the refractive index of the material of the prism is $\cot\left(\frac{A}{2}\right)$. The angle of minimum deviation of the prism is

1) $\pi+2A$ 2) $\pi-2A$ 3) $\frac{\pi}{2}+A$ 4) $\frac{\pi}{2}-A$

23. A ray of light incident on the hypotenuse of a right angled prism after travelling parallel to the base inside the prism is incident on second refracting surface. If ' μ' is the refractive index of material of prism, the maximum value of the base angle for which light is totally reflected from the hypotenuse is

1) $\sin^{-1}\left(\frac{1}{\mu}\right)$ 2) $\tan^{-1}\left(\frac{1}{\mu}\right)$ 3) $\sin^{-1}\left(\frac{\mu-1}{\mu}\right)$ 4) $\cos^{-1}\left(\frac{1}{\mu}\right)$

24. A prism of glass $(\mu = 1.5)$ is dipped in to water as shown in the figure. If the refractive index of water is 4/3, then the incident ray will be totally reflected if

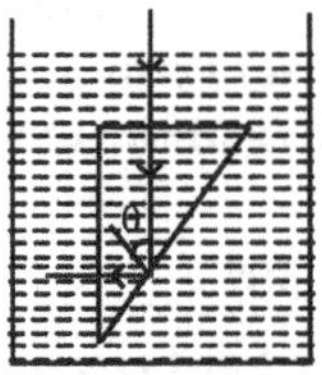

1) $\sin\theta > \frac{8}{9}$ 2) $\sin\theta < \frac{8}{9}$

3) $\sin\theta = \frac{9}{8}$ 4) $\sin\theta = \frac{8}{9}$

25. ABC is a right angled glass prism of refractive index 1.5. $\angle A$, $\angle B$ and $\angle C$ are 60^0, 30^0 and 90^0 respectively.A thin layer of liquid is on the face AB. For a ray of light which is incident normally on AC to be totally reflected at AB, the refractive index of the liquid on AB should be

1) 1.5 2) 1.4 3) 1.3 4) 1.2

26. The principle section of a glass prism is an isosceles triangle ABC with AB = AC, the face AC is silvered. A ray of light is incident normally on the face AB and after two reflections, it emerges from the base BC perpendicular to the base. Angle of BAC of the prism is

1) 30^0 2) 36^0 3) 18^0 4) 72^0

27. **The ratio of angle of minimum deviation of a thin prism in air and when dipped in water will be**

$\left({}_a\mu_g = \frac{3}{2} \text{ and } {}_a\mu_w = \frac{4}{3}\right)$

1) 1/8 2) 1/4 3) 3/4 4) 4/1

28. **Two prisms A and B are in contact with each other have angular dispersions of 2^0 and 4^0 respectively. The dispersive power of 'A' is 0.002. If the combination produces dispersion without deviation, the dispersive power of 'B' is**

1) 0.001 2) 0.004 3) 0.002 4) 0.006

29. **Two prisms A and B have dispersive powers of 0.012 and 0.018 respectively. The two prisms are in contact with each other. The prism 'A' produces a mean deviation of 1.2^0, the mean deviation produced by 'B' if the combination is achromatic is**

1) 3.6^0 2) 0.8^0 3) 0.4^0 4) 1.8^0

30. **Two prisms A and B in contact with each other. The dispersive power of A and B is 0.04 and 0.03 respectively. The angular dispersion produced by 'A' is 8^0 . The angular dispersion produced by the combination, if the combination does not produce a net deviation is**

1) 14^0 2) 8^0 3) 6^0 4) 2^0

Lens maker's formula & Aberrations :

31. **Two plano concave lenses of glass of refractive index 1.5 have radii of curvature 20 cm and 30 cm respectively. They are placed in contact with the curved surface towards each other and the space between them is filled with a liquid of refractive index 5/2. The focal elngth of the combination is (in cm)**

1) 6 2) – 92 3) 108 4) 12

32. **The two surfaces of a biconvex lens has same radii of curvatures. This lens is made of glass of refractive index 1.5 and has a focal length 10 cm in air. The lens is cut into two equal halves along a plane perpendicular to its principal axis to yield two plano–convex lenses. The two pieces are glued such that the convex surfaces touch each other. If this combination lens is immersed in water of refractive index 4/3 its focal length (in cm) is (2006 E)**

1) 5 2) 10 3) 20 4) 40

33. **The effective focal length of the lens combination shown in figure is – 60 cm. The radii of curvature of the curved surfaces of the plano–convex lenses are 12 cm each and refractive index of the material of the lens is 1.5. The refractive index of the liquid is (2006 M)**

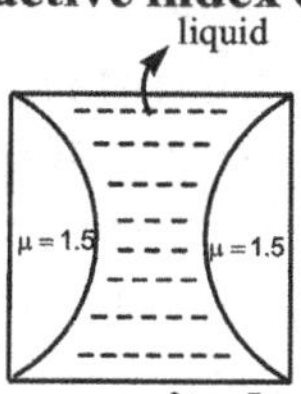

1) 1.33
2) 1.42
3) 1.53
4) 1.60

34. **A convex lens is in contact with concave lens. The magnitude of the ratio of their focal length is 2/3. Their equivalent focal length is 30 cm. Their individual focal lengths are**

1) – 75 cm, 50 cm 2) – 10 cm, 15 cm 3) – 50 cm, 75 cm 4) – 15 cm, 10 cm

35. **Two converging glass lenses 'A' and 'B' have focal lengths in the ratio 2 : 1. The radius of curvature of first surface of lens 'A' is 1/4th of the second surface where as the radius of curvature of first surface of lens 'B' is twice that of second surface. Then the ratio between the radii of the first surfaces of A and B is**

1) 5 : 3 2) 3 : 5 3) 1 : 2 4) 5 : 6

36. **A double convex lens is made of glass which has refractive index 1.55 for violet rays and 1.50 for red rays. If the focal length for violet rays is 20 cm, the focal length for red rays will be**

1) 9 cm 2) 28 cm 3) 20 cm 4) 22 cm

37. Two thin symmetrical lenses one converging and other of diverging nature are made from different material have equal radii of curvature R = 15 cm, the lenses are put in contact and immersed in water ($\mu_W = 4/3$). The focal length of the system in water is 30 cm. Then the difference between refractive indices of the two lenses is

1) 1/2 2) 1/4 3) 1/3 4) 3/4

38. A plano convex lens fits exactly in to a plano concave lens. Their plane surfaces are parallel to each other. If the lenses are made of different materials of refractive indices μ_1 and μ_2 and 'R' is the radius of curvature of the curved surfaces of the lenses then the focal length of the combination is

1) $\frac{R}{\mu_1 - \mu_2}$ 2) $\frac{2R}{\mu_1 - \mu_2}$ 3) $\frac{R}{2(\mu_1 - \mu_2)}$ 4) $\frac{R}{\mu_1 + \mu_2}$

39. The refractive index of a lens material is 1.5 and focal length f. Due to some chemical changes in the material, its refractive index has increased by 2%. The percentage change in its focal length is

1) +4.5% 2) –4.5% 3) +5.67% 4) –5.67%

40. Two lenses of power – 15D and 5D are in contact with each other. The focal length of the combination is (in cm)

1) 20 2) 10 3) –20 4) –10

41. A thin glass (refractive index 1.5) lens has optical power of –8D in air. Its optical power in a liquid medium with refractive index 1.6 will be

1) 25D 2) –25D 3) 1D 4) –1D

42. A candle placed 25 cm from a lens, forms an image on a screen 75 cm on the other side of the lens. The focal length and type of the lens should be

1) 18.75 cm, convex lens 2) 18.75 cm, concave lens
3) 20.25 cm, convex lens 4) 20.25 cm, concave lens

43. The dispersive powers of the materials of the two lenses forming an achromatic doublet is 4 : 3. If the focal length of the combination is 60 cm, focal lengths of the two lenses are

1) – 15 cm, 20 cm 2) 15 cm , – 20 cm 3) 20 cm, – 15 cm 4) 20 cm, – 25 cm

44. The dispersive power of the material of lens of focal length 20 cm is 0.08. The longitudinal chromatic aberration of the lens is (in cm)

1) 0.08 2) 0.004 3) 1.6 4) 0.16

45. The focal length of the field lens (which is an achromatic combination of two lenses) of telescope is 90 cm. The dispersive powers of the two lenses in the combination are 0.024 and 0.036. The focal length of two lenses are

1) 30 cm, 60 cm 2) 30 cm, – 45 cm 3) 45 cm, 90 cm 4) 15 cm, – 45 cm

Optical Instruments :

46. The focal lengths of the objective and eye–lens of a microscope are 1 cm and 5 cm respectively. If the magnifying power for the relaxed eye is 45, then the length of the tube is

1) 30 cm 2) 25 cm 3) 15 cm 4) 12 cm

47. If the focal length of objective and eye lens are 1.2 cm and 3 cm respectively and the object is put 1.25 cm away from the objective lens and the final image is formed at infinity. The magnifying power of the microscope is

1) 150 2) 200 3) 250 4) 400

48. **The length of a compound microscope is 14cm and its magnifying power when final image is formed at near point is 25. If the focal length of eyepiece is 5cm. The distance of object from the objective and the focal length of objective lens are**

1) $\frac{59}{25}$cm, $\frac{59}{31}$cm   2) $\frac{59}{31}$cm $\frac{59}{25}$cm   3) 3cm, 2cm   4) 4cm, 1cm

49. **The focal length of objective and eye lens of a microscope are 4 cm and 8 cm respectively. If the least distance of distinct vision is 24 cm and object distance is 4.5 cm from the objective lens, then the magnifying power of the microscope will be**

1) 18   2) 32   3) 64   4) 20

50. **In a compound microscope the objective and eyepiece have focal length of 0.95cm and 5cm and are kept at a distance of 20cm. If the final image is formed at least distance of distant vision, the position of object and the total magnification is**

1) $\frac{95}{94}$cm ,94   2) $\frac{94}{95}$cm ,95   3) 0.94cm, 94   4) 0.95cm, 95

51. **The focal length of objective and the eyepiece of a compound microscope are 1cm and 5cm. An object is placed at a distance of 1.1cm from the objective and if the final image is formed at infinity the magnifying power and distance between the lenses are respectively**

1) 60, 16cm   2) 50, 16cm   3) 50, 15.7cm   4) 60, 16.7cm

52. **In a compound microscope, the focal lengths of two lenses are 1.5 cm and 6.25 cm an object is placed at 2 cm from objective and the final image is formed at 25 cm from the eye lens. The distance between the two lenses is (in cm)**

1) 6   2) 7.75   3) 9.25   4) 11

53. **The length of the compound microscope is 14 cm. The magnifying power for relaxed eye is 25. If the focal length of eye lens is 5 cm, then the object distance for objective will be (in cm)**

1) 1.8   2) 1.5   3) 2.1   4) 2.4

54. **The focal lengths of the objective and the eye – piece of a compound microscope are 2.0 cm and 3.0 cm respectively. The distance between the objective and the eye – piece is 15.0 cm. The final image formed by the eye – piece is at infinity. The two lenses are thin. The distances in centimeters of the objeect and image produced by the objective measured from the objective respectively.**

1) 2.4 and 12   2) 2.4 and 15   3) 2.3 and 3   4) 2 and 12

55. **The magnifying power of an astronomical telescope is 8 and the distance between the two lenses is 54 cm. The focal length of eye lens and objective will be respectively.**

1) 48 cm, 6 cm   2) 8 cm, 64 cm   3) 64 cm, 8 cm   4) 6 cm, 48 cm

56. **In an astronomical telescope in normal adjustment, straight black line of length L is drawn on the objective. The eye piece forms a real image of this line. The length of this image is . The magnification of the telescope is**

1) $\frac{L}{\ell}$ 2) $\frac{L}{\ell}+1$ 3) $\frac{L}{\ell}-1$ 4) $\frac{L+1}{\ell-1}$

57. **focal length of the objective of a terrestrial telescope is 80 cm and it is adjusted for parallel rays, then its magnifying power is 20. If the focal length of erecting lens is 20 cm, then full length of telescope will be (in cm)**

1) 84   2) 100   3) 124   4) 164

48. **The length of a compound microscope is 14cm and its magnifying power when final image is formed at near point is 25. If the focal length of eyepiece is 5cm. The distance of object from the objective and the focal length of objective lens are**

1) $\frac{59}{25}$cm, $\frac{59}{31}$cm 2) $\frac{59}{31}$cm $\frac{59}{25}$cm 3) 3cm, 2cm 4) 4cm, 1cm

49. **The focal length of objective and eye lens of a microscope are 4 cm and 8 cm respectively. If the least distance of distinct vision is 24 cm and object distance is 4.5 cm from the objective lens, then the magnifying power of the microscope will be**

1) 18 2) 32 3) 64 4) 20

50. **In a compound microscope the objective and eyepiece have focal length of 0.95cm and 5cm and are kept at a distance of 20cm. If the final image is formed at least distance of distant vision, the position of object and the total magnification is**

1) $\frac{95}{94}$cm ,94 2) $\frac{94}{95}$cm ,95 3) 0.94cm, 94 4) 0.95cm, 95

51. **The focal length of objective and the eyepiece of a compound microscope are 1cm and 5cm. An object is placed at a distance of 1.1cm from the objective and if the final image is formed at infinity the magnifying power and distance between the lenses are respectively**

1) 60, 16cm 2) 50, 16cm 3) 50, 15.7cm 4) 60, 16.7cm

52. **In a compound microscope, the focal lengths of two lenses are 1.5 cm and 6.25 cm an object is placed at 2 cm from objective and the final image is formed at 25 cm from the eye lens. The distance between the two lenses is (in cm)**

1) 6 2) 7.75 3) 9.25 4) 11

53. **The length of the compound microscope is 14 cm. The magnifying power for relaxed eye is 25. If the focal length of eye lens is 5 cm, then the object distance for objective will be (in cm)**

1) 1.8 2) 1.5 3) 2.1 4) 2.4

54. **The focal lengths of the objective and the eye – piece of a compound microscope are 2.0 cm and 3.0 cm respectively. The distance between the objective and the eye – piece is 15.0 cm. The final image formed by the eye – piece is at infinity. The two lenses are thin. The distances in centimeters of the objeect and image produced by the objective measured from the objective respectively.**

1) 2.4 and 12 2) 2.4 and 15 3) 2.3 and 3 4) 2 and 12

55. **The magnifying power of an astronomical telescope is 8 and the distance between the two lenses is 54 cm. The focal length of eye lens and objective will be respectively.**

1) 48 cm, 6 cm 2) 8 cm, 64 cm 3) 64 cm, 8 cm 4) 6 cm, 48 cm

56. **In an astronomical telescope in normal adjustment, straight black line of length L is drawn on the objective. The eye piece forms a real image of this line. The length of this image is . The magnification of the telescope is**

1) $\frac{L}{\ell}$ 2) $\frac{L}{\ell}+1$ 3) $\frac{L}{\ell}-1$ 4) $\frac{L+1}{\ell-1}$

57. **focal length of the objective of a terrestrial telescope is 80 cm and it is adjusted for parallel rays, then its magnifying power is 20. If the focal length of erecting lens is 20 cm, then full length of telescope will be (in cm)**

1) 84 2) 100 3) 124 4) 164

58. An astronomical telescope and a Galilean telescope are identical objective lenses. They have the same magnification, when both are in normal adjustment. The eye piece of the astronomical telescope has a focal length f.

1) The tube lengths of the two telescopes differ by f

2) The tube lengths of the two telescopes differ by 2f

3) The tube lengths of the two telescopes differ by 3f

4) The tube lengths of the two telescopes differ by 4f

59. The focal lengths of the lenses of an astronomical telescope are 50 cm and 5 cm. The length of the telescope when the image is formed at the least distance of distinct vision is (in cm)

1) 45 2) 55 3) $\frac{275}{6}$ 4) $\frac{325}{6}$

Eye Pieces :

60. If the effective focal length of Ramsden's eye piece is equal to the effective focal length of Huygen's eye piece, then the ratio of focal lengths of field lens of Ramsden's to that of Huygen's eye piece is

1) 2 : 1 2) 2 : 3 3) 1 : 1 4) 1 : 3

61. The equivalent focal length of lenses used in Ramsden's eye piece is 6 cm. If they are put in contact with their flat surfaces together, the focal length of new arrangement becomes

1) 8 cm 2) 12 cm 3) 2 cm 4) 4 cm

62. The equivalent focal length of the combination of lenses in a Ramsden's eye piece is 4.2 cm. Then the focal length of each lens and distance between two lenses are respectively

1) 5.6 cm, 3.73 cm 2) 6.5 cm, 4.73 cm 3) 2.8 cm, 7.5 cm 4) 4.8 cm, 4.5 cm

63. The distance of the cross wires from eye lens in a Ramsden's eye piece is 22/3 cm, then the equivalent focal length of the eye piece is

1) 12 cm 2) 6 cm 3) 4 cm 4) 8 cm

64. In Ramsden's eye piece, the two plano convex lenses each of focal length 'f' are separated by a distance 12 cm. The equivalent focal length (in cm) of the eye piece is

1) 10.5 2) 12 3) 13.5 4) 15.5

ANSWERS

1) 2	2) 4	3) 3	4) 3	5) 4	6) 1	7) 1	8) 4	9) 3	10) 4
11) 1	12) 2	13) 3	14) 1	15) 2	16) 4	17) 4	18) 2	19) 2	20) 1
21) 1	22) 2	23) 4	24) 1	25) 4	26) 2	27) 4	28) 2	29) 2	30) 4
31) 4	32) 4	33) 4	34) 4	35) 4	36) 4	37) 3	38) 1	39) 4	40) 4
41) 3	42) 1	43) 2	44) 3	45) 2	46) 3	47) 2	48) 1	49) 2	50) 1
51) 2	52) 4	53) 1	54) 1	55) 4	56) 1	57) 4	58) 2	59) 4	60) 2
61) 4	62) 1	63) 2	64) 3						

COMPREHENSION

Comprehension – I

A transparent cube of 24 cm edge contains a smaller air bubble. Its apparent depth when viewed through first face is 10 cm and when viewed through the opposite second face is 8 cm

1. The refractive index of the material of the cube is

1) 1.25 2) 1.33 3) 1.66 4) 2

2. Critical angle of the material of the cube is

1) $\sin^{-1}(4/3)$ 2) $\cos^{-1}(3/4)$ 3) $\cos^{-1}\left(\sqrt{7}/4\right)$ 4) $\tan^{-1}(4/3)$

3. **Actual distance of air bubble from the first face is**

1) 12.5 cm 2) 13.3 cm 3) 14.8 cm 4) 15 cm

4. **Actual distance of air bubble from the second face is**

1) 8.2 cm 2) 9.6 cm 3) 10.6 cm 4) 12 cm

Comprehension – II : In an astronomical telescope the focal lengths of the objective and the eye piece are 100 cm and 5 cm respectively. If the telescope is focussed on a scale 2m from the objective, the final image is formed at 25 cm from the eye then

5. **Magnification of eye piece is**

1) 2 2) 4 3) 6 4) 8

6. **Magnification of Telescope is**

1) 6 2) 12 3) 18 4) 24

7. **Length of the telescope is**

1) 154.5 cm 2) 172 cm 3) 196.4 cm 4) 204.2 cm

Comprehension – III : ABC is right angled prism kept in air. A ray (1) is incident on the face AB along the normal. Another ray (2) is incident on the face AB such that it emerges normally form the face AC. Then

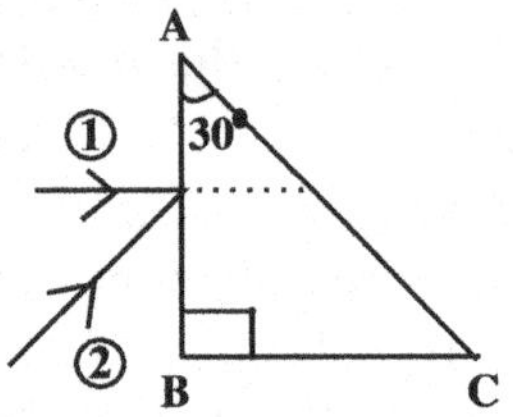

8. **The minimum value of refractive index of the material of the prism for which the ray (1) under goes total internal reflection on the face AC is**

1) $\sqrt{2}$ 2) 2 3) 1.5 4) 2.2

9. **The angle of incidence of the ray (2) on the face AB is**

1) 0^0 2) 45^0 3) 30^0 4) 90^0

10. **The deviction suffered by the light ray (2) is**

1) 60^0 2) 45^0 3) 30^0 4) 90^0

Comprehension – IV

A crown glass prism of angle 5^0 is to be combined with a flint glass prism in such a way that the mean ray passes undeviated. Refractive indices for red, yellow and violet are 1.514, 1.517 and 1.523 respectively for crown glass and 1.613, 1.620 and 1.632 for flint glass then

11. **The angle of flint glass needed is**

1) 3.6^0 2) 4.2^0 3) 6.4^0 4) 8.2^0

12. **The net angular dispersion produced by the combination when white light goes is,**

1) 0.0348^0 2) 0.0369^0 3) 0.0422^0 4) 0.0448^0

13. **Dispersive power of crown glass is**

1) 0.012 2) 0.014 3) 0.017 4) 0.019

Comprehension – V

A convex lens of focal length 20cm and concave lens of focal length 10cm one placed 20 cm apart. In between them an object is placed at a distance 'x' from the convex lens

14. **What is the value of x in cm so that the images formed by both th lenses coincide ?**

1) $20(\sqrt{3}-1)$ 2) $\frac{20\sqrt{3}-1}{\sqrt{3}}$ 3) $\frac{20}{\sqrt{3}}$ 4) $10(\sqrt{3}-1)$

15. **What will be the linear magnification produced by convex lens ?**

1) $\sqrt{3}+1$ 2) $\sqrt{3}$ 3) $\sqrt{3}+2$ 4) $2\sqrt{3}$

16. **What will be the linear magnification produced by concave lens ?**

1) $\frac{1}{\sqrt{3}}$ 2) $\frac{1}{\sqrt{2}}$ 3) $\sqrt{3}-1$ 4) $2\sqrt{3}-3$

Comprehension – VI

Two parallel rays are travelling in a medium of refractive index $\mu_1 = \frac{4}{3}$. One of the rays passes through a parallel glass slab of thickness 't' and refractine index $\mu_2 = \frac{3}{2}$.

17. **The path difference between the two rays due to the glass slab will be**

1) 4t /3 2) 3t / 2 3) t / 8 4) t / 6

18. **The critical angle of the pair of media is**

1) $\text{Sin}^{-1}(8/9)$ 2) $\sin^{-1}(3/4)$ 3) $\text{Sin}^{-1}(2/3)$ 4) $\text{Sin}^{-1}(1/2)$

19. **Instead of being paralled rays, if the rays are diverging through an angle θ in medium of $\mu_1 = \frac{4}{3}$, on passing through slab as described above, the angle of divergence**

1) Increases 2) Decreases 3) Remains the same 4) May increase or decrease

Comprehension – VII

Ray strikes normally on the face AB of prism of $\mu = \frac{3}{2}$. A liquid of refractine index μ_1 exists on AC.

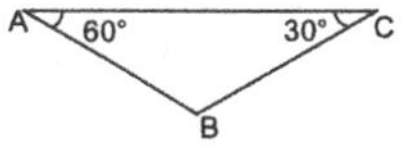

20. **The μ_1 for which T.I.R takes place at AC is**

1) $> \frac{\sqrt{3}}{2}$ 2) $> \frac{3\sqrt{3}}{4}$ 3) $> \sqrt{3}$ 4) $< \frac{\sqrt{3}}{2}$

21. **The μ_1 for which the ray grazes the face AC is**

1) $> \frac{\sqrt{3}}{2}$ 2) $\frac{3\sqrt{3}}{4}$ 3) $> \sqrt{3}$ 4) $< \frac{\sqrt{3}}{2}$

22. **The μ_1 for which the ray emerges from AC is**

1) $> \frac{3\sqrt{3}}{2}$ 2) $\frac{3\sqrt{3}}{4}$ 3) $> \sqrt{3}$ 4) $< \frac{\sqrt{3}}{2}$

Comprehension – VIII

A ray incident at an angle of 60° enters a glass sphere of refractine index $\mu = \sqrt{3}$. This ray is reflected and refracted at the farther surface of sphere

23. **The angle between reflecred and refracted rays at this surface is**

1) 90° 2) 60° 3) 70° 4) 140°

24. **The deviation between incident and emergent rays is**

1) 30° 2) 60° 3) 90° 4) 120°

25. **Instead if the ray enters at angle 60° to one of its diameters and the ray emerges parallel to the same diametre the refractive index of material of sphere is**

1) $\sqrt{2}$ 2) $\sqrt{3}$ 3) 3/2 4) 2

ANSWERS

1) 2	2) 3	3) 2	4) 3	5) 3	6) 1	7) 4	8) 2	9) 4	10) 1
11) 2	12) 1	13) 3	14) 2	15) 2	16) 4	17) 3	18) 1	19) 3	20) 2
21) 2	22) 1	23) 1	24) 2	25) 2					

EXERCISE – IV(A)

1. Two plane mirrors are at 45^0 to each other. If an object is placed between them, then the number of images will be

1) 5 2) 9 3) 7 4) 8

2. A ray reflected successively from two plane mirrors inclined at a certain angle undergoes a deviation of 300^0. The number of observable images is

1) 60 2) 12 3) 11 4) 5

3. To get five images of a single object, one should have two plane mirrors at an angle of

1) 30^0 2) 72^0 3) 90^0 4) 150^0

4. A small object is placed 10 cm infront of a plane mirror. If you stand behind the object 30 cm from the mirror and look at its image, the distance focused for your eye will be

1) 60 cm 2) 20 cm 3) 40 cm 4) 80 cm

5. A watch shows time as 3 : 25 when seen through a mirror, time appeared will be

1) 8 : 35 2) 9 : 35 3) 7 : 35 4) 8 : 25

6. A man runs towards mirror at a speed of 15 m/s. The speed of image with respect to man is

1) 7.5 m/s 2) 15 m/s 3) 30 m/s 4) 45 m/s

7. A person standing in the centre of a room, 12 m high looks into a plane mirror fixed on the wall. Then the minimum length of the plane mirror for him to see the full length image of the wall behind him, is equal to

1) 12 m 2) 8 m 3) 4 m 4) 6 m

8. A ray of light is incident at 50^0 on the middle of one of the two mirrors arranged at an angle of 60^0 between them. The ray then touches the second mirror, get reflected back to the first mirror, making an angle of incidence of

1) 50^0 2) 60^0 3) 70^0 4) 80^0

9. An object is placed at 20cm from a convex mirror of focal length 10 cm. The image formed by a mirror is

1) Real and at 20cm from the mirror 2) Virtual and at 20cm from the mirror 3) Virtual and at 20/3cm from the mirror 4) Real and at 20/3cm from the mirror

10. A point object is placed at a distance of 30cm from a convex mirror of focal length 30cm. The image will form at

1) Infinity 2) Focus 3) Pole 4) 15cm behind the mirror

11. With a concave mirror, an object is placed at a distance x_1 from the principal focus, on the principal axis. The image is formed at a distance x_2 from the principal focus. The focal length of the mirror is

1) $x_1 x_2$ 2) $\frac{x_1 + x_2}{2}$ 3) $\sqrt{\frac{x_1}{x_2}}$ 4) $\sqrt{x_1 x_2}$

12. An object 1 cm tall is placed 4 cm infront of a mirror. In order to produce an upright image of 3 cm height one needs a

1) Convex mirror of radius of curvature 12cm 2) Concave mirror of radius of curvature 12 cm 3) Concave mirror of radius of curvature 4 cm 4) Plane mirror of height 12 cm

13. Radius of concave mirror is 40 cm and the size of image (real) is twice as that of object, then the object distance is

1) 60 cm 2) 20 cm 3) 40 cm 4) 30 cm

14. A plano - convex lens of refractive index 1.5 and radius of curvature 30cm is silvered at the curved surface. Now this lens has been used to form the image of an object. At what distnace from this lens and object be placed in order to have a real image of the size of the object

1) 20cm 2) 30cm 3) 60cm 4) 10cm

15. **A glass sphere (μ= 1.5) of radius 20cm has a small air bubble 4 cm below its centre. The sphere is viewed from outside and along a vertical line through the bubble. The apparent depth of the bubble below the surface of sphere is (in cm)**

1) 13.33 2) 26.67 3) 15 4) 30

16. **In Gallilean telescope, if the powers of an objective and eye lens are respectively + 1.25 D and – 20 D, then for relaxed vision, the length and magnification will be**

1) 21.25 cm and 16 2) 75 cm and 20 3) 75 cm and 16 4) 8.5 cm and 21.25

17. **Two point white dots are 1 mm apart on a black paper. They are viewed by eye of pupil diameter 3 mm. Approximately, the maximum distance at which these dots can be resolved by eye (wavelength of light = 500 nm)**

1) 5 m 2) 3 m 3) 6 m 4) 1 m

18. **An opera glass (Gallilean telescope) measures 9 cm from the objective to the eye piece. The focal length of the objective is 15 cm. Its magnifying power is**

1) 2.5 2) 2/5 3) 5/6 4) 0.4

19. **Two plane mirrors are arranged at right angles to each other as shown in fig0ure. A ray of light is incident on the horizontal mirror at an angle 'θ'. For what value of θ the ray emerges parallel to the incoming ray after reflection from the vertical mirror ?**

1) 60^0
2) 30^0
3) 45^0
4) all of these

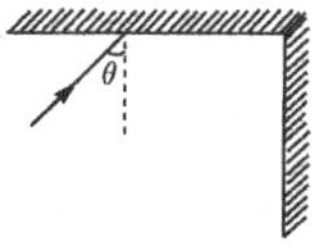

20. **An infinitely long rod lies along the axis of a concave mirror of focal length 'f'. The near end of the rod is at a distance u > f from the mirror. Its image will have a length.**

1) $\frac{uf}{u-f}$ 2) $\frac{uf}{u+f}$ 3) $\frac{f^2}{u+f}$ 4) $\frac{f^2}{u-f}$

21. **A plane mirror is placed at origin parallel of y–axis, facing the positive x – axis. An object starts from (2m, 0, 0) with a velocity of $\left(2\hat{i}+2\hat{j}\right)m/s$. The relative velocity of image with respect to object is along**

1) Positive x – axis 2) negative x – axis 3) positive y – axis 4) negative y–axis

22. A ray incident at an angle of incidence 60^0 **enters a glass sphere of refractive index $\mu=\sqrt{3}$. This ray is reflected and refracted at the farther surface of the sphere. The angle between reflected and refracted rays at this surface is**

1) 90^0 2) 60^0 3) 70^0 4) 40^0

23. **An object 'O' is placed in front of a small plane mirror M_1 and a large convex mirror M_2 of focal length 'f'. The distnace between 'O' and M_1 is x and the distance between M_1 and M_2 is y. The images of 'O' formed by M_1 and M_2 coincide. The magnitude of 'f' is**

1) x–y

2) $\frac{x^2-y^2}{2y}$

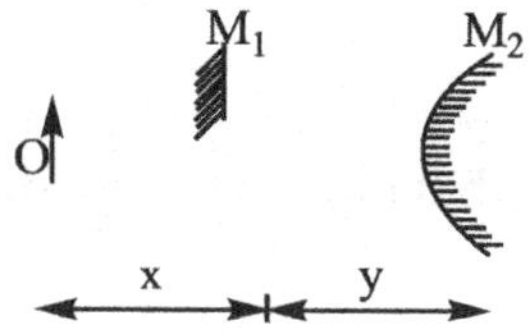

3) $\frac{x^2+y^2}{2y}$ 4) $\frac{x^2+y^2}{x+y}$

24. **A ray of light is incident on a glass sphere of refractive index 3/2. What should be the angle of incidence so that the ray which enters the sphere does not come out of the sphere ?**

1) $\tan^{-1}(2/3)$ 2) $\sin^{-1}(2/3)$ 3) 90^0 4) $\cos^{-1}(1/3)$

25. **Two plane mirrors A and B are aligned parallel to each other, as shown in the figure. A light ray is incident at an angle 30^0 at a point just inside one end of A. The plane of incidence coincides with the plane of the figure. The maximum number of times the ray undergoes reflections (including the first one) before it emerges out is**

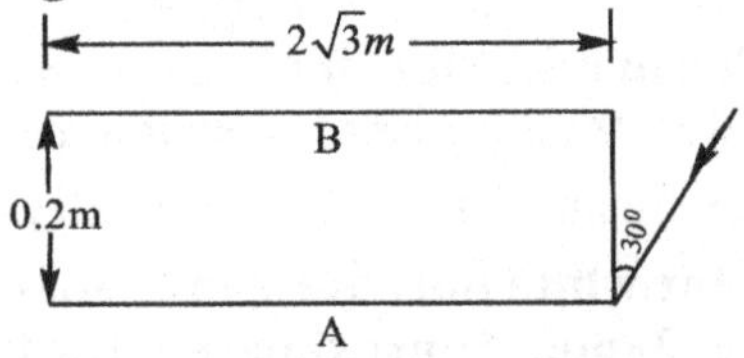

1) 28
2) 30
3) 32
4) 34

26. **A rod of length 10 cm lies along the principal axis of a concave mirror of focal length 10 cm in such a way that the end closer to the pole is 20 cm away from it. The length of the image is**

1) 5 cm 2) 10 cm 3) 15 cm 4) 20 cm

27. **A 2 cm high object is placed on the principal axis of a concave mirror at a distance of 12 cm from the pole. If the image is inverted, real and 5 cm high. The location of the image and the focal length of the mirror is**

1) 30 cm, 8.6 cm 2) 8.6 cm, 30 cm 3) 30 cm, 10 cm 4) 10 cm, 30 cm

ANSWERS

1) 3	2) 3	3) 2	4) 3	5) 1	6) 3	7) 3	8) 3	9) 3	10) 4
11) 4	12) 2	13) 4	14) 1	15) 2	16) 3	17) 1	18) 1	19) 4	20) 4
21) 2	22) 1	23) 2	24) 3	25) 2	26) 1	27) 1			

EXERCISE - IV(B)

1. **A light bulb is placed between two plane mirror inclined at an angle 60^0. The number of images formed are [AIIMS97, AIEEE2002, PMT2004]**

1) 6 2) 2 3) 5 4) 4

2. **Two plane mirrors are inclined to each other at an angle θ. A ray of light is reflected first at one mirror and then at the other. The total deviation of the ray is**

1) $360^\circ - 2\theta$ 2) $240^\circ - 2\theta$ 3) 2θ 4) $180^\circ - \theta$

3. **To get three images of a single object, one should have two plane mirrors at an angle of**

1) 30^0 2) 60^0 3) 90^0 4) 150^0

4. **An object is at a distance of 0.5 m in front of a plane mirror. Distance between the object and image is**

1) 0.5 m 2) 1 m 3) 0.25 m 4) 1.5 m

5. **A clock hung on a wall has marks instead of numerals in its dial. On the adjoining wall, there is a plane mirror and the image of the clock in the mirror indicates the time 4:20. Then the time on the clock is**

1) 7 : 40 2) 4 : 20 3) 2 : 40 4) 4 : 07

6. **A child walks towards a fixed plane mirror at a speed of 5 kmh^{-1}. The velocity of the image is (in kmh^{-1})**

1) 5 2) – 5 3) 10 4) – 10

7. **A man of length h requires a mirror, to see his own complete image of length at least equal to**

1) h/4 2) h/3 3) h/2 4) h

8. **Two vertical plane mirrors are inclined at an angle of 60^0 with each other. A ray of light travelliing horizontally is reflected first from one mirror and then from the other mirror. Then the resultant deviation is**

1) 60^0 2) 120^0 3) 180^0 4) 240^0

9. **A convex mirror has a focal length f. A real object is placed at a distance f in front of it from the pole produces an image at**
   1) Infinity 2) f 3) f/2 4) 2f
10. **A convex mirror has a focal length of 20 cm. A real object is placed at a distance of 20 cm in front of the mirror from the pole. The mirror produces an image at**
   1) infinity 2) 20 cm 3) 40 cm 4) 10 cm
11. **A concave mirror has radius of curvature of 1m. Light from a distant star is incident on the mirror. The distance of the image of the star from the mirror is**
   1) 0.5m 2) 1m 3) 2m 4) 0.25m
12. **A concave mirror gives an image three time as large as the object placed at a distance of 20 cm from it. For the image to be real, the focal length should be**
   1) 10 cm 2) 15 cm 3) 20 cm 4) 30 cm
13. **A concave mirror of focal length f (in air) is immersed in water (μ = 4/3). The focal length of the mirror in water will be**
   1) f 2) 4/3 f 3) 3/4 f 4) 7/3 f
14. **A thin plano – convex lens acts like a concave mirror of focal length 0.2 m when silvered from its plane surface. The refractive index of the material of the lens is 1.5. The radius of curvature of the convex surface of the lens will be**
   1) 0.4 m 2) 0.2 m 3) 0.1 m 4) 0.75 m
15. **A mark on the surface of a glass sphere(μ = 1.5) is viewed from a diametrically opposite position. It appers to be at a distance 10cm from its actual position. The radius of the sphere is**
   1) 5cm 2) 10cm 3) 15cm 4) 25cm
16. **A Galileo telescope has an objective of focal length 100 cm and magnifying power 50. The distance between the two lenses in normal adjustment will be**
   1) 96 cm 2) 98 cm 3) 102 cm 4) 104 cm
17. **Wave length of light used in an optical instrument are λ_1=4000 A^0 and λ_2=5000 A^0, then ratio of their respective resolving power is**
   1) 16:25 2) 25:16 3) 4:5 4) 5:4
18. **A telescope of diameter 2 m uses light of wavelength 5000 A^0 for viewing stars. The minimum angular seperation between two stars whose image is just resolved by this telescope is**
   1) 4×10^{-4} rad 2) 0.25×10^{-6} rad 3) 0.31×10^{-6} rad 4) 5.0×10^{-3} rad
19. **Two plane mirrors are inclined at angle 'θ as shown in figure. If a ray parallel to OB strikes the other mirror at P and finally emerges parallel to OA after two reflections then θ is equal to**
   1) 90^0
   2) 60^0
   3) 45^0
   4) 30^0

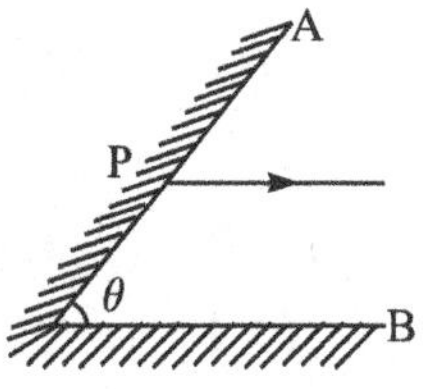

20. **A short linear object of length 'b' lies along the axis of a concave mirror of focal length 'f' at a distance 'u' from the pole of the mirror. The size of the image is approximately equal to**
   1) $b\left(\frac{u-f}{f}\right)^{1/2}$ 2) $b\left(\frac{f}{u-f}\right)^{1/2}$ 3) $b\left(\frac{u-f}{f}\right)$ 4) $b\left(\frac{f}{u-f}\right)^2$

21. **An object is placed infront of a concave mirror of focal length 'f'. A virtual image is formed with a magnification of 2. To obtain a real image of same magnification, the object has to moved by a distance**

1) f 2) f/2 3) 3f/2 4) 2f/2

22. **A ray of light falls on a transparent sphere with centre at C as shown in figure. The ray emerges from the sphere parallel to line AB. The refractive index of the sphere is**

1) $\sqrt{2}$
2) $\sqrt{3}$
3) 3/2
4) 1/2

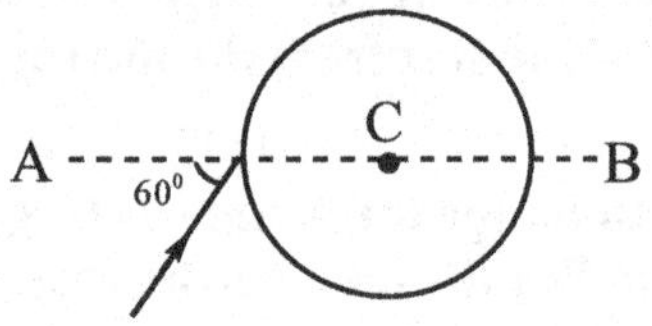

23. **A concave mirror has a focal length 20cm. The distance between the two positions of the object for which the image size is double of the object size is**

1) 20 cm 2) 40 cm 3) 30 cm 4) 60 cm

24. **A spherical surface of radius of curvature R, separates air (refractive index 1.0) from glass (refractive index 1.5). The centre of curvature is in the glass. A point object P placed in air is found to have a real image Q in the galss. The line PQ cuts the surface at a point O and PO = OQ. The distnace PO is equal to**

1) 5R 2) 3R 3) 2R 4) 1.5 R

25. **An object is placed on the principal axis of a concave mirror of focal length 10 cm at a distance of 8 cm from the pole. Find the position and the nature of the image.**

1) 40 cm, virtual 2) 40 cm, real 3) 20 cm, virtual 4) 20 cm, real

26. **At what distance from a convex mirror of focal length 2.5 m should a boy stand so that his image has a height equal to half the original height ?**

1) 2.5 m from the mirror
2) 5 m from the mirror
3) 7.5 m from the mirror
4) 10 m from the mirror

27. **There is a small air bubble inside a glass sphere $(\mu = 1.5)$ of radius 10 cm. The bubble is 4 cm below the surface and is viewed normally from the outside. The apparent depth of the bubble is**

1) 3 cm below the surface
2) 5 cm below the surface
3) 8 cm below the surface
4) 10 cm below the surface

ANSWERS

1) 3	2) 1	3) 3	4) 2	5) 1	6) 2	7) 3	8) 4	9) 3	10) 4
11) 1	12) 2	13) 1	14) 2	15) 1	16) 2	17) 4	18) 3	19) 2	20) 4
21) 1	22) 2	23) 1	24) 1	25) 1	26) 1	27) 1			

Chapter - 11

WAVE OPTICS

WAVE OPTICS

SYNOPSIS

In geometrical optics, light is represented as a ray which travels in a straight line in a homogeneous medium. The phenomenon like Interference and Dffraction cannot be explained on the basis of particle nature of light. These phenomenon can only be explained on the basis of wave nature of light. This part of optics is called physical optics or wave optics. The wave theory of light was presented by Christian Huygen. It should be pointed out that Huygen did not know whether the light waves were longitudinal or transverse and also how they propagate through vaccum. It was then explained by Maxwell by introducing electromagnetic wave theory in nineteenth century.

Geometrical optics

1. In geometrical optics, light is assumed to be travelling in a straight line. This property is known as rectilinear propagation.
2. By using rectilinear propagation of light, laws of reflection, refraction, total internal reflection etc. are explained geometrically.

Physical optics or wave optics:

1. In physical optics, light is considered as a wave
2. Huygen's wave principle and principle of superposition are used to explain interference and diffraction
3. Electromagnetic wave nature of light is used to explain the concept of polarisation.

Condition for applicability of geometrical optics and wave optics: When the size of the object incteracting with light, is much larger than the wavelength of light, we can apply geometrical optics.

When the wavelength of light is comparable to or less than the size of the object interacting with light, we can apply wave optics.

If 'b' is the size of the object interacting with light, 'l' is the distance between the object and the screen and 'λ' is the wavelengh of light then,

i) The condition for applicability of geometrical optics is $\frac{b^2}{l\lambda} >> 1$

ii) The condition for applicability of wave optics is

$$\frac{b^2}{l\lambda} \approx 1 \; or \; \frac{b^2}{l\lambda} << 1$$

Note: The object interacting with light may be a mirror, a lens, a prism, an aperture (pin hole), a slit and a straight edge.

WAVE FRONT

According to wave theory of light, a source of light sends out disturbance in all directions. In a homogeneous medium, the disturbance reaches to all those particles of the medium in phase, which are located at the same distance from the source of light and hence at any instant, all such particles must be vibrating in phase with each other.

The locus of all the particles of the medium, which at any instant are vibrating in the same phase, is called the wavefront.

Depending upon the shape of the source of light, wavefront can be of the following types

1. **Spherical wavefront:** A spherical wavefront is produced by a point source of light. It is because, the locus of all such points, which are equidistant from the point source, is a sphere (a)

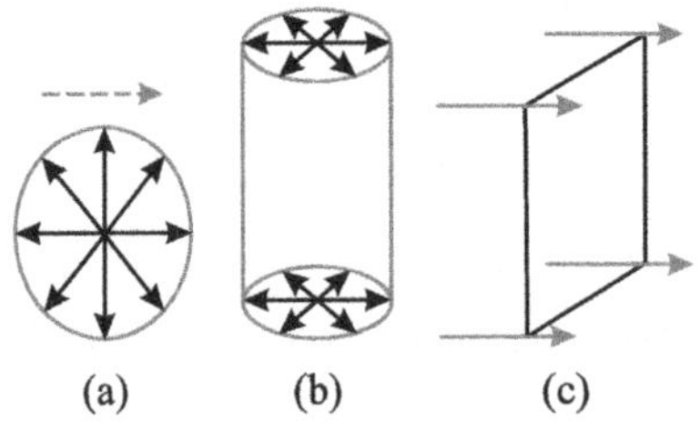

(a) (b) (c)

2. **Cylindrical wavefront:** When the source of light is linear in shape (such as a slit), a cylinderical wavefront is produced. It is because, all the points, which are equidistant from the linear source, lie on the surface of a cylinder (b).
3. **Plane wavefront:** A small part of a spherical or a cylindrical wavefront originating from a distant source will appear plane and hence it is called a plane wavefront (c).

Huygen's Principle: Every point on the wave front becomes a source of secondary disturbance and generates wavelets which spread out in the medium with the same velocity as that of light in the forward direction only.

- The envelope of these secondary waves at any instant of time gives the position of the new wave front at that instant.
- The wave front in medium is always perpendicular to the direction of wave propagation.

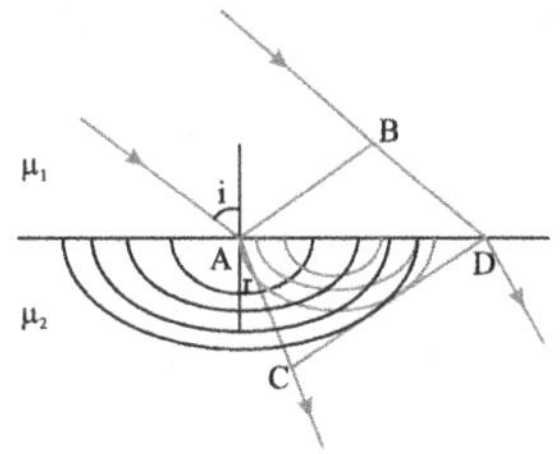

AB is width of incident beam

CD is width of refracted beam

$$\frac{\text{width of incident beam}}{\text{width of refracted beam}} = \frac{\cos i}{\cos r}$$

The Doppler Effect:

i) When any source emitting light (like sun, moon, star, atom etc) is approaching or receding from the observer then the frequency or wavelength of light appears to be changing to the observer. This apparent change in frequency or waveelength of light is called Doppler effect in light.

Blue Shift: When the distance between the source and observer is decreasing (i.e. the source is approaching the observer) then frequency of light appears to be increasing or wavelength appears to be decreasing i.e. the spectral line in electromangetic spectrum gets displaced towards blue end, hence it is known as blue shift.

Red Shift: When the distance between the source and observer is increasing (i.e. the source is receding from the observer) then frequency of light appears to be decreasing or wavelength appears to be increasing i.e. the spectral line in electromangetic spectrum gets displaced towards red end, hence it is known as red shift.

Doppler Shift, $\frac{\Delta v}{v} = \frac{V}{C}$ (where V is the speed of source and C is the speed of light)

W. E-1 ***What speed should a galaxy move with respect to us so that the sodium line at 589.0 nm is observed at 589.6 nm?***

Sol. $\frac{\Delta\lambda}{\lambda} = \frac{V}{C}$;

$$V = +c\left(\frac{0.6}{589.0}\right) = 3\times10^{8}\left(\frac{0.6}{589.0}\right) = +3.06\times10^{5}\,ms^{-1}$$

Therefore. the galaxy is moving away from us with speed 306 km/s.

Principle of superposition of waves:

If two or more waves meet at a place simultaneously in the same medium, the particles of the medium undergo displacements due to all the waves simultaneously. The resultant wave is due to the resultant displacement of the particles.

Principle of superposition of waves states that when two or more waves are simultaneously impressed on the particles of the medium, the resultant displacement of any particle is equal to the sum of displacements of all the waves. (or)

"When two or more waves overlap, the resultant displacement at any point and at any instant is the vector sum of the instantaneous displacements that would be produced at the point by individual waves, if each wave were present alone".

If $y_1, y_2, ...y_n$ denote the displacements of 'n' waves meeting at a point, then the resultant displacement is given by $y = y_1 + y_2 + ... + y_n$.

a) **Superposition of coherent waves:** Consider two waves travelling in space with an angular frequency ω. Let the two waves arrive at some point simultaneously. Let y_1 and y_2 represent the displacements of two waves at this point.

$\therefore\ y_1 = A_1 \sin(\omega t + \phi_1)\ \&\ y_2 = A_2 \sin(\omega t + \phi_2)$

Then according to the principle of superposition the resultant displacement at the point is given by,

$$y = y_1 + y_2 \text{ or } y = A_1 \sin(\omega t + \phi_1) + A_2 \sin(\omega t + \phi_2)$$

$$= A_1(\sin \omega t \cos \phi_1 + \cos \omega t \sin \phi_1)$$

$$+ A_2(\sin \omega t \cos \phi_2 + \cos \omega t \sin \phi_2)$$

$$= A \cos \phi . \sin \omega t + A \sin \phi . \cos \omega t$$

$$= A \sin(\omega t + \phi)$$

where $A \cos \phi = A_1 \cos \phi_1 + A_2 \cos \phi_2$.......(1)

and $A isn\phi = A_1 \sin \phi_1 + A_2 \sin \phi_2$.......(2)

Here A and ϕ are respectively the amplitude and initial phase of the resultant displacement

Squaring and adding equations (1) & (2), we get

$$A = \sqrt{A_1^2 + A_2^2 + 2A_1A_2 \cos(\phi_1 - \phi_2)}$$

$$= \sqrt{A_1^2 + A_2^2 + 2A_1A_2 \cos \phi}\text{.....(3)}$$

Where $\phi = \phi_1 - \phi_2$, phase difference between the two waves.

Dividing equation (2) by equation (1), we get

$$\tan \phi = \frac{A_1 \sin \phi_1 + A_2 \sin \phi_2}{A_1 \cos \phi_1 + A_2 \cos \phi_2}\text{........(4)}$$

Since the intensity of a wave is proportional to square of the amplitude, the resultant intensity I of the wave from equation (3) may be written as

$$I = I_1 + I_2 + 2\sqrt{I_1 I_2} \cos \phi \text{ ....(5)}$$

where I_1 and I_2 be the intensities of the two waves.

It can be seen that the amplitude (intensity) of the resultant displacement varies with phase difference of the constituent displacements.

Case I : When $\phi = \phi_1 - \phi_2 = 0, 2\pi, 4\pi ...... 2n\pi$

where $n = 0, 1, 2, ...$

$\Rightarrow\ \cos \phi = 1$

$\therefore\ A = A_1 + A_2$ from (3)

and $\sqrt{I} = \sqrt{I_1} + \sqrt{I_2}$ from (5)

Hence the resultant amplitude is the sum of the two individual amplitudes. This condition refers to the constructive interference.

Case II: When $\phi = \phi_1 - \phi_2 = \pi, 3\pi, 5\pi.....\ (2n-1)\pi$ where $n = 1, 2, 3, ...$;

$\Rightarrow \cos\phi = -1$

$\therefore\ A = |A_1 - A_2|$ and $\sqrt{I} = |\sqrt{I_1} - \sqrt{I_2}|$

Hence the resultant amplitude is the difference of the individual amplitudes and is referred to as destructive interference.

b) Supersposition of incoherent waves:

Incoherent waves are the waves which do not maintain a constant phase diference. The phase of the waves fluctuates irregularly with time and independently of each other. In case of light waves the phase fluctuates randomly at a rate of about 10^8 per second. Light detectors such as human eye, photographic film etc, cannot respond to such rapid changes. The detected intensity is always the average intensity, averaged over a time interval which is very much larger than the time of fluctuations. Thus

$I_{av} = I_1 + I_2 + 2\sqrt{I_1 I_2} < \cos\phi >$. The average value of the $\cos\phi$ over a large time interval will be zero and hence $I_{av} = I_1 + I_2$

This implies that the superposition of incoherent waves gives uniform illumination at every point and is simply equal to the sum of the intensities of the component waves.

Interference:

- The variation in intensity occurs due to the redistribution of the total energy of the interfering waves is called interference.
- Interference of light is a wave phenomenon.
- The source of light emitting wave of same frequency and travelling with either same phase or constant phase difference are called Coherent Sources.

  Ex: Two virtual sources derived from a single source can be used as Coherent Sources.
- The source producing the light wave travelling with rapid and random phase changes are called Incoherent Sources.

  Ex: 1. Light emitted by two candles.
  2. Light emitted by two lamps.

Conditions for Steady Interference

- The two sources must be coherent.
- Two sources must be narrow.
- Two sources must be close together.

NOTE: The two sources must be mono chromatic, otherwise the fringes of different colours overlap and hence interference cannot be observed.

Young's Double Slit Experiment

- Young with his experiment measured the most important characteristic of the light wave i.e wavelength (λ)
- Young's experiment conclusively established the wave nature of light.

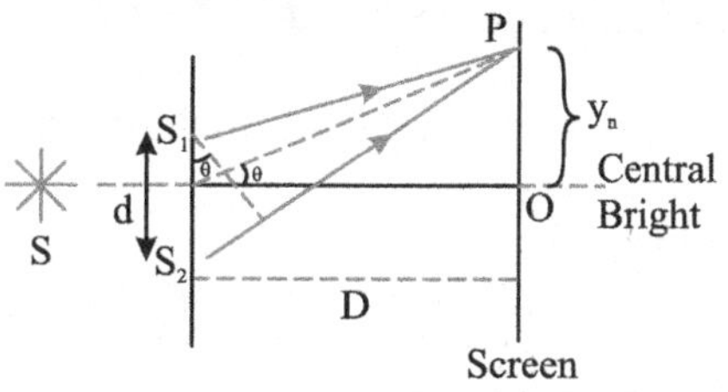

$s_1 s_2 = d$

- When source illuminates the two slits, the pattern observed on the screen consists of large number of equally spaced bright and dark bands called "interference fringes"

a) Bright fringes :

Bright fringes occur whenever the waves from S_1 and S_2 interfere constructively. i.e. on reach ing 'P', the waves with crest (or trough) superimpose at the same time and they are said to be in phase.

The condition for finding a bright fringe at 'P' is that $S_2P - S_1P = n\lambda$,

Where $n = 0, \pm 1, \pm 2, \pm 3, ...$ and n is called the order of bright fringe. Hence for n^{th} order bright fringe, the path difference is

$$d\sin\theta = n\lambda$$

$$\Rightarrow d\left(\frac{y_n}{D}\right) = n\lambda$$

$$\therefore \; y_n = \frac{n\lambda D}{d}$$

Where y_n is the position of nth maximum from O.

The bright fringe corresponding to n = 0, is called the zero - order fringe or central maximum. It means it is the fringe with zero path difference between two waves on reaching the point P.

The bright fringe corresponding to n = 1 is called first order bright fringe i.e., if the path difference between the two waves on reaching 'P' is λ. Similarly second order bright fringe $n = 2$ is located where the path difference is 2λ and so on.

From $I = 4I_0\cos^2\left(\frac{\phi}{2}\right)$

For maximum intensity $\cos\frac{\phi}{2} = 1$

$$i.e. \frac{\phi}{2} = 0, \pm, \pi, \pm 2\pi ....$$

(or) Phase difference between the waves $\phi = \pm 2\ \pi n$ with n = 0, 1, 2, 3 ....

The corresponding path difference, $\Delta x = n\lambda$

Hence $I_{max} = 4I_0$.

b) Dark fringes :

Dark fringes occur whenever the waves from S_1 and S_2 interfere destructively. i.e., on reaching 'P' one wave with its crest and another wave with its trough superimpose. Then the phase difference between the waves is π and the waves are said to be in opposite phase.

Destructive interference occurs at P, if S_1P and S_2P differ by a odd integral multiple of $\frac{\lambda}{2}$.

Thus the condition for finding dark fringe at P is that $S_2P - S_1P = (2n-1)\frac{\lambda}{2}$.

Where $n = \pm 1, \pm 2, \pm 3, ...$, and n is called order of dark fringe. Hence for n^{th} order dark fringe, the path difference, $d\sin\theta = (2n-1)\frac{\lambda}{2}$

$$\Rightarrow d\left(\frac{y_n}{D}\right) = (2n-1)\frac{\lambda}{2} \therefore y_n = \left(\frac{2n-1}{2}\right)\frac{\lambda D}{d}$$

Where y_n is the position of nth minima from O.

The first dark fringe occurs when

$S_2P - S_1P = \frac{\lambda}{2}$. This is called first order dark (n = 1) fringe and similarly for $S_2P - S_1P = \frac{3\lambda}{2}$ second order dark fringe (n = 2) occurs and so on.

From $I = 4I_0 \cos^2\left(\frac{\phi}{2}\right)$

For minimum intensity $\cos\frac{\phi}{2} = 0$

i.e., $\frac{\phi}{2} = \pm\frac{\pi}{2}, \pm\frac{3\pi}{2}, \pm\frac{5\pi}{2}......$

(or) $\phi = \pm\pi, \pm 3, \pm 5\pi......$

(or) $\phi = \pm(2n-1)\pi$ with n = 1, 2, 3 ......

The corresponding path difference, $\Delta x = (2n-1)\frac{\lambda}{2}$

Hence $I_{\min} = O$

c) **Fringe width (β):**

The distance between two adjacent bright (or dark) fringes is called the fringe width. It is denoted by β.

The nth order bright fringe occurs from the central maximum at $y_n = \frac{n\lambda D}{d}$

The $(n+1)^{th}$ order bright fringe occurs from the central maximum at $y_{n+1} = \frac{(n+1)\lambda D}{d}$

$\therefore$ The fringe separation, β is given by

$$\beta = y_{n+1} - y_n = \frac{\lambda D}{d}$$

In a similar way, the same result will be obtained for the dark fringes also.

$\therefore$ Fringe width, $\beta = \frac{\lambda D}{d}$

Thus fringe width is same every where on the screen and the width of bright fringe is equal to the width of dark fringe.

$$\therefore \; \beta_{bright} = \beta_{dark} = \beta = \frac{\lambda D}{d}$$

d) The locus of the point P lying in the xy-plane such that $S_2P - S_1P = (\Delta x)$ (path difference) is a constant, is a hyperbola. If the distance D is very large compared to the fringe width, the fringes will be very nearly straight lines.

Note:

Constructive Interference

i) a) If the phase difference is $\phi = (2n)\pi$ (even multiples of π). Where n = 0, 1,2, 3,......

i.e. when $\phi = 0, 2\pi, 4\pi........2n\pi$

b) If the path difference $x = 2n\left(\frac{\lambda}{2}\right)$ (even multiples of half wavelength).

i.e when $x = 0, \lambda, 2\lambda........n\lambda$

The amplitude and intensity are maximum.

$$\boxed{A_{max} = (A_1 + A_2)}$$

$$\boxed{I_{max} = \left(\sqrt{I_1} + \sqrt{I_2}\right)^2 = (A_1 + A_2)^2}$$

Note: If $A_1 = A_2 = a$ then $\boxed{A_{max} = 2a}$

If $I_1 = I_2 = I_0$ then $\boxed{I_{max} = 4I_0}$

Destructive Interference

ii) a) If the phase difference $\phi = (2n-1)\pi$ (odd multiples of π) where n = 1, 2, 3.....

i.e. when $\phi = \pi, 3\pi, 5\pi.......(2n-1)\pi$

b) If the path difference $x = (2n-1)\lambda/2$ (odd multiples of $\lambda/2$)

i.e. when $x = \frac{\lambda}{2}, \frac{3\lambda}{2}, \frac{5\lambda}{2}........\frac{(2n-1)\lambda}{2}$

The amplitude and Intensity are minimum.

$$\boxed{A_{min} = (A_1 - A_2)}$$

$$\boxed{I_{min} = \left(\sqrt{I_1} - \sqrt{I_2}\right)^2 = (A_1 - A_2)^2}$$

Note: If $A_1 = A_2 = a$ then $\boxed{A_{min} = 0}$

If $I_1 = I_2 = I_0$ then $\boxed{I_{min} = 0}$

$$\boxed{\frac{I_{max}}{I_{min}} = \frac{\left(\sqrt{I_1} + \sqrt{I_2}\right)^2}{\left(\sqrt{I_1} - \sqrt{I_2}\right)^2} = \frac{(A_1 + A_2)^2}{(A_1 - A_2)^2}}$$

iii) phase difference $= \frac{2\pi}{\lambda}$ (path difference).

$$\boxed{\phi = \frac{2\pi}{\lambda}x}$$

iv) Since $\beta \alpha \lambda, \beta_{Red} > \beta_{voilet}, as\, \lambda_{red} > \lambda_{voilet}$

v) In YDSE, if blue light is used instead of red light then β decreases ($\because \lambda_B < \lambda_R$)

vi) If YDSE is conducted in vaccum instead of air, then β increases ($\because \lambda_{vaccum} > \lambda_{air}$)

vii) In certain field of view on the screen, if n_1 fringes are formed when light of wavelength λ_1 is used and n_2 fringes are formed when light of wavelength λ_2 is used, then

$$y = \frac{n\lambda D}{d} = \text{constant} \Rightarrow n\lambda = \text{constant}$$

$$\therefore\ n_1\lambda_2 = n_2\lambda_2 \text{ (or) } n_1\beta_1 = n_2\beta_2$$

viii) The distance of nth bright fringe from central maximum is $(y_n)_{bri} = \frac{n\lambda D}{d} = n\beta$

The distance of mth dark fringe from central maximum is

$$(y_m)_{dark} = \frac{(2m-1)}{2}\frac{\lambda D}{d} = \frac{(2m-1)}{2}\beta$$

$\therefore$ The distance between n^{th} bright and m^{th} dark fringes is

$$(y_n)_{bri} - (y_m)_{dark} = n\beta - \frac{(2m-1)}{2}\beta$$

ix) When white light is used in YDSE the inteference patterns due to different component colours of white light overlap (incoherently). The central bright fringes for different colours are at the same position.

Therefore, the central fringe is white. For a point P for whih $S_2P - S_1P = \frac{\lambda_b}{2}$ where $\lambda_b (\approx 4000 A^0)$ represents the wavelength for the blue colour, the blue component will be absent and the fringe will appear red in colour.

Slightly farther away where $S_2Q - S_1Q = \frac{\lambda_r}{2}$ where $\lambda_r (\approx 8000\, A^0)$ is the wavelength for the red colour, the fringe will be predominantly blue.

Thus, the fringe closest on either side of the central white fringe is red and fathest will appear blue. After a few fringes, no clear fringe pattern is seen.

x) To know maximum number of possible maxima on the screen

If $d\sin\theta = n\lambda\,(or)\sin\theta = \frac{n\lambda}{d}$

As $\sin\theta \le 1, \frac{n\lambda}{d} \le 1 \qquad \therefore n \le \frac{d}{\lambda}$

Therefore the maximum number of complete maxima on the screen will be $2(n)+1$

Ex: If $d = 3\lambda\ then\ \sin\theta = \frac{n\lambda}{3\lambda} = \frac{n}{3}$ As $\sin\theta \le 1$,

n can take values $-3, -2, -1, 0, 1, 2, 3$

$\therefore$ Maximum number of maxima is 7.

xi) Fringe visibility (or) band visibility (V) :

It is the measure of contrast between the bright and dark fringes

Fringe visibility, $V = \frac{I_{max} - I_{min}}{I_{max} + I_{min}}$

where $I_{max} = I_1 + I_2 + 2\sqrt{I_1 I_2}$

and $I_{min} = I_1 + I_2 - 2\sqrt{I_1 I_2}$

$$\therefore\ V = \frac{4\sqrt{I_1 I_2}}{2(I_1 + I_2)} = \frac{2\sqrt{I_1 I_2}}{(I_1 + I_2)}$$

V has no unit and no dimensional formula.

Generally, $0 < V < 1$.

Fringe visibility is maximum, if $I_{min} = 0$, then

$V = 1$

For poor visibility, $I_{max} = I_{min}$, then $V = 0$

i.e., if $V = 1$, then the fringes are very clear and contrast is maximum and if $V = 0$, then there will be no fringes and there will be uniform illumination i.e., the contrast is poor.

xii) When one slit is fully open and another one is partially open then the contrast between the fringes decreases. i.e., if the slit widths are unequal, the minima will not be completely dark.

xiii) Missing wavelengths in front of one slit in YDSE:

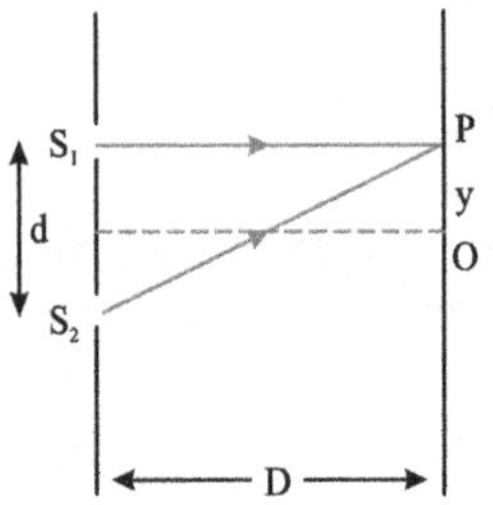

Suppose P is a point of observation in front of slit S_1 as shown in figure. Path difference between the two waves from S_1 and S_2 is

$$\Delta x = S_2P - S_1P = \sqrt{D^2 + d^2} - D$$

$$= D\left(1 + \frac{d^2}{D^2}\right)^{1/2} - D = D\left(1 + \frac{d^2}{2D^2}\right) - D = \frac{d^2}{2D}$$

$$\therefore \Delta x = \frac{d^2}{2D} \text{.........(1)}$$

But for missing wavelengths, intensity will be zero. i.e., the corresponding path difference,

$$\Delta x = (2n-1)\frac{\lambda}{2} \qquad \text{...... (2)}$$

From equations (1) and (2)

$$\frac{d^2}{2D} = (2n-1)\frac{\lambda}{2}$$

$$\therefore \text{ Missing wavelength, } \lambda = \frac{d^2}{(2n-1)D}$$

By putting $n = 1, 2, 3, ...$, the wavelengths at P are

$$\lambda = \frac{d^2}{D}, \frac{d^2}{3D}, \frac{d^2}{5D}, ...$$

In the above case, if **bright fringes** are to be formed exactly opposite to S_1 then

$$\frac{d^2}{2D} = n\lambda \Rightarrow \lambda = \frac{d^2}{2Dn}$$

By putting n = 1, 2, 3 ,,, the possible wavelengths at P are

$$\lambda = \frac{d^2}{2D}, \frac{d^2}{4D}, \frac{d^2}{6D}, \ldots.$$

xiv) Lateral displacement of fringes:

To determine the thickness of a given thin sheet of transparent material such as glass or mica, that transparent sheet is introduced in the path of one of the two interfering beams. The fringe pattern gets displaced towards the beam in whose path the sheet is introduced. This shift is known as lateral displacement or lateral shift.

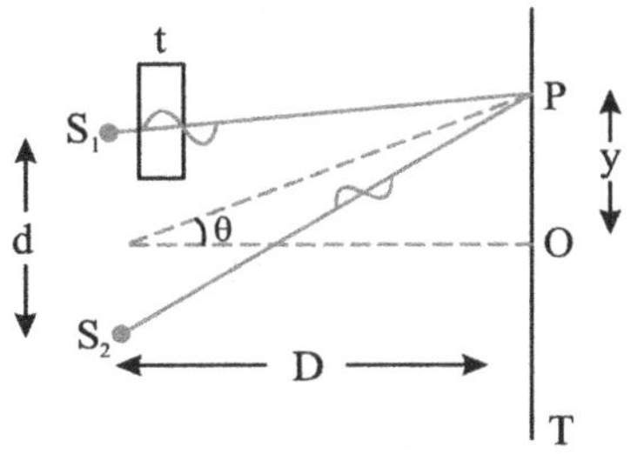

The optical path from S_1 to $P = (S_1P - t) + \mu t$. The optical path from S_2 to $P = S_2P$.

To get central zero fringe at P, $\Delta_{s_1p} = \Delta_{s_2p}$

$\Rightarrow S_1P - t + \mu t = S_2P$

$\therefore\ S_2P - S_1P = (\mu - 1)t$

Since $\mu > 1$, this implies $S_2P > S_1P$ hence the fringe pattern must shift towards the beam from S_1.

But $S_2P - S_1P = d\sin\theta = d\frac{y}{D}$, where 'y' is the lateral shift.

$$\therefore\ (\mu - 1)t = d\frac{y}{D}$$

$$\therefore \text{Lateral shift } (y) = \frac{D}{d}(\mu - 1)t = \frac{\beta}{\lambda}(\mu - 1)t$$

(or) Thickness of sheet

$$t = \frac{yd}{(\mu - 1)D} = \frac{y\lambda}{(\mu - 1)\beta}$$

From the above it is clear that

a) For a given colour, shift is independent of order of the fringe i.e. shift in zero order maximum = shift in 9th minima (or) shift in 6th maxima = shift in 2nd minima. Since the refractive index depends on wavelength hence lateral shift is different for different colours.

b) The number of fringes shifted = $\frac{\text{lateral shift}}{\text{fringe width}}$

$$\therefore n = \frac{y}{\beta} = \frac{(\mu - 1)t}{\lambda} \text{ (or) } n\lambda = (\mu - 1)t$$

Therefore, number of fringes shifted is more for shorter wavelength.

c) If a transparent sheet of thickness 't' and its relative refractive index μ_r (w.r.t. surroundings) be introduced in one of the beams of interference, then

1) the lateral shift $y = \frac{(\mu_r - 1)tD}{d}$

2) the number of fringes shifted $n = \frac{(\mu_r - 1)t}{\lambda}$

d) Due to the presence of transparent sheet, the phase difference between the interfering waves at a given point is given by $= \frac{2\pi}{\lambda}(\mu - 1)t$.

e) If YDSE is performed with two different colours of light of wavelengths λ_1 & λ_2 but by placing the same transparent sheet in the path of one of the interfering waves then $n_1\lambda_1 = n_2\lambda_2$.

where n_1 and n_2 are the number of fringes shifted with wavelengths λ_1 & λ_2.

vi) When two different transperent sheets of thickness t_1, t_2 and refractive index μ_1, μ_2 are placed in the paths of two interfering waves in YDSE, if the central bright fringe position is not shifted, then $(\mu_1 - 1)t_1 = (\mu_2 - 1)t_2$.

Important Concepts :

➢ **Formation of colours in thin films** :

a) Interference due to reflected light

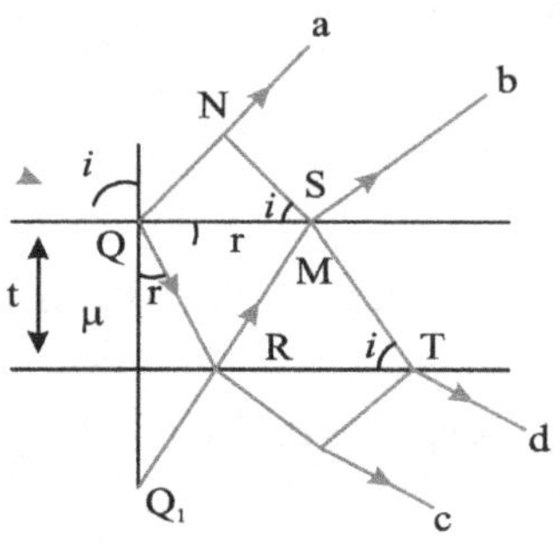

Reflected system :

➢ Path difference between the rays Qa and QRSb. (PD) = QRS in medium - QN in air

$\therefore P.D = 2\mu t \cos r$ This is the path lag

due to reflection on film additional path lag of $\lambda/2$ exists. (stoke's theorem)

Total path difference $= 2\mu t \cos r + \frac{\lambda}{2}$

Condition for maximum

➢ $2\mu t \cos r + \frac{\lambda}{2} = n\lambda$

OR $\boxed{2\mu t \cos r = (2n-1)\frac{\lambda}{2}}$ For all values of n is equal to 1, 2, 3.............. n.

➢ **Condition for Minimum**

$2\mu t \cos r + \frac{\lambda}{2} = (2n-1)\frac{\lambda}{2}$

$\boxed{2\mu t \cos r = n\lambda}$ for values of n = 0,1,2,3... n = 0 gives the central minima.

For normal incident $i = o = r$

$2\mu t = n\lambda$ for dark ; $2\mu t = (2n-1)\frac{\lambda}{2}$ for bright.

Transmitted system

- Interference of two rays Rc and Td. By symmetry it can be concluded that the path difference between the rays in $2\mu t\cos r$.

  But there would not be any extra phase lag because either of the two rays suffers reflection at denser surface.

- **Condition for maxima :** $\boxed{2\mu t\cos r = n\lambda}$

- **Condition for minimum :**

$$\boxed{2\mu t\cos r = (2n-1)\frac{\lambda}{2}}$$

If YDSE is conducted with white light,

- Central fringe is always achromatic (white)
- When path difference is small, then some coloured fringes are obtained on two sides of the central fringe. The outer edge of the fringe is violet and inner edge is red.
- The fringe width is different for different colours
- The number of fringes obtained is less than that with monochromatic light source.

WE-2: ***Light waves from two coherent sources having intensity ratio 81 : 1 produce interference. Then, the ratio of maxima and minima in the interference pattern will be***

Sol. Given, $\frac{I_1}{I_2} = \frac{A_1^2}{A_2^2} = \frac{81}{1}$

$\therefore \frac{A_1}{A_2} = \frac{9}{1}$ **or** $A_1 = 9A_2$ .... (1) $\therefore \frac{I_{\max}}{I_{\min}} = \frac{(A_1 + A_2)^2}{(A_1 - A_2)^2}$

From Eq. (i), we get

$$\frac{I_{\max}}{I_{\min}} = \frac{(9A_2 + A_2)^2}{(9A_2 - A_2)^2} = \frac{(10)^2}{(8)^2} = \frac{25}{16}$$

WE-3: ***Two slits are made one millimetre apart and the screen is placed one metre away. When blue-green light of wavelength 500 nm is used, the fringe separation is***

Sol. Fringe separation, $\beta = \frac{D\lambda}{d}$

Given, $D = 1m, \lambda = 500\,nm = 5\times10^{-7}\,m$ **and**

$d = 1mm = 1\times10^{-3}\,m$

$\therefore$ Fringe separation, $\beta = \frac{1\times5\times10^{-7}}{1\times10^{-3}}\,m$

$= 5\times10^{-4}\,m = 0.5\,mm$

WE-4: ***In YDSE, the two slits are separated by 0.1 mm and they are 0.5 m from the screen. The*** wavelengh of light used is 5000 $\mathring{A}$. ***What is the distance between 7th maxima and 11th minima on the screen?***

Sol. Here, d = 0.1 mm = 10^{-4} m,

$D = 0.5\,m, \lambda = 5000\,\mathring{A} = 5.0\times10^{-7}\,m$

$\therefore \Delta x = (X_{11})_{dark} - (X_7)_{bright} = \frac{(2\times11-1)\lambda D}{2d} - \frac{7\lambda D}{d}$

$\Delta x = \frac{7\lambda D}{2d} = \frac{7\times5\times10^{-7}}{2\times10^{-4}}$

$= 8.75 \times 10^{-3} m$

$= 8.75 mm$

WE-5: ***In Young's double slit experiment interference fringes 1^0 apart are produced on the screen, the slit separation is*** $(\lambda = 589\ nm)$

Sol. The fringe width, $\beta = \dfrac{\lambda D}{d}$

The angualr separation of the fringes is given by

$$\theta = \frac{\beta}{D} = \frac{\lambda}{d}$$

Given, $\theta = 1^0 = \dfrac{\pi}{180} rad$

$\lambda = 589\ nm$

$$\therefore\ d = \frac{\lambda}{\theta} = \frac{589 \times 180 \times 10^{-9}}{\pi}$$

$= 0.0337\ mm$

WE-6: ***In Young's double slit experiment, the wavelength of red light is 7800 Å and that of blue light is 5200 Å. The value of n for which nth bright band due to red light coincides with (n + 1)th bright band due to blue light, is***

Sol. $\dfrac{n_R \lambda_R D}{d} = \dfrac{n_B \lambda_B D}{d}$ **or** $\dfrac{n_R}{n_B} = \dfrac{\lambda_B}{\lambda_R} = \dfrac{5200}{7800} = \dfrac{2}{3}$

Therefore 2nd of red coincides with 3rd of blue.

WE-7: ***Young's double slit experiment is made in a liquid. The 10th bright fringe in liquid lies where 6th dark fringe lies in vacuum. The refractive index of the liquid is approximately***

Sol. Fringe width $\beta = \frac{\lambda D}{d}$. When the apparatus is immersed in a liquid, λ and hence β is reduced μ (refractive index) times.

$$10\beta' = (5.5)\beta \text{ or } 10\lambda'\left(\frac{D}{d}\right) = (5.5)\frac{\lambda D}{d}$$

$$\text{or } \frac{\lambda}{\lambda'} = \frac{10}{5.5} = \mu \text{ or } \mu = 1.8$$

WE-8: ***In Young's double slit experiment, how many maximas can be obtained on a screen (including the central maximum) on both sides of the central fringe if*** $\lambda = 2000 Å$ ***and*** $d = 7000\ Å$***?***

Sol. For maximum intensity on the screen $d \sin\theta = n\lambda$ or $\sin\theta = \dfrac{n\lambda}{d}$; $= \dfrac{(n)(2000)}{(7000)} = \dfrac{n}{3.5}$

maximum value of $\sin\theta = 1$

$\therefore n = -3, -2, -1, 0, 1, 2, 3$; $\therefore$ 7 maximas.

WE-9: ***In a double slit experiment the angular width of a fringe is found to be*** 0.2^0 ***on a screen placed I m away. The wavelength of light used in 600 nm. What will be the angular width of the fringe if the entire experimental apparatus is immersed in water? Take refractive index of water to be 4/3.***

Sol. Angular fringe separation,

$\theta = \frac{\lambda}{d} ord = \frac{\lambda}{\theta}$; In water, $d = \frac{\lambda'}{\theta'}$

$\therefore \frac{\lambda}{\theta} = \frac{\lambda'}{\theta'} or \frac{\theta'}{\theta} = \frac{\lambda'}{\lambda} = \frac{1}{\mu} = \frac{3}{4}$

or $\theta' = \frac{3}{4}\theta = \frac{3}{4} \times 0.2^0 = 0.15^0$

WE-10: ***In a Young's experiment, one of the slits is covered with a transparent sheet of thickness $3.6 \times 10^{-3} cm$ due to which position of central fringe shifts to a position orginally occupied by 30th fringe.If $\lambda = 6000$ Å, then find the refractive index of the sheet.***

Sol. The position of 30th bright fringe,

$y_{30} = \frac{30\lambda D}{d}$ Now position shift of central fringe is

$y_0 = \frac{30\lambda D}{d}$; But we know, $y_0 = \frac{D}{d}(\mu - 1)t$

$\frac{30\lambda D}{d} = \frac{D}{d}(\mu - 1)t$

$\Rightarrow \quad (\mu - 1) = \frac{30\lambda}{t} = \frac{30 \times (6000 \times 10^{-10})}{(3.6 \times 10^{-5})} = 0.5$

$\therefore \quad \mu = 1.5$

WE-11: ***The maximum intensity in the case of n identical incoherent waves each of intensity $2\frac{W}{m^2}$ is $32\frac{W}{m^2}$ the value of n is***

Sol. $I = n I_0$, $32 = n 2$, $n = 16$

WE-12: ***Compare the intensities of two points located at respective distance $\frac{\beta}{4}$ and $\frac{\beta}{3}$ from the central maxima in a interference of YDSE (β is the fringe width)***

Sol. $\Delta\theta = \frac{2\pi}{\lambda}\Delta x = \frac{2\pi}{\lambda}\left(\frac{d}{D}\frac{\beta}{4}\right) = \frac{2\pi}{\lambda}\left(\frac{d}{D}\frac{\lambda D}{4d}\right)$

$\therefore \quad \Delta\theta = \frac{2\pi}{4} = \frac{\pi}{2} \Rightarrow I = 4I_0 \cos^2\left(\frac{\pi}{4}\right)$

Similarly $\Delta\theta = \frac{2\pi}{3} \Rightarrow I = 4I_0 \cos^2\left(\frac{2\pi}{2\times 3}\right) = I_0$

$\therefore$ required ratio $= \boxed{2:1}$

WE-13: ***In Young's double slit experiment intensity at a point is (1/4) of the maximum intensity. Angular position of this points is***

Sol: $I = I_{max}\cos^2\left(\frac{\phi}{2}\right)$; $\therefore \frac{I_{max}}{4} = I_{max}\cos^2\left(\frac{\phi}{2}\right)$

$\cos\frac{\phi}{2} = \frac{1}{2} or \frac{\phi}{2} = \frac{\pi}{3}$

$\therefore \phi = \frac{2\pi}{3} = \left(\frac{2\pi}{\lambda}\right).\Delta x$ where $\Delta x = d\sin\theta$

$$\frac{\lambda}{3} = d\sin\theta, \sin\theta = \frac{\lambda}{3d}, \theta = \sin^{-1}\left(\frac{\lambda}{3d}\right)$$

WE-14: ***In Young's double slit experiment the y co-ordinates of central maxima and 10th maxima are 2 cm and 5 cm respectively. When the YDSE apparatus is immersed in a liquid of refractive index 1.5 the corresponding y co-ordinates will be***

Sol. Fringe width $\beta \propto \lambda$. Therefore, λ and hence β will decrease 1.5 times when immersed in liquid. The distance between central maxima and 10th maxima is 3 cm in vacuum. When immersed in liquid it will reduce to 2 cm. Position of central maxima will not change while 10th maxima will be obtained at y=4 cm.

WE-15: ***In YDSE, bi-cromatic light of wavelengths 400 nm and 560 nm are used. The distance between the slits is 0.1 mm and the distance between the plane of the slits and the screen is 1m. The minimum distance between two successive regions of complete darkness is:***

Sol. Let nth minima of 400 nm coincides with mth minima of 560nm, then

$$(2n-1)\left(\frac{400}{2}\right) = (2m-1)\left(\frac{560}{2}\right) \text{ or}$$

$$\frac{2n-1}{2m-1} = \frac{7}{5} = \frac{14}{10} = ...$$

i.e., 4th minima of 400 nm coincides with 3rd minima of 560 nm. Location of this minima is,

$$Y_1 = \frac{(2\times4-1)(1000)(400\times10^{-9})}{2\times0.1} = 14\,mm$$

Next 11th minima of 400 nm will coincide with 8th minima of 560 nm.

Location of this minima is,

$$Y_2 = \frac{(2\times11-1)(1000)(400\times10^{-9})}{2\times0.1} = 42\,mm$$

$\therefore$ Required distance $Y_2 - Y_1 = 28\ mm$.

WE-16: ***An interference is observed due to two coherent sources S_1 placed at origin and S_2 placed at $(0, 3\lambda, 0)$. Here λ is the wavelength of the sources. A detector D is moved along the positive x-axis. Find x-coordinates on the x-axis (excluding x = 0 and $x=\infty$) where maximum intensity is observed.***

Sol: At x = 0, path difference is 3λ. Hence, third order maxima will be obtained. At $x=\infty$, path difference is zero. Hence, zero order maxima is obtained. In between first and second order maxima will be obtained.

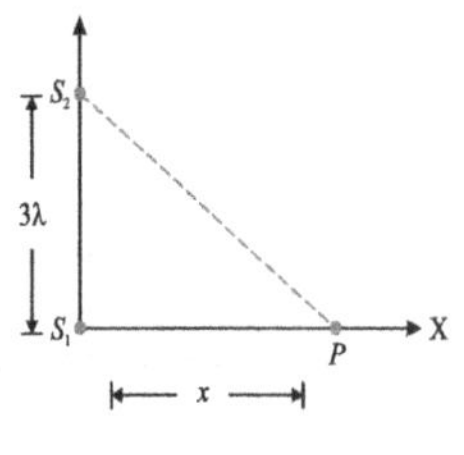

First order maxima:

$S_2P - S_1P = \lambda$ (or) $\sqrt{x^2+9\lambda^2} - x = \lambda$

or $\sqrt{x^2+9\lambda^2} = x+\lambda$ Squaring both sides, we get $x^2+9\lambda^2 = x^2+\lambda^2+2x\lambda$. Solving this, we get $x = 4\lambda$. Second order maxima:

$S_2P - S_1P = 2\lambda$; (or) $\sqrt{x^2+9\lambda^2} - x = 2\lambda$ (or)

$\sqrt{x^2+9\lambda^2}=(x+2\lambda)$ Squaring both sides, we get

$x^2+9\lambda^2=x^2+4\lambda^2+4x\lambda$

Solving, we get $x=\frac{5}{4}\lambda=1.25\lambda$

Hence, the desired x coordinates are,

$x=1.25\lambda$ and $x=4\lambda$.

WE-17: ***Two coherent light sources A and B with separation 2λ are placed on the x-axis symmetrically about the origin. They emit light of wavelength λ. Obtain the positions of maxima on a circle of large radius, lying in the x-y plane and with centre at the origin.***

Sol:

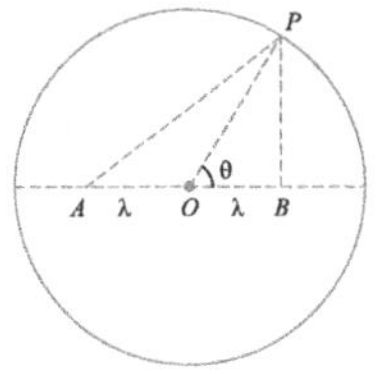

For P to have maximum intensity, $d\cos\theta=n\lambda$

$2\lambda\cos\theta=n\lambda$ $\cos\theta=\frac{n}{2}$ where n is integer

For $n=0,\ \theta=90^0,270^0$

$n=\pm1,\theta=60^0,120^0,\ 240^0,300^0$

$n=\pm2,\ \theta=0^0,180^0$

So, positions of maxima are at

$\theta=0^0,60^0,90^0,120^0,180^0,240^0,270^0$ and 300^0 ; i.e., 8 positions will be obtained.

Short cut : In $d=n\lambda$ then number of maximum on the circle is 4n.Note: For minima; $\Delta x=(2n-1)\frac{\lambda}{2}$

WE-18: ***Two coherent point sources S_1 and S_2 vibrating in phase emit light of wavelength λ. The separation between the sources is 2λ. Consider a line passing through S_2 and perpendicular to the line S_1S_2. Find the position of farthest and nearest minima***

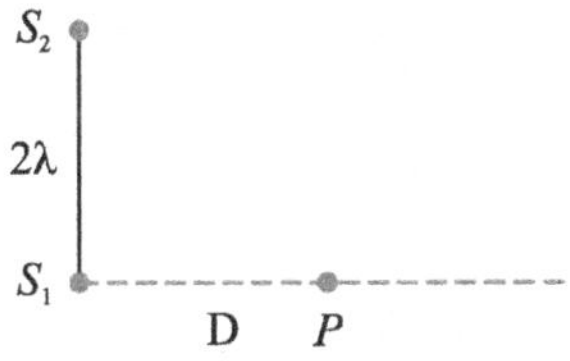

Sol: $\Delta x_{\min}=(2n-1)\frac{\lambda}{2}$ The farthest minima has path difference $\lambda/2$ while nearest minima has path difference $(3/2)\lambda$. For the nearest minima.

$S_1P-S_2P=\frac{3}{2}\lambda$; [as maximum path difference is 2λ]

$\Rightarrow\sqrt{(2\lambda)^2+D^2}-D=\frac{3}{2}\lambda\Rightarrow(2\lambda)^2+D^2=\left(\frac{3}{2}\lambda+D\right)^2$

$$\Rightarrow 4\lambda^2 + D^2 = \frac{9}{4}\lambda^2 + D^2 \times 2 \times \frac{3}{2}\lambda \times D$$

$$\Rightarrow 3D = 4\lambda - \frac{9\lambda}{4} = \frac{7\lambda}{4} \Rightarrow D = \frac{7}{12}\lambda$$

For the farthest minima, $S_1P - S_2P = \frac{\lambda}{2}$

$$\Rightarrow \sqrt{4\lambda^2 + D^2} - D = \frac{\lambda}{2} \Rightarrow 4\lambda^2 + D^2 = \frac{\lambda^2}{4} + D^2 + D\lambda \Rightarrow D = 4\lambda - \lambda/4 = \frac{15\lambda}{4}$$

WE 19: ***A ray of light of intensity I is incident on a parallel glass slab at a point A as shown. It undergoes partial reflection and refraction. At each reflection 20% of incident energy is reflected. The rays AB and A' B' undergo interference. The ratio I_{max}/I_{min} is***

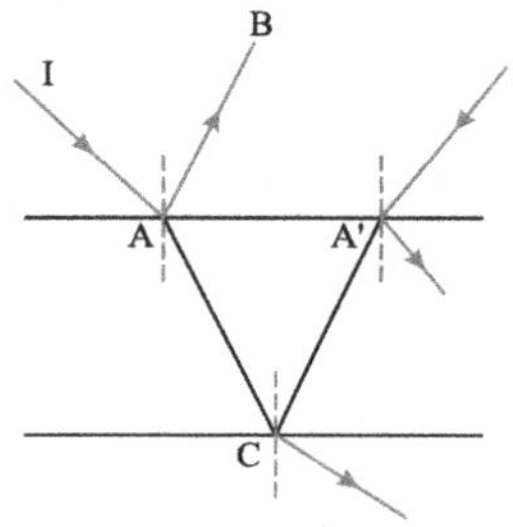

Sol: According to the question, Intensity of ray AB, $I_1 = \frac{I_0}{5}$ and Intensity of ray $A'B'$,

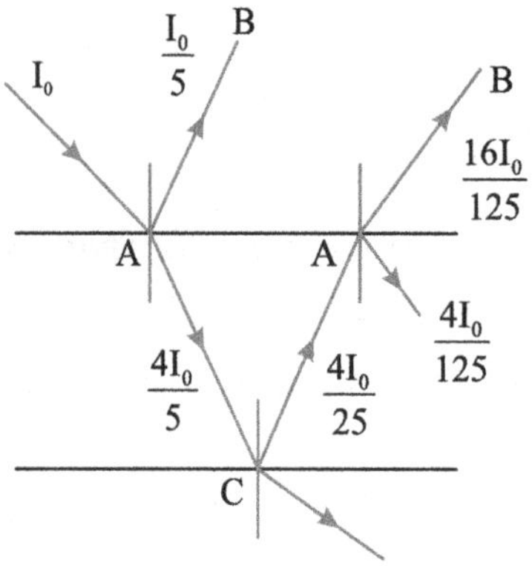

$$I_2 = \frac{16I_0}{125}, I_{\max} = (\sqrt{I_1} + \sqrt{I_2})^2 = \frac{81}{125}I_0,$$

$$I_{\min} = (\sqrt{I_1} - \sqrt{I_2})^2 = \frac{I_0}{125}, \frac{I_{\max}}{I_{\min}} = 81.$$

WE 20: ***In a YDSE experiment if a slab whose refractive index can be varied is placed in front of one of the slits, then the variation of resultant intensity at mid-point of screen with 'μ' will be best represented by ($\mu \geq 1$).***

[Assume slits of equal width and there is no absorption by slab]

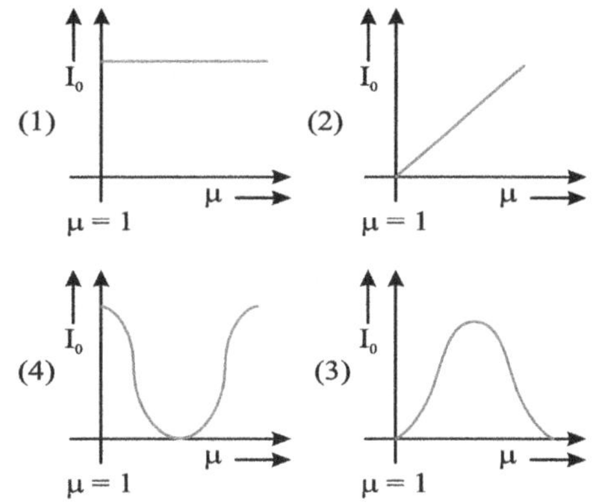

Sol. $\Delta x = (\mu - 1)t$; For $\mu = 1, \Delta x = 0$

$\therefore$ I = maximum = I_0; As μ increases path difference Δx also increases.; For $\Delta x = 0$ to $\frac{\lambda}{2}$, intensity will decrease from I_0 to zero.

Then for $\Delta x = \frac{\lambda}{2}$ *to* λ, intensity will increase from zero to I_0.

Hence option 3 is correct

WE 21: ***Consider the optical system shown in fig. The point source of light S is having wavelength equal to λ. The light is reaching screen only after reflection. For point P to be 2nd maxima, the value of λ would be*** $(D >> d$ & $d >> \lambda)$

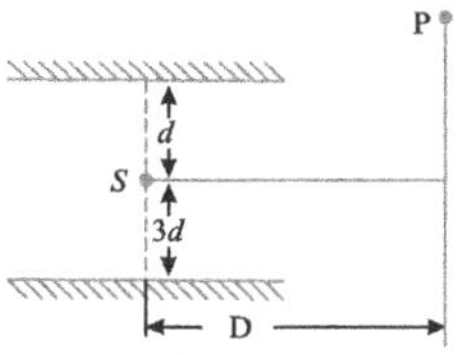

1) $\frac{12d^2}{D}$ 2) $\frac{6d^2}{D}$ 3) $\frac{3d^2}{D}$ 4) $\frac{24d^2}{D}$

Sol: a. At P, $\Delta x = \frac{(8d) \times 3d}{D}$;

For 2nd maxima, $\Delta x = 2\lambda$; $\Rightarrow \frac{24d^2}{D} = 2\lambda$

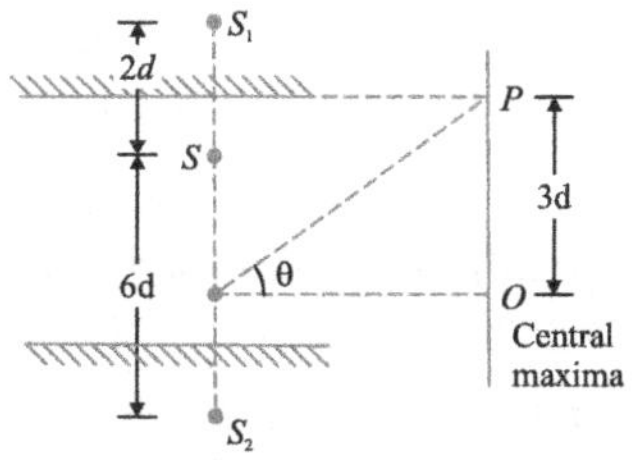

$\Rightarrow \lambda = \frac{12d^2}{D}$

WE 22: ***Two coherent point sources S_1 and S_2 vibrating in phase emit light of wavelength λ. The separation between them is 2λ as shown in figure. The first bright fringe is formed at 'P' due to interference on a screen placed at distance 'D' from $S_1 (D >> \lambda)$, then OP is***

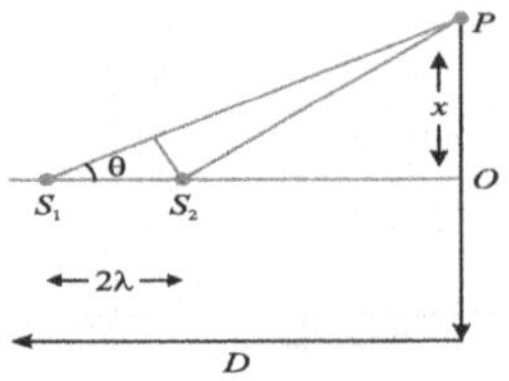

1) $\sqrt{3}$ D 2) 1.5 D 3 $\sqrt{2}$ D 4) 2 D

Sol: $\Delta x = d\cos\theta = \lambda\,;\cos\theta = \frac{\lambda}{d} = \frac{\lambda}{2\lambda} = \frac{1}{2}$

$\theta = 60^0 \tan 60 = \frac{x}{D}\; x = \sqrt{3}D$

Diffraction

- The bending of light around edges of an obstacle on the enchroachment of light within geometrical shadow is known as *"diffraction of light"*
- Diffraction is a characteristic wave property.
- Diffraction is an effect exhibited by all electro-magnetic waves, water waves and sound waves
- Diffraction takes place with very small moving particles such as atoms, neutrons and electrons which show wavelike properties.
- When light passes through a narrow aperture some light is found to be enchroached into shadow regions.
- When slit width is larger, the enchroachment of light is small and negligible.
- When slit width is comparable to wavelength of light the enchroachment of light is more
- If the size of obstacle or aperture is comparable with the wavelength of light, light deviates fromrectilinear propagation near edges of obstacle or aperture and enchroaches into geometrical shadow.
- Diffraction phenomenon is classified into two types, a) Fresnel diffraction b) Fraunhoffer diffraction

Fresnel Diffraction

- The source or screen or both are at finite distances from diffracting device (obstacle or aperture)
- In Fresnel diffraction, the effect at any point on the screen is due to exposed wave front which maybe spherical or cylindrical in shape.
- Fresnel diffraction does not require any lens to modify the beam.
- Fresnel diffraction can be explained in terms of "half period zones or strips"

Fraunhoffer Diffraction:

The source and the screen are at infinite distance from diffracting device (aperture or obstacle).

- In Fraunhofer diffraction the wave front meeting the obstacle is plane wave front.
- Fraunhofer diffraction requires lenses to modify the beam.

Diffraction Due to Single Slit

- Diffraction is supposed to be due to interference of secondary wavelets from the exposed portion of wavefront from the slit.

Whereas in interference, all bright fringes have same intensity. In diffraction, bright bands are ofdecreasing intensity.

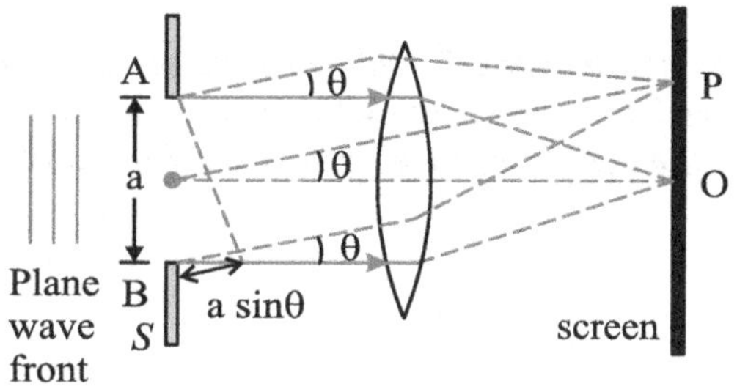

i) Condition for minimum intensity is

$\boxed{a\sin\theta = n\lambda}$ $(n = 1, 2, 3, ...)$

Where 'a' is the width of the slit, θ is the angle of diffraction

ii) Condition for maximum intensity

$\boxed{a\sin\theta = (2n+1)\frac{\lambda}{2}}$ $(n = 1, 2, 3, ...)$

The intensity decreases as we go to successive maxima away from the centre, on either side. The width of central maxima is twice as that of secondary maxima.

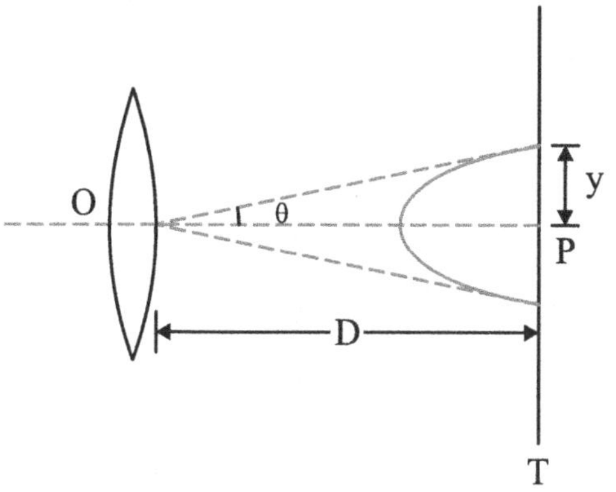

For first minia $a\sin\theta = \lambda$

$$a\frac{y}{D} = \lambda \ (\because\ \sin\theta \approx \tan\theta) \therefore y = \frac{\lambda D}{a}$$

Width of central maxima $w = 2y = \dfrac{2\lambda D}{a}$

Note: If lens is placed close to the slit, then D = f. Hence 'f' be the focal length of lens, then width of the central maximum $w = \dfrac{2f\lambda}{a}$.

Note: If this experiment is performed in liquid other than air, width of diffraction maxima will decrease and becomes $\dfrac{1}{\mu}$ times. With while light, the central maximum is white and the rest of the diffraction bands are coloured.

➢ **Interference and diffraction bands**

If N interference bands are contained by the width of the central bright.

$$\text{width} = N\beta = N\left(\frac{D\lambda}{d}\right); \therefore \frac{2D\lambda}{a} = \frac{ND\lambda}{d}$$

therefore width of the slit $a = \frac{2d}{N}$

WE-23: ***A parallel beam of light of wavelength 500 nm falls on a narrow slit and the resulting diffraction pattern is observed on screen 1 m away. It is observed that the first minimum is at a distance of 2.5 mm from the centre of the screen. Find the width of the slit.***

Sol: $\theta = \frac{y}{D}, \theta = \frac{2.5\times10^{-3}}{1} radian$

Now, $a\sin\theta = n\lambda$

Since θ is very small, therefore $\sin\theta = \theta$.

or $a = \frac{n\lambda}{\theta} = \frac{1\times500\times10^{-9}}{2.5\times10^{-3}} m$

$= 2\times10^{-4} m = 0.2\,mm$

WE-24: ***A screen is placed 50 cm from a single slit, which is illuminated with 6000 Å light, If distance between the first and third minima in the diffraction pattern is 3.00 mm, what is the width of the slit?***

Sol: In case of diffraction at single slit, the position of minima is given by $a\sin\theta = n\lambda$. Where d is the aperture size and for small θ:

$\sin\theta = \theta = (y/D)$

$\therefore a\left(\frac{y}{D}\right) = n\lambda,\ i.e., y = \frac{D}{a}(n\lambda)$

So that, $y_3 - y_1 = \frac{D}{a}(3\lambda - \lambda) = \frac{D}{a}(2\lambda)$ and hence, $a = \frac{0.50\times(2\times6\times10^{-7})}{3\times10^{-3}} = 2\times10^{-4} m$

$= 0.2\,mm$

WE-25: ***In a single slit diffraction experiment first minimum for $\lambda_1 = 660\,nm$ coincides with first maxima for wavelength λ_2. Calculate λ_2.***

Sol: Position of minima in diffraction pattern is given by; $a\sin\theta = n\lambda$

For first minima of λ_1, we have

$a\sin\theta_1 = (1)\lambda_1$ or $\sin\theta_1 = \frac{\lambda_1}{a}$ .....(i)

The first maxima approximately lies between first and second minima. For wavelength λ_2 its position will be

$a\sin\theta_2 = \frac{3}{2}\lambda_2 \therefore \sin\theta_2 = \frac{3\lambda_2}{2a}$ ...... (ii)

The two will coincide if,

$\theta_1 = \theta_2$ or $\sin\theta_1 = \sin\theta_2$

$\therefore \frac{\lambda_1}{a} = \frac{3\lambda_2}{2a}$ or

$\lambda_2 = \frac{2}{3}\lambda_1 = \frac{2}{3}\times660\,nm = 440\,nm$

WE-26: ***Two slits are made one millimeter apart and the screen is placed one meter away. What should the width of each slit be to obtain 10 maxima of the double slit pattern within the central maximum of the single slit pattern.***

Sol: We have $a\theta = \lambda$ (or) $\theta = \frac{\lambda}{a}$

(a = width of each slit)

$$10\frac{\lambda}{d} = 2\frac{\lambda}{a}$$

$$\therefore \quad a = \frac{d}{5} = \frac{1}{5} = 0.2\,mm$$

The Validity of Ray Optics:

The distance of the screen from the slit, so that spreading of light due to diffraction from the centre of screen is just equal to size of the slit, is called Fresnel distance. It is denoted by Z_F.

The diffraction pattern of a slit consists of secondary maximum and minima on the two sides of the central maximum. Therefore, one can say that on diffraction from a slit, light spreads on the screen in the form of central maximum. The angular position of first secondary minimum is called half angular width of the central maximum and it is given by

$$\theta = \frac{\lambda}{a} \text{ (provided } \theta \text{ is small)}$$

If the screen is placed at a distance D from the slit, then the linear spread of the central maximum is given by

$$y = D\theta = \frac{D\lambda}{a}$$

It is, in fact, the distance of first secondary minimum from the centre of the screen. It follows that as the screen is moved away (D is increased), the linear size of the central maximum i.e., spread distance, when $D = Z_F$,

y = a (size of the slit)

Setting this condition in the above equation, we have

$$a = \frac{Z_F \lambda}{a} \quad \text{or} \quad \boxed{Z_F = \frac{a^2}{\lambda}}$$

It follows that if screen is placed at a distance beyond Z_F, the spreading of light due to diffraction will be quite large as compared to the size of the slit. The above equation shows that the ray-optics is valid in the limit of wavelength tending to zero.

WE-27: ***For what distance is ray optics a good approximation when the aperture is 3 mm wide and the wavelength is 500 nm?***

Sol: For distance $Z \le Z_F$,

ray optics is the good appropriate

$$\text{Fresnel distance } Z_F = \frac{a^2}{\lambda} = \frac{(3\times10^{-3})^2}{5\times10^{-7}} = 18\,m$$

Limit of resolution:

➢ The smallest linear or angular separation between two point objects at which they can be just separately seen or resolved by an optical instrument is called the limit of resolution of the instrument.

Resolving Power:

➢ The resolving power of an optical instrument is reciprocal of the smallest linear or angular separation between two point objects, whose images can be just resolved by the instrument.

Resolving power = $\frac{1}{Limit\ of\ resolution}$

The resolving power of an optical instument is inversely propotional to the wavelength of light used.

Diffraction as a limit on resolving power:

- All optical instuments like lens, telescope, microscope, etc, act as apertures. Light on passing through them undergoes diffraction. This puts the limit on their resolving power.

Rayleigh's criterion for resolution:

- The images of two point objects are resolved when the central maximum of the diffraction pattern of one falls over the first minimum of the diffraction pattern of the other.

Resolving Power of a Microscope:

- T he resolving power of a microscope is defined as the reciprocal of the smallest distance d between two point objects at which they can be just resolved when seen in the microscope.

- Resolving power of microscope $= \frac{1}{d} = \frac{2\mu\sin\theta}{1.22\lambda}$

Clearly, the resolving power of a microscope depends on:

i) the wave length (λ) of the light used

ii) Half the angle (θ) of the cone of light from each point object.

iii) the refractive index (μ) of the medium between the object and the objective of the microscope

Resolving Power of a Telescope:

- The resolving power of a telescope is defined as the reciprocal of the smallest angular separation '$d\theta$' between two distant objects whose images can be just resolved by it.

- Resolving power of telescope $\frac{1}{d\theta} = \frac{D}{1.22\lambda}$

Clearly, the resolving power of telescope de pends on : (i) the diameter (D) of the telescope objective (ii) The wavelength (λ) of the light used.

WE- 28: ***Assume that light of wavelength 6000 Å is coming from a star. What is the limit of resolution of a telescope whose objective has a diameter of 100 inch?***

Sol: A 100 inch telescope implies that

a = 100 inch = 254 cm. Thus if,

$\lambda \approx 6000$ Å $= 6\times10^{-5}cm$ then,

$$\Delta\theta = \frac{1.22\lambda}{a} \approx 2.9\times10^{-7} \text{ radians}$$

POLARIZATION

- The properties of light, like interference and diffraction demonstrate the wave nature of light.
- Both longitudinal and transverse waves can exhibit interference and diffraction effects.
- The properties like polarization can be exhibited only by transverse waves.
- The peculiar feature of polarized light is that human eye cannot distinguish between polarised and unpolarised light.
- As light is an electromagnetic wave, among its electric and magnetic vectors only electric vector is mainly responsible for optical effects.
- The electric vector of wave can be identified as a "light vector"
- Ordinary light is unpolarised light in which electric vector is oriented randomly in all directions perpendicular to the direction of propagation of light.

- The phenomena of confining the vibrations of electric vector to a particular direction perpendcular to the direction of propagation of light is called "Polarization". Such polarised light is called linearly polarised or plane polarised light.
- The plane in which vibrations are present is called "plane of polarization."

(a) (b)

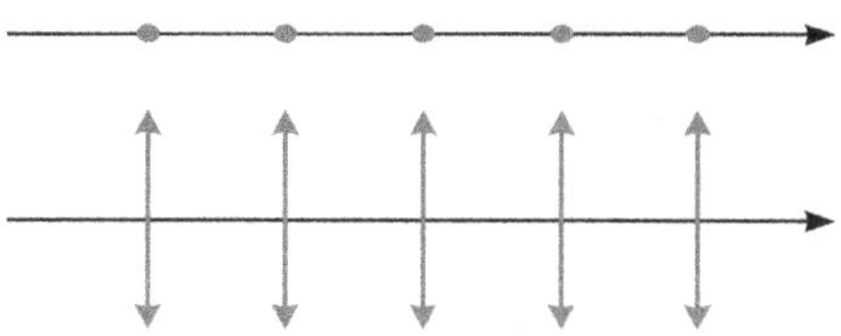

Polarized light

(a)

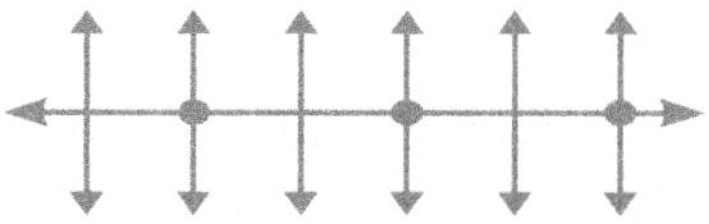

Partially polarized light

(b)

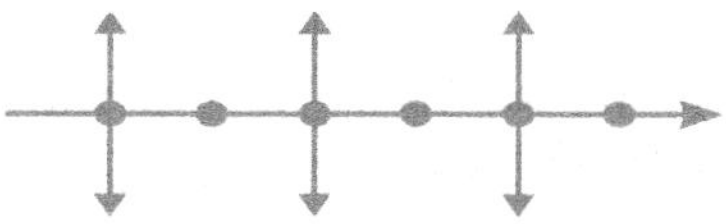

Partially polarized light

(c)

Partially polarized light

- Plane polarised light can be produced by different methods like
  i. Reflection ii. Refraction
  iii. Double refraction iv. Polaroids.

Polarization by Reflection

- The ordinary light beam is incident on transparent surface like glass or water. Both reflected and refracted beams get partially polarised.
- The degree of polarization changes with angle of incidence.
- At a particular angle of incidence called "polarising angle" the reflected beam gets completely plane polarised. The reflected beam has vibrations of electric vector perpendicular to the plane of paper.
- The polarising angle depends on the nature of reflecting surface.

Brewster's Law: When angle of incidence is equal to "polarising angle" the reflected and refracted rays will be perpendicular to each other.

- Brewster's law states that "The refractive index of a medium is equal to the tangent of polarising angle θ_p".

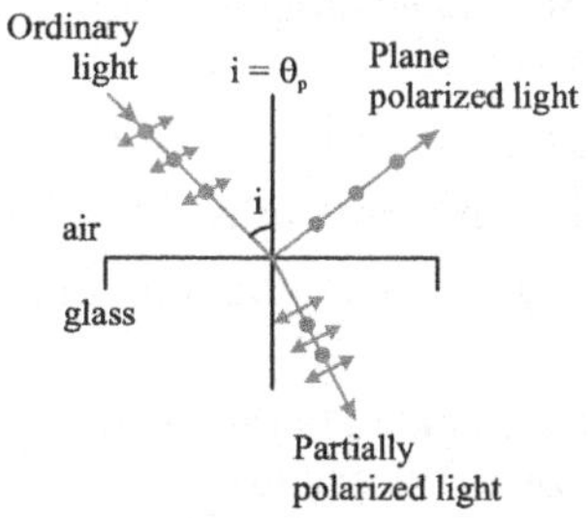

- The refractive index of the medium changes with wavelength of incident light and so polarising angle will be different for different wavelengths.
- The complete polarization is possible when incident light is monochromatic.

$$\mu = \frac{\sin\theta_p}{\sin r} = \frac{\sin\theta_p}{\sin(90^0 - \theta_p)} = \frac{\sin\theta_p}{\cos\theta_p} = \tan\theta_p$$

- From **Brewster's law,** $\mu = \tan\theta_p$.
- If $i=\theta_p$, the reflected light is completely polarised and the refracted light is partially polarised.
- If $i<\theta_p$ or $i>\theta_p$, both reflected and refracted rays get partially polarised.
- For glass $\theta_p = \tan^{-1}(1.5) \approx 57^0$

  For water $\theta_p = \tan^{-1}(1.33) \approx 53^0$

WE-29: ***When light of a certain wavelength is incident on a plane surface of a material at a glancing angle 30^0, the reflected light is found to be completely plane polarized determine.***

a) refractive index of given material and

b) angle of refraction.

Sol: a) Angle of incident light with the surface is 30^0. The angle of incidence = $90^0 - 30^0 = 60^0$. Since reflected light is completely polarized, therefore incidence takes place at polarizing angle of incidence θ_p.

$\therefore \theta_p = 60^0$

Using Brewster's law

$\mu = \tan\theta_p = \tan 60^0 \quad \mu = \sqrt{3}$

b) From Snell's law

$$\mu = \frac{\sin i}{\sin r} \quad \therefore \sqrt{3} = \frac{\sin 60^0}{\sin r}$$

or $\sin r = \frac{\sqrt{3}}{2} \times \frac{1}{\sqrt{3}} = \frac{1}{2}$, $r = 30^0$.

Polarisation by Refraction

- The unpolarised light when incident on a glass plate at an angle of incidence equal to the polarising angle, the reflected light is completely plane polarised, but the refracted light is partially polarised.
- The refracted light gets completely plane polarised if incident light is allowed to pass through number of thin glass plates arranged parallel to each other. Such an arrangement of glass plates is called "pile of plates".

Polarisation by Double Refraction (Additional)

- Bartholinus discovered that when light is incident on a calcite crystal two refracted rays are produced. It is called "double refraction" or "birefringence"
- An ink dot made on the paper when viewed through calcite crystal two images are seen due to double refraction. On rotating the crystal one image remains stationary and the other image rotates around the stationary image.
- The rotating image revolves round the stationary image in circular path.
- The stationary image is formed due to ordinary ray and revolving image is formed by extraordinary ray.
- A plane which contains the optic axis and is perpendicular to the two opposite faces is called the principal section of crystal.
- The ordinary ray emerging from the calcite crystal obey the laws of refraction and vibrations are perpendicular to the principal section of the crystal.

- The extra ordinary ray does not obey the laws of refraction and the vibrations are in the plane of principal section of crystal.
- Both ordinary and extraordinary rays are plane polarised.

Polaroid : Polaroid is an optical device used to produce plane polarised light making use of the phenomenon of "selective absorption".

- More recent type of polaroids are H-polaroids.
- H-polaroids are prepared by stretching a film of polyvinyl alcohol three to eight times to original length.

Effect of polarizer on natural light:

If one of waves of an unpolarized light of intensity I_0 is incident on a polaroid and its vibration amplitude A_0 makes an angle θ with the transmission axis, then the component of vibration parallel to transmission axis will be $A_0 \cos\theta$ while perpendicular to it $A_0 \sin\theta$. Now as polaroid will pass only those vibrations which are parallel to its transmission axis, the intensity I of emergent light wave will be

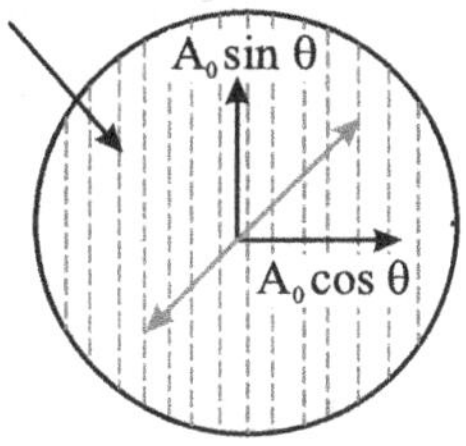

$I = KA_0^2 \cos^2\theta$ (or)

$I = I_0 \cos^2\theta [as\ I_0 = KA_0^2]$ In unpolarized light, all values of θ starting from 0 to 2π are equally probable, therefore

$$I = I_0 < \cos^2\theta > \Rightarrow I = \frac{I_0}{2\pi}\int_0^{2\pi} \cos^2\theta d\theta = \frac{I_0}{2}$$

$$\therefore\ I = \frac{I_0}{2}$$

Thus, if unpolarized light of intensity I_0 is incident on a polarizer, the intensity of light transmitted through the polarizer is $\frac{I_0}{2}$. The amplitude of polarized light is $\frac{A_0}{\sqrt{2}}$.

Effect of Analyser on plane polarized light:

When unpolarized light is incident on a polarizer, the transmitted light is linearly polarized. If this light further passes through analyser, the intensity varies with the angle between the transmission axes of polarizer and analyser.

Malus states that "the intensity of the polarized light transmitted through the analyser is proportional to cosine square of the angle between the plane of transmission of analyser and the plane of transmission of polarizer." This is known as **Malus law**.

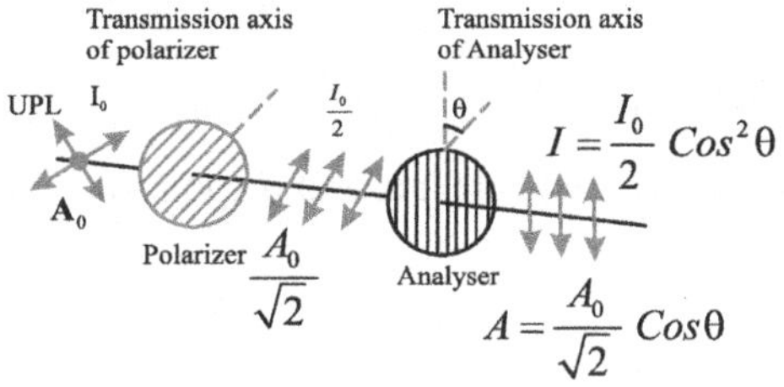

Therefore the intensity of polarized light after passing through analyser is $I = \frac{I_0}{2}\cos^2\theta$

Where I_0 is the intensity of unpolarized light. The amplitude of polarized light after passing through analyser is $A = \frac{A_0}{\sqrt{2}} \cos\theta$.

Case (i) : If $\theta = 0^0$ axes are parallel then $I = \frac{I_0}{2}$

Case (ii): If $\theta = 90^0$ axes are perpendicular, then $I = 0$.

Case (iii): If $\theta = 180^0$ axes are parallel then $I = \frac{I_0}{2}$

Case (iv): If $\theta = 270^0$ axes are perpendicular then $I = 0$ Thus for linearly polarized light we obtain two positions of maximum intensity and two positions of minimum (zero) intensity, when we rotate the axis of analyser w.r.t to polarizer by an angle 2π. In the above cases if the polariser is rotated with respect to analiser then there is no change in the outcoming intensity.

Note: In case of three polarizers P_1, P_2 and P_3: If θ_1 is the angle between transmission axes of P_1 and P_2, θ_2 is the angle between transmission axes of P_2 and P_3. Then the intensity of emerging light from P_3 is

$$I = \frac{I_0}{2} \cos^2\theta_1 \cos^2\theta_2.$$

WE-30: ***Unplarized light falls on two polarizing sheets placed one on top of the other. What must be the angle between the characteristic directions of the sheets if the intensity of the tramsitted light is one third of intensity of the incident beam?***

Sol: Intensity of the light transmitted through the first polarizer $I_1 = I_0/2$, where I_0 is the intensity of the incident unpolarized light.

Intensity of the light transmitted through the second polarizer is $I_2 = I_1 \cos^2\theta$ where θ is the angle between the characteristic directions of the polarizer sheets.

But $I_2 = I_0/3$ (given)

$$\therefore\ I_2 = I_1 \cos^2\theta = \frac{I_0}{2}\cos^2\theta = \frac{I_0}{3}$$

$$\therefore \cos^2\theta = 2/3 \Rightarrow \theta = \cos^{-1}\sqrt{\frac{2}{3}}$$

WE-31: ***Unpolarized light of intensity 32 Wm^{-2} passes through three polarizers such that the transmission axis of the last polarizer is crossed with the first. If the intensity of the emerging light is 3 Wm^{-2}, what is the angle between the transmission axes of the first two polarizers? At what angle will the transmitted intensity be maximum?***

Sol: If θ is the angle between the transmission axes of first polaroid P_1 and second P_2 while ϕ between the transmission axes of second polaroid P_2 and third P_3, then according to given problem.

$$\theta + \phi = 90^0 \quad \text{or} \quad \phi = (90^0 - \theta) .....(1)$$

Now if I_0 is the intensity of unpolarized light incident on polaroid P_1, the intensity of light transmitted through it,

$$I_1 = \frac{1}{2} I_0 = \frac{1}{2}(32) = 16\frac{W}{m^2} ......(2)$$

Now as angle between transmission axes of polaroids P_1 and P_2 is θ, in a accordance with Malus law, intensity of light transmitted through P_2 will be

$I_2 = I_1 \cos^2\theta = 16\cos^2\theta$ ....(3)

And as angle between transmission axes of P_2 and P_3 is ϕ, light transmitted through P_3 will be

$I_3 = I_2 \cos^2\phi = 16\cos^2\theta\cos^2\phi$.........(4)

According to given problem, $I_3 = 3W/m^2$

So, $4(\sin 2\theta)^2 = 3$ *i.e.*, $\sin 2\theta = (\sqrt{3}/2)$ or

$2\theta = 60^0$, i.e., $\theta = 30^0$.

WE-32: ***Discuss the intensity of transmitted light when a polaroid sheet is rotated between two crossed polaroids?***

Sol: Let I_0 be the intensity of polarised light after passing through the first polariser P_1. Then the intensity of light after passing through second polariser P_2 will be

$I = I_0\cos^2\theta$, where θ is the angle between pass axes of P_1 and P_2. Since P_1 and P_2 are crossed the angle between the pass axes of P_2 and P_3 will be $(\pi/2 - \theta)$. Hence the intensity of light emerging from

P_3 will be $I = I_0\cos^2\theta\cos^2\left(\dfrac{\pi}{2} - \theta\right)$

$= I_0\cos^2\theta\sin^2\theta = (I_0/4)\sin^2 2\theta$

Therefore, the transmitted intensity will be maximum when $\theta = \pi/4$.

CONCEPTUAL

INTERFERENCE

1. **A plane wave front falls on a convex lens. The emergent wave front is**
   1) Plane 2) Cylindrical 3) Spherical diverging 4)Spherical converging
2. **When two light waves meet at a place**
   1) their displacements add up 2) their intensities add up
   3) both will add up 4) Energy becomes zero
3. **The following phenomena which is not explained by Huygens' construction of wave front is**
   1) refraction 2) reflection 3) diffraction 4) origin of spectra
4. **A wavefront is an imaginary surface where**
   1) phase is same for all points
   2) phase changes at constant rate at all points along the surface.
   3) constant phase difference continuously changes between the points
   4) phase changes all over the surface
5. **Huygen's wave theory is used**
   1) to determine the velocity of light 2) to find the position of the wave front
   3) to determine the wavelength of light 4) to find the focal length of a lens.
6. **In a Laser beam the photons emitted are**
   1) same wavelength 2) coherent 3) of same velocity 4) All the above
7. **The amplitudes of two interfering waves are 4 cm and 3 cm respectively. If the resultant amplitude is 1 cm then the interference becomes**
   1) constructive 2) Destructive
   3) Both constructive and destructive 4) given data is insufficient
8. **Two coherent waves are represented by $y_1 = a_1\cos\omega t$ and $y_2 = a_2\sin\omega t$. The resultant intensity due to interference will be**
   1) $(a_1^2 - a_2^2)$ 2) $(a_1^2 + a_2^2)$ 3) $(a_1 - a_2)$ 4) $(a_1 + a_2)$

9. **Two light waves are represented by $y_1 = a\sin\omega t$ and $y_2 = a\sin(\omega t + \delta)$. The phase of the resultant wave is**

1) 2δ 2) $\frac{\delta}{2}$ 3) $\frac{\delta}{3}$ 4) $\frac{\delta}{4}$

10. **Laser light is considered to be coherent because it consists of**

1) many wavelengths 2) uncoordinated wavelengths
3) coordinated waves of exactly the same wavelength
4) divergent beams

11. **Two waves having the same wave length and amplitude but having a constant phase difference with time are known as**

1) identical waves 2) incoherent waves 3) coherent waves 4) collateral waves

12. **Light waves spreading from two sources produce steady interference only if they have**

1) congruence 2) coherence 3) same intensity 4) same amplitude

13. **Four different independent waves are represented by**

a) $y_1=a_1 \sin \omega_1 t$ b) $y_2=a_2 \sin \omega_2 t$ c) $y_3=a_3 \sin \omega_3 t$ d) $y_4=a_4\sin(\omega_4 t+\frac{\pi}{3})$

The sustained interference is possible due to

1) a & c 2) a & d 3) c & d
4)not possible with any combination

14. **Interference fringes in Young's double slit experiment with monochromatic light are**

1) always equispaced 2) always unequally spaced
3) both equally and unequally spaced 4) formed by a portion of the wave front.

15. **The necessary condition for an interference by two sources of light is that:**

1) two light sources must have the same wavelength
2) two point sources should have the same amplitude and same wavelength
3) two sources should have the same wavelength, nearly the same amplitude and have a constant phase angle difference
4) the two point sources should have a randomly varying phase difference

16. **For the sustained interference of light, the necessary condition is that the two sources should**

1) have constant phase difference only 2) be narrow
3) be close to each other 4) of same amplitude with constant phase difference

17. **When interference of light takes place**

1) Energy is created in the region of maximum intensity
2) Energy is destroyed in the region of maximum intensity
3) Conservation of energy holds good and energy is redistributed
4) Conservation of energy doesn't hold good

18. **Which of the following is conserved when light waves interfere**

1) momentum 2) amplitude 3) energy 4) intensity

19. **The path difference between two interfering waves at a point on screen is 70.5 times the wave length. The point is**

1) Dark 2) Bright 3) Not possible 4) Green in colour

20. **Interference is produced with two coherent sources of same intensity. If one of the sources is covered with a thin film so as to reduce the intensity of light coming out of it to half, then**

1) Bright fringes will be less bright and dark fringes will be less dark
2) Bright fringes will be more bright and the dark fringes will be more dark
3) Brightness of both types of the fringes will remain the same
4) Dark region will spread completely

21. **For constructive interference between two waves of equal wavelength, the phase angle δ should be such that**

1) $\cos^2\frac{\delta}{2}=-1$ 2) $\cos^2\frac{\delta}{2}=0$ 3) $\cos^2\frac{\delta}{2}=1$ 4) $\cos^2\frac{\delta}{2}=\text{infinite}$

22. **Two coherent waves each of amplitude 'a' traveling with a phase difference δ when superpose with each other the resultant intensity at a given point on the screen is**

1) $a^2(1+\cos\delta)$ 2) $4a^2(1+\cos\delta)$ 3) $2a^2(1+\cos\delta)$ 4) $(1+\cos\delta)$

23. **In the set up shown, the two slits S_1 *and* S_2 *are* not equidistant from the slit S.**

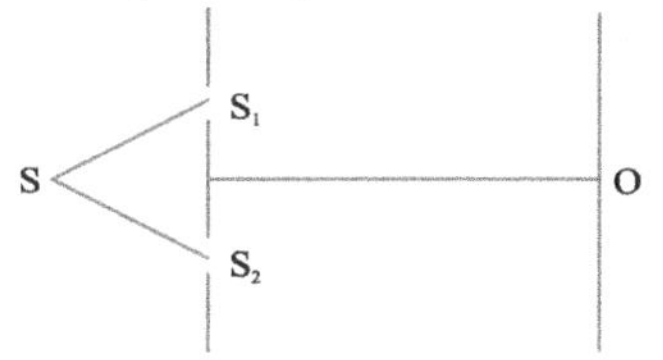

The central fringe at O is then

1) always bright 2) always dark

3) either dark or bright depending on the position of S. 4) neither dark nor bright

24. **In young's experiment of double slit, the number of times the intensity of the central bright band greater than the individual intensity of the interfering waves**

1) 2 2) 4 3) 6 4) 16

25. **A young's double slit experiment uses a monochromatic source. The shape of the interference fringes formed on the screen is a**

1) straight line 2) parabola 3) hyperbola 4) circle

26. **The contrast in the fringes in any interference pattern depends on :**

1) fringe width 2) wave length

3) intensity ratio of the sources 4) distance between the sources.

27. **If monochromatic red light is replaced by green light the fringe width becomes**

1) increase 2) remain same 3) we cannot say 4) decrease

28. **Interference was observed in interference chamber, when air was present. Now the chamber is evacuated, and if the same light is used, a careful observer will see**

1) no interference 2) interference with central bright band

3) interference with central dark band

4) interference in which breadth of the fringe will be slightly increased.

29. **In young's experiment with white light central fringe is white. If now a transparent film is introduced in the upper beam coming from the top slit, the white fringe**

1) moves down ward 2) moves upward

3) remains at the same place 4) totally disappears

30. **In young's double slit experiment the slits are of different length and widths. The amplitude of the light waves is directly proportional to the**

1) length of the slit 2) distance between the slits

3) area of the slits 4) width of slits

31. **In a double slit experiment, instead of taking slits of equal widths, one slit is made twice as wide as the other. Then, in the interference pattern.**

1) The intensities of both the maxima and the minima increase

2) The intensity of the maxima increases and the minima has zero intensity.

3) The intensity of the maxima decreases and that of the minima increases.

4) The intensity of the maxima decreases and the minima has zero intensity.

32. **When the width of slit aperture is increased by keeping 'd' as constant in Young's experiment**

1) Fringe width will increase 2)Fringe width will decrease and then increase

3)Fringe width first increases then decreases 4) Gradually the fringes will be merge

SR-Physics

33. **When viewed in white light, soap bubble show colours because of**
1) Interference 2) Scattering 3) Diffraction 4) Dispersion

34. **When petrol drops from a vehicle fall over rain water on road surface colours are seen because of**
1) Dispersion of light 2) Interference of light
3) Scattering of light 4) Absorption of light

35. **Coherent light is incident on two fine parallel slits S_1 and S_2 as shown in fig. If a dark fringe occurs at P, which of the following gives possible phase differences for the light waves arriving at P from S_1 and S_2?**

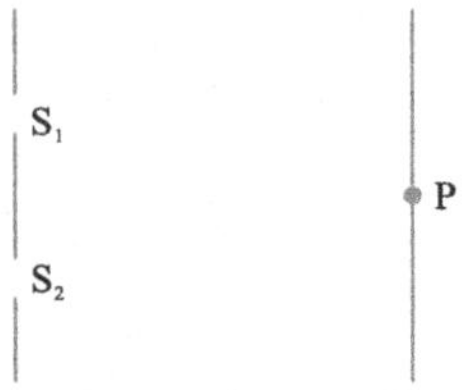

1) $2\pi, 4\pi, 6\pi$ ... 2) $1/2\pi, 5/2\pi, 9/2\pi$ ...
3) $\pi, 3\pi, 5\pi$ ..... 4) $1/2\pi, 3/2\pi, 5/2\pi$ ....

36. **In young's double slit experiment, the distance of the n-th dark fringe from the centre is**
1) $n\left(\frac{\lambda D}{2d}\right)$ 2) $n\left(\frac{2d}{\lambda D}\right)$ 3) $(2n-1)\frac{\lambda D}{2d}$ 4) $(2n-1)\frac{4d}{\lambda D}$

37. **When a thin film of thickness t is placed in the path of light wave emerging out of the slit, then increase in the length of optical path will be**
1) $(\mu-1)t$ 2) $(\mu+1)t$ 3) μt 4) $\frac{\mu}{t}$

38. **If the Young's double slit experiment is performed with white light, then**
1) the central maximum will be dark 2) there will not be completely dark fringe
3) the fringe next to the central will be red 4) the fringe next to the central will be violet

39. **Which of the following decides about the contrast between bright and dark fringes in an interference experiment?**
1) wavelength 2) distance between two coherent sources
3) fringe width 4) intensity ratio

40. **If torch is used in place of monochromatic light in Young's experiment, what will happen?**
1) Fringe will occur as from monochromatic source
2) Fringe will appear for a moment and then it will disappear
3) No fringes will appear 4) Only bright fringe will appear

41. **At a finite distance from the source, a point source of light produces**
1) spherical wave front 2) plane wavefront
3) cylindrical wavefront 4) both spherical and plane wavefronts

42. **Nature of wave front depends on**
1) shape of source 2) distance of source 3) both 1 and 2 4) none of these

43 **If Young's double slit apparatus is shifted from air to water, then**
1) Fringe width decreases 2) Fringe width increases
3) Fringe width remains same 4) Fringe system disappears

44. **In Young's double slit experiment the phase difference between the waves reaching the central fringe and fourth bright fringe will be**
1) zero 2) 4π 3) 6π 4) 8π

45. Instead of using two slits as in young's experiment, if we use two separate but identical sodium lamps, which of the following occur

a) uniform illumination is observed
b) widely separate interference
c) very bright maximum
d) very minimum

1) a only 2) a, b only 3) c, d only 4) b, d only

46. Alternate bright and dark fringes appear in Young's double slit experiment due to the phenomenon of

1) Polarisation 2) Diffraction 3) Interference 4) Dispersion

DIFRACTION OF LIGHT

47. The bending of light about corners of an obstacle is called

1) Dispersion 2) Refraction 3) Deviation 4) Diffraction

48. To observe diffraction, the size of an obstacle

1) Should be of the same order as wave length
2) Should be much larger than the wave length
3) Has no relation to wave length
4) May be greater or smaller than the wave length

49. In diffraction pattern

1) The fringe widths are equal
2) The fringe widths are not equal
3) The fringes can not be produced
4) The fringe width may or may not be equal

50. Sun light filtering through a tree leaves often makes circular patches on the ground because

1) The sun is round
2) The space through which light penetrates is round
3) Light is transverse in nature
4) Of diffraction effects

51. In studying diffraction pattern of different obstacles, the effect of

1) full wave front is studied
2) portion of a wave front is studied
3) waves from two coherent sources is studied
4) waves from one of the coherent source is studied.

52. Both light and sound waves produce diffraction. It is more difficult to observe the diffraction with light waves because.

1) Light wave do not require medium
2) Wavelength of light waves is far smaller
3) Light waves are transverse
4) Speed of light is far greater

53. In Young's double slit experiment

1) only interference occurs
2) only diffraction occurs
3) both interference and diffraction occurs
4) polarisation occurs

54. Light travels in a straight line because

1) it is not absorbed by atmosphere
2) its velocity is very high
3) diffraction effect is negligible
4) due to interference

55. One of the following statements is correct. Pick out the one

1) Diffraction can not take place without interference
2) Interference will not take place with out diffraction.
3) Interference and diffraction are the result of polarization
4) The fringe width in Young's double slit experiment does not depends on the wave length.

56. Diffraction of light is

1) the bending of light at the surface of separation when it travels from rarer medium of denser medium
2) the bending of light at the surface of separation when it travels from denser medium to rarer medium
3) encroachment of light into the geometrical shadow of the obstacle placed in its path
4) emergence of a light ray grazing the surface of separation when it travels from denser to rarer medium

57. Pick out the correct statements
1) diffraction is exhibited by all electromagnetic waves but not by mechanical waves
2) diffraction cannot be observed with a plane polarized light
3) the limit of resolution of a microscope decreases with increase in the wavelength of light used
4) the width of central maximum in the diffraction pattern due to single slit increases as wavelength increases

58. A lens of focal length f gives diffraction pattern of Fraunhoffer type of a slit having width a. If wavelength of light is λ, the distance of first dark band and next bright band from axis is given by

1) $\frac{a}{\lambda}$ f   2) $\frac{\lambda}{a}$ f   3) $\frac{\lambda}{af}$ 4) aλf

59. The class of diffraction in which incident and diffracted wave fronts are planar is called
1) Fresnel diffraction   2) Fraunhoffer diffraction
3) Huygen's diffraction   4) Newton's diffraction

60. Neutron diffraction pattern is used to determine
1) Density of solids   2) Atomic number of elements
3) Crystal structure of solid   4) Refractive index of liquid

61. Geometrical shadow is formed due to the phenomenon of
1) Diffraction of light   2) Polarisation of light
3) Interference of light   4) Rectilinear propagation of light

62. The surface of crystals can be studied using
1) diffraction of visible light   2) diffraction of x-rays
3) interference of sound waves   4) refraction of radio waves

63. The diffraction bands observed in the case of straight edge producing diffraction effects are
1) equally spaced like the interference bands but with less contrast
2) unequally spaced with increasing width as we move away from the edge of geometric shadow
3) unequally spaced with decreasing width as we move away from the edge of geometric shadow
4) equally spaced like the interference bands but with more contrast

64. A we move away from the edge into the geometrical shadow of a straight edge, the intensity of illumination
1) Decreases   2) Increases
3) Remains unchanged   4) Increase and then decreases

65. In Fresnel's diffraction, wavefront must be
1) spherical   2) cylindrical   3) plane   4) both 1 and 2

RESOLVING POWER

66. The resolving power of human eye is
1) $\approx 1'$ 2) $\approx 1^0$ 3) $\approx 10'$ 4) $\approx 5'$

67. Resolving power of a telescope increases with
1) Increase in focal length of eye piece   2) Increase in focal length of objective
3) Increase in aperture of eye piece   4) Increase in aperture of objective

68. To increase both the resolving power and magnifying power of a telecscope
1) Both the focal length and aperture of the objecive has to be increased
2) The focal length of the objective has to be increased
3) The aperture of the objective has to be increased
4) The wavelength of light has to be decreased

POLARISATION

69. Waves that cannot be polarised are
1) Longitudinal   2) Transverse   3) Electromagnetic   4) Light

70. Human eye
1) Can detect polarised light
2) Can not detect polarisation of light
3) Can detect only circularly polarised light
4) Can detect only linearly polarised light

71. Polarisation of light was first successfully explained by
1) Corpuscular theory
2) Huygens' wave theory
3) Electromagnetic wave theory
4) Planck's theory

72. Plane of polarisation is
1) The plane in which vibrations of the electric vector takes place
2) A plane perpendicular to the plane in which vibrations of the electric vector takes place
3) Is perpendicular to the plane of vibration
4) Horizontal plane

73. In the propagation of polarised light waves, the angle between the plane of vibration and the plane of polarization is
1) 0^0
2) 90^0
3) 45^0
4) 180^0

74. Transverse wave nature is established by
1) Interference
2) Diffraction
3) Polarization
4) All the above

75. Choose the correct statement.
1) the Brewster's angle is independent of wavelength of light.
2) the Brewster's angle is independent of the nature of reflecting surface
3) the Brewster's angle is different for different wavelengths
4) Brewster's angle depends on wavelength but not on the nature of reflecting surface.

76. The polarising angle for glass is
1) same for different kinds of glass
2) different for different kinds of glass
3) same for lights of all colours
4) varies with time

77. When an unpolarised light is polarized, then the intensity of light of the polarized wave
1) remains same
2) doubled
3) halved
4) depends on the colour of the light.

78. Unpolarising light falls on two polarizing sheets so oriented that no light is transmitted. If a third polarizing sheet is placed between them; not parallel to either of the above two sheets in question
1) no light is transmitted
2) some light is transmitted
3) light may or may not be transmitted
4) certainly 50% light is transmitted.

79. When light falls on two polaroid sheets, one observes complex brightness then the two polaroids axes are
1) Mutually perpendicular
2) Mutually parallel
3) Angle between their two axes is 45^0
4) None of the above

80. Polaroid are used
1) to eliminate head light glare in automobile
2) in production of 3-D motion pictures
3) in sun glasses
4) all the above

81. A diffraction pattern is obtained using a beam of redlight. What happens if the red light is replaced by blue light
1) no change
2) diffraction bands become narrower and crowded together
3) bands beome broader and farther apart
4) bands disappear

82. In a diffraction pattern the width of any fringe is
1) directly proportional to slit width
2) inversely proportional to slit width
3) Independent of the slit width
4) None of the above

83. **Yellow light is used in a single slit of diffraction experiment with slit width 0.6 mm. If yellow light is replaced by X-rays then the observed pattern will reveal**

1) that the central maximum is narrower 2) more number of fringes
3)less number of firnges 4) no diffraction patterns

84. **A beam of light AO is incident on a glass slab ($\mu = 1.54$) in the direction shown. The reflected ray OB is passed through a Nicol prism. On viewing through a Nicol prism, we find on rotating the prism that**

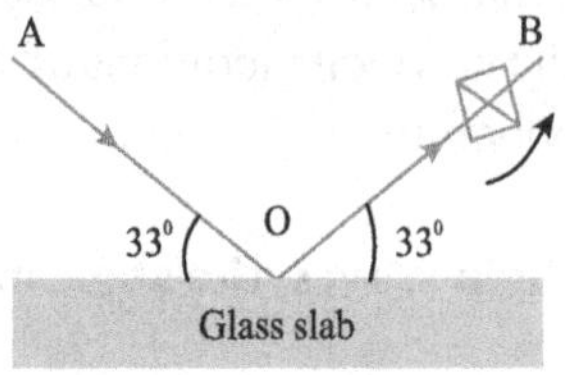

1) the intensity is reduced down to zero and remains zero
2) the intensity reduces down somewhat and rises again
3) there is no change in intensity
4) the intensity gradually reduces to zero and then again increases

85. **A star is going away from the earth. An observer on the earth will see the wavelength of light coming from the star**

1) decreased 2) increased
3) neither decreased nor increased
4) decreased or increased depending upon the velocity of the star

86. **Red shift is an illustration of**

1) low temperature emission 2) high frequency absorption
3) Doppler effect 4) unknown phenomenon

87. **If the shift of wavelength of light emitted by a star is towards violet, then this shows that star is**

1) stationary 2) moving towards earth
3) moving away from earth 4) Information is incomplete

88. **When there is a relative motion of an observer from a source of light, the apparent change in its wavelength is termed as**

1) Raman effect 2) Seebeck effect 3) Doppler's effect 4) Gravitational effect

89. **In the context of Doppler effect in light, the term red shift signifies**

1) decrease in frequency 2) increase in frequency
3) decrease in intensity 4) increase in intensity

90. **As we change the colour of light from Red to Blue, which of the following is correct for the polarizing angle and critical angle of glass?**

1) the former increases, the latter decreases 2) the former decreases, the latter increases
3) the former increases, the latter increases 4) the former decreases, the latter decreases

91. **Atmospheric refraction is due to**

1) changing pressure in the atmosphere 2) varying density of atmosphere
3) varying temperature of the atmosphere 4) both (2) and (3).

92. **Which of the following phenomenon is not common to sound and light waves**

1) Interferenec 2) Diffraction 3) Polarisation 4) Reflection

93. **Polarisation can be produced by**

1) Reflection 2) Refraction 3) Scattering 4) All of the above

94. An unpolarised light is incident on a surface separating two transparent media of different optical densities at the polarizing angle. Then the reflected ray and refracted ray are
1) parallel to each other
2) perpendicular to each other
3) inclined to each other making an angle 45^0
4) none of the above

95. The intensity of the polarized light transmitted through the analyzer is given by
1) Brewster's law
2) Malus Law
3) Fresnel's assumptions
4) law of superposition

96. Statement A: In the interference pattern the intensity is same at all points in a brightband
Statement B: In Young's double slit experiment, as we move away from the central maximum, the third maximum always comes before the third minimum.
1) Both A and B are true
2) Both A and B are false
3) A is true but B is false
4) A is false and B is true

97. A light of wavelength λ is incident on an object of size b. If a screen is at a distance D from the object. Identify the correct condition for the observation of different phenomenon

a) if $b^2 = D\lambda$, Fresnel diffraction is observed

b) if $b^2 >> D\lambda$, Fraunhoffer diffraction is observed

c) $b^2 << D\lambda$, Fraunhoffer diffraction is observed

d) $b^2 >> D\lambda$, the approximation of geometrical optics is applicable

1) a, b and d are true
2) a,c and d are true
3) a and c are true
4) a and d are true

DIRECTIONS Q.NO: 98 TO 127

In each of the following questions, a statement of Assertion (A) is given followed by a corresponding statement of reason (R) just below it. Of the statement mark the correct answer.
1) Both 'A' and 'R' are true and 'R' is the correct explanation of 'A'
2) Both 'A' and 'R' are true and 'R' is not the correct explanation of 'A'
3) 'A' is true and 'R' is false
4) A' is false and 'R' is true

98. Assertion (A) : In Young's double slit experiment the band width for red colour is more
Reason (R) : Wavelength of red is small

99. Assertion (A) : Thin films such as soap bubble or a thin layer of oil on water show beautiful colours when illuminated by sunlight
Reason (R) : The colours are obtained by dispersion of light only

100. Assertion (A) : When tiny circular obstacle is placed in the path of light from some distance, a bright spot is seen at the centre of the shadow of the obstacle.
Reason (R) : Destructive interference occurs at the centre of the shadow.

101. Assertion (A) : Coloured spectrum is seen when we look through a fine cotton cloth
Reason (R) :It is due to the diffraction of white light on passing through fine slits.

102. Assertion (A) : Diffraction is common in sound but not common in light waves
Reason (R) : Wavelength of light is more than the wavelength of sound

103. Assertion (A) : We cannot observe diffraction pattern from a wide slit illuminated by monochromatic light
Reason (R) : In diffraction pattern, all the bright bands are not of the same intensity.

104. Assertion (A) : Transverse wave nature of light is proved by polarisation
Reason (R) : According to Maxwell, light is an electromagnetic wave but not mechanical wave

105. Assertion (A) : Coloured spectrum is seen when we look through a cloth
Reason (R) : Diffraction of light takes place when light is travelling through the pores of cloth

106. **Assertion (A) : Young's double slit experiment can be performed using a source of white light.**
Reason (R) : The wavelength of red light is less than the wavelength of other colors in white light.

107. **(A) : The unpolarised light and polarized light can be distinguished from each other by using Polaroid.**
(R) : A Polaroid is capable of producing plane polarized beams of light.

108. **Assertion (A) : Illumination of the sun at noon is maximum because**
Reason (R): The sun rays are incident almost normally

109. **Assertion (A): The phase difference between any two points on a wave front is zero**
Reason (R): Light from the source reaches every point of the wave front at the same time

110. **Assertion (A) : In Young's double slit experiment white light is used and slits are covered with red and blue filters respectively. The phase difference at any point on the screen will continuously change and uniform illumination is produced on the screen**
Reason (R): Two independent sources of light would no longer act as coherent sources

111. **Assertion (A) : In interference pattern intensity of successive fringes due to achromatic light is not same**
Reason (R): In interference, only redistribution of energy takes place

112. **Assertion (A): Light from two coherent sources is reaching the screen. If the path difference at a point on the screen for yellow light is $3\lambda/2$, then the fringe at the point will be coloured.**
Reason (R): Two coherent sources always have constant phase relationship

113. **Assertion (A): No interference pattern is detected when two coherent sources are very close to each other. (i.e separation almost zero)**
Reason (R): The fringe width is inversely proportional to the distance between the two slits

114. **Assertion (A): In Young's double slit experiment interference pattern disapperars when one of the slits is closed**
Reason (R): Interference occurs due to superimposition of light wave from two coherent sources

115. **Assertion (A): The maximum intensity in interference pattern is four times the intensity due to each slit of equal width.**
Reason (R): Intensity is directly proportional to square of amplitude.

116. **Assertion (A): The fringe obtained at the centre of the screen is known as zeroth order fringe, or the central fringe**
Reason (R): Path difference between the waves from S_1 *and* S_2, reaching the central fringe (or zero order fringe) is zero

117. **Assertion (A) : If the phase difference between the light waves emerging from the slits of the Young's experiment is π-radian, the central fringe will be dark**
Reason (R) : Phase difference is equal to $(2\pi/\lambda)$ times the path difference.

118. **Assertion (A) : At sunrise or at sunset the sun appears to be reddish while at mid-day the sun looks white**
Reason (R) : Scattering due to dust particles and air molecules

119. **Assertion (A) : If the whole apparatus of Young's experiment is immersed in liquid, the fringe width will decrease.**
Reason (R) : The wavelength of light in water is more than that in air

120. **Assertion (A) : The soap film in sun light is colourful**
Reason(R):Thin films produce interference of light

121. Assertion (A): Coloured spectrum is seen when we look through a cloth
Reason (R): Diffraction of light takes place when light is travelling through the pores of cloth

122. Assertion (A) : Radio waves diffract pronouncedly around the sharp edges of the buildings than visible light waves.
Reason (R) : Wave length of radio waves is comparable to the dimension of the edges of the building.

123. Assertion (A) : When an unpolarised light is incident on a glass plate at Brewster angle, the reflected ray and refracted ray are mutually perpendicular
Reason (R) :The refractive index of glass is equal to sine of the angle of polarisation.

124. Assertion (A) : If two waves of same amplitude produce a resultant wave of same amplitude, then the phase difference between them may be 120°.
Reason (R) : The resultant amplitude of two waves is equal to algebraic sum of amplitude of two waves.

125. Assertion (A) : Although the surfaces of a goggle lens are curved, it does not have any power.
Reason (R) : In case of goggles, both the curved surfaces have equal radii of curvature.

126. (A) : For best contrast between maxima and minima in the interference pattern of Young's double slit experiment, the intensity of light emerging out of the two slits should be equal
(R) : The intensity of interference pattern is proportional to square of amplitude.

127. (A) : In Young's double slit experiment the fringes become indistinct if one of the slits is converd with cellophane paper.
(R) : The cellophane paper decreases the wavelength of light.

MATCHING TYPE QUESTIONS

128. Match list A and list B accurately

LIST - A	LIST - B
a) spherical wave front	e) linear source
b) plane wave front	f) point light source
c) cylindrical wave front	g) at infinite distance

1) (a, f); (b, g); (c, e)
2) (a, f); (b, e); (c, g)
3) (a, g); (b, f); (c, e)
4) (a, e); (b, g); (c, f)

129. Match the following

PART-A	PART-B
a) achromatic light	d) $\frac{\lambda D}{d}$
b) monochromatic	
e) distance between bright two successive bright bands	
c) fringe width two successive dark	f) distance between bands
g) central fringe is lways bright	h) central fringe is always achromatic

1) $a \to g \quad b \to e, f, g, \quad c \to e, f, g$
2) $a \to g,h \quad b \to h,g \quad c \to d,e,f$
3) $a \to e, f, g \quad b \to g \quad c \to e, f, g$
4) $a \to e \quad b \to h, \quad c \to g, h$

130. Match the following

PART-A	PART-B
a) Polarisation	e) All types of waves
b) interference	f) longitudinal waves
c) diffraction	g) transverse waves
d) reflection	h)only with transverse waves
	i) stationary waves produced in stretched strings

1) $a\to g$ $b\to e,f,g,i$ $c\to e,f,g$ $d\to e,f,g$ 2) $a\to h,g$ $b\to f,g$ $c\to g$ $d\to h$
3) $a\to e,f,g$ $b\to g$ $c\to e,f,g$ $d\to g$ 4) $a\to e$ $b\to h,i$ $c\to g,h$ $d\to e$

131. Match the following :

List-I	List-II
a) Silver lining of mountains	e) polarization by refraction
b) Rectilinear propagation light	f) transverse nature of light
c) Polarization	g) diffraction
d) Pile of plates	h) ray optics

1) $a-h,b-g,c-f,d-e$ 2) $a-g,b-h,c-e,d-f$
3) $a-f,b-h,c-h,d-e$ 4) $a-g,b-h,c-f,d-e$

132. Match the following

List-I	List-II
a) coherent, monochromatic highly unidirectionally	e) Malus law
b) $I=I_0\cos^2\theta$	f) Polariod
c) Selective absorption is exhibited by	g) Spherical wave front
d) Fresnel diffraction	h) LASER

1) $a-h,b-e,c-f,d-g$ 2) $a-g,b-h,c-e,d-f$
3) $a-h,b-g,c-e,d-f$ 4) $a-g,b-h,c-f,d-e$

133. Match the following

List-I	List-II
a) Interference	e) Thamos young
b) Polarisation by reflection	f) Bartholinus
c) Diffraction	g) Grimaldi
d) Polarisation by refraction	h) Malus

1) $a-e,b-g,c-f,d-h$ 2) $a-h,b-f,c-g,d-e$
3) $a-e,b-h,c-g,d-f$ 4) $a-h,b-g,c-f,d-e$

134. In Young's double slit experiment, what will be the effects of the following

Column - I	Column - II
A) A thin translucent plate is inserted in front of one of the slits	p) Fringe width changes
B) A thin transparent glass plate is inserted in front of one of the slits	q) Fringe width remains unaltered
C) The entire set up is immersed in water	r) Fringe width remains unaltered
D) Both slits are covered with translucent paper	s) Brightness of fringe remains unaltered
	t) Fringe width and brightness are directly related

1) A – q,r, B – q, s, C – p, r, D – q, r 2) A – q,r, B – q, s, C – p, s, D – q, r
3) A – p,r, B – q, s, C – p, s, D – q, r 4) A – q,s, B – q, t, C – p, s, D – q, r

CONCEPTUAL - KEY

1) 4	2) 2	3) 4	4) 1	5) 2	6) 4	7) 2	8) 2	9) 2	10) 3
11) 3	12) 2	13) 4	14) 1	15) 3	16) 4	17) 3	18) 3	19) 1	20) 1
21) 3	22) 3	23) 3	24) 2	25) 3	26) 3	27) 4	28) 2	29) 2	30) 3
31) 1	32) 2	33) 1	34) 2	35) 3	36) 3	37) 1	38) 3	39) 4	40) 3
41) 1	42) 3	43) 1	44) 4	45) 1	46) 3	47) 4	48) 1	49) 2	50) 4
51) 2	52) 2	53) 3	54) 3	55) 1	56) 3	57) 4	58) 2	59) 2	60) 3
61) 4	62) 2	63) 3	64) 1	65) 4	66) 1	67) 4	68) 4	69) 1	70) 2
71) 3	72) 1	73) 2	74) 3	75) 3	76) 2	77) 3	78) 2	79) 2	80) 4
81) 2	82) 2	83) 4	84) 4	85) 2	86) 3	87) 2	88) 3	89) 1	90) 1
91) 4	92) 3	93) 4	94) 2	95) 2	96) 1	97) 2	98) 3	99) 3	100)3
101)1	102)3	103)2	104)2	105)1	106)3	107)1	108)1	109)1	110)1
111)1	112)4	113)1	114)1	115)2	116)1	117) 2	118)1	119)3	120)1
121)1	122)1	123)3	124)3	125)1	126)2	127)3	128)1	129) 2	130)1
131)4	132)1	133)3	134) 2						

LEVEL - I (CW)

INTERFERENCE

1. **The displacements of two interfering light waves are $y_1 = 4\sin\omega t$ and $y_2 = 3\cos(\omega t)$. The amplitude of the resultant wave is (y_1 and y_2 are in CGS system)**
   1) 5cm 2) 7cm 3) 1 cm 4) zero
2. **Two coherent sources of different intensities send waves that interfere. The ratio of maximum to minimum intensity is 25. The intensity ratio of the sources is**
   1) 25 : 1 2) 5 : 1 3) 9 : 4 4) 625 : 1
3 **Two sources of intensity $2I$ and $8I$ are used in an interference experiment. The intensity at a point where the waves from two sources superimpose with a phase difference of (a) zero (b) $\pi/2$ $and (c) \pi$ is**
   1) $18I, 10I, 2I$ 2) $5I, 4I, I$ 3) $2I, I, \frac{I}{2}$ 4) $2I, 10I, 18I$
4. **The intensity of interference waves in an interference pattern is same as I_0. The resultant intensity at the point of observation will be**
   1) $I = 2I_0[1+\cos\phi]$ 2) $I = I_0[1+\cos\phi]$ 3) $I = \frac{[1+\cos\phi]}{I_0}$ 4) $I = \frac{[1+\cos\phi]}{2I_0}$
5. **In Young's double slit experiment, the constant phase difference between two sources is $\frac{\pi}{2}$. The intensity at a point equidistant from the slits in terms of maximum intensity I_0 is**
   1) I_0 2) $I_0/2$ 3) $3I_0/4$ 4) $3I_0$
6. **The path difference between two interfering waves at a point on the screen is $\lambda/6$ from central maximum. The ratio of intensity at this point and that at the central fringe will be**
   1) 0.75 2) 7.5 3) 85.3 4) 853
7. **In a Young's double slit experiment, 12 fringes are observed to be formed in a certain region of the screen when light of wavelength 600 nm is used. If the light of wavelength 400 nm is used, the number of fringes observed in the same region of the screen will be**
   1) 12 2) 18 3) 24 4) 8

8. A double slit apparatus is immersed in a liquid of refractive index 1.33. It has slit separation of 1mm and distance between the plane of slits and screen 1.33 m. The slits are illuminated by a parallel beam of light whose wavelength in air is $6300 A^0$. The fringe width is

1) $(1.33\times 0.63)mm$ 2) $\frac{0.63}{1.33}mm$ 3) $\frac{0.63}{(1.33)^2}mm$ 4) $0.63mm$

9. The fringe width at a distance of 50cm from the slits in young's experiment for light of wavelength 6000A^0 is 0.048cm. The fringe width at the same distance for $\lambda = 5000\text{A}^0$ will be

1) 0.04cm 2) 0.4cm 3) 0.14cm 4) 0.45cm

10. In young's double slit experiment the two slits are illuminated by light of wavelength 5890^0A and the distance between the fringes obtained on the screen is 0.2^0. If the whole apparatus is immersed in water then the angular fringe width will be, if the refractive index of water is 4/3

1) 0.30^0 2) 0.15^0 3) 15^0 4) 30^0

11. A plate of thickness t made of material of refractive index μ is placed in front of one of the slits in a double slit experiment. What should be the minimum thickness t which will make the intensity at the centre of the fringe pattern zero?

1) $(\mu-1)\frac{\lambda}{2}$ 2) $(\mu-1)\lambda$ 3) $\frac{\lambda}{2(\mu-1)}$ 4) $\frac{\lambda}{(\mu-1)}$

12. In Young's double slit experiment the separation between the slits is halved and the distance between the slits and screen is doubled. The fringe width is

1) unchanged 2) halved 3) doubled 4) quadrupled

13. The maximum number of possible interference maxima for slit separation equal to twice the wavelength in Young's double slit experiment is

1) infinite 2) five 3) three 4) zero

14. Two identical coherent sources produce a zero order bright fringe on a screen. If β is the band width, the minimum distance between two points on either side of the bright fringe where the intensity is half that of maximum intensity is

1) $\beta/2$ 2) $\beta/4$ 3) $\beta/3$ 4) $\beta/6$

15. In Young's double slit experiment, the 8th maximum with wavelength λ_1 is at a distance d_1 from the central maximum and the 6th maximum with wavelength λ_2 is at a distance d_2 from central maximum. Then, d_1/d_2 is equal to

1) $\frac{4}{3}\left(\frac{\lambda_2}{\lambda_1}\right)$ 2) $\frac{4}{3}\left(\frac{\lambda_1}{\lambda_2}\right)$ 3) $\frac{3}{4}\left(\frac{\lambda_2}{\lambda_1}\right)$ 4) $\frac{3}{4}\left(\frac{\lambda_1}{\lambda_2}\right)$

DIFFRACTION

16. First diffraction minima due to a single slit diffraction is at $\theta = 30^0$ for a light of wavelength $6000 A^0$. The width of slit is

1) $1\times10^{-6}cm$ 2) $1.2\times10^{-6}m$ 3) $2\times10^{-6}cm$ 4) $2.4\times10^{-6}m$

17. In a single slit diffraction, the width of slit is 0.5 cm, focal length of lens is 40cm and wavelength of light is $4890 A^0$. The distance of first dark fringe is

1) $2\times10^{-5}m$ 2) $4\times10^{-5}m$ 3) $6\times10^{-5}m$ 4) $8\times10^{-5}m$

18. Angular width of central maxima is $\pi/2$. When a slit of width 'a' is illuminated by a light of wavelength $7000 A^0$ then a =

1) $9\times10^{-9}m$ 2) $8.9\times10^{-7}m$ 3) $9\times10^{-7}m$ 4) $9.8\times10^{-7}m$

RESOLVING POWER

19. The sun subtends an angle of $(1/2)^0$ on earth. The image of sun is obtained on the screen with the help of a convex lens of focal length 100 cm the diameter of the image obtained on the screen will be

1) 18 cm 2) 1 mm 3) 50 cm 4) 8.73 mm

20. The limit of resolution of microscope, if the numerical aperture of microscope is 0.12, and the wavelength of light used is 600 nm, is

1) 0.3 μm 2) 1.2 μm 3) 2.5 μm 4) 3 μm

21. The least resolvable angle by a telescope using objective of aperture 5 m is nearly $(\lambda = 4000A^0)$

1) $\frac{1}{50^0}$ 2) $\frac{1}{50}$ *minute* 3) $\frac{1}{50}$ sec 4) $\frac{1}{500}$ sec

22. Wavelength of light used in an optical instrument are $\lambda_1 = 4000A^0$ and $\lambda_2 = 5000A^0$, then ratio of their respective resolving powers (corresponding to λ_1 and λ_2) is

1) 16 : 25 2) 9 : 1 3) 4 : 5 4) 5 : 4

POLARISATION

23. The angle of incidence at which reflected light is totally polarised for reflection from air to glass (refractive index *n*) is

1) $\sin^{-1}(n)$ 2) $\sin^{-1}(1/n)$ 3) $\tan^{-1}(1/n)$ 4) $\tan^{-1}(n)$

24. A light ray is incident on a transparent medium of μ = 1.732 at the polarising angle. The angle of refraction is

1) 60^0 2) 30^0 3) 45^0 4) 90^0

25. A ray of light in air is incident on a glass plate at polarising angle of incidence. It suffers a deviation of 22^0 on entering glass. The angle of polarization is

1) 90^0 2) 56^0 3) 68^0 4) Zero

26. The critical angle for total internal reflection for a substance is 45^0. The polarizing angle for this substance is $\left(\tan 54^0 44' = \sqrt{2}\right)$

1) $46^0 16^1$ 2) $54^0 44^1$ 3) $46^0 44^1$ 4) $54^0 16^1$

27. Unpolarized light of intensity I_0 is incident on a polarizer and the emerging light strikes a second polarizing filter with its axis at 45^0 to that of the first. The intensity of the emerging beam

1) $\frac{I_0}{2}$ 2) $\frac{I_0}{4}$ 3) I_0 4) $\frac{I_0}{3}$

28. The axes of the polariser and analyser are inclined to each other at 60^0. If the amplitude of polarised light emergent through analyser is A. The amplitude of unpolarised light incident on polariser is

1) $\frac{A}{2}$ 2) A 3) $2A$ 4) $2\sqrt{2}A$

29. Unpolarised light of intensity I is incident on a polarizer and the emerging light strikes a second polarizing filter with its axis at 45^0 to that of the first. Determine

a) the intensity of the emerging beam and

b) its state of polarization

1) $\frac{I}{4}$ and parallel to second filter 2) $\frac{I}{4}$ and perpendicular to second filter

3) $\frac{I}{8}$ and parallel to second filter 4) $\frac{I}{8}$ and perpendicular to second filter

HUYGEN'S PRINCIPLE

30. A parallel beam of width 'a' is incident on the surface of glass slab $(\mu = 3/2)$ at an angle 'i' and the angle of refraction in glass is 'r'. The width of the refracted parallel beam will be

1) equal to a 2) less than a 3) more than a 4) exactly 2a/3

31. When a parallel beam of monochromatic light suffers refraction while going from a rarer medium into a denser medium, which of the following are correct?

a) the width of the beam decreases b) the width of the beam increases

c) the refracted beam makes more angle with the interface

d) the refracted beam makes less angle with the interface

1) a, c true 2) b, d true 3) a, d true 4) b, c true

32. A parallel beam of light in incident on a liquid surface such that the wave front makes an angle 30^0 with the surface and has a width of $\sqrt{3}$ m, the width of the refracted beam is____ $\left({}_a\mu_L = \sqrt{3}\right)$

1) 3 m 2) $\sqrt{3}$ m 3) $\frac{\sqrt{11}}{3}$ m 4) $\sqrt{\frac{11}{3}}$ m

LEVEL - I (C. W) - KEY

1) 1	2) 3	3) 1	4) 1	5) 2	6) 1	7) 2	8) 4	9) 1	10) 2
11) 3	12) 4	13) 2	14) 2	15) 2	16) 2	17) 2	18) 4	19) 4	20) 4
21) 3	22) 4	23) 4	24) 2	25) 2	26) 2	27) 2	28) 4	29) 1	30) 3
31) 4	32) 4								

LEVEL - I (C. W) - HINTS

1. $A = \sqrt{a_1^2 + a_2^2 + 2a_1a_2\cos\theta}$

2 $\frac{I_{max}}{I_{min}} = \frac{\left(\sqrt{I_1} + \sqrt{I_2}\right)^2}{\left(\sqrt{I_1} - \sqrt{I_2}\right)^2}$

3. $I = I_1 + I_2 + 2\sqrt{I_1}\sqrt{I_2}\cos\theta$

4. $I = I_1 + I_2 + 2\sqrt{I_1}\sqrt{I_2}\cos\phi$ $\quad I = I_0 + I_0 + 2I_0\cos\phi = 2I_o(1+\cos\phi)$

5. $I = I_0\cos^2\left(\frac{\theta}{2}\right)$; $I = I_0\cos^2\left(\frac{\pi}{4}\right)$

6. $I = I_0\cos^2\left(\frac{\theta}{2}\right)$; $\theta = \frac{2\pi}{\lambda}x$

7. $n_1\lambda_1 = n_2\lambda_2$

8. $\beta = \frac{\beta}{\mu} = \frac{\lambda D}{\mu d}$ 9. $\beta = \frac{\lambda D}{d}$ 10. $\theta^1 = \frac{\theta}{\mu}$

11. $(\mu - 1)t = (2n-1)\lambda/2$ 12. $\beta = \frac{\lambda D}{d}$

13. $d\sin\theta = n\lambda$

14. $I = I_0\cos^2\left(\frac{\theta}{2}\right)$, $d\tan\theta = \Delta x$

15. Position of nth maxima from central maxima is given by $x_n = \frac{n\lambda D}{d}$

For 8th maxima $x_8 = \frac{8\lambda_1 D}{d_1}$

and for 6th maxima $x_6 = \frac{6\lambda_2 D}{d_2}$

Now, $x_6 = x_8$; $\Rightarrow \frac{d_1}{d_2} = \frac{n_1\lambda_1}{n_2\lambda_2} = \frac{4}{3}\left(\frac{\lambda_1}{\lambda_2}\right)$

16. $\sin\theta = \frac{x}{d} = \frac{\lambda}{b}$; $a\sin\theta = n\lambda$, $a = \frac{n\lambda}{\sin\theta}$

17. $a\sin\theta = n\lambda$ $a.\frac{Y}{D} = n\lambda$; But = D = f

$a.\frac{Y}{D} = n\lambda \quad Y = \frac{n\lambda f}{a}$

18. $a\sin\theta = n\lambda$; $a = \frac{\lambda}{\sin\theta} = w = 2\theta \quad \theta = \frac{w}{2}$

19. $D = f \times \theta$

20. $X = \frac{1.22\lambda}{2\mu\sin\theta}$ where $\mu\sin\theta$ is numarical aperture

21. $\theta = \frac{1.22\lambda}{a}$

22. $R \propto \frac{1}{\lambda}$

23. $Tan i_p = \mu$; $i_p + r = 90^0$

24. $\mu = Tan I_p \quad I_p + r = 90$

25. $i + r = 90^0$ $r = i - d$

26. $\mu = \frac{1}{\sin c} \quad \mu = Tan i$

27. $I = \frac{I_0}{2}, I_2 = \frac{I_0}{2}\cos^2\theta$

28. $A_1 = A_0\cos 45 = \frac{A_0}{\sqrt{2}}$; $A_2 = \frac{A}{\sqrt{2}}\cos\theta$

29. $I_2 = \frac{I_0}{2}\cos^2\theta$

30. $\frac{W_i}{W_r} = \frac{\cos i}{\cos r}$

31. $\frac{W_i}{W_r} = \frac{\cos i}{\cos r}, i > r$

32. $\mu_1\sin i = \mu_2\sin r$, $\frac{W_i}{W_r} = \frac{\cos i}{\cos r}$

LEVEL - I (HW)

INTERFERENCE

1. The displacements of two interfering light waves are $y_1 = 2\sin\omega t$ and $y_2 = 5\sin\left(\omega t + \frac{\pi}{3}\right)$ the resultant amptitude is

1) 39 cm 2) $\sqrt{39}\,cm$ 3) 7 cm 4) $\sqrt{29}\,cm$

2. The intensity ratio of two waves is 9 : 1. If they produce interference, the ratio of maximum to minimum intensity will be

1) 4 : 1 2) 2 : 1 3) 9 : 1 4) 3 : 2

3. Two beams of light having intensities I and 4I interfere to produce a fringe pattern on a screen. The phase difference between the beams is $\pi/2$ at point A and π at point B. then the difference between the resultant intensities at A and B is

1) $2I$ 2) $4I$ 3) $5I$ 4) $7I$

4. The maximum intensity in Young's double slit experiment is I_0. What will be the intensity of light in front of one of the slits on a screen where path difference is $\frac{\lambda}{4}$?

1) $\frac{I_0}{2}$ 2) $\frac{3}{4}I_0$ 3) I_0 4) $\frac{I_0}{4}$

5. In Young's double slit experiment, we get 60 fringes in the field of view of monochromatic light of wavelength $4000\,A^o$. If we use monochromatic light of wavelength $6000\,A^o$, then the number of fringes obtained in the same field of view are

1) 60 2) 90 3) 40 4) 1.5

6. The separation between successive fringes in a double slit arrangement is x. If the whole arrangement is dipped under water, what will be the new fringe separation? [The wavelength of light being used is 5000 Å]

1) 1.5 x 2) x 3) 0.75 x 4) 2x

7. In the Young's double slit experiment, a mica slip of thickness t and refractive index μ is introduced in the ray from first source S_1. By how much distance fringes pattern will be displaced? (d=distance between the slits and D is the distance between slits and screen)

1) $\frac{d}{D}(\mu-1)t$ 2) $\frac{D}{d}(\mu-1)t$ 3) $\frac{d}{(\mu-1)D}$ 4) $\frac{D}{d}(\mu-1)$

8. In young's double slit experiment, the 10th maximum of wave length λ_1 is at a distance of y_1 from the central maximum. When the wavelength of the source is changed to λ_2, 5th maximum is at a distance of y_2 from its central maximum. Then $\frac{y_1}{y_2}$ is

1) $\frac{2\lambda_1}{\lambda_2}$ 2) $\frac{2\lambda_2}{\lambda_1}$ 3) $\frac{\lambda_1}{2\lambda_2}$ 4) $\frac{\lambda_2}{2\lambda_1}$

9. Two coherent monochormatic light sources are located at two vertices of an equilateral triangle. If the intensity due to each of the source independently is $1\,Wm^{-2}$ at the third vertex. The resultant intensity due to both the sources at that point (i.e at the third vertex) is (in Wm^{-2})

1) zero 2) $\sqrt{2}$ 3) 2 4) 4

DIFFRACTION

10. Light of wavelength $6000\,A^o$ is incident on a single slit. The first minimum of the diffraction pattern is obtained at 4 mm from the centre. The screen is at a distance of 2 m from the slit. The slit width will be

1) 0.3 mm 2) 0.2 mm 3) 0.15 mm 4) 0.1 mm

11. A plane wave of wavelength 6250Å is incident normally on a slit of width $2\times10^{-2}cm$. The width of the principal maximum on a screen distant 50 cm will be

1) $312.5\times10^{-2}cm$ 2) $312.5\times10^{-4}cm$ 3) $312cm$ 4) $312.5\times10^{-5}cm$

12. The distance between the first and the sixth minima in the diffraction pattern of a single slit is 0.5 mm. The screen is 0.5 m away from the slit. If the wavelength of light used is 5000 Å. Then the slit width will be

1) 5 mm 2) 2.5 mm 3) 1.25 mm 4) 1.0 mm

RESOLVING POWER

13. The diameter of an objective of a telescope, which can just resolve two stars situated at angular displacement of 10^{-4} degee, should be ($\lambda = 5000\ A^0$)

1) 35 mm 2) 35 cm 3) 35 m 4) 3.5 cm

14. A telescope is used to resolve two stars separated by 4.6×10^{-6} rad. If the wavelength of light used is 5460 A^0, what should be the aperture of the objective of the telescope?

1) 0.448 m 2) 0.1448 m 3) 1.1448 m 4) 0.011 m

15. Two point sources distant 0.1 meter away viewed by a telescope. The objective is covered by a screen having a hole of 1 mm width. If the wavelength of the light used is 6500 A^0, then the maximum distance at which the two sources are seen just resolved, will be nearly

1) 125.0 m 2) 164 m 3) 131 m 4) 144 m

POLARISATION

16. Two polaroids are kept crossed to each other. Now one of them is rotated through an angle of 45^0. The percentage of incident light now transmitted through the system is

1) 15% 2) 25% 3) 50% 4) 60%

17. The amplitude of polarised light transmitted through a polariser is A. The amplitude of unpolarised light incident on it is

1) $A/2$ 2) $A/\sqrt{2}$ 3) 2A 4) $\sqrt{2}A$

18. Unpolarised lgiht of intensity 32 W/m² passes through a polariser and analyser which are at an angle of 30^0 with respect to each other. The intensity of the light coming from analyser is

1) $16\sqrt{3}\,W/m^2$ 2) $12\ W/m^2$ 3) $16W/m^2$ 4) $14W/m^2$

19. The critical angle of a transparent crystal is 60^0. Then its polarizing angle is

1) $\theta = \tan^{-1}\left(\frac{2}{\sqrt{3}}\right)$ 2) $\theta = \sin^{-1}(\sqrt{2})$ 3) $\theta = \cos^{-1}\left(\frac{1}{\sqrt{2}}\right)$ 4) $\theta = \cot^{-1}(\sqrt{2})$

20. When an unpolarised light of intensity I_0 is incident on a polarising sheet, the intensity of the light which does not get transmitted is

1) $1/2I_0$ 2) $1/4I_0$ 3) zero 4) I_0

LEVEL - I (H. W)-KEY

1) 2	2) 1	3) 2	4) 1	5) 3	6) 3	7) 2	8) 1	9) 4	10) 1
11) 1	12) 2	13) 2	14) 2	15) 1	16) 2	17) 4	18) 2	19) 1	20) 1

LEVEL - I (H. W)-HINTS

1. $A = \sqrt{a_1^2 + a_2^2 + 2a_1a_2\cos\theta}$

2 $\frac{I_{max}}{I_{min}} = \frac{\left(\sqrt{I_1}+\sqrt{I_2}\right)^2}{\left(\sqrt{I_1}-\sqrt{I_2}\right)^2}$

3. Here, $I_1 = I, I_2 = 4I, \theta_1 = \pi/2, \theta_2 = \pi$

Resultant intensity $I_{\theta_1} = I_1 + I_2 + 2\sqrt{I_1 I_2}\cos\theta_1$

$= I + 4I + 2\sqrt{I \times 4I}\cos\pi/2 = 5I$

Resultant intensity $I_{\theta_2} = I_1 + I_2 + 2\sqrt{I_1 I_2}\cos\theta_2$

$= I + 4I + 2\sqrt{1 \times 4I}\cos\pi$

$= 5I - 4I = I \quad \therefore \; I_{\theta_1} - I_{\theta_2} = 5I - I = 4I$

4. Corresponding phase difference will be

$$\phi = \left(\frac{2\pi}{\lambda}\right)(\Delta x) = \left(\frac{2\pi}{\lambda}\right)\left(\frac{\lambda}{4}\right) = \frac{\pi}{2}$$

or $\frac{\phi}{2} = \frac{\pi}{4}$; $\therefore \; I = I_0 \cos^2\left(\frac{\phi}{2}\right)$

$$= I_0 \cos^2\left(\frac{\pi}{4}\right) = \frac{I_0}{2}$$

5. As, $x = n_1\beta_1 = n_2\beta_2 = n_1\lambda_1 = n_2\lambda_2$

$$\therefore \; n_2 = \frac{n_1\lambda_1}{\lambda_2} = \frac{60 \times 4000}{6000} = 40$$

6. When the arrangement is dipped in water,

$$\beta' = \frac{\beta}{\mu} = \frac{x}{4/3} = \frac{3}{4}x = 0.75x$$

7. For a path difference $(\mu - 1)t$, the shift is

$$x = (\mu - 1)t\frac{D}{d}$$

8. $y_2 = 5\left(\frac{\lambda_2 D}{d}\right)$; $y_1 = 10\left(\frac{\lambda_1 D}{d}\right)$

9. at the third vertex path diff $= 0$, hence intensity is max.

10. From, $a\sin\theta = n\lambda$,

$$a\frac{x}{D} = n\lambda \text{ or } a = \frac{n\lambda D}{x} = \frac{1 \times 6000 \times 10^{-10} \times 2}{4 \times 10^{-3}}$$

$$\therefore \quad a = 3 \times 10^{-4} = 0.3\,mm$$

11. Here, $\lambda = 6250$ Å $= 6520 \times 10^{-10}$ m

$a = 2 \times 10^{-2}\,cm = 2 \times 10^{-4}\,m$

$D = 50\,cm = 0.5\,m$

$\therefore$ Width of central maxima $= \frac{2\lambda D}{a}$

$$= \frac{2 \times 6250 \times 10^{-10} \times 0.5}{2 \times 10^{-4}} = 312.5 \times 10^{-3}\,cm$$

12. Distance between first and sixth minima $x = \frac{5\lambda D}{a}$

13. $\theta = \frac{1.22\lambda}{a}$

14. The aperture (a) of the telescope is given as $a = \frac{1.22\lambda}{\theta}$

15. $\theta = \frac{1.22\lambda}{a}$

16. $I = \left(\frac{I_0}{2}\right)(\cos 45^0)^2 = \frac{I^0}{2}\left(\frac{I}{\sqrt{2}}\right)^2 = \frac{I_0}{4}$

$\Rightarrow \quad I = 25\% \; of \; I_0$

17. $\frac{A_0}{\sqrt{2}} = A$ 18. $I = \frac{I_0}{2}\cos^2\theta$

19. $\mu = \frac{1}{\sin c}$ and $\mu = \tan\theta$ 20. $I = I_0 \cos^2\theta$

Intensity of polarised light $= \frac{I_0}{2}$

$\therefore$ Intensity of untransmitted light $= I_0 - \frac{I_0}{2} = \frac{I_0}{2}$

LEVEL - II (CW)

INTERFERENCE

1. **In Young's double slit experiment the intensity of light at a point on the screen where the path difference is λ is K. The intensity of light at a point where the path difference is $\frac{\lambda}{3}$ [λ is the wavelength of light used] is**

1) K/4 2) K/3 3) K/2 4) K

2. **In Young's double slit experiment, Let β be the fringe width and I_0 be the intensity at the central bright fringe. At a distance 'x' from the central bright fringe, the intensity will be**

1) $I_0 \cos\left(\frac{x}{\beta}\right)$ 2) $I_0 \cos^2\left(\frac{x}{\beta}\right)$ 3) $I_0 \cos^2\left(\frac{\pi x}{\beta}\right)$ 4) $\frac{I_0}{4}\cos^2\left(\frac{\pi x}{\beta}\right)$

3. **In Young's double slit experiment the fringe pattern is observed on a screen placed at a distance D. The slits are illuminated by light of wavelength λ. The distance from the central point where the intensity falls to half the maximum is**

1) $\frac{\lambda D}{3d}$ 2) $\frac{\lambda D}{2d}$ 3) $\frac{\lambda D}{d}$ 4) $\frac{\lambda D}{4d}$

4. **In a double slit experiment, the slit separation is 0.20 cm and the slit to screen distance is 100 cm. The positions of the first three minima, if wavelength of the source is 500 nm is**

1) $\pm 0.125cm, \pm 0.375cm, \pm 0.625cm$ 2) $\pm 0.025cm, \pm 0.075cm, \pm 0.125cm$

3) $\pm 12.5cm, \pm 37.5cm, \pm 62.5cm$ 4) $\pm 1.25cm, \pm 3.75cm, \pm 6.25cm$

5. **In a Young's double slit experiment, the fringes are displaced by a distance x when a glass plate of refractive index 1.5 is introduced in the path of one of the beams. Then this plate is replaced by another plate of the same thickness, the shift of fringes is $3/2\,x$. The refractive index of the second plate is**

1) 2.25 2) 2.0 3) 1.75 4) 1.25

6. **A double slit experiment is performed with light of wavelength 500 nm. A thin film of thickness $2\mu m$ and refractive index 1.5 is introduced in the path of the upper beam. The location of the central maximum will**

1) remain unshifted 2) shift downward by nearly two fringes

3) shift upward by nearly two fringes 4) shift downward by 10 fringes

7. In Young's double slit experiment, an interference pattern is obtained on a screen by a light of wavelength $6000A^0$ coming from the coherent sources S_1 *and* S_2. At certain point p on the screen third dark fringe is formed. Then the path difference $S_1p - S_2p$ in micron is

1) 0.75 2) 1.5 3) 3.0 4) 4.5

8. In double slit experiment fringes are obtained using light of wavelength 4800 A^0 One slit is covered with a thin glass film of refractive index. 1.4 and another slit is covered by a film of same thickness but refractive index 1.7. By doing so, the central fringe is shifted to fifth bright fringe in the original pattern. The thickness of glass film is

1) 2×10^{-3} mm 2) 4×10^{-3} mm 3) 6×10^{-3} mm 4) 8×10^{-3} mm

9. In Young's double slit experiment, 5th dark fringe is obtained at a point. If a thin transparent film is placed in the path of one of waves, then 7th bright fringe is obtained at the same point. The thickness of the film in terms of wavelength λ and refractive index μ will be

1) $\frac{1.5\lambda}{(\mu-1)}$ 2) $1.5(\mu-1)\lambda$ 3) $2.5(\mu-1)\lambda$ 4) $\frac{2.5\lambda}{(\mu-1)}$

10. The Young's double slit experiment is performed with blue light and green light of wavelengths 4360 A^0 and 5460 A^0 respectively. If y is the distance of 4th maxima from the central one, then

1) $y_b = y_g$ 2) $y_b > y_g$ 3) $y_b < y_g$ 4) $\frac{y_b}{y_g} = \frac{5460}{4360}$

11. In double slit experiment, the distance between two slits is 0.6 mm and these are illuminated with light of wavelength $4800\,A^0$. The angular width of dark fringe on the screen at a distance 120 cm from slits will be

1) 8×10^{-4} *radian* 2) 6×10^{-4} *radian* 3) 4×10^{-4} *radian* 4) 16×10^{-4} *radian*

12. Fig shows a double slit experiment, P and Q are the two coherent sources. The path lengths PY and QY are $n\lambda$ and $(n+4)\lambda$ respectively where n is whole number and λ is wavelength. Taking the central bright fringe as zero, what is formed at Y?

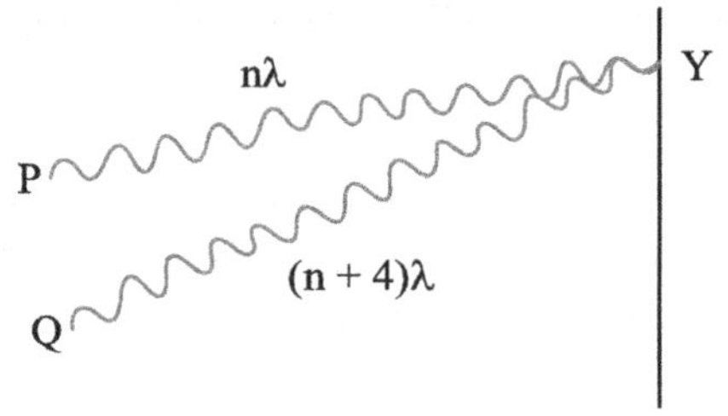

1) First Bright 2) First Dark 3) Fourth Bright 4) Second Dark

13. White light is used to illuminate two slits in Young's double slit experiment. Separation between the slits is b and the screen is at a distance d (>> b) from the slits. Then wavelengths missing at a point on the screen directly infront of one of the slits are

1) $\frac{b^2}{d}, \frac{b^2}{3d}$ 2) $\frac{b^2}{d}, \frac{b^2}{4d}$ 3) $\frac{b^2}{2d}, \frac{b^2}{3d}$ 4) $\frac{b^2}{2d}, \frac{b^2}{4d}$

DIFFRACTION

14. The I[st] diffraction mininum due to single slit diffraction is θ, for a light of wave length $5000A^0$. If the width of the slit is $1\times10^{-4}cm$ then the value of θ

1) 30^0 2) 45^0 3) 60^0 4) 15^0

15. Light of wavelength $5000\times10^{-10}m$ is incident normally on a slit. The first minimum of the diffraction pattern is observed to lie at a distance of 5mm from the central maximum on a screen placed at a distance of 3m from the slit. Then the width of the slit is

1) 3 cm 2) 0.3 cm 3) 0.03 cm 4) 0.01 cm

16. **A small aperture is illuminated with a parallel beam of $\lambda = 628nm$. The emergent beam has an angular divergence of 2^0. The size of the aperture is**

1) $9\mu m$ 2) $18\mu m$ 3) $27\mu m$ 4) $36\mu m$

17. **A beam of light of wavelength 600 nm from a distant source falls ona single slit 1.00 mm wide and the resulting diffraction pattern is observed on a screen 2m away. Then distance between the first dark fringes on either side of the central fringe is**

1) 1.2 mm 2) 2.4 mm 3) 3.6 mm 4) 2.4 cm

RESOLVING POWER

18. **A person wants to see two pillars distant 11 km, separately. The distance between the pillars must be approximately**

1) 3m 2) 1m 3) 0.25 m 4) 0.5 m

19. **Two point white dots are 1 mm apart on a black paper. They are viewed by eye of pupil diameter 3mm. Approximately, what is the maximum distance at which these dots can be resolved by the eye? [Take wavelength of light = 500 nm]**

1) 6m 2) 3 m 3) 5 m 4) 1 m

POLARISATION

20. **A horizontal beam of vertically polarized light of intensity 43 W/m^2 is sent through two polarizing sheets. The polarizing direction of the first is 60° to the vertical, and that of the second is horizontal. The intensity of the light transmitted by the pair of sheets is (nearly)**

1) 8.1 W/m^2 2) 7.3 W/m^2 3) 6.4 W/m^2 4) 3.8 W/m^2

21. **Unpolarised light of intensity $32Wm^{-2}$ passes through three polarisers such that the transmission axis of the last polariser is crossed with first. If the intensity of the emerging light is $3Wm^{-2}$, the angle between the axes of the first two polarisers is**

1) 45^0 2) 60^0 3) 30^0 4) zero

22. **Two polaroids are oriented with their transmision axes making angle of 30^0 with each other. The fraction of incident un polarised light is transmitted**

1) 37% 2) 37.5% 3) 3.36% 4) 36.5%

23. **The polaroids P_1, P_2 & P_3 are arranged coaxially. the angle between P_1 and P_2 is 37^0. The angle between P_2 and P_3 is, if intensity of emerging light is one quarterth of intensity of unpolarized light**

1) $\theta = \cos^{-1}\left(\frac{5}{4}\right)$ 2) $\theta = \cos^{-1}\left(\frac{4}{5}\right)$ 3) $\theta = \cos^{-1}\left(\frac{4}{5\sqrt{2}}\right)$ 4) $\theta = \cos^{-1}\left(\frac{5}{4\sqrt{2}}\right)$

24. **A ray of light is going from air to glass such that the reflected light is found to be completely plane polarized. Also the angle of refraction inside the glass is found exactly equal to the angle of deviation suffered by the ray. The refractive index of the glass is**

1) 1.5 2) $\sqrt{2}$ 3) $\sqrt{3}$ 4) 4/3

25. **A plane polarized beam of intensity I is incident on a polariser with the electric vector inclined at 30° to the optic axis of the polariser passes through an analyzer whose optic axis is inclined at 30° to that of polariser. Intensity of light coming out of the analyzer is**

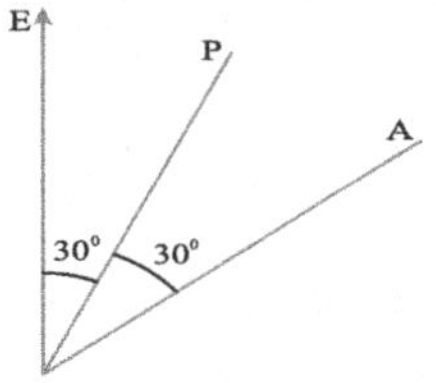

1) (9/16)I 2) (3/4)I 3) (1/4)I 4) $\left(\sqrt{3}/2\right)I$

LEVEL - II (C. W) - KEY

1) 1	2) 3	3) 4	4) 1	5) 3	6) 3	7) 2	8) 4	9) 4	10) 3
11) 1	12) 3	13) 1	14) 1	15) 3	16) 4	17) 2	18) 1	19) 3	20) 1
21) 3	22) 2	23) 4	24) 3	25) 1					

LEVEL - II (C. W) - HINTS

1. $K = I_0\cos^2(\pi) \rightarrow (1)$

   $I = I_0\cos^2\left(\frac{\pi x}{3}\right) \rightarrow (2) \quad \theta = \frac{2\pi x}{3}$

2. $I = I_0\cos^2\frac{\theta}{2}; \qquad \theta = \frac{2\pi}{\lambda}x$

3. $I = I_0\cos^2\frac{\theta}{2}\ ; \quad y = \frac{(n\lambda)D}{d}$

   $y = x.\frac{D}{d} = \frac{\lambda}{4}\frac{D}{d}$

4. $y = \frac{n\lambda D}{d}$; n=1,2,3,4 ...

5. $x = (\mu - 1)t$

6. $(\mu - 1)t = n\lambda$

7. $x = (2n-1)\frac{\lambda}{2}$

8. $n\lambda = (\mu_2 - \mu_1)t\ ;\ 5\beta = (\mu_2 - \mu_1)t\frac{\beta}{\lambda}$

9. $x = (2n-1)\frac{\lambda D}{2d}$

10. $n_1\lambda_1 = n_2\lambda_2$

11. $\theta = \frac{\lambda}{d}$

12. $\Delta x = n\lambda$

13. $x = (2n-1)\frac{\lambda D}{2d}$

14. $a\sin\theta = n\lambda$

15. $a\sin\theta = n\lambda;\ a\frac{Y}{D} = n\lambda \quad a = \frac{n\lambda D}{Y}$

16. $a\sin\theta = n\lambda;\ \theta = \frac{\lambda}{a}\frac{n\pi}{180}rad$

17. $Y = \frac{\lambda D}{a}$; w = 2Y

18. Resolving power of eye $= \left(\frac{1}{60}\right)^0$

    $\frac{1}{60}\times\frac{\pi}{180} = \frac{d}{11000}$

19. $\sin\theta \simeq \theta = \frac{1.22\lambda}{d} = \frac{y}{D}$

20. $I_2 = \frac{I_0}{2}\cos^2\theta$

21. $I_1 = \frac{I_0}{2};\ I_2 = I_1\cos^2\theta;\ \ I_3 = \frac{I_0}{2}\cos^2\theta\sin^2\theta$

22. $I_1 = \frac{I_0}{2};\ I_2 = I_1\cos^2\theta$; Fraction $\frac{I_2}{I} = \frac{3}{8}$

23. $I^1 = \frac{I}{2}.\cos^2\theta_1\cos^2\theta_2$

24. $i = 90 - r, r = d, d = i - r, \mu = \tan i$

25. Use malus law

LEVEL - II (HW)

INTERFERENCE

1. **In Young's double slit experiment the intensity of light at a point on the screen where the path difference λ is K. The intensity of light at a point where the path difference is $\frac{\lambda}{6}$ [λ is the wavelength of light used] is**

   1) K/4 2) K/3 3) 3K/4 4) K

2. **In a Young's double slit experiment, D equals the distance of screen and d is the separation between the slits. The distance of the nearest point to the central maximum where the intensity is same as that due to a single slit is equla to**

   1) $\frac{D\lambda}{d}$ 2) $\frac{D\lambda}{2d}$ 3) $\frac{D\lambda}{3d}$ 4) $\frac{2D\lambda}{d}$

3. **With two slits spaced 0.2 mm apart and a screen at a distance of 1 m, the third bright fringe is found to be at 7.5 mm from the central fringe. The wavelength of light used is**

   1) 400 nm 2) 500 nm 3) 550 nm 4) 600 nm

4. **The central fringe of the interference pattern produced by the light of wavelength 6000 A is found to shift to the position of 4th dark fringe after a glass sheet of refractive index 1.5 is introduced. The thickness of glass sheet would be**

   1) 4.8 μm 2) 4.2 μm 3) 5.5 μm 4) 3.0 μm

5. **In Young's double slit inteference experiment the wavelength of light used is 6000 Å. If the path difference between waves reaching a point P on the screen is 1.5 microns, then at that point P**

   1) Second bright band occurs 2) Second dark band occur
   3) Third dark band occur 4) Third bright band occur

6. **When a mica plate of thickness 0.1 mm is introduced in one of the interfering beams, the central fringe is displaced by a distance equal to 10 fringes. If the wavelength of the light is 6000 Å, the refractive index of the mica is**

   1) 1.06 2) 1.6 3) 2.4 4) 1.3

7. **In Young's experiment inteference bands are produced on the screen placed at 1.5 m from the two slits 0.15 mm apart and illuminated by light of wavelength 6000 Å. If the screen is now taken away from the slit by 50 cm the change in the fringe width will be**

   1) $2\times10^{-4}m$ 2) $2\times10^{-3}m$ 3) $6\times10^{-3}m$ 4) $7\times10^{-3}m$

8. **When a thin transparent plate of Refractive Index 1.5 is introduced in one of the interfearing beams produces 20 fringes shift. If it is replaced by another thin plate of half that thickness and of refractive index 1.7, the number of fringes that undergo displacement is**

   1) 23 2) 14 3) 28 4) 7

9. **In young's double slit experiment one of the slits is wider than other, so that amplitude of the light from one slit is double of that from other slit. If I_m be the maximum intensity, the resultant intensity I, when they interfere at phase difference ϕ is given by**

   1) $\frac{I_m}{9}(4+5\cos\phi)$ 2) $\frac{I_m}{3}\left(1+2\cos^2\frac{\phi}{2}\right)$
   3) $\frac{I_m}{5}\left(1+4\cos^2\frac{\phi}{2}\right)$ 4) $\frac{I_m}{9}\left(1+8\cos^2\frac{\phi}{2}\right)$

10. In young's doulbe slit experiment, the two slits act as coherent sources of waves of equal amplitude A and wavelength λ. In another experiemnt with the same arrangement the two slits are made to act as incoherent sources of waves of same amplitude and wavelength. If the intensity at the middle point of the screen in the first case is I_1 and in the second case I_2, then the ratio $\frac{I_1}{I_2}$ is

1) 4 2) 2 3) 1 4) 0.5

11. A mixture of light, consisting of wavelength 590 nm and an uknown wavelength, illuminates Young's double slit and gives rise to two overlapping interference patterns on the screen. The central maximum of both light coincide. Further, it is observed that the third bright fringe of known light coincides with the 4th bright fringe of the unknown light. The wavelength of the unknown light is

1)393.4 nm 2)885.0 nm 3)442.5 nm 4)776.8 nm

12. In Young's experiment using monochromatic light, the fringe pattern shifts by a certain distance on the screen when a mica sheet of refractive index 1.6 and thickness 2 micron is introduced in the path of one of the interfering waves. The mica sheet is then removed and the distance between the slits and the screen is doubled. It is found that the distance between successive maxima now is the same as the observed fringe shift upon the introduction of the mica sheet. The wavelength of light is

1) 5762 Å 2) 5825 Å 3) 6000 Å 4) 6500 Å

DIFFRACTION

13. Plane microwaves are incident on a long slit having a width of 5.0 cm. The wavelength of microwaves if the first diffraction minimum is formed at $\theta = 30^0$ is

1) 2.5 cm 2) 5 cm 3) 7.5 cm 4) 10 cm

14. A screen is placed 50 cm from a single slit, which is illuminated with 6000 A^0 light. if distance between the first and third minima in the diffraction pattern is 3.0 mm, then the width of the slit is

1) 0.1 mm 2) 0.2 mm 3) 0.4 mm 4) 0.8 mm

15. A slit of width 'd' is placed infront of a lens of focal length 0.5 m and is illuminated normally with light of wavelength 5.89×10^{-7} m. The first diffraction minima on either side of the central diffraction maximum are separated by $2\times10^{-3}m$. The width of the slit is

1) $1.47\times10^{-4}m$ 2) $2.94\times10^{-4}m$ 3) $1.47\times10^{-7}m$ 4) $2.92\times10^{-7}m$

POLARISATION

16. Un polarised light passes through a polariser and analyser which are at an angle of 45^0 with respect to each other. The intensity of polarised light coming from analyser is 5W / m^2. The intensity of unpolarised light incident on polariser is

1) $5\sqrt{3}\,W/m^2$ 2) $10\,W/m^2$ 3) $20\,W/m^2$ 4) $5\frac{\sqrt{3}}{4}W/m^2$

17. A beam of ordinary light is incident on a system of four polaroids which are arranged in succession such that each polaroid is turned through 30^0 with respect to the preceding one. The percentage of the incident intensity that emerges out from the system is approximately

1) 56% 2) 6.25% 3) 21% 4) 14%

18. Two polaroid sheets are placed one over the other with their axes inclined to each other at an angle θ. If only 12.5% of the intensity of the light incident on the first sheet emerges out from the second sheet, the value of θ is

1) 30^0 2) 60^0 3) 45^0 4) 90^0

19. An unpolarised light is incident on a plate of refractive index $\sqrt{3}$ and the reflected light is found to be completely plane polarised. The angles of incidence and refraction are respectively

1) $60^0, 30^0$ 2) $30^0, 60^0$ 3) $Sin^{-1}\left(\frac{1}{\sqrt{3}}\right), 45^0$ 4) $Tan^{-1}\left(\frac{\sqrt{3}}{2}\right), 30^0$

LEVEL - II (H. W) -KEY

1) 3	2) 3	3) 2	4) 2	5) 3	6) 1	7) 2	8) 2	9) 4	10) 2
11) 3	12) 3	13) 1	14) 2	15) 2	16) 3	17) 3	18) 2	19) 1	

LEVEL - II (H. W)- HINTS

1. $K = I_0 \cos^2(\pi) \rightarrow (1)$

$I = I_0 \cos^2\left(\frac{\pi}{6}\right) \rightarrow (2)$

2. $I = 4I_0 \cos^2 \frac{\phi}{2} \Rightarrow I_0 = 4I_0 \cos^2 \frac{\phi}{2}$

$\phi = \frac{2\pi}{3}$, $\phi = \frac{2\pi}{\lambda}\Delta x$; But $\Delta x = \frac{yd}{D}$

$\Rightarrow \frac{yd}{D} = \frac{\lambda}{3}$ $\Rightarrow y = \frac{\lambda D}{3d}$

3. $y = \frac{n\lambda D}{d}$

4. $(\mu - 1)t = (2n-1)\frac{\lambda}{2}$

5. Path difference $= (2n-1)\frac{\lambda}{2}$

6. Shift $= \frac{\beta}{\lambda}(\mu - 1)t$

7. $\beta = \frac{\lambda D}{d}$

8. Shift $= \frac{\beta}{\lambda}(\mu - 1)t$

9. $I = I_1 + I_2 + 2\sqrt{I_1 I_2}\cos\phi$ $I_1 = 4I_2$, $I_{max} = 9I_2$

10. For coherent sources

$I_1 = 4I_0 \cos^2 \phi / 2 = 4I_0$ For incoherent sources

$I_2 = I_0 + I_0 = 2I_0$; $\therefore$ $\frac{I_1}{I_2} = 2$

11. $3\lambda_1 = 4\lambda_2$

$\Rightarrow \lambda_2 = \frac{3}{4}\lambda_1 = \frac{3}{4} \times 590$; $= \frac{1770}{4} = 442.5$ nm

12. $y = \frac{D}{d}(\mu - 1)t$ .... (1)

When the distance between the plane of slits and screen is changed from D to 2D, then

$\beta = \frac{2D}{d}\lambda$; $\frac{D}{d}(\mu - 1)t = \frac{2D(\lambda)}{d}$ ... (2)

$\Rightarrow \lambda = \frac{1}{2}(\mu - 1)t$

13. $a\sin\theta = n\lambda$

14. Position of first minima on a single slit diffraction pattern is given by $d\sin\theta = n\lambda$

For small value of θ, $\sin\theta \approx \theta = \dfrac{y}{D}$

$\therefore \dfrac{y.a}{D} = n\lambda$ or $y = \dfrac{n\lambda D}{a}$

$\therefore$ Distance between third order minima and first order minima will be

$$\Delta y = y_3 - y_1 = \frac{(3-1)(\lambda D)}{a} = \frac{2\lambda D}{a}$$

15. $w = 2y = \dfrac{2\lambda D}{a} = \dfrac{2f\lambda}{a}$

16. $I = \dfrac{I_0}{2}\cos^2\phi$

17. $I = \dfrac{I_o}{2}\left(\cos^2\theta\right)^{n-1}$ where n= number of polaroids

18. $I = \dfrac{I_0}{2}\cos^2\phi$

19. $\mu = \tan i_p$; $r = 90^0 - i_p$

LEVEL - III

INTERFERENCE OF LIGHT

1. A ray of light of intensity I is incident on a parallel glass slab at a point A as shown. It undergoes partial reflection and refraction. At each reflection 25% of incident energy is reflected. The rays AB and $A'B'$ undergo interference. The ratio I_{max} / I_{min} is

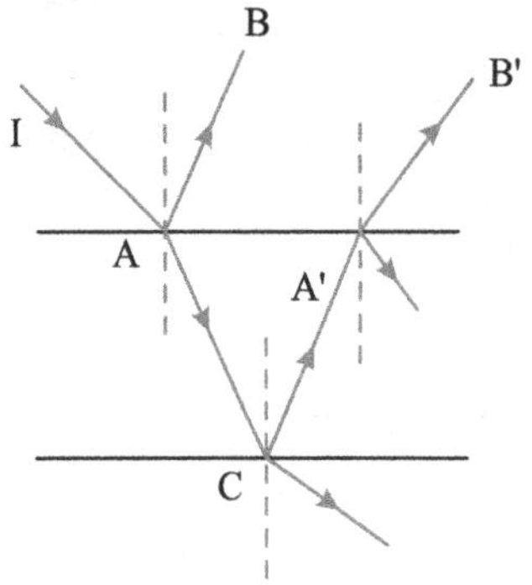

1) 4 : 1 2) 8 : 1 3) 7 :1 4) 49 : 1

2. Two coherent sources of intensity ratio β interfere, then $\dfrac{I_{max} - I_{min}}{I_{max} + I_{min}}$ is

1) $\dfrac{\beta}{1+\beta}$ 2) $\dfrac{2\sqrt{\beta}}{1+\beta}$ 3) $\dfrac{2\sqrt{\beta}}{1+\sqrt{\beta}}$ 4) $\dfrac{2\beta}{1+\sqrt{\beta}}$

3. Monochromatic green light of wavelength 550 nm illuminates two parallel narrow slits $7.7\,\mu m$ apart. The angular deviation θ of third order (for m = 3) bright fringe a) in radian and b) in degree

1) 21.6, 12.4^0 2) 0.216, 1.24^0 3) 0.216, 12.4^0 4) 216, 1.24^0

4. A source emitting wavelength 480 nm and 600 nm is used in YDSE. The separation between the slits is 0.25 mm. the interference is observed 1.5 m away from the slits. The linear separation between first maxima of two wavelengths is

1) 0.72 mm 2) 0.72 cm 3) 7.2 cm 4) 7.2 mm

5. **In the Young's double slit experiment, maximum number of bright bands observed (inclusive of the central bright band) is found to be 11. If λ is the wavelength of the monochromatic light used, the distance between the slits is**

1) 5λ 2) 6λ 3) 10λ 4) 11λ

6. **In a double slit experiment, interference is obtained from electron waves produced in an electron gun supplied with voltage V. If λ is wavelength of the beam, D is the distance of screen, d is the spacing between coherent source, h is Planck's constant, e is charge on electron and m is mass of electron, then fringe width is given as**

1) $\frac{hD}{\sqrt{2meV}d}$ 2) $\frac{2hD}{\sqrt{meV}d}$ 3) $\frac{hd}{\sqrt{2meV}D}$ 4) $\frac{2hd}{\sqrt{meV}D}$

7. **Two identical narrow slits S_1 and S_2 are illuminated by light of wavelength λ from a point source P. If, as shown in the diagram above the light is then allowed to fall on a screen, and if n is a positive integer, the condition for destructive interference at Q is that**

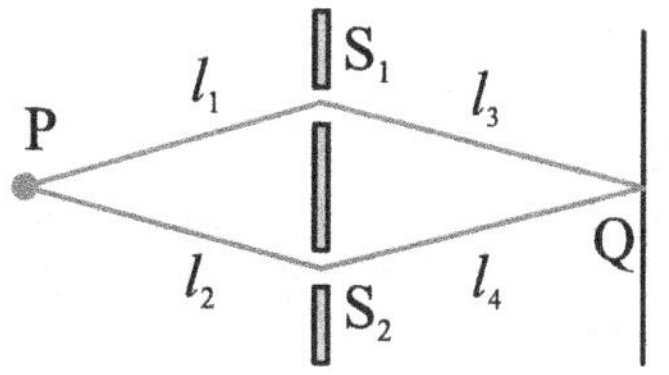

1) $(l_1 - l_2) = (2n+1)\lambda/2$ 2) $(l_3 - l_4) = (2n+1)\lambda/2$

3) $(l_3 + l_3) - (l_2 + l_4) = n\lambda$ 4) $(l_1 + l_3) - (l_2 + l_4) = (2n+1)\lambda/2$

8. **Fig., here shows P and Q as two equally intense coherent sources emitting radiations of wavelength 20m. The separation PQ is 5m, and phase of P is ahead of the phase of Q by 90°. A, B and C are three distant points of observation equidistant from the mid-point of PQ. The intensity of radiations of A, B, C will bear the ratio**

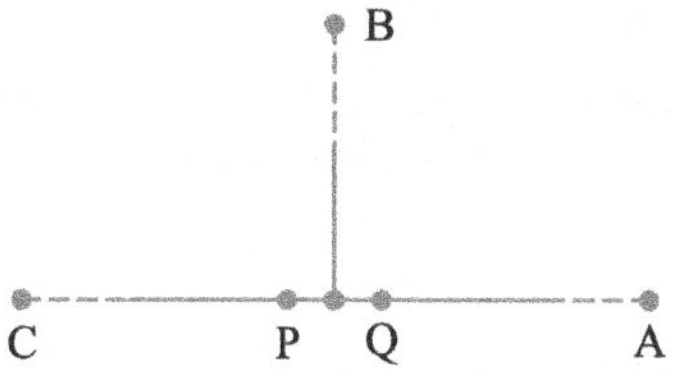

1) 0 : 1 : 4 2) 4 : 1 : 0 3) 0 : 1 : 2 4) 2 : 1 : 0

9. **Two coherent sources separated by distance d are radiating in phase having wavelength λ. A detector moves in a big circle around the two sources in the plane of the two sources. The angular position of n = 4 interference maxima is given as**

1) $\sin^{-1}\frac{n\lambda}{d}$ 2) $\cos^{-1}\frac{4\lambda}{d}$ 3) $\tan^{-1}\frac{d}{4\lambda}$ 4) $\cos^{-1}\frac{\lambda}{4d}$

10. **In a double slit experiment, the separation between the slits is d and distance of the screen from slits is D. If the wavelength of light used is λ and I is the intensity of central bright fringe, then intensity at distance x from central maximum is**

1) $I\cos^2\left(\frac{\pi^2 xd}{\lambda D}\right)$ 2) $I^2\sin^2\left(\frac{\pi xd}{2\lambda D}\right)$ 3) $I\cos^2\left(\frac{\pi xd}{\lambda D}\right)$ 4) $I\sin^2\left(\frac{\pi xd}{\lambda D}\right)$

11. **The polaroids are placed in the path of unpolarized beam of intensity I_0 such that no light is emitted from the second polaroid. If a third polaroid whose polarization axis makes an angle θ with the polarization axis of first polaroid, is placed between these polariods then the intensity of light emerging from the last polaroid will be**

1) $\left(\frac{I_0}{8}\right)\sin^2 2\theta$ 2) $\left(\frac{I_0}{4}\right)\sin^2 2\theta$ 3) $\left(\frac{I_0}{2}\right)\cos^2\theta$ 4) $I_0\cos^4\theta$

12. **Two coherent point sources S_1 and S_2 vibrating in phase emit light of wavelength λ. separation between the sources is 2λ. The smallest distance from s_2 on a line passing through The S_2 and perpendicular to s_1s_2, where a minimum of intensity occurs is**

1) $\frac{7\lambda}{12}$ 2) $\frac{15\lambda}{4}$ 3) $\frac{\lambda}{2}$ 4) $\frac{3\lambda}{4}$

13. **In Young's double slit experiment S_1 and S_2 are two slits. Films of thickness t_1 and t_2 and refractive indices μ_1 and μ_2 are placed in front of S_1 and S_2 respectively. If $\mu_1 t_1 = \mu_2 t_2$, then the central maximum will**

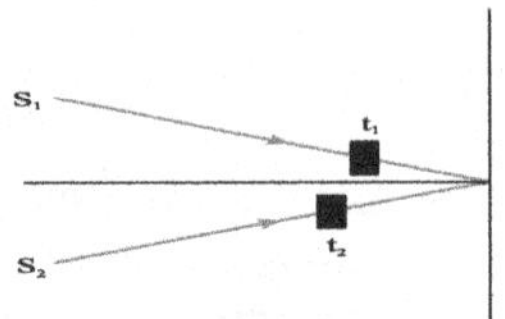

1) Not shift 2) Shift towards S_2 irrespective of amounts of t_1 and t_2

3) Shift towards S_2 irrespective of amounts of t_1 and t_2

4) Shift towards S_1 if $t_2 > t_1$ and towards S_2 if $t_2 < t_1$

14. **A monochromatic beam of light is used for the formation of fringes on a screen by illuminating the two slits in the Young's double slit interference experiment. When a thin film of mica is interposed in the path of one of the interfering beams**

1) the fringe-width increases 2) the fringe-width decreases

3) the fringe pattern disappears 4) fringe-width remains the same but the pattern shifts

15. **What happens to the fringe pattern when the Young's double slit experiment is performed in water instead of air?**

1) Shrinks 2) Disappears 3) Unchanged 4) Enlarged

16. **Two periodic waves of intensities I_1 and I_2 pass through a region at the same time in the same direction. The sum of the maximum and minimum intensities is:**

1) $\left(\sqrt{I_1} - \sqrt{I_2}\right)^2$ 2) $2(I_1 + I_2)$ 3) $I_1 + I_2$ 4) $\left(\sqrt{I_1} + \sqrt{I_2}\right)^2$

17. **What is the minimum thickness of a soap bubble needed for constructive interference in reflected light if the light incident on the film has wavelengths 900 nm? Assume the refractive index for the film is $\mu = 1.5$**

1) 100 nm 2) 150 nm 3) 200 nm 4) 250 nm

18. **Two identical coherent sources are placed on a diameter of a circle of radius R at separation x (<<R) symmetrically about the centre of the circle. The sources emit identical wavelength λ each. The number of points on the circle with maximum intensity is $(x = 5\lambda)$**

1) 20 2) 22 3) 24 4) 26

19. **In a YDSE shown in Fig a parallel beam of light is incident on the slits from a medium of refractive index n_1. The wavelength of light in this medium is λ_1. A transparent slab of thickness 't' and refractive index n_3 is put in front of one slit. The medium between the screen and the plane of the slits is n_2. The phase difference between the light waves reaching point "O". (symmetrical, relative to the slits) is**

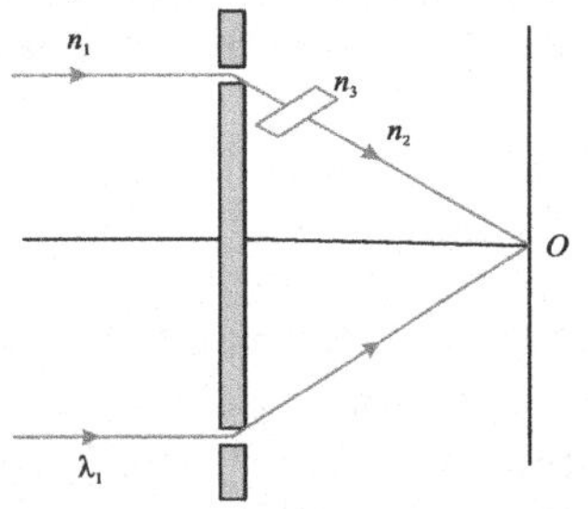

1) $\frac{2\pi}{n_1\lambda_1}(n_3 - n_2)t$ 2) $\frac{2\pi}{\lambda_1}(n_3 - n_2)t$ 3) $\frac{2\pi n_1}{n\,\lambda}\left(\frac{n_3}{n} - 1\right)t$ 4) $\frac{2\pi n_1}{\lambda_1}(n_3 - n_2)t$

20. In Young's double slit experiment, the 10th bright fringe is at a distance x from the central fringe. Then

a) the 10th dark fringe is at a distance of $19x/20$ from the central fringe

b) the 10th dark fringe is at a distance of $21x/20$ from the central fringe

c) the 5th dark fringe is at a distance of x/2 from the central fringe.

d) the 5th dark fringe is at a distance of $9x/20$ from the central fringe.

1) a,b,c only 2) b,c,d only 3) a,d only 4) a,b,c,d only

DIFFRACTION

21. If I_0 is the intensity of the principle maximum in the single slit diffraction pattern then, with doubling the slit width, the intensity becomes

1) I_0 2) $I_0/2$ 3) $\frac{1}{2}I_0$ 4) $4I_0$

22. Light of wavelength 6000 A^0 from a distant source falls on a slit 0.5mm wide. The distance between two dark bands on each side of the central bright band of the diffraction pattern observed on a screen placed at a distance 2m from the slit is

1) 1.2nm 2) 2.4nm 3) 3.6nm 4) 4.8mm

23. The ratio of radii of Fresnel's fourth to ninth zone is

1) 1 : 4 2) 4 : 0 3) 9 : 4 4) 2 : 3

24. A parallel beam of wavelength $\lambda = 4500 A^0$ passes through a long slit of width $2\times10^{-4} m$. The angular divergence for which most of the light is diffracted is (in $\times10^{-5}$ radian)

1) $\frac{2\pi}{3}$ 2) $\frac{5\pi}{4}$ 3) $\frac{3\pi}{4}$ 4) $\frac{\pi}{3}$

POLARISATION

25. A polariser and an analyser are oriented so that the maximum amount of lights is transmitted. Fraction of its maximum value is the intensity of the transmitted light reduced when the analyzer is rotated through (intensity of incident light = I_0)

a) 30^0 b) 45^0 c) 60^0

1) 0.375 I_0, 0.25 I_0, 0.125 I_0 2) 0.25 I_0, 0.375 I_0, 0.125 I_0

3) 0.125 I_0, 0.25 I_0, 0.0375 I_0 4) 0.125 I_0, 0.375 I_0, 0.25 I_0

26. A polaroid examines two adjacent plane polarised beams A and B whose planes of polarisation are mutually perpendicular. In the first position of the analyser, beam B shows zero intensity. From this position a rotation 30^0 shows that the two beams have same intensity. The ratio of intensity of the two beams I_A and I_B

1) 1:3 2) 3:1 3) $\sqrt{3}:1$ 4) $1:\sqrt{3}$

27. An analyser is inclined to a polariser at an angle of 30^0. The intensity of light emerging from the analyser is $\frac{1}{n}$th of that is incident on the polariser. Then n is equal to

1) 4 2) 4/3 3) 8/3 4) 1/4

28. When a beam of light wavelength λ is incident on the surface of a liquid at an angle ϕ, the reflected ray in completely polarized. The wavelength of light in the liquid medium is

1) $\lambda Tan\phi$ 2) $\frac{\lambda}{Tan\phi}$ 3) $\frac{\lambda}{Cos\phi}$ 4) $\frac{\lambda}{Sin\phi}$

LEVEL - III - KEY

1) 4	2) 2	3) 3	4) 1	5) 1	6) 1	7) 4	8) 4	9) 2	10) 3
11) 1	12) 1	13) 4	14) 4	15) 1	16) 2	17) 2	18) 1	19) 1	20) 3
21) 1	22) 4	23) 4	24) 2	25) 1	26) 1	27) 3	28) 2		

LEVEL - III - HINTS

1. $\frac{I_{max}}{I_{min}} = \left(\frac{\sqrt{I_1}+\sqrt{I_2}}{\sqrt{I_1}-\sqrt{I_2}}\right)^2$; $I_1 = \frac{I}{4} : I_2 = \frac{9I}{64}$

2. $\frac{I_1}{I_2} = \beta$; $I_{max} = \left(\sqrt{I_1}+\sqrt{I_2}\right)^2$ $I_{min} = \left(\sqrt{I_1}-\sqrt{I_2}\right)^2$

3. $\theta = \frac{\lambda}{d}$ 4. $\frac{dy}{D} = n\lambda$ 5. $d\sin\theta = n\lambda$

6. $\beta = \frac{\lambda D}{d}$, $\lambda = \frac{h}{\sqrt{2mVe}}$

8. $\Delta\phi = \theta \pm \frac{2\pi}{\lambda}(\Delta x)$, $I = I_0\cos^2\frac{\theta}{2}$

9. $d\sin\theta = n\lambda$ 10. $I = I_0\cos^2\frac{\theta}{2}$

11. $I = \frac{I_0}{2}\cos^2\theta$ 12. $\Delta x = (2n-1)\frac{\lambda}{2}$

13. Shift $= (\mu-1)\frac{t\beta}{\lambda}$ 14. Shift $= (\mu-1)\frac{t\beta}{\lambda}$

15. $\beta' = \frac{\beta}{\mu}$ 16. $\left(\sqrt{I_1}+\sqrt{I_2}\right)^2 + \left(\sqrt{I_1}-\sqrt{I_2}\right)^2$

17. $2\mu t\cos r = (2n+1)\frac{d}{2}$ for minimum thickness

$$t = \frac{\lambda}{4\mu\cos r} = 150nm$$

18. Path diff at p is ; $\Delta = 2(x/2\ \cos\theta) = x\cos\theta$

For intensity to be maximum

$\Delta x = n\lambda$; $x\cos\theta = n\lambda$

$\frac{n\lambda}{x} > 1$; $x = 5\lambda$; n = 1,2,3,4,5

Therefore in all for quadrants these can be 20 maximum

19. Optical path difference between the waves $= (n_3 - n_2)t$

Phase difference $= 2\lambda(n_3 - n_2)t$; $\frac{2\pi(n_3-n_2)t}{n_1\lambda}$

20. $y_{bright} = y_n = n\left(\frac{\lambda D}{d}\right)$

$y_{dark} = y_n = (2n-1)\frac{\lambda D}{2d}$; Where n = 1,2,3,4

21. $I = I_0\left(\frac{\sin\phi}{\phi}\right)^2$ Where ϕ is path difference for principal maxima, ϕ=0 Hence, intensity will remain same (=I_0) Increasing the width of the slit will make the central peak more narrow.

22. Required distance $= \frac{2D\lambda}{d}$

23. $r_a : r_b = \sqrt{a} : \sqrt{b}$

24. $a\sin\theta = n\lambda$; $\theta = \frac{\lambda}{a}\frac{n\pi}{180} rad$

25. $I = \frac{I_0}{2}\cos^2\theta$

26. $\frac{I_A}{I_B} = \frac{\cos\theta_B}{\cos\theta_A}$

27. $I = \frac{I_0}{2}\cos^2\theta$

28. $\mu = \tan i, \mu_1\lambda_1 = \mu_2\lambda_2$

LEVEL - IV

1. **Huygen's principle of secondary wavelets can be used to**
a) deduce the laws of reflection of light **b) deduce the laws of refraction of light**
c) explain the transverse nature of light waves
d) predict the location of a wavefront as time passes
1) a, b only 2) a, c only 3) a, b, d only 4) b, c only

2. **Following statements which are true for light waves but not for sound waves are/is**
(I) The speed of waves is greater in vacuum than in a medium
(II) Waves of different frequencies travel with different speeds in a medium
(III) Waves travel with different speeds in different media.
1) (I) only 2) (I) and (III) 3) (II) and (III) 4) (I), (II) and (III)

3. **If white light is used in Young's double-slit experiment.**
a) bright white fringe is formed at the centre of the screen
b) fringes the different colours are observed on both sides of central fringe clearly only in the first order.
c) the first order violet fringe's are closer to the centre of the screen than the first order red fringes
d) The first order red fringes are closer to the centre of the screen than the first order violet fringes
1) Only a and d are true 2) Only a and b are true
3) Only a,b and c are true 4) All are true

4. **In a double slit experiment, instead of taking slits of equal widths, one slit is made twice as wide as the order. Then in the interference pattern, the intensity.**
a) of maxima will increases **b) of maxima will decrease**
c) of minima will increase **d) of minima will decrease**
1) a,b only 2) b, c only 3) a, c only 4) c, b only

5. **In Young's double slit experiment for producing interference pattern, the fringe width depends on**
i) wave length **ii) distance between the two slits**
iii) distance between the screen and the slits **iv) distance between source and the slits**
1) i only 2)i, ii only 3)i, ii and iii 4) i, ii and iv

6. **Both in interference and diffraction phenomena, alternate dark and bright fringes are obtained on screen**
i) generally fringe width is same in interference and not same in diffraction
ii) the central fringe in interference has maximum brightness and the intensity gradually decreases on either side
iii) in interference the intensity of all bright fringes in same
iv) both the phenomena are produced from same coherent sources
1) i only 2) i and ii 3) i,ii and iv 4) i, ii and iii

7. **When light is incident on a glass block at polarizing angle**
a) reflected ray is plane polarized
b) reflected and refracted rays are perpendicular
c) reflected and refracted rays are partially polarized
d) refracted ray is partially polarised
1) a, c and d are correct
2) a, b and d are correct
3) b, c and d are correct
4) a, b and c are correct

8. **Consider the following statements A and B identify the correct answer**
(A) Polarized light can be used to study the helical structure of nucleic acids
(B) Optic axis is a direction and not any particular line in the crystal.
1) A and B are correct
2) A and B are wrong
3) A is correct and B is wrong
4) A is wrong and B is correct

COMPREHENSION -I

Wave property of electrons implies that they will show diffraction effects. Davission and Germer demonstrated this by diffracting electrons from crystals. The law governing the diffraction from a crystal is obtained by requiring that electron waves reflected from the planes of atoms in a crystal interfere constructively (see figure)

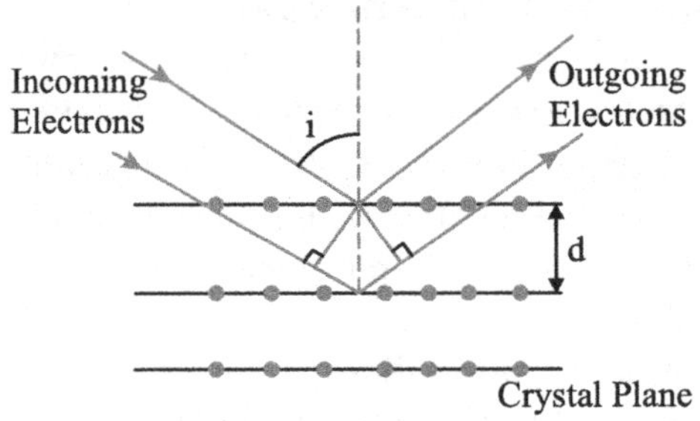

9. **Electrons accelerated by potential V are diffracted from a crystal. If $d = 1A^0$ and $i = 30^0$, V should be about ($h = 6.6\times10^{-34} Js,\ m = 9.1\times10^{-31} kg,\ e = 1.6\times10^{-19} C$)**
1) 2000 V
2) 50 V
3) 500 V
4) 1000 V

10. **If a strong diffraction peak is observed when electrons are incident at an angle 'i' from the normal to the crystal planes with distance 'd' between them (see figure) de-Broglie wavelength λ_{dB} of electrons can be calculated by the relationship (n is an integer)**
1) $d \sin i = n\lambda_{dB}$
2) $2d \cos i = n\lambda_{dB}$
3) $2d \sin i = n\lambda_{dB}$
4) $d \cos i = n\lambda_{dB}$

11. **In an experiment, electrons are made to pass through a narrow slit of width 'd' comparable to their de-Broglie wavelength. They are detected on a screen at a distance 'D' from the slit (see figure)**

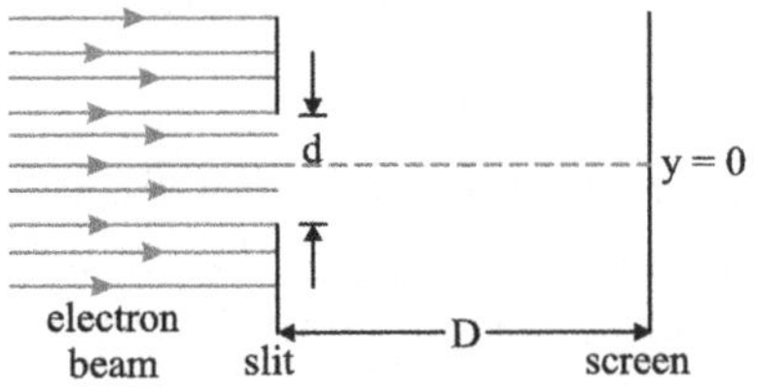

The following graphs that can be expected to represent the number of electrons 'N' detected as a function of the detector position 'y' (y=0 corresponds to the middle of the slit) is

1) 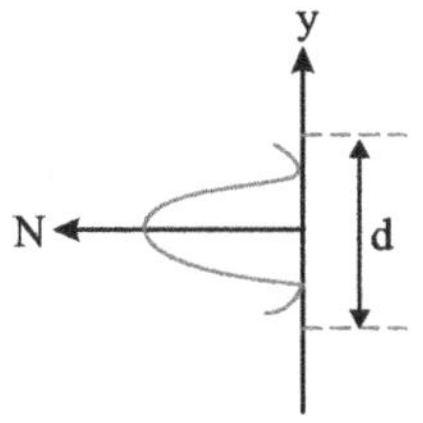

2) 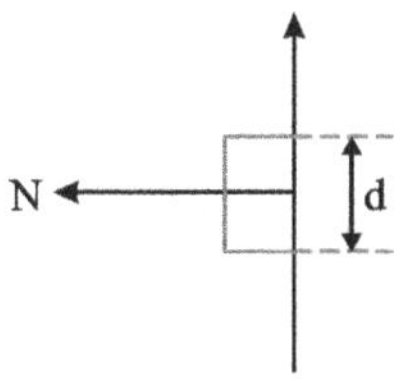

3) 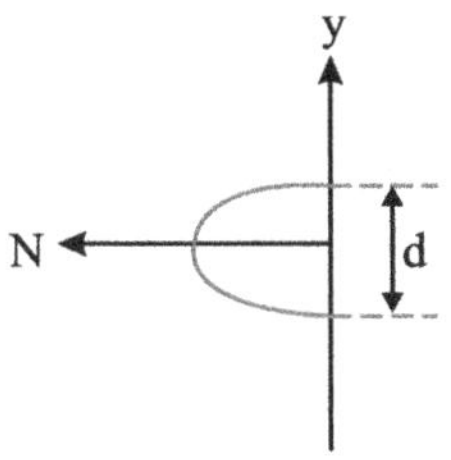

4) 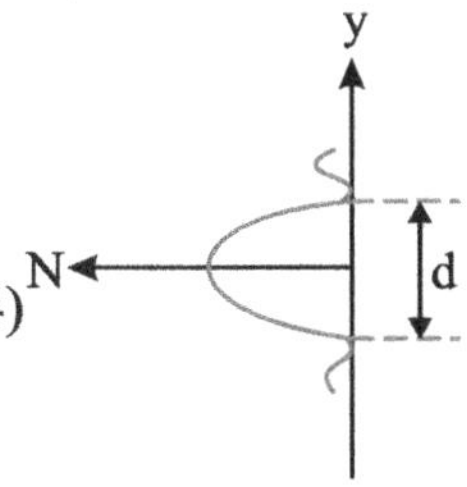

COMPREHENSION -II

A Young's double slit experiment is conducted with slit separation 10mm, where the screen is 2m away from the slits. If wavelength of light used is $6000A^0$, answer the following

12. **Fringe width in mm is**
    1) 0.12 2) 0.24 3) 0.36 4) 0.48
13. **distance of 4th dark band from central fringe in mm is**
    1) 0.14 2) 0.28 3) 0.42 4) 0.56

14 **If the wavelength is increased by $1000A^0$, and placed the whole apparatus in water of refractive index 4/3, the new fringe width in mm is**
   1) 0.210 2) 0.105 3) 0.315 4) 0.420

LEVEL - IV - KEY

1) 3 2) 4 3) 3 4) 3 5) 3 6) 4 7) 2 8) 1 9) 2 10) 2
11) 4 12) 1 13) 3 14) 2

Made in United States
Orlando, FL
06 November 2026

90299267R00208